Occupational Hazards in the Health Professions

Editors

Dag K. Brune, Ph. D.
Chairman
A & J Brune Memorial Foundation
Oslo, Norway

Christer Edling, Ph.D.
Professor of Occupational Medicine
Department of Occupational Medicine
University Hospital
Uppsala, Sweden

CRC Press, Inc.
Boca Raton, Florida

Library of Congress Cataloging-in-Publication Data

Occupational hazards in the health professions / editors, Dag K. Brune, Christer Edling.
p. cm.
Bibliography; p.
Includes index.
ISBN 0-8493-6931-2
1. Medical personnel--Health and hygiene. I. Brune. Dag, 1931-
II. Edling, Christer.
RC965.M39023 1989
363.1'1961--dc19 88-23217
CIP

This book represents information obtained from authentic and highly regarded sources. Reprinted material is quoted with permission, and sources are indicated. A wide variety of references are listed. Every reasonable effort has been made to give reliable data and information, but the author and the publisher cannot assume responsibility for the validity of all materials or for the consequences of their use.

Direct all inquiries to CRC Press, Inc., 2000 Corporate Blvd., N.W., Boca Raton, Florida, 33431.

International Standard Book Number 0-8493-6931-2

Library of Congress Card Number 88-23217
Printed in the United States

PREFACE

An increasing concern about occupational hazards in the health profession has prompted this book. Different topics covered include chemical, biological, and radiation hazards as well as ergonomics, eczema, psychosomatic risks, and indoor climate in the hospital environment.

The veterinary surgery safety aspects for health professionals are similar to those described within the field of human medicine and specific hazards in the veterinary profession have not been surveyed in this book.

Questions and concerns about HIV and AIDS are assigned a separate chapter.

Radiation hazards have been paid particular attention not solely in view of common problems inherent with X-ray therapy and diagnosis, but also in topics dealing with lasers, microwave equipment, as well as ultra- and infrasound equipments.

A dentist who is concerned about the safety aspects of dental surgeries will find that the book includes various topics pertinent to this matter, such as ergonomics specifically devoted to muscular strain, hazards during amalgam treatment, as well as safety recommendations related to curing lights and the handling of vibrating tools.

Due to a broad knowledge of the frequency of injuries inherent with various occupations as well as comprehensive governmental legislation in the appropriate fields in Sweden, such matters have been emphasized according to Swedish experience.

The aim has been to give the book a wide presentation of various safety aspects for different categories of health professionals, and it might be useful for physicians, nurses, dentists, veterinarians, etc. The book may be of particular interest to professionals within the occupational safety area in hospitals and dental clinics.

Several authors have contributed to this book. The editors take the opportunity to forward their sincere thanks to the contributors for fruitful discussions and cooperation in the editing of the book.

The Editors

EDITORS

Dag Brune, Ph.D., was appointed Docent(Associate Professor) in Nuclear Chemistry in 1967 at Chalmers University of Technology, Gothenburg, Sweden. From 1957 to 1975 Dr. Brune served as a research associate at the Swedish Atomic Energy Commission at Studsvik, working with basic research in nuclear physics and chemistry as well as implementing nuclear technique in various biomedical, metallurgical, and environmental fields. During this period much attention was devoted to studies of trace elements in the human body, and health related to various occupational conditions. During 1976 to 1982 Dr. Brune served as Head of the Physical/Chemical Division at the Scandinavian Institute of Dental Materials in Oslo, and was mainly active in biomaterial research and in studies of pollution of internal working environments. Special attention was paid to man's internal exposure to various heavy metals from biomaterials. Further, Dr. Brune has been assigned during various periods by the United Nations, working as an expert for the International Atomic Energy Agency in Colombia, Peru, Ecuador, and Turkey, implementing nuclear technology in medical and industrial fields. He has been a visiting scientist at Centre d'Etudes Nucleaires, Grenoble, France, and at the Mayer-Leibnitz Institute in Garching, Munich.

Dr. Brune is author/coauthor of about 120 publications and author/editor of 3 books. He is also chairman of the A & J Brune Memorial Foundation, and has been involved in social and humanitarian activities in Latin America. His current research interest is in the area of trace element research and human ecology.

Christer Edling, M.D., Ph.D., is a Professor of Occupational Medicine and Head of the Department of Occupational Medicine, University Hospital, Uppsala, Sweden.

Dr. Edling obtained a B.A. degree in psychology from Stockholm University in 1962 and his M.D. degree from Karolinska Institute in 1969. He received his Ph.D. degree in occupational medicine from Linköping University in 1983.

He has been the organizer, chairman, and speaker of several national and international symposia held in Europe, Africa, Australia, and North America. Since 1985 he has been Chairman of the "Section for Occupational Medicine and Environmental Health" at the Swedish Medical Society and since 1984 Chairman of the "Board on Postgraduate Medical Training in Occupational Medicine" at the National Board of Health and Welfare. Furthermore, he is a member of the "Criteria Group" at the National Institute of Occupational Health. He serves on the editorial board of the *Scandinavian Journal of Work Environment & Health* and is a member of several national and international professional societies. He has authored or coauthored more than 100 papers dealing with research areas such as occupational epidemiology, radon and lung cancer, solvents and neurotoxic effects, as well as irritants and nasal disturbances.

His current research interests are in the areas of occupational cancer and toxic effects of gases and vapors.

CONTRIBUTORS

Elisabet Broberg, M.S.
Senior Statistician
National Board of Occupational Safety & Health
Solna, Sweden

Lars G. Burman, M.D., Ph.D.
Professor
Department of Bacteriology
National Bacteriological Laboratories
Stockholm, Sweden

Lars Gerhardsson, M.D., Ph.D.
Department of Occupational Medicine
University Hospital
Umea, Sweden

Catharina Hagberg, D.D.S., Ph.D.
Department of Orthodontics
Karolinska Institute
Huddinge, Sweden

Mats Hagberg, M.D., Ph.D.
Professor
National Institute of Occupational Health
Solna, Sweden

Tore Kalager, M.D.
Chief Physician
Department of Internal Medicine
Buskerud sentraalsykehus
Drammen, Norway

Ole Didrik Laerum, M.D., Ph.D.
Professor
The Gade Institute
Department of Pathology
Haukeland Hospital
University of Bergen
Bergen, Norway

Elisabeth Lagerlöf, M.A.
Work Environment Attache
Swedish Embassy
Washington, D.C.

Arne Lervik, D.D.S.
Occupational Dentist
Oslo, Norway

Egil Lingaas, M.D.
Department of Infection Control
Rikshospitalet
Oslo, Norway

Per Lundberg, Ph.D.
Scientific Secretary
Department of Toxicology
National Institute of Occupational Health
Solna, Sweden

Ronnie Lundström, D.Med.Sc.
Associate Proffessor
National Institute of Occupational Health
Umeå, Sweden

Andreas O. Myking, M.D., Ph.D.
Professor
The Gade Institute
Department of Pathology
Haukeland Hospital
University of Bergen
Bergen, Norway

Eskil Nilsson, M.D., Ph.D.
Associate Professor
Department of Dermatology
Sundsvall Hospital
Sundsvall, Sweden

Roland Örtengren, Ph.D.
Professor
Department of Industrial Ergonomics
University of Linköping
Linköping, Sweden

Bertil R. R. Persson, Ph.D.
Professor
Radiation Physics Department
University Hospital
Lund, Sweden

Bodil Persson, M.D.
Senior Registrar
Department of Occupational Medicine
University Hospital
Linköping, Sweden

Eirik Røynstrand
Chief Techician
Department of Histopathology
Haukeland Hospital
Bergen, Norway

Töres Theorell, M.D., Ph.D.
Professor
National Institute for Psychosocial Factors & Health
Stockholm, Sweden

J. D. G. Troup, Ph.D.
Honorary Senior Research Fellow
Department of Orthopedic & Accident Surgery
University of Liverpool
Liverpool, England
and Consultant and Visiting Professor
Institute of Occupational Health
Helsinki, Finland

John Widström, M.Hs.
Occupational Hygienist
Department of Occupational Medicine
University Hospital
Uppsala, Sweden

D. P. Wyon, Ph.D.
Human Criteria Laboratory
National Swedish Institute for Building Research
Gävle, Sweden

TABLE OF CONTENTS

Chapter 1

GENERAL CONCEPTS

Per Lundberg

TABLE OF CONTENTS

I. INTRODUCTION

In this chapter, a few examples are given of substances which might involve a risk for the health professionals. The hazards of these chemicals will be dealth with in more detail elsewhere in this book, as will the hazards of, for example, inorganic mercury in dental offices and formaldehyde in pathological departments. The examples given are merely to illustrate different ways of providing occupational standards and to point out that there exist some national differences in regulating the work enviroment.

The basis for providing occupational safety for health professionals is the same as for occupational safety for the working environment as such. The regulatory authorities provide occupational standards which may be recommendations or which may have a legal status. Usually, in the area of occupational health, one thinks of standards in terms of exposure limit values for workplace contaminants, but standards may, for example, involve requirements of the best available technology.

II. EXPOSURE LIMIT VALUES AND RISK MANAGEMENT

Laws, recommendations, stipulations, and engineering standards, etc. for acceptable and permitted levels of chemicals in workroom air have been published during the last 50 years. Such lists exist in more than 70 countries. However, few countries generate or update their lists independently. The expression ''exposure limit value'' is used here as a general term and will therefore cover the various expressions employed in the national lists, such as ''maximum allowable concentration'', ''threshold limit value'', ''permissible level'', ''limit value'', ''average limit value'', ''permissible limit'', ''time-weighted average'', ''industrial hygiene standard'', etc. The definition of these various expressions is given in the separate national lists, but it should be noted that the same terms are used with different meanings. As an example, the expression ''maximum permissible concentration'' in certain countries indicates ceiling values, whereas in others it defines average concentrations.[1]

The risk management underlying the occupational exposure limit values can be divided into four steps:

1. Risk identification
2. Risk estimation
3. Risk evaluation
4. Risk control

At least the first two steps are scientific issues and the last two are political and administrative in nature. It is important to realize that risk management is a focus for the driving forces of society. This fact also makes a simple comparison of exposure limit values from different countries partly misleading. Within the context of different health policies and differences in technical development and economic capacities, national exposure limit values will differ. However, all countries can strive to achieve a common scientific approach in a standard setting.

In the first phase in setting exposure limit values, experts evaluate the current knowledge on a toxic compound. Knowledge of the adverse effects of many toxic substances is, however, limited. Furthermore, the conclusions drawn are often based on the effects of short-term experiments only. Consequently, long-term effects such as carcinogenicity and neurobehavioral effects may not have been considered. In the second phase of the establishment of an exposure limit value, technological and economical consequences are considered. The final value is thus usually a compromise between the medical/toxicological, technical, and economical points of view which means it is the result of political decision making. Exposure limit values should therefore be considered primarily as administrative norms. They may have a legal status or be recommendations. They should never be used as fine lines between safe and dangerous concentrations. They rather reflect exposures accepted by the society.

The scientific parts of the risk management should deal with published scientific literature only. The assessment should aim to describe dose-effect relationships, dose-response relationships, and the critical effect. These concepts can be defined as follows. *Dose-effect* relationship means a factor-specified relationship between dose or exposure and the degree of effect on the individual level. *Dose-response* relationship describes a factor-specific relation between dose or exposure and the frequency of affected individuals. A dose-response relationship can be obtained for different types of effects, such as irritation, peripheral neuropathy, or cancer. A *critical effect*, finally, may be defined as that particular effect which appears earliest or at the lowest exposure level. The critical effect could be any adverse and unwanted effect and is not merely the most dramatic effect. With this use of the concept, the best prevention to any acute or chronic change will be obtained.

The documentations of the medical/toxicological considerations are published in some countries. For example, the American Conference of Governmental Industrial Hygienists (ACGIH),[2] the Federal Republic of Germany,[3] and Sweden[4] yearly publish a documentation for new or revised exposure limit values. The techological and economical considerations are usually not publicly documented. The ACGIH is an association of scientists and hygienists within the occupational health sciences. They publish annually an unofficial list of exposure limit values, called threshold limit values (TLVs). The ACGIH values are only intended as recommendations and do not represent the official U.S. occupational exposure limits, although some of their values have been adopted by the Occupational Safety and Health Administratio (OSHA) as official values for the U.S.

Except for the documents specifically written in connection with national exposure limit values, evaluated information on health hazards is produced by some international bodies. The International Program on Chemical Safety (IPCS) is a joint venture of the United Nations Environmental Program, the International Labor Organization (ILO), the the World Health Organization (WHO). WHO, under this joint sponsorship, publishes environmental health criteria, where ad hoc expert groups evaluate the health effects of chemicals. The WHO also publishes technical reports where some of them also give recommendations for health-based occupational exposure limits. The International Agency for Research on Cancer (IARC), within the WHO, evaluates the carcinogenic risk of chemicals to humans, presented in a series of monographs. Here, too, the evaluation is made by ad hoc expert working groups. (Some addresses to distributors of the different documents mentioned here are given in the Appendix to this chapter.)

A large body of scientific articles and reviews has, during the last decades, been published concerning interpretation and management of exposure limit values; some examples are given in the reference list.[5-25] Whatever numerical value an exposure limit is given, it should never be regarded as a borderline between safe and dangerous concentrations. The best practice is to maintain concentrations as low as possible. The differences between different countries can be exemplified by a couple of substances occurring in the work environment of health professionals.

III. HALOTHANE, EXPOSURE LIMIT VALUES

Halothane (2-bromo-2-chloro-1,1,1-trifluoroethane) is used as an anesthetic gas, usually together with other halogenated anesthetic gases or nitrous oxide. The introduction of the 1986-1987 ACGIH TLV Booklet[26] states: "Threshold limit values refer to airborn concentrations of substances and represent conditions under which it is believed that nearly all workers may be repeatedly exposed day after day without adverse effect. Because of wide variation in individual susceptibility, however, a small percentage of workers may experience discomfort from some substances at concentrations at or below the threshold limit''. The time-weighted average exposure limit in the ACGIH 1986-1987 TLV booklet[26] for halothane is given the value of 50 ppm (400 mg/m^3). Halothane is placed in the list under ''Notice of Intended Changes'', where it has been since the 1984-1985 list. It means that it is not yet adopted, but is considered as a trial limit. The documentation behind this trial limit has not been published. It should be noticed in this context that the lists given by the ACGIH are used in many countries as the basis for their own national exposure limit levels.

In Sweden the National Board of Occupational Safety and Health (NBOSH) gives a list of exposure limit values.[27] The list has a legal status through the Occupational and Safety Act. The scientific considerations are handled by the Criteria Group for Occupational Standards. This group consists of, besides occupational health scientists from the Institute of Occupational Health (former Research Department of NBOSH) and from the universities, experts from the employers' and employees' central organizations. For each substance discussed, the Criteria Group publishes a consensus report.[4] Another group, within the Supervision Department of NBOSH, takes care of the technical and economical issues and proposes exposure limit values which are finally promulgated by the Laymen Board of NBOSH.

Halothane is, in the Swedish list of exposure limits, given a value of 5 ppm (40 mg/m^3). Furthermore, it is said that ''the same value expressed in parts per million is also applied to similar halogenated anesthetizing gases which do not have their own set of limit values.''[27] In their consensus report on halothane,[28] the Criteria Group concluded that the critical effect of exposure to halothane seems to be its effect on the central nervous system (CNS). The Criteria Group also concludes that ''in estimating the risks of halothane exposure the possible effects on pregnancy outcome should be taken into account.''[28]

In the Federal Republic of Germany (FRG), a special Commission for the Investigation of Health Hazards of Compounds in the Work Area has been established in order to deal with exposure limit values. The commission annually produces a list of exposure limit values, which has official status, and also, since 1972, prepares scientific documentation with an argumentation to justify the magnitude of the exposure limit value. The commission consists of scientists from universities and industry.

Halothane, in the list from the FRG,[29] is given an exposure limit value of 5 ppm (40 mg/m^3). In the scientific documentation[3] on halothane (written in 1979), the commission notes that an indirect risk estimation must be used, as the epidemiological and experimental data are inadequate. Halothane gives trifluoroacetic acid as a metabolite, which is easily accumulated in the body. This metabolite interferes possibly with tricholoroacetic acid in binding to plasma proteins. The commission therefore proposes that the blood level of trifluoroacetic acid should not exceed 2.5 μg/ml and that an exposure limit value of 5 ppm would satisfy that. They also stress the uncertainty in this risk estimation.

Although the exposure limit value for halothane is the same in Sweden and the FRG, the scientific basis differs. One reason, of course, is that the FRG documentation was prepared in 1979 and the Swedish documentation was prepared in 1985. One other difference between the two countries is that the exposure limit values in the FRG are officially purely health-based while the Swedish exposure limit values are administrative norms based on technological and economical as well as scientific considerations.

The exposure limit values for halothane given above are all time-weighted average concentrations for a normal 8-h workday and a 40-h workweek. However, there also exist *short-term* exposure limits, which usually are 15-min time-weighted average concentrations. The short-term limits are not separate, independent exposure limits, but should be regarded as supplemental where acute health effects are recognized.

Again taking halothane as an example, the ACGIH[26] in their list of ''Notice of Intended

Changes'' has not given any short-term value. In Sweden, halothane is given a short-term value of 10 ppm (80 mg/ml^3). However, as said in the Swedish list,[27] the short-term values should only be regarded as approximate guidelines and they do not have the same legal status as the 8-h time-weighted values. In the list from the FRG,[29] the short-term value for halothane is expressed as twice the 8-h average exposure limit value as a 30-min time-weighted average exposure.

IV. NITROUS OXIDE, EXPOSURE LIMIT VALUES

Nitrous oxide (N_2O) is another substance commonly used for anesthetic purposes. Of the three examples given for halothane above, ACGIH, Sweden, and the FRG, only Sweden has adopted an exposure limit value for nitrous oxide, 100 ppm (180 mg/m^3).[27] The Swedish Criteria Group, in a consensus report,[30] concludes that the critical effect of irreversible nature seems to be the teratogenic effect and the critical effect of reversible nature is the disturbance of mental function. In the case of mental disturbances, it is further said that as there is a large variation in the quality of the studies published and contradictory results, the question of the lowest exposure level which can interfere with mental function remains unanswered. The teratogenic effects are seen in animal studies where rats were exposed to 1000 ppm of N_2O for 19 d.[31]

The U.S. National Institute for Occupational Safety and Health (NIOSH) in 1977 recommended[32] that ''occupational exposure to nitrous oxide shall be controlled so that no worker is exposed at a TWA (time-weighted average) concentration greater than 25 ppm during anesthetic administration.'' The choice of this value was based mainly on observations that N_2O might have a potential to impair functional capacities of exposed workers. In a study,[33] measurable decrements in audiovisual tasks of volunteers exposed during testing at concentrations as low as 50 ppm of N_2O were shown.

V. OCCUPATIONAL STANDARDS OTHER THAN EXPOSURE LIMIT VALUES

A. Anesthetic Gases

In most cases in the practical situation, exposure levels less than the exposure limit values for anesthetic gases are attainable. This does not mean that the risk is negligible for adverse effects among health professionals working with anesthetic gases in hospitals, dental offices, or elsewhere. To further minimize the risk, the regulatory authorities may provide standards other than exposure limit values. These standards may be recommendations or may have a legal status.

In the case of anesthetic gases, NIOSH can be taken as an example of proposing a recommended standard.[32] This standard recommends control procedures and work practices which include handling of anesthetic gas machines, use of a face mask, pressure tests, and anesthetic equipment maintenance. There are also recommendations of medical surveillance for all employees subject to exposure to anesthetic gases and of information to the employees of hazards from anesthetic gases.

In Sweden the occupational standard for anesthetic gases has a legal status. It is issued by NBOSH,[34] and contains regulations about anesthetic equipment maintenance, ventilation, and control of the workroom air. According to the Swedish Occupational and Safety Act, it is the responsibility of employers to inform employees about the hazards. In contrast to the NIOSH recommendations, nothing special is said about medical surveillance. More than anything else, this reflects a difference of principles. The NIOSH criteria documents invariably define exposure levels at which certain actions are initiated as fractions of the proposed exposure value (e.g., 40%). In Sweden, as well as in other Nordic countries, medical or other preventive measures should be instituted on the basis of assessment of the actual health risk.

B. Cytotoxic Drugs

Cytotoxic drugs are preparations which are used mainly in the treatment of malignant growing cells. They may damage growth and reproduction of normal cells as well. The potential for harmful effects developing over the longer term is, however, well known. In several countries, the authorities have provided occupational standards for handling cytotoxic drugs. These standards, whether guidelines or mandatory, all aim to minimize the risk of exposure for health professionals. In the U.S., OSHA has given guidelines for cytotoxic drugs which recommend controls and work practice techniques to reduce the risk of a hazard.[35] The guidelines describe practices in preparation, usage, and disposal of cytotoxic drugs, including personal protective equipment, preparation equipment, administration equipment, medical surveillance, etc. About preparation equipment, they say that "work with cytotoxics must be carried out in a BSC" (Biological Safety Cabinet) "but where one is not currently available, a respirator with a high efficiency filter provides the best protection."[35]

In the Swedish ordinance,[36] with legal status, it is said that the preparation of cytotoxic drugs shall be carried out in a BSC or by using other techniques which give at least the same degree of protection. One example of a suitable BSC is given, with special requirements for the air filtration capacity. One step further is taken by the Standards Association of Australia in its standard on cytotoxic drug safety cabinets.[37] It has determined the construction and function of the cabinet in more detail, including the exhaust filter that should be used.

The medical surveillance recommended in the OSHA guidelines[35] includes full information to all employees, a preplacement physical examination, and a registry of all staff who routinely prepare and administer cytotoxic drugs. The medical staff may recommend group screening for urine mutagenesis or for the presence of certain cytostatic drugs in the urine. They note that, at present, no techniques exist for screening individual employers. It is also recommended that staff members who are pregnant or breast-feeding should be transferred to comparable duties that do not involve handling of cytotoxic drugs, if they so request.

In the context of medical surveillance, the Swedish ordinance[36] concludes that, at present, screening methods indicating exposure to mutagenic or carcinogenic environmental factors are under development. They cannot yet be used for individual screening. Nothing special is said about pregnant or breast-feeding employees in the ordinance, but if any uncertainty is at hand, the occupational medicine clinics should be consulted.

As an example of an adverse effect, it could be mentioned that an increased frequency of sister-chromatid exchanges (SCE) has been found in lymphocytes of nurses handling cytotoxic drugs compared to office workers.[38] Patients under therapy, however, had a five to six times higher frequency of SCEs than the oncology nurses. On the other hand, the SCE frequency of the oncology nurses and the nurses from other hospital departments did not differ statistically significantly. No health effect is known to be associated with SCEs as such, but SCEs should be considered to be an adverse sign of exposure to mutagenic and potentially carcinogenic environmental agents, for the population studied.[39]

These are only a few examples and details of occupational standards for work with cytotoxic drugs. Standards of similar content exist in several countries and, furthermore, some hospitals and other institutions have special guidelines of their own. The standards for handling cytotoxic drugs are also an example of where use of the best available technology seems to be the way of minimizing the health hazard risk and of protecting employees of adverse effects.

VI. MONITORING EXPOSURE LEVELS

In order to ensure that the air levels of contaminants in the occupational environment are well below the exposure limit values given, analytical methods should be used. Most common is air sampling during the work day and analysis of the samples. In many countries,[40,41] the authorities provide descriptions of methods for sampling and analysis. With stationary sampling, the analytical result will tell if the level of the contaminant in the workroom is acceptable or not. With personal sampling, a measurement of the concentration inhaled by an individual is recorded. Another way of evaluating the individual exposure is by biological monitoring, which includes measurements of a substance or its metabolic product(s) in tissues, fluids, or exhaled air of exposed workers. There are, however, generally some problems that hinder wider use of biological measurements:

1. The relatively wide range in individual response and the wide range of "normal"
2. The lack of simple specific analytical methods of sufficient sensitivity
3. The different breathing rates during physical work
4. The complexity of metabolism

In a few cases, a biological exposure limit value has been established. One example is halothane in the FRG.[29] The value is given to 250 μg of tricholoroacetic acid per deciliter of blood, in samples taken at the end of the workweek. In the scientific documentation[42] given 1982, it is said that the value given corresponds to the exposure limit value for halothane of 5 ppm (see also Section III).

Air monitoring and/or biological monitoring can thus be used to control the exposure levels of hazardous substances in the occupational environment. Air monitoring is mainly utilized to ensure that an acceptable level is maintained. Biological monitoring can furthermore be used in a medical surveillance program and thereby detect individuals at risk.

VII. CONCLUDING REMARKS

It is not uncommon for health care professionals to regard themselves as immune from any harm arising from their work. During the course of treating their patients, they may inadvertently exposure themselves and their staffs to hazardous substances. As "employers" it is their duty to fully inform their staffs of the hazards and of the standards given for handling a harmful substance, whether the standards have a legal status or are recommendations. Thus, in the practical situation, reliable information is one way of minimizing the risk.

The exposure may be derived from direct contact with the hazardous substances, via contaminated material, or via handling biological fluids or excreta. Work practices of highest hygienic quality, whether stated in an occupational standard or not, are another way of minimizing the risk and thereby having a good occupational environment in hospitals, dental offices, or elsewhere.

APPENDIX: USEFUL ADDRESSES

INTERNATIONAL DOCUMENTATION FOR OCCUPATIONAL EXPOSURE LIMIT VALUES

World Health Organization
Distribution and Sales Service
CH-1211 Geneva, Switzerland

NATIONAL DOCUMENTATION FOR OCCUPATIONAL EXPOSURE LIMIT VALUES

Federal Republic of Germany
Verlag Chemie GmbH
D-6940 Weinheim, FRG

The Netherlands
Direktoraat-Generaal van de Arbeid
Postbus 69
NL-2270 MA Voorburg, The Netherlands

Sweden
Arbetarskyddsstyrelsen
Publication Service
S-171 84 Solna, Sweden

United Kingdom
Her Majesty's Stationery Office
P.O. Box 276
London SW8 5DT, England

United States
Occupational Safety and Health Administration
Superintendent of Documents
U.S. Government Printing Office
Washington, D.C. 20401 U.S.
National Board of Occupational Safety and Health
NIOSH Publications
4676 Columbia Parkway
Cincinnati, Ohio 45226 U.S.
American Conference of Governmental Industrial Hygienists
6500 Glenway Avenue
Building D-5
Cincinnati, Ohio 45211 U.S.

REFERENCES

1. ILO, Occupational Exposure Limits for Airborne Toxic Substances, 2nd ed., Occupational Safety and Health Ser. No. 37, International Labor Organization, Geneva, 1980.
2. ACGIH, *Documentation of the Threshold Limit Values and Biological Exposure Indices,* 5th ed., American Conference of Governmental Industrial Hygienists, Inc., Cincinnati, Ohio, 1986.
3. **Henschler, D.,** Toxikologisch-arbeitsmedizinische Bergründung von MAK-Werten, 11th ed., Verlag Chemie, Weinheim, 1985/1986 (in German).
4. **Lundberg, P., Ed.,** Scientific basis for Swedish occupational standards. VII. Arbete och Hälsa, 35, National Board of Occupational Safety and Health, Solna, 1986.
5. **Stokinger, H. E.,** Industrial air standards — theory and practice, *J. Occup. Med.,* 15, 249, 1973.
6. **Zielhuis, R. L.,** Permissible limits for occupational exposure to toxic agents, *Int. Arch. Arbeitsmed.,* 33, 1, 1974.
7. **Roschin, A. V. and Timoteevskaja, L. A.,** Chemical substances in the work environment. Some comparative aspects of USSR and US hygienic standards, *Ambio,* 4, 30, 1975.
8. **Dinman, B. D.,** Development of workplace environment standards in foreign countries. I. Historical perspectives; criteria of response utilized in the USSR, *J. Occup. Med.,* 18, 409, 1976.

9. **Dinman, B. D.,** Development of workplace environment standards in foreign countries. II. Concepts of higher nervous function in the USSR, *J. Occup. Med.*, 18, 477, 1976.
10. **Dinman, B. D.,** Development of workplace environment standards in foreign countries. III. Procedures for development of MAC values in the USSR, *J. Occup. Med.*, 18, 556, 1976.
11. **Zielhuis, R. L. and Notten, W. R. F.,** Permissible levels for occupational exposure: basic concepts, *Int. Arch. Occup. Environ. Health*, 42, 269, 1979.
12. **Holmberg, B.,** Setting of exposure standards, in *Advances in Medical Oncology, Research and Education*, Birch, J. M., Ed., Pergamon Press, Oxford, 1979, 115.
13. **Mastromatteo, E.,** On the concept of threshold, *Am. Ind. Hyg. Assoc. J.*, 42, 763, 1981.
14. **Rantanen, J., Aitio, A., Hemminki, K., Järvisalo, J., Lindström, K., Tossavainen, A., and Vainio, H.,** Exposure limits and medical surveillance in occupational health, *Am. J. Ind. Med.*, 3, 363, 1982.
15. **Henschler, D.,** Exposure limits: history, philosophy, future developments, *Ann. Occup. Hyg.*, 28, 79, 1984.
16. **Cook, W. A.,** History of ACGIH TLVs, *Ann. Am. Conf. Ind. Hyg.*, 12, 3, 1985.
17. **Henschler, D.,** Development of occupational limits in Europe, *Ann Am. Conf. Ind. Hyg.*, 12, 37, 1985.
18. **El Batawi, M. A. and Goelzer, B. I. F.,** Internationally recommended health-based occupational exposure limits: a programme in the World Health Organization, *Ann. Am. Conf. Ind. Hyg.*, 12, 49, 1985.
19. **Bardodej, Z.,** Occupational exposure limits for protection of working people in modern society, *Ann. Am. Conf. Ind. Hyg.*, 12, 73, 1985.
20. **Holmberg, B. and Lundberg, P.,** Exposure limits for mixtures, *Ann. Am. Conf. Ind. Hyg.*, 12, 111, 1985.
21. **Damgaard Nielsen, G. and Bakbo, J. C.,** Exposure limits for irritants, *Ann. Am. Conf. Ind. Hyg.*, 12, 119, 1985.
22. **Vainio, H. and Tomatis, L.,** Exposure to carcinogens: scientific and regulatory aspects, *Ann. Am. Conf. Ind. Hyg.*, 12, 135, 1985.
23. **Lundberg, P. and Holmberg, B.,** Occupational standard setting in Sweden — procedure and criteria, *Ann. Am. Conf. Ind. Hyg.*, 12, 249, 1985.
24. **Xue-Qu, G. and You-Xin, L.,** Standard setting for occupational health in China, *Ann. Am. Conf. Ind. Hyg.*, 12, 253, 1985.
25. **Guest, I. G.,** The control of occupational exposure in the United Kingdom, *Ann. Am. Conf. Ind. Hyg.*, 12, 259, 1985.
26. ACGIH, *Threshold Limit Values and Biological Exposure Indices* for 1986-87, American Conference of Governmental Industrial Hygienists, Inc., Cincinnati, Ohio, 1986.
27. NBOSH, Occupational Limit Values, Ordinance AFS 1984:5, National Board of Occupational Safety and Health, Liber Tryck, Stockholm, 1984.
28. **Lundberg, P., Ed.,** Scientific basis for Swedish occupational standards. VI. Arbete och Hälsa, 32, National Board of Occupational Safety and Health, Solna, 1985, 103.
29. Deutsche Forschungsgesellschaft, Maximale Arbeitsplatzkonzentrationen und Biologische Arbeitsstofftoleranzwerte 1986, Mitteilung XXII der Senatskommission zur Prüfung gesundheitsschädlicher Arbeitsstoffe, Verlag Chemie, Weinheim, 1986 (in German).
30. **Lundberg, P., Ed.,** Scientific basis for Swedish occupational standards. III. Arbete och Hälsa, 24, National Board of Occupational Safety and Health, Solna, 1982, 36.
31. **Vieira, E., Cleaton-Jones, P., Austin, J. C., Moyes, D. G., and Shaw, R.,** Effects of low concentrations of nitrous oxide on rat fetuses, *Anesth. Analg.*, 59, 175, 1980.
32. NIOSH, Criteria for a Recommended Standard. Occupational Exposure to Waste Anesthetic Gases and Vapors, DHEW (NIOSH) Publ. No. 77-140, Department of Health, Education and Welfare, Cincinnati, Ohio, 1977.
33. **Bruce, D. L. and Bach, M. J.,** Trace Effects of Anesthetic Gases on Behavioral Performance of Operating Room Personnel, DHEW Publ. No. (NIOSH) 76-169, Department of Health, Education and Welfare, Cincinnati, Ohio, 1976.
34. NBOSH, Anesthetic Gases, Ordinance AFS 1983:11, National Board of Occupational Safety and Health, Liber Tryck, Stockholm, 1983 (in Swedish).
35. U.S. Department of Labor, Guidelines for Cytotoxic (Antineoplastic) Drugs, Occupational Safety and Health Administration Instruction Publ. 8-1.1, Washington, D.C., 1986.
36. NBOSH, Cytotoxic Drugs, Ordinance AFS 1984:8, National Board of Occupational Safety and Health, Liber Tryck, Stockholm, 1984 (in Swedish).
37. Standards Association of Australia, Australian Standard for Cytotoxic Drug Safety Cabinets, AS 2567-1982, North Sydney, New South Wales, 1982.
38. **Norppa, H., Sorsa, M., Vainio, H., Gröhn, P., Heinonen, E., Holsti, L., and Nordman, E.,** Increased sister chromatid exchange frequencies in lymphocytes of nurses handling cytostatic drugs, *Scand. J. Work Environ. Health*, 6, 299, 1980.

39. **Sorsa, M.,** Monitoring of sister chromatid exchange and micronuclei as biological endpoints, in *Monitoring Human Exposure to Carcinogenic and Mutagenic Agents,* Berlin, A., Draper, M., Hemminki, K., and Vainio, H., Eds., IARC Sci. Publ. No. 59, International Agency for Research on Cancer, Lyon, 1984, 339.
40. NIOSH, Manual of Analytical Methods, DHHS (NIOSH) Publ. No. 84-100, National Institute for Occupational Health and Safety, Cincinnati, Ohio, 1984.
41. NBOSH, Principles and Recommendations for the Sampling and Analysis of Substances Included in the List of Limit Values, Arbete och Hälsa, 20, National Board of Occupational Safety and Health, Solna, 1984 (in Swedish).
42. **Henschler, D. and Lehnert, G.,** Biologische Arbeitsstoff-Toleranz-Werte (BAT-Werte), Arbeitsmedizinisch-toxikologische Begründungen, 3rd ed., Verlag Chemie, Weinheim, 1986 (in German).

Chapter 2

OCCUPATIONAL INJURIES AND DISEASES

Elisabeth Lagerlöf and Elisabet Broberg

TABLE OF CONTENTS

I. BACKGROUND

Every year, all occupational injuries and diseases, as well as commuting injuries, are reported via the social security offices to the Swedish Information System on Occupational Injuries and Diseases (ISA).[1,2] The notification forms give a detailed description of the injury as well as information about occupation, branch of industry, etc. The statistics on occupational injuries and diseases for health professionals presented here are obtained through the ISA system. Every year, roughly 6000 occupational injuries and 1700 occupational diseases are reported for this group.

Every fifth year, a census of the population is taken in Sweden. The last one was taken in 1985 and data showed that 430,000 persons (360,000 women and 70,000 men) worked in medical, dental, or other health services in the following occupations:

Health care (nurses, nursing assistants, physiotherapists, dental nurses, etc.)	64%
Medical workers (physicians, dentists)	7%
Clerical workers (office work)	8%
Kitchen workers	5%
Cleaning personnel	5%
Technicians, laboratory personnel	3%
Repairmen, construction workers	1%
Support staff	1%
Psychologists, curators, etc.	3%
Other	3%

A. The Swedish Work Injury Act

The Swedish Work Injury Act came into force in July 1977. Unlike previous legislation, work injury insurance presents a general description of occupational injuries. Occupational injuries are taken to comprise injuries resulting from accidents or other harmful influences at work. No list of occupational diseases is thus referred to by the act. As examples of other types of influence besides accidents, the act refers to chemical substances, energy radiation, continuous repeated unusual pressure, vibrations from tools or machinery, noise, infections, low or high temperatures, and damp and heavy drafts. Harmful factors also include all monotonous or unusually strenuous work operations, which means that they also include strenuous work postures. Mentally strenuous conditions directly associated with work can also have such harmful effects as to come into the scope of the new act. Special rules apply, however, to infectious diseases.

Work injury insurance is coordinated with general health insurance for the first 90 d after an injury occurs, which means that during this period the injured person receives the same benefits as if he had been ill with a cold, e.g., about 90% of his earned income. After 90 d, the injury insurance provides in principle 100% of the person's earnings if the claim is approved. All medical care costs are covered. A person whose work capacity has been permanently reduced will receive an annuity which in principle provides full compensation for his loss of earnings. The work injury insurance employs an economic concept of disability. Thus, the law does not regulate questions concerning pain and suffering or disfigurement and disadvantage. In these respects, the injured employee can instead obtain compensation under a special social security insurance agreement concluded between employer's associations and the trade union (i.e., the social partners). The agreement is called the Job Security Insurance. The insurance is financed by premiums paid by employers.

The Job Security Insurance is a no-fault liability insurance, i.e., the injured person does not have to prove that an employer or a colleague caused the injury. The principle followed here is that the compensation shall correspond to what would have been paid in accordance with general indeminification regulation. Concurrent with this construction, it is prohibited for an injured person to sue the employer or another person embraced by the insurance for

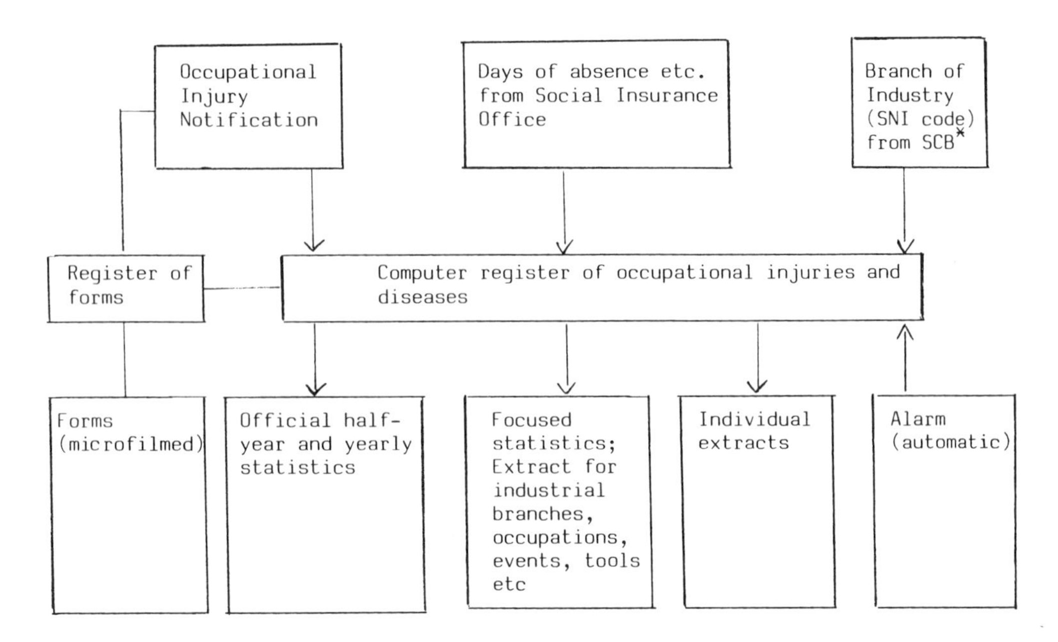

FIGURE 1. The information in the Swedish Information System on Occupational Injuries and Diseases (ISA). *Statistics, Sweden.

damages. In this manner, the social partners have succeeded in doing away with the frequently long-winded and time-consuming proceedings in a court of law.

B. The Information System on Occupational Injuries and Diseases (ISA)

ISA was started in 1979 as a way to present statistics on occupational injuries and diseases, so that they could be used as a preventive tool to diminish the number of accidents and diseases. Figure 1 shows roughly what information is collected and what can be obtained from the system. As seen from Figure 1, the base for collecting information about occupational injuries and diseases is a notification based on the Work Injury Act. All injuries with at least 1 d of absence from work are included in the system.

The information is collected in a data base, which not only enables one to get statistics in the form of fixed tables, but also allows a free search among a lot of different variables. The ISA system collects information of more than 80 variables that characterize an injury or disease such as:

1. Age, sex, and occupation of the injured person
2. Branch of industry
3. Company
4. Part of Sweden, i.e., county
5. Description of the injury or disease
6. Nature of injury, body part injured
7. Employment and pay conditions
8. Work schedule as well as time and place of the injury

All injuries are classified in the form of a single ''casual chain'', so that both the accident events and the external agencies that are involved can be presented. The diseases are classified according to the International Classification of Diseases (ICD), 8th revision, but also according to the pattern of exposure, i.e., inhalation or skin penetration, as well as according to the type of exposure.

Information can be retrieved through a standard software package, which enables one to search for almost any variables in the notification form. Also the different codes can be searched for, in all about 3000 different code values. It is thus possible not only to make yearly standard statistics, but also to do in-depth studies like this one, as well as special reports on machines, certain types of events, etc. Also, statistics on companies, individual or for a whole industry branch, can be produced for the Labor Inspectorate in order to make priorities.

II. OCCUPATIONAL INJURIES AND DISEASES FOR THE MEDICAL AND HEALTH SERVICES

A. Overview and Trends

In 1984, 6276 injuries at work were reported, as well as 1934 occupational diseases and 2272 commuting injuries. Figure 2 gives an overview of the injuries and diseases. Among the occupational injuries, musculoskeletal injuries (overexertion of body parts; 46%) and falls (21%) are the most frequent ones. Among occupational diseases, musculoskeletal (56%) and chemically induced diseases (24%) are most frequent, while commuting injuries occur while walking (falls; 49%) or bicycling (26%). This overall picture, however, varies a great deal among different occupations in the health services. Every year about 325,000 d are lost due to occupational injuries and diseases in the health services. Of these sick leave days, 43% are due to occupational injuries, 36% are due to diseases, and 21% are due to commuting injuries.

The injury rate for health services is low compared with most other branches of industry (19.0 per 1000 full-time employees compared with 29.8 which is the average for all industries). Looking at a 5-year trend, however, it is noticeable that the number of injuries for health services has increased since 1982, while the injuries decreased in most other branches. This also applies to the frequency rates. The trend from 1980 to 1984 is shown in Table 1. As can be seen from the table, the mean number of sick leave days (severity rate) has also increased.

The rates for injuries as well as diseases in the following sections are based on the number of cases per 1000 employees in the different occupations in the health services according to the census of 1980. No data on part- or full-time occupations are available. The occupations shown in different tables cover 97% of the persons working in the health services. The occupations with the highest rates are not the groups working in direct contact with patients, but rather other groups in the health services. The rates for occupational injuries vary as well as the rates for men and women. Among women, laundry personnel have the highest rate, with 73.8 cases per 1000 laundry personnel, which is more than 3 times as high as the average for all women in the health services (21.3). The second highest rates are found among bathing personnel (41.2) and kitchen personnel (41.2), followed by cleaning personnel (33.7) and nursing assistants (28.6).

The highest rate for men is found among kitchen personnel, with 81.2 cases per 1000, which is more than 3 times as high as the average for all men (23.2). The second highest rate exists among repairmen and construction workers (74.1), followed by ambulance drivers (60.0), cleaning personnel (59.2), and hospital support personnel (39.6). Among the occupational groups with direct contact with patients, the mental health personnel have the highest rate, with 28.1 cases per 1000 employed in the occupation.

As the background and genesis of occupational injuries and diseases are different, they will be presented separately. In one area, however, the occupational injuries and diseases are similar and hard to separate from each other, namely, in the musculoskeletal area. That area will be presented under a separate section, since musculoskeletal injuries and diseases constitute the dominating safety risk in the health services.

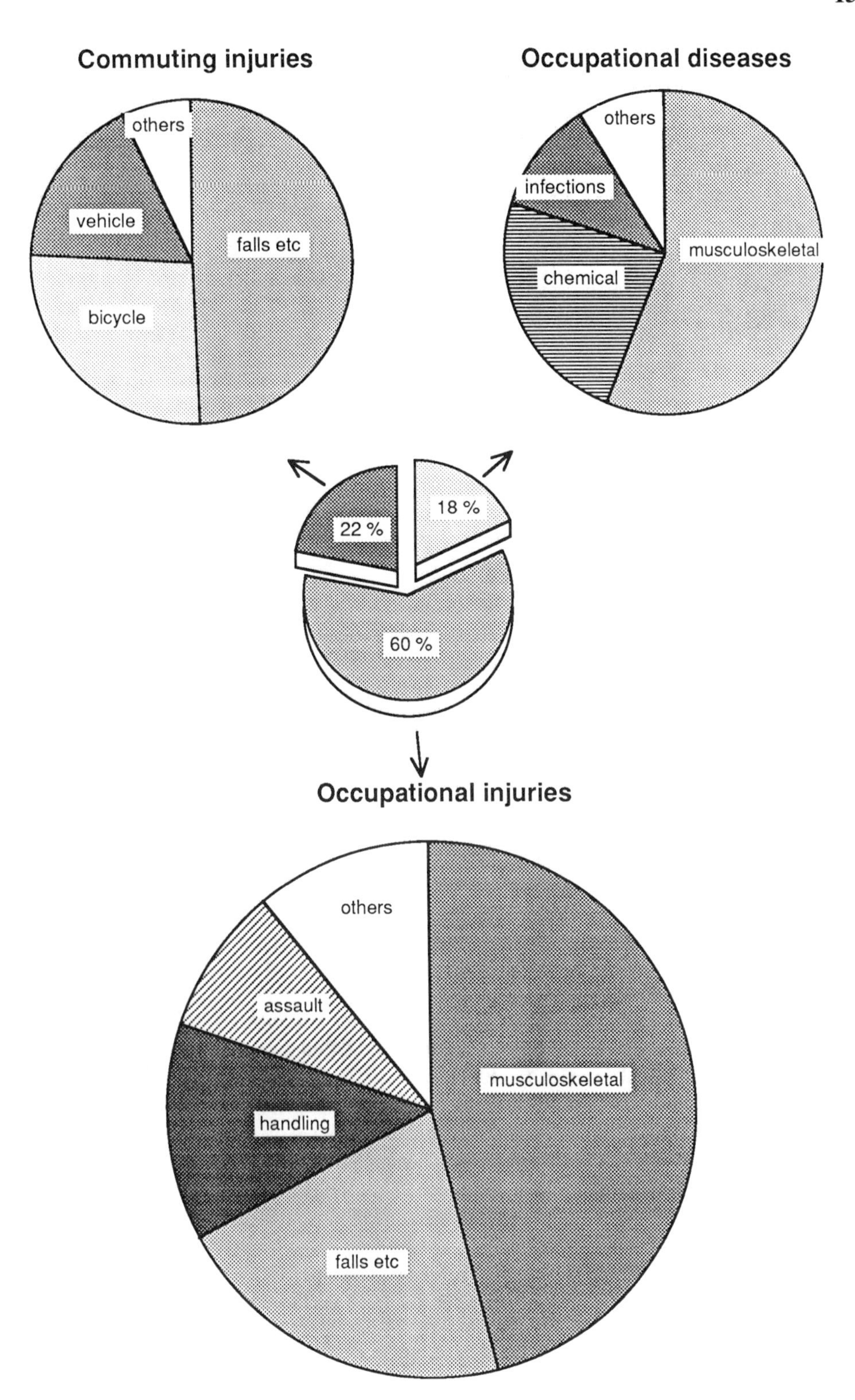

FIGURE 2. Reported occupational injuries and diseases in the health services in Sweden for 1984.

Table 1
TRENDS IN HEALTH SERVICE IN SWEDEN (1980 TO 1984)

Year	Full-time employed	Average age (years)	Occupational accidents			Occupational diseases		
			Number	Rate	Sick leave (d)	Number	Rate	Sick leave (d)
1980	295,483	35.6	5,123	17.3	20	1,774	6.0	40
1981	310,904	35.9	5,004	16.1	21	1,690	5.4	45
1982	310,720	36.3	5,335	17.2	24	1,690	5.4	55
1983	317,975	36.6	5,513	17.3	25	1,891	5.9	61
1984	330,711	36.7	6,276	19.0	27	1,934	5.8	66

Note: Number of cases, number of cases per 1000 full-time employed, as well as the average number of days of sick leave per injury and disease are provided for both occupational accidents and occupational diseases.

Table 2
OCCUPATIONAL INJURIES IN THE HEALTH SERVICE (1984) ACCORDING TO MAIN EVENT COMPARED WITH ALL BRANCHES IN SWEDEN

Event	Health Services		All branches (%)
	Number	%	
Fall, stumbling, etc.	1331	21	30
Contact with flying or falling objects	223	4	12
Contact with mobile objects, machine parts, also during handling	833	13	27
Injury with vehicle when traveling, hit by vehicle	180	3	2
Assaults (including unintentional)	534	9	2
Overexertion of part of the body (musculoskeletal injuries)	2894	46	19
Other	281	2	8
Total	6276	100	100

B. Occupational Injuries

1. Injury Pattern and Frequency Rates for Different Occupations

Between 1981 and 1984 occupational injuries increased by 25%. The increase refers mainly to musculoskeletal injuries (+40%), assaults (unintentional blows by patients are also included; +46%), and vehicle injuries (+66%), as well as injuries when handling tools and by contact with mobile machines (+26%). Other types of injuries are more constant in number over the years. The distribution of occupational injuries for 1984 by different events is shown in Table 2. A comparison between all industrial branches in Sweden shows that the health services have a much higher percentage of musculoskeletal injuries and assaults than in all industrial branches.

The number of occupational injuries for women and men per 1000 professionals in different occupations in the health services is found in Tables 3 and 4. The accident pattern differs very much between different occupations. For the largest female occupational group, nursing assistants, musculoskeletal injuries dominate with 15 cases per 1000 nursing assistants. For the group with the highest total rate, cleaning personnel, falls are the most frequent accidents with injury (14). The highest rate of injuries due to assaults (unintentional blows by patients are included) is found among mental health personnel (9).

The largest male occupational group is physicians. They have the lowest rate of all

Table 3
NUMBER OF OCCUPATIONAL INJURIES PER 1000 EMPLOYED IN THE HEALTH SERVICES (1984) — WOMEN

Occupation	Fall, stumbling, etc.	Hit by flying or falling object	Contact with mobile machine parts	Vehicle injuries	Assault (including unintentional)	Over-exertion	Others	Total	Number of Cases
Physicians	1.5	0.2	0.2	—	0.2	0.2	—	2.3	11
Nurses	2.0	0.2	0.6	0.3	0.8	3.4	0.3	7.5	349
Midwives	1.4	—	0.7	0.3	—	2.1	0.3	4.8	14
Mental health personnel	3.4	0.5	0.8	0.3	8.6	8.3	0.6	22.5	337
Nursing assistants	3.6	0.5	1.6	0.5	1.4	15.2	0.6	23.4	2912
Medical technical assistants	3.8	0.5	2.5	1.0	0.5	3.0	1.0	12.3	49
Laboratory personnel	1.7	0.1	2.1	0.1	0.1	0.9	0.7	5.8	50
Dentists	1.9	—	2.3	0.4	0.4	0.4	1.2	6.6	17
Dental nurses	2.0	0.2	1.3	0.1	0.2	0.7	0.4	4.8	79
Dental technicians	—	1.5	6.6	—	—	0.7	1.5	10.2	14
Physiotherapists, masseurs	3.3	0.5	1.0	0.5	0.4	5.5	0.4	11.6	114
Bathing personnel	8.2	—	—	—	3.5	15.3	1.2	28.2	24
Office workers	2.1	0.1	0.5	0.2	—	0.6	0.1	3.5	100
Pedagogic personnel	2.6	—	—	0.5	1.0	1.5	—	5.6	11
Psychologists, curators, etc.	1.6	—	—	0.5	0.4	0.2	0.2	2.9	16
Kitchen personnel	7.6	2.1	14.0	0.3	0.4	4.9	4.4	33.7	520
Cleaning personnel	7.0	0.9	2.5	0.4	0.5	4.8	0.9	16.9	276
Laundry personnel	14.1	0.9	7.9	3.5	0.9	7.0	2.6	36.9	42
Average	3.4	0.5	1.9	0.4	1.3	8.0	0.7	16.1	—
Number of cases	1069	147	602	122	394	2512	225	—	5071

Table 4
NUMBER OF OCCUPATIONAL INJURIES PER 1000 EMPLOYED IN THE HEALTH SERVICES (1984) — MEN

Occupation	Fall, stumbling, etc.	Hit by flying or falling object	Contact with mobile machine parts	Vehicle injuries	Assault (including unintentional)	Over-exertion	Others	Total	Number of cases
Physicians	0.4	—	0.2	0.2	0.2	0.4	0.2	1.6	21
Nurses	2.2	0.4	1.1	1.5	0.4	3.3	0.7	9.6	26
Mental health personnel	3.4	0.5	0.9	0.8	12.2	6.3	0.5	24.7	195
Nursing assistants	2.9	0.2	1.6	0.2	2.3	14.6	0.1	22.0	192
Dentists	1.2	0.2	1.6	—	0.2	0.8	0.4	4.5	22
Dental technicians	0.4	—	2.5	—	—	0.5	1.0	4.5	9
Physiotherapists, masseurs	3.9	1.0	2.9	—	3.9	2.9	1.0	15.3	16
Ambulance drivers	16.3	2.6	1.7	1.7	1.7	37.7	—	61.7	72
Technicians, etc.	2.3	0.9	3.3	—	—	0.9	0.9	8.4	18
Office workers	1.6	—	0.4	1.6	—	1.2	—	4.7	12
Psychologists, curators, etc.	1.1	—	—	1.7	1.1	1.1	—	5.1	9
Support personnel	7.2	2.6	7.6	2.4	0.5	10.5	1.4	32.2	135
Repairmen, construction workers	13.2	7.6	17.9	2.6	0.7	10.3	4.6	56.9	172
Kitchen personnel	9.4	7.7	33.4	0.9	0.9	11.1	8.6	71.9	84
Cleaning personnel	15.1	2.2	14.0	3.2	1.1	9.7	2.2	47.4	44
Average	4.1	1.2	3.6	0.9	2.2	6.0	0.9	18.9	—
Number of cases	262	76	231	58	140	382	56	—	1205

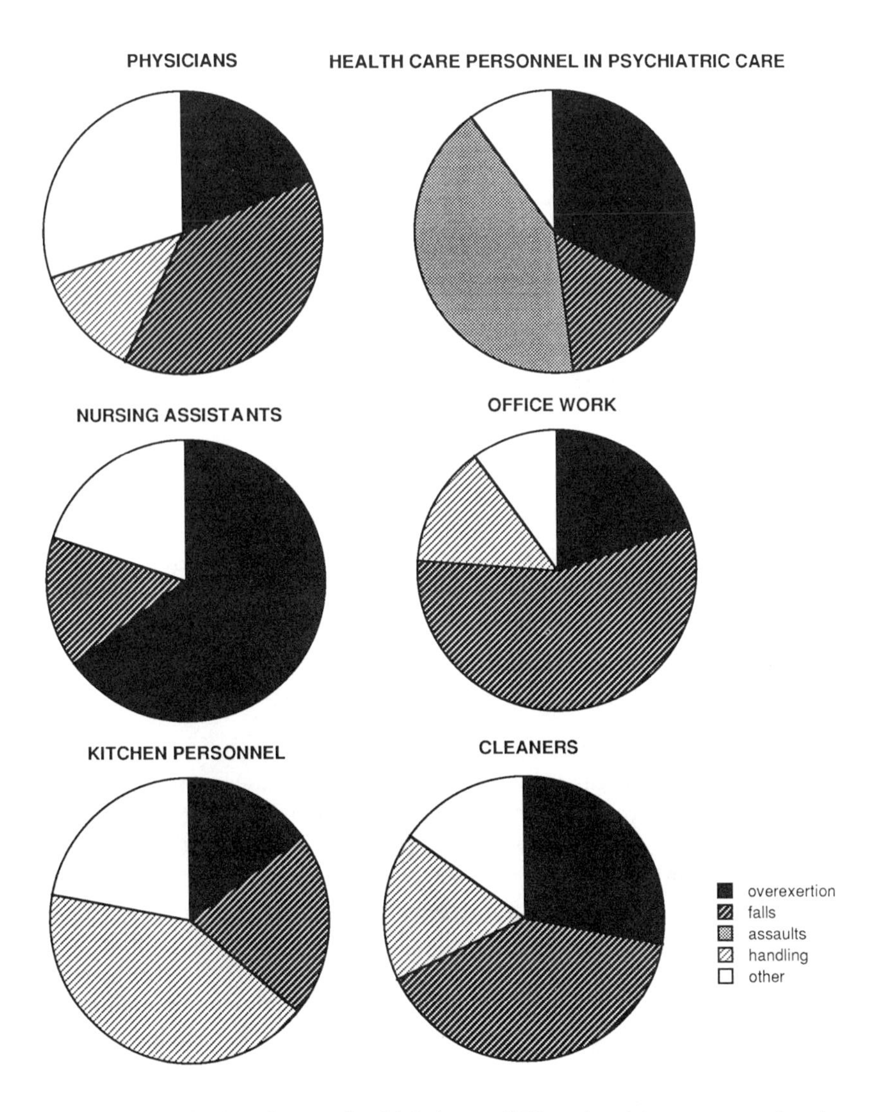

FIGURE 3. Number of occupational injuries per 1000 employed in some occupations in the health services distributed on type of event in Sweden for 1984.

categories with 2 injuries per 1000 physicians. The highest rate is found among male kitchen personnel and the most frequent type of injury is handling of tools and machinery. A comparison between Tables 3 and 4 shows that there are larger differences between occupations then between men and women in the same occupation. Figure 3 summarizes the different injury pattern for some of the occupations.

For nurses over 40% of the injuries are musculoskeletal. One fourth of the injuries are from falls or stumbling. Mental health personnel suffer over 40% of the injuries by assaults or other injuries caused by patients. One third of their injuries are musculoskeletal. A little over one fourth of the injuries among dentists are from accidents when handling tools and machines, and one fourth are caused by falls or stumbling. Over 40% of the dental nurses' injuries are from falls, stumbling, and a little over one fourth are caused by handled objects and tools. Bathing personnel have a high injury rate and more than half of the injuries are musculoskeletal — 30% of the injuries are from falls and stumbling. Ambulance drivers are also a high-risk group and most frequent are musculoskeltal injuries (60%). Falls and stumbling which constitute one fourth of the injuries are second. Repairmen and construction

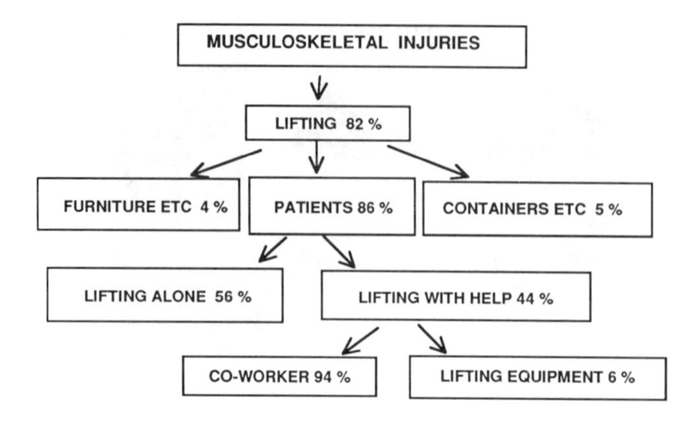

FIGURE 4. Accident scenario on musculoskeletal injuries in the health services in Sweden for 1984.

workers also have high injury rates. One third of the injuries happen during contact with tools and machines. One fourth are from falls and stumbling and one fifth are musculoskeletal. Cleaning personnel constitute a large group in the health services (5%) and 40% of their injuries are from falls, often on wet and slippery floors. Over one fourth of the injuries are musculoskeletal.

2. Some Accident Scenarios in the Health Services

The ISA system also allows us to present a more detailed description of the injuries. In the following, the three most significant accident events will be presented, i.e., overexertion (musculoskeletal injuries), falls, and assaults. The accident scenario for injuries due to overexertion can be seen in Figure 4. As seen from the figure, most of the injuries occur while lifting patients, especially when lifting the patient without help. In the health services, it is recommended that lifting without help should be avoided. As seen from the figure, lifting equipment is involved in 6% of the injuries when the injured person is lifting with help. Data on the use of lifting equipment are not available, but most hospitals, etc. are equipped with lifting equipment. Among the injured persons, 17% have less than a year of experience, while 15% have worked more than 15 years in the health services. Of the injuries, 21% occurred during the evening or at night (6 p.m. to 6 a.m.).

Falls are, to a great extent, caused by slipping (50%) or tripping (10%). Other causes are "treading on air" (4%), loss of support of underlaying surface (3%), getting stuck (3%), being pushed (2%), etc. It is even possible to retrieve information on the type of surface the person has slipped or tripped on through the ISA system. Often only a rough description is given, such as floor (20%), staircase (12%), or road or court (15%). Sometimes more specific information is obtained, such as wet floors (9%; not included in the 20% caused by floor). Falls (4%) have also been caused by slipping on ice or snow.

The accident scenario for assaults shows that roughly 70% of the assaults are intentional and of these 13% occurs when the injured person is working alone. Roughly one third of the assaults have occurred during evenings or nights (6 p.m. to 6 a.m.), and 20% of the injured persons have less job experience than 1 year, while 20% have more than 15 years of experience. All together, the persons injured due to assaults are older than those injured due to overexertion. This could be explained by the higher average age of the personnel employed in psychiatric care.

C. Occupational Diseases

Table 5 shows occupational diseases according to physicians' diagnosis. Most frequently reported are the musculoskeletal diseases, which represent 50% of the reported cases. Fre-

Table 5
OCCUPATIONAL DISEASES IN THE HEALTH SERVICES ACCORDING TO PHYSICIAN'S DIAGNOSIS (1984)

Diagnoses	Number	%
Infectious diseases	189	10
Occupational infections	96	
Sepsis	7	
Chicken pox	8	
Lab hepatitis	18	
Tumors	3	
Psychiatric diseases	36	2
Mental stress	22	
Nervous system diseases	104	5
Brachial neuritis	10	
Carpal tunnel syndrome, etc.	15	
Hearing impairment	46	
Circulatory system diseases	8	
Respiratory system diseases	100	5
Acute respiratory infections	9	
Pleural plaques	39	
Digestive diseases	18	
Skin and subcutaneous tissues	369	19
Cellulite	11	
Eczema	196	
Due to detergents	70	
Musculoskeletal diseases	960	50
Chronic back pain	295	
Lumbago	58	
Lumbago ischia	57	
Cervical	29	
Cervical brachialgia	54	
Rhizopathia cervicalis	40	
Tendon inflammation	239	
Tendovaginitis	130	
Epicondylitis	88	
Arthroses, etc.	43	
Myalgia	129	
Not otherwise specified symptoms	124	6
Shoulder pain	27	
Joint pain	27	
Injuries due to violence and poisoning	13	
Other diagnosis groups	6	
Information unavailable	4	
Total	1934	100

uent diagnoses are chronic lower back pain, such as lumbago, and tendovaginitis or tenlonitis. The second most frequently diagnosed group includes skin diseases and diseases in he subcutaneous tissues, especially dermatitis. The third diagnosis group includes infectious liseases, especially occupational infections in the health services. Tables 6 and 7 present he cases attributed to a suspected cause among different occupations for women and men.

Laundry personnel have the highest rate among women with 37 cases per 1000. They have the highest rate both regarding musculoskeletal diseases and diseases caused by chemicals. The most frequent diagnoses are contact dermatitis (38%), mainly caused by detergents,

Table 6
NUMBER OF OCCUPATIONAL DISEASES PER 1000 EMPLOYED IN THE HEALTH SERVICES (1984) — WOMEN

Occupation	Musculoskeletal diseases	Caused by chemical products	Infectious diseases	Caused by physical factors	Caused by psychosocial factors	Total	Number of cases
Physicians	0.2	—	0.6	—	0.2	1.0	5
Nurses	0.8	0.6	0.5	0.1	0.2	2.1	99
Midwives	1.7	0.3	—	—	—	2.1	6
Mental health personnel	1.5	0.7	0.9	0.1	0.5	3.7	56
Nursing assistants	2.8	1.3	0.9	0.0	0.1	5.2	648
Medical technical assistants	3.0	3.0	0.5	0.3	0.3	7.0	28
Laboratory personnel	2.3	0.6	0.7	—	0.3	3.9	34
Dentists	11.2	1.2	1.2	0.4	—	14.3	37
Dental nurses	3.1	0.8	0.1	0.1	—	4.2	68
Dental technicians	1.5	2.9	—	—	—	4.4	6
Physiotherapists, masseurs	1.2	0.3	0.3	—	0.1	1.9	19
Bathing personnel	7.0	2.3	7.0	—	—	16.4	14
Office workers	3.4	0.8	—	0.3	0.3	5.1	143
Psychologists, curators, etc.	0.2	0.4	—	—	0.7	1.3	7
Telephone operators	1.6	0.5	—	0.5	—	2.7	5
Kitchen personnel	4.2	2.3	0.1	0.2	0.6	7.5	116
Cleaning personnel	12.7	3.4	0.4	0.1	0.1	16.8	274
Laundry personnel	21.1	14.9	—	0.9	—	36.9	42
Average	3.1	1.2	0.6	0.1	0.2	5.2	—
Number of cases	976	382	182	26	57	—	1648

Table 7
NUMBER OF OCCUPATIONAL DISEASES PER 1000 EMPLOYED IN THE HEALTH SERVICES (1984) — MEN

Occupation	Musculoskeletal diseases	Caused by chemical products	Infectious diseases	Caused by physical factors	Caused by psychosocial factors	Total	Number of cases
Physicians	0.1	0.2	0.4	—	0.2	0.8	10
Nurses	0.4	1.5	—	—	—	2.6	7
Mental health personnel	0.9	0.6	1.1	0.1	0.5	3.4	27
Nursing assistants	1.8	1.4	1.0	—	0.1	4.3	38
Dentists	2.7	0.2	0.4	1.2	0.2	4.7	23
Dental technicians	2.0	1.0	—	1.0	—	4.0	8
Physiotherapists, masseurs	—	1.9	1.9	—	—	3.9	4
Ambulance drivers	3.4	—	0.9	—	—	4.3	5
Technicians, etc.	1.4	0.9	0.5	1.4	0.8	4.6	10
Office workers	0.4	0.4	0.4	—	0.4	1.6	4
Support personnel	4.1	2.4	0.2	1.2	—	8.1	34
Repairmen, construction workers	4.0	7.6	0.7	5.0	—	17.2	52
Kitchen personnel	5.1	3.4	—	0.9	—	9.4	11
Cleaning personnel	8.6	1.1	1.1	—	1.1	11.8	11
Average	1.6	1.3	0.6	0.6	0.2	4.5	—
Number of cases	105	85	38	40	14	—	286

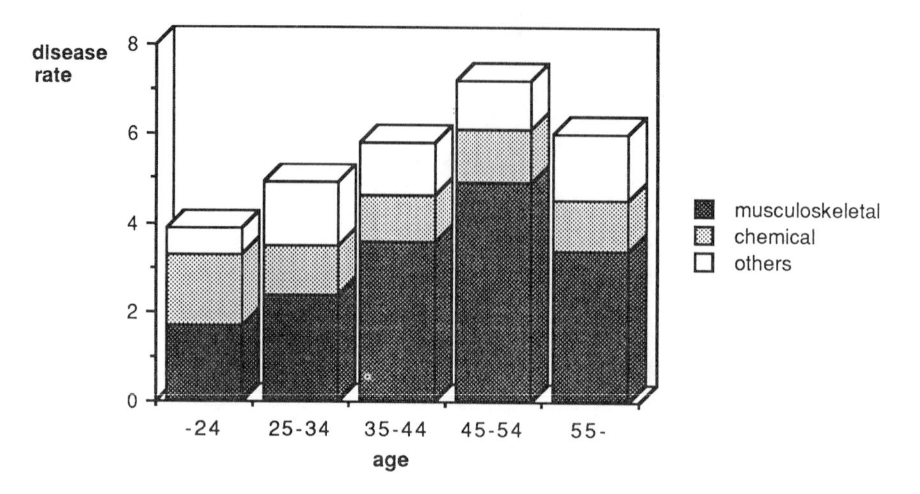

FIGURE 5. Number of occupational diseases per 1000 women employed in the health services, distributed by age and cause of disease (in Sweden, 1984).

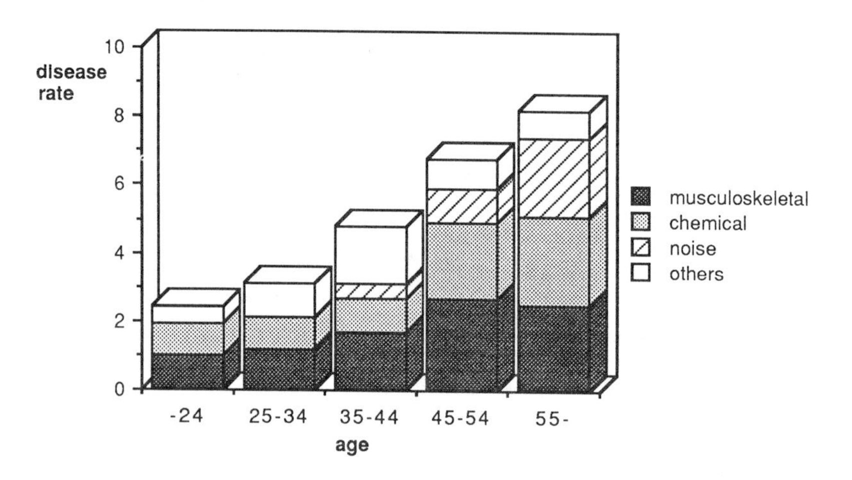

FIGURE 6. Number of occupational diseases per 1000 men employed in the health services, distributed by age and cause of disease (in Sweden, 1984).

myalgia (10%), cervicalis (8%), supraspinatus tendonitis (8%), and epicondylitis (6%). Repairment and construction workers have the highest rate among men. The high rate can be explained by chemicals, especially asbestos-related thickening of the pleura (pleural plaques) and hearing impairment caused by noise. The most frequent diagnosis for physicians is infectious diseases, while frequent diagnoses for nurses are chronic back pain, eczemas, and occupational infections. Nursing assistants suffer the same illnesses as nurses. The most frequent illnesses are chronic lower back pain, tendon inflammation, myalgia, contact dermatitis, and occupational infections. For dentists the most frequent diagnoses are tendonitis and chronic back pain syndrome, and the same applies to dental nurses, who also have a high rate of contact dermatitis. Office workers have high numbers of tendon inflammation and chronic lower back pain. Kitchen personnel have a high rate of tendonitis, chronic back pain, and eczemas. The same applies to cleaning personnel, who also have a high rate of myalgia.

Occupational diseases can also be related to the number of employees in different age groups. Figures 5 and 6 show age-related rates, distributed by suspected cause, separately for men and women. The youngest age group among women has a total of 3.9 cases per

1000 employed, of which 1.7 cases are due to musculoskeletal diseases and 1.6 cases are due to exposure to chemicals. The highest rate occurs in the age group from 45 to 54 years, with a total of 7.2 cases per 1000 of which 4.9 cases are musculoskeletal and 1.2 cases are caused by chemicals.

Also, men's occupational diseases are age related, with an increase from 2.4 cases per 1000 for the youngest age group to 8.2 cases for the oldest. Musculoskeletal diseases are increasing from 1.0 case for the youngest age group to 2.5 cases per 1000 for the oldest men. The chemically induced occupational diseases have increased from 0.9 cases to 2.6 cases. Also, the type of chemically induced disease varies. The dominating disease for the younger age group is eczemas and for the older age group, asbestos-related diseases (pleural plaques) dominate. For the oldest age groups, a high rate of hearing impairment due to noise is found.

D. Musculoskeletal Injuries and Diseases

Figure 2 shows that 56% of the occupational diseases and 46% of the occupational injuries are musculoskeletal. The musculoskeletal diseases have an average of 87 sick leave days per case, and the acute injuries have an average of 26 sick leave days per case. All together, about 60% of the absences for injuries and diseases in the health services are due to musculoskeletal injuries and diseases.

The reason why we have chosen to present a special section for musculoskeletal injuries and diseases is that the notification form does not make clear the extent to which the injury involved is due to work, has deteriorated because of work, stems from constitutional factors or age-related changes in the tissues, can be attributed to leisure activities, etc.

The boundary line between accident-related injuries and injuries due to physical wear and tear or to overexertion tends to blur. An injury onset, which is perceived to be acute by the claimant, in connection with a heavy lift consequent upon overexertion of a part of the body, may very well have been caused by several years of unremitting heavy labor. The appearance of the injury just when this particular lift is being made could be a purely random occurrence. On the other hand, an inflammation of a tendon sheath may arise after only a brief period elapses if the claimant is wholly inexperienced or untrained. In the following, a musculoskeletal injury is classified when some acute event has occurred that has contributed to the injury or if the injury can be attributed to a particular untoward occurrence. Any gradually increasing complaint, as well as a complaint which cannot be related to a specific occurrence, but where it may be considered likely that the workload on the job has contributed to the genesis for the injury, is classified as an occupational disease.

Most of the musculoskeletal cases are those with an acute debut. This applies to all age groups and to both men and women, although musculoskeletal diseases are more frequent in the higher age groups. Back injuries clearly dominate in all age groups, both for men and women, and the highest rate is found among the youngest (6.9 injuries per 1000). The second most frequent injuries are to the shoulder and the arm and can be caused by both accidents and diseases. The rates for diseases are higher among older persons than among younger persons. Table 8 shows the number of cases per 1000 employed distributed on different parts of the body. The tendencies are the same for men and women, but the levels are different. Women have higher rates for all ages and all parts of the body except for injuries to the hip and leg.

The highest rate for musculoskeletal injuries and diseases among women is found for laundry personnel, with 28 injuries per 1000 workers in the occupation, followed by bathing personnel (18), nursing assistants (18), and dentists (13) (see Tables 3 and 6 for other occupations). For men the highest rates are among ambulance drivers (41 cases per 1000 employed), cleaning personnel (18), nursing assistants (16), kitchen personnel (16), hospital support personnel (15), and repairmen and construction workers (14) (see Tables 4 and 7

Table 8
NUMBER OF MUSCULOSKELETAL INJURIES AND DISEASES (1984) PER 1000 EMPLOYED DISTRIBUTED BY BODY PART

	Musculoskeletal injuries		Musculoskeletal diseases	
Injured part of the body	**Women**	**Men**	**Women**	**Men**
Neck	0.7	0.3	0.9	0.3
Back	5.2	4.1	0.7	0.4
Arm, shoulder	1.3	0.7	1.3	0.7
Hip, leg	0.6	0.6	0.1	0.2
Total	8.0	6.0	3.1	1.6
Number of cases	2512	382	976	105

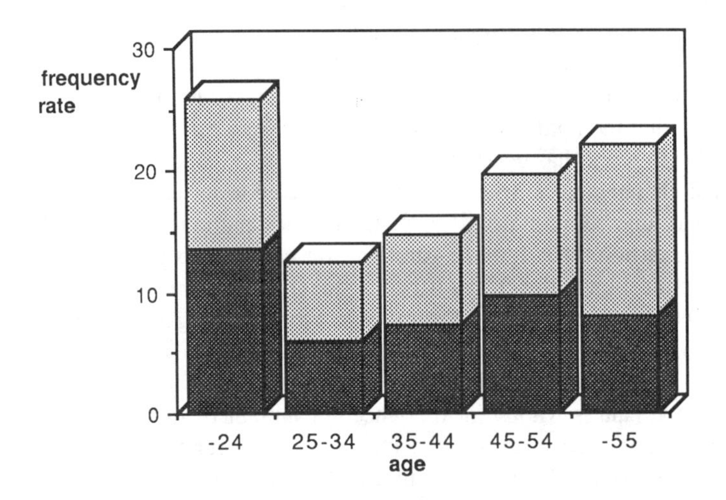

FIGURE 7. Musculoskeletal injuries and diseases compared with the total frequency of injuries and diseases among health care occupations (in Sweden, women).

for other occupations). For the largest occupational group in the health services, nurses and nursing assistants, the relation between musculoskeletal injuries and diseases and the total rates for different age groups is presented in Figures 7 and 8.

As seen from the figure, the highest rate is found among women in the youngest age group. A more detailed analysis pinpoints the absolutely youngest age group as having the highest risk, namely, those who are less than 20 years old. These women have just entered their profession and account for 5% of the personnel among nursing assistants.

III. CONCLUDING OBSERVATIONS

Musculoskeletal injuries and diseases are the most frequent health and safety problems among health professionals. These problems also account for 60% of the absences for injuries and diseases in the health services. The highest rates are found among laundry personnel and bathing personnel, but the largest occupational group in the health services, nurses and nursing assistants, where 67% of the workforce is employed, also has a very high rate. Most of the occupational musculoskeletal injuries are acute and they usually occur when the

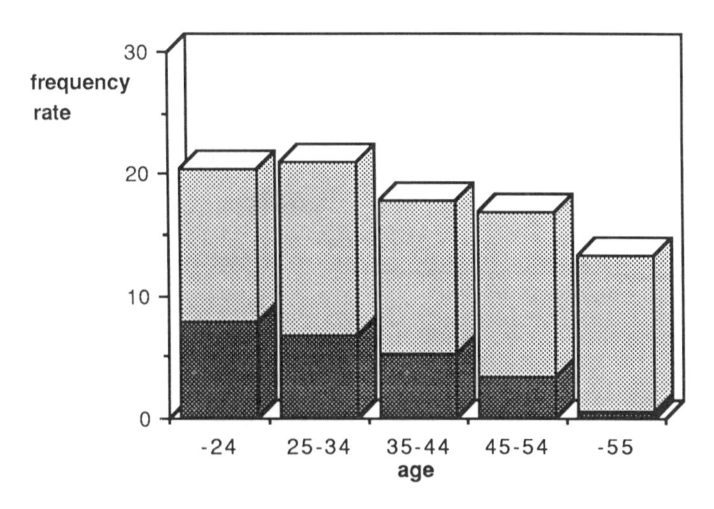

FIGURE 8. Musculoskeletal injuries and diseases compared with the total frequency of injuries and diseases among health care occupations (in Sweden, men).

injured person is lifting without help. Very often this occurs when applying inappropriate work postures, such as twisting and lifting at the same time. Among the nurses and nursing assistants, the highest risk for these kinds of injuries and diseases is found among the youngest age group, women less than 20 years old. Studies have shown[3] that individual susceptibility to musculoskeletal disorders must play a role for the outcome. The importance of muscle strength and other physical characteristics has been emphasized by Keyserling et al.[4] In a recent follow-up study, muscle strength has been demonstrated as a predictor for occupational cervicobrachial disorders.[5] Thus, muscle strength has to be considered when young women enter the occupation. Training in lifting techniques is another important item to reduce the musculoskeletal disorders.

Injuries due to falls, contact with tools and handled objects, and assaults are also frequent among the health professionals. Assaults have their highest rate among mental health personnel, where they account for 42% of the injuries. One third of the assaults occurs during the evening or night, when the working staff usually is reduced.

Occupational diseases, apart from the musculoskeletal ones, are often causes by chemicals or infections. The number of different types of skin diseases such as eczemas, etc. is high and many of them are due to detergents. Among the infectious diseases, so-called ''hospital-infectious'' diseases (staph infections, etc.) are numerous. More information on occupational diseases and other health risks for health professionals will be presented in the following chapters.

REFERENCES

1. Arbetsskador 1983 (Occupational Injuries in 1983), Sveriges officiella statistik, Arbetarskyddsstyrelsen, Statistiska Centralbyrån, Stockholm, 1986 (Swedish with summary in English).
2. **Andersson, R. and Lagerlöf, E.,** Accident data in the new Swedish Information System on occupational injuries, *Ergonomics*, 26, 33, 1983.
3. **Kilbom, A., Lagerlöf, E., Liew, M., and Broberg, E.,** An ergonomic study of notified cases of occupational musculoskeletal diseases, in Proc. 1984 Int. Conf. Occupational Ergonomics, Human Factors Conference, Canada, 1984.
4. **Keyserling, W. M., Herrin, C. D., and Chaffin, D. B.,** Isometric strength testing as a means of controlling medical incidents on strenuous jobs, *J. Occup. Med.*, 22, 332, 1980.
5. **Kilbom, Å.,** Assymetric straight and occupational muscle disorders, *Eur. J. App. Physiol.*, 57, 322, 1988.

Chapter 3

INDOOR CLIMATE

David P. Wyon

TABLE OF CONTENTS

I. INDOOR CLIMATE IN THE HOSPITAL ENVIRONMENT

A. Introduction to Thermal Hygiene

1. Physiological Basis: Heat Balance and Metabolism of Human Body

Heat is produced continuously in the human body at a rate that is strongly dependent on the activity being performed. Fuel is required for producing this heat, and air is required to burn it. In these respects, the human body does not differ from the furnace that maintains the temperature of a building at the required temperature. The fuel we burn, or metabolize, is food. The calorific value of the food eaten during a 24-h period, less the mechanical work performed, is equal to the heat produced in the body. It must all be dissipated during the 24-h period — all, but not more: we can tolerate only small and relatively brief changes in temperature, unlike houses that can store heat from warm days to cold nights or even over periods of several days. Houses need not maintain their internal temperature so exactly as we do, neither are they subject to the fivefold alterations in rate of heat production that we impose on our bodies as we engage in different activities.

The average rate of heat production for a 24-h period has been calculated as 102 W for women and 126 W for men. These figures are based on food intake for an average-sized person, not engaging in any of the more active occupations or pursuits. Large people will eat more food and produce more heat. It is usual to normalize these figures by dividing by the body surface area, a good measure of the size of a person and a critical factor in heat balance, as the outside wall area is for a house. Since women in northern Europe have an average body surface area of 1.6 m^2 (men, 1.8 m^2), the 24-h average rates of heat production are 64 and 70 W/m^2, respectively. Ignoring sex differences gives an average of 67 W/m^2.

The 24-h average is a useful concept of calculating the heat contribution from occupants to a house, but it conceals large and systematic differences at different times of day and is therefore inadequate even for calculating contributions to the heat balance of a room. However, as we shall see below, it provides a surprisingly good basis for calculating the average required room temperature in different clothing ensembles. In most cases, however, it is necessary to allow for different rates of heat production for different activities: sleeping, 40 W/m^2; sitting quietly, 55 W/m^2; standing still working, 78 W/m^2; light house work, 110 W/m^2. Much higher rates of working occur occasionally in the hospital, corresponding to medium heavy industrial work, 150 W/m^2, and heavy manual work, 200 W/m^2, but they are seldom maintained for long. Rate of heat production is governed by the activity in which we engage. It is therefore to a very large extent under conscious control during leisure time in the home. This is not the case during working hours nor for the majority of hospital patients. Rate of heat loss is governed by five principal factors: clothing insulation, air temperature, radiant temperature, air velocity, and air humidity. These have been listed in descending order of importance for heat balance in normal dwellings and for normal activities. Humidity is in last place because it affects mainly the rate at which sweat can evaporate. Sweating is something we try to avoid, for hygienic reasons as much as for the fact that it is the last line of defense. Similarly, shivering is a heat balance mechanism we try to avoid. Hospitals should be planned thermally so that only in freak weather conditions is it necessary for the body to use these strategies. In hot countries, this may be an unrealistic goal, and if so, humidity at once becomes an important factor in heat balance. Sweating can save enormous amounts of the energy now used for summer air conditioning in advanced hot countries, but this rather specialized form of energy conservation will not be dealt with here.

2. Indoor Temperature

The permissible range of temperature is given by the restraints imposed upon the other five factors governing the heat balance of the body — activity, clothing, radiant temperature, air velocity, and humidity. Under nonsweating conditions, even very large differences in

Table 1
PERMISSIBLE TEMPERATURE RANGE

A. Minimum Temperature Without Shivering

Activity	Metabolism (W/m^2)	Clothing (clo) 0	0.5	1.0	1.5	2.0	3.0
Sleeping	40	30	28	26	24	21	17
Sitting	55	27	24	21	18	15	—
24-h average	67	25	22	18	14	—	—
Standing work	78	23	19	15	—	—	—
Active nursing	110	20	14	—	—	—	—

B. Maximum Temperature Without Sweating

Activity	Metabolism (W/m^2)	Clothing (clo) 0	0.5	1.0	1.5	2.0	3.0
Sleeping	40	32	30	28	26	23	19
Sitting	55	30	27	24	21	18	—
24-h average	67	29	25	21	18	—	—
Standing work	78	27	23	19	—	—	—
Active nursing	110	26	29	—	—	—	—

Note: Table 1 is based on an equation for dry heat balance by Humphreys,[7] assuming that 0.7 of the total heat loss is nonevaporative, that the air velocity is 0.1 m/s, except in active nursing, where 0.3 m/s relative air velocity is assumed, and that the permissible temperature range is given by the limits of vasoconstriction and vasodilation, beyond which shivering and sweating occur, respectively.

relative humidity, between the extremes at which humidity becomes a nuisance in itself, 20 to 70%, can be compensated thermally by a change of only 1°C in the air temperature. Air velocity becomes a nuisance by causing local cooling of the body above 0.2 m/s and blows paper about. For this reason, it is usual to limit air velocity to 0.1 m/s in rooms occupied by sedentary people. Under these conditions, the air temperature and the radiant temperature affect heat to about the same extent. The "operative temperature" can be taken as the arithmetic mean of the two and used to characterize the temperature of the space. Table 1 sets out the maximum and minimum air temperatures for the activities whose heat production rate has been discussed above, as a function of clothing (or bed clothing) insulation value. At these limits, sweating or shivering, respectively, must ocur in the steady state. They have been calculated on the basis of the heat flow from the center of the body at 37°C, to the room, taking account of the minimum and maximum insulation afforded by the body tissues when fully vasodilated and vasoconstricted, respectively. The insulation value of the clothing worn is a parameter of Table 1. The insulation value of the air is a function of its relative velocity, taken here to be 0.1 m/s except in the case of active nursing where 0.3 m/s is assumed. A small amount of uncertainty is introduced by the dependence of clothing insulation on air velocity, but since effective clothing insulation can be altered much more in an adaptive way, by opening or closing buttons, for instance, this can probably be neglected.

Clothing and activity, the parameters of Table 1, may be seen to have very large effects on the permissible range of room temperature. They also naturally affect optimum temperatures, which for the present purpose can be taken to mean optimally comfortable temperatures and to lie midway between the limits set out in Table 1. It is a common observation that preferred temperatures vary seasonally, diurnally, and between countries. Many field studies have documented these differences, without normalizing clothing and activity. Had they done so, it is probable that the undoubtedly large and systematic differences in clothing and activity that occur between seasons, between times of day, and between countries would have accounted for the differences in preferred temperature.

Fanger has performed extensive studies of preferred temperature in standard clothing, sitting still in a laboratory, and has never found any variation due to the above factors.[1] It seems therefore adequate to assume that by taking account of the clothing and activity likely to occur in a given place at a given time, Table 1 values may be used to give the permissible range of room temperature without further correction for the so-called seasonal, diurnal, and geographical bias. The only exception, already excluded from the present treatment, but worth reiterating, is when sweating is socially acceptable or economically necessary in hot countries. Table 1 upper limits do not then apply, other heat stress equations must be used, and heat acclimatization must be considered qualitatively. This will introduce marked seasonal and geographical variation.

Thermal gradients occur in both space and time. Horizontal thermal gradients are experienced as temperature swings in time by occupants who move through them. Temperature swings and the factors influencing their perception and their effects on comfort, performance, and behavior have been studied extensively by Wyon et al.[2-5] Their results may be summarized for the present purposes as showing a surprising lack of hysteresis during the relatively slow temperature swings occurring in hospitals — dynamic effects may be neglected and occupants may be assumed to respond to the actual temperature, regardless of its past variation.

Vertical thermal gradients have been insufficiently studied. A correct treatment of criteria should be based on a quantitative understanding of physiological response. This is available for total heat balance in a uniform environment, but not yet for an environment that varies between different parts of the body. It has been found good practice, however, to limit vertical thermal gradients to 2°C/m. This requirement is included in building regulations in several countries. Larger gradients are, in practice, likely to occur only with such marked convective circulation of air that drafts and cold floors will cause problems rather than the thermal gradient per se. It is worth noting that such thermal gradients are always positive, the temperature increasing with height above the floor. Although very few thermal comfort studies have asked for multiple judgements of thermal comfort referred to different parts of the body, Wyon et al.[6] showed that thermal preference is for a negative thermal gradient. (In other words, people prefer to keep a cool head rather than get cold feet.) Most heating systems can achieve at best a small positive thermal gradient in air temperature. However, radiant systems can easily create a negative gradient in operative temperature. The limits applicable in this case are discussed in the following section.

3. Surface Heat Radiation

The operative temperature in a room is affected to the same extent by the surface temperatures as the air temperature. The mean radiant temperature at any point in the occupied zone must be calculated in order to be able to estimate the overall heat balance of the human body. Table 1 may then be used to give the permissible range of operative temperature. It is necessary to define the occupied zone quite closely in this connection, for the solid angle subtended by a given surface area whose temperature differs from the air temperature determines its contribution to the mean radiant temperature. This becomes unreasonably large if the occupied zone is taken to extend right up to walls and windows.

Building regulations often define the occupied zone as extending to within 1 m of the walls or window and to 0.6 m of an outside wall without a window for the special purpose of calculating the plane radiant temperature. High-density occupation will require that the occupied zone in practice extends into the last meter excluded above. Real health risks can thereby occur, for example, when occupants must sleep right up against solid outside walls. Increasing insulation standards are reducing these risks, but it is worth considering whether extra floor space would not in some situations be justified economically by the high alternative cost of meeting building regulations for thermal environment if the occupied zone must extend to the walls.

In addition to satisfying the above requirements for overall heat balance, surface temperatures must not cause excessive local heat loss from the body. If radiation exchange with a cold surface causes a particular part of the body to be colder than it would be in the absence of the surface, the sensation and effect is identical to that caused by a cold draft or excessive air velocity. As stated above, the underlying physiological model for response to such asymmetric thermal loads is lacking. It has been found good practice to require in building regulations that the plane radiant temperature must be calculated or measured in the occupied zone and to introduce the concept of the plane operative temperatures, analogous to operative temperature. This quantity should not be less than 18°C anywhere nor should differences in plane operative temperature within the occupied zone exceed 5°C. In practice the extreme points are usually 1 m from the center of a radiator below the window and 1 m from the center of the window. Observe that the plane operative temperature at the latter point will be lower than at the former. An occupant seated or standing in front of the window will experience a negative gradient in plane operative temperature, provided that the radiator is fulfilling its other function of preventing down-draft by creating upward convection. Whereas the radiant heat exchange of the body as a whole with window and radiator, as reflected by the conventional calculation of mean radiant temperature or measurement of globe temperature, may show acceptable operative temperature values, care must still be taken to ensure that the asymmetric radiant field does not cause too much local heat gain or loss from different parts of the body. The above criteria have been found adequate even under severe winter conditions with present-day windows. With increased insulation standards for outside walls and windows leading to reduced radiator size and surface temperature, the asymmetry will be reduced and it will be easy to meet the above criteria except in special cases of very large windows or high-temperature radiant sources.

4. Air Velocity (Draft)

Air velocities that are too high can cause excessive local cooling of parts of the body. They can cause discomfort by drying the mucous membranes of the eyes, nose, and mouth, particularly in dry winter conditions. Some people are particularly sensitive to the latter effect. Older women often fall into this category and account for a large proportion of complaints of draft. Even the former effect is highly dependent on the clothing worn and the exact way clothing insulation is distributed over the body surface. Sex differences in clothing lead to sex differences in complaints of draft. Women tend to wear less clothing on the legs, ankles, shoulders, and arms than do men and are therefore more likely to be affected by drafts. Thus, draft susceptibility is a highly individual matter. For this reason, although work has been done on minimum temperatures for avoiding complaints of drafts at a given air velocity, it is better to provide the means for individual adjustment of air velocities. They can then be increased in hot weather to provide increased cooling, local or general, but reduced to avoid drafts or to raise the effective temperature in cold weather. Raising the temperature to avoid drafts is not conducive to energy conservation. It is good practice to limit the general air velocity in an occupied zone to 0.1 m/s, but to provide the means to increase air velocity when required. The general effect of air velocity on the heat transfer coefficient for whole body heat balance is discussed below, but in hospitals it is unlikely to be a major source of variation.

5. Combined Effects of Air Temperature, Radiation, and Draft on the Human Body

There have been many attempts to predict quantitatively the effects of the thermal environment on the human body. Over 20 different indices of thermal stress have been used in different connection. All have their limitations: some take account only of certain thermal factors, neglecting others in order to simplify the measurement of thermal climate in a particular situation. They lead to widely different predictions of thermal stress even in theory and therefore cannot predict reliably any relevant measure of thermal strain in practice, except in the particular conditions under which they were derived. Most such indices attempt to predict equivalent combinations of extreme thermal stress that produce the same limiting physiological strain, e.g., maximum permitted heart rate, sweat rate, or central body temperature. They do not lead explicitly to a predicted numerical value for these relevant parameters, however, but lead to an arbitrary value of ''thermal stress''. Strain parameters must be linked to this dimension experimentally. Since there is always a statistical distribution of physiological response even under identical conditions, the selection of a reassuringly exact number on the thermal stress dimension conceals a very large degree of uncertainty on the strain dimension. Only when the measure of physiological strain is the proportion observed to die of heat stroke has the exact shape of the distribution been studied with any close attention.

These indices of extreme thermal stress are of limited use in assessing conditions in dwellings. They are all based on experiments where sweating was profuse and discomfort was considerable. Another set of indices are based on subjective responses to comfort questionnaires. They equate thermal conditions in terms of the thermal discomfort they produce. Their weakness is that complaints of discomfort depend very much on what people are trying to do, in spite of the heat or the cold, and in almost all of the experiments on which comfort indices are based, the subjects were not trying to do anything. Although quite successful in equating different combinations of thermal conditions giving optimum comfort for sitting quietly, they provide a rather inadequate basis for the two kinds of assessment required for optimizing the hospital environment, which are (1) to determine the limits of the reasonably comfortable zone beyond which it is justifiable to use energy to improve the thermal environment and (2) to provide a thermal environment in which it is possible for the body to maintain thermal balance in a way that does not hinder the performance of specified activities. Most comfort indices take no account of clothing, activity, or the adaptive mechanisms of the body. They therefore neglect the most important factors that can extend the comfort zone. The only basis they could provide for (1) above is to predict some arbitrary ''unacceptable'' degree of discomfort or percentage uncomfortable, while giving no information at all on (2). Table 1 shows how inadequate this simplification would be. Fanger made an important step forward by taking account of clothing and activity level.[1] His equation can show the conditions providing ideal comfort for a given activity level and clothing, ideal comfort being a particular state of thermal balance as defined by the mean skin temperature and sweat rate of subjects in exact thermal comfort. A comfort zone defined by acceptable clothing changes would be much narrower than required for (1), as it would not utilize deviations from ideal comfort on a statistical basis, predicting the number of people not in ideal comfort rather than the consequences for an individual in terms of the adaption his body would be required to make. It is therefore not possible to address (2) at all by using Fanger's equation.

Humphreys introduced a much simpler equation for the heat balance of the human body which enables assessments (1) and (2) to be made.[7] It is an equation for the heat flow from the central body core at 37°C to the surroudings, taking account of the three insulating layers that govern this flow: that of the body tissues, of the clothing, and of the surrounding air. For continuous occupation, the heat flow must be equal to the heat produced in the body, otherwise a stable central body temperature could not be maintained. The temperature differences across these three layers of thermal resistance are expressed by the three terms of the equation and together equal the temperature difference between body core and air temperature:

$$T_b - T_a = \frac{M}{A} R_b + K \frac{M}{A} R_c + K \frac{M}{A} (4.2 + 13u^{0.5})^{-1}$$

where T_b = body core temperature, 37°C; T_a = air temperature, °C; M/A = metabolic rate of heat production per square meter of body surface, W/m^2; K = proportion of metabolic heat dissipated by means other than evaporation, about 0.7 indoors; R_b = thermal resistance of body tissues, $m^{2}{}^{\circ}C/W$; R_c = thermal resistance of clothing, $m^{2}{}^{\circ}C/W$ (popular unit: 1 clo = 0.155 $m^{2}{}^{\circ}C/W$); u = air speed, m/sec; and $(4.2 + 13u^{0.5})^{-1}$ = thermal resistance between clothing and surroundings, $m^{2}{}^{\circ}C/W$.

Using Humphrey's[7] limiting values for body tissue resistance, R_b = 0.04 $m^{2}{}^{\circ}C/W$ at onset of sweating (0.09 shivering), this equation allows the calculation of conditions that permit stable adjustment of skin temperature to the prevailing air temperature. It is convenient to select fixed values of metabolic rate appropriate to a given activity and fixed clothing resistance, in order to allow the calculation of maximum and minimum room temperatures. Table 1 was prepared in this way. Air velocity should not be assumed to be less than 0.1 m/s. Thermal radiation can be included by using the operative temperature instead of the air temperature, i.e., by replacing T_a by the arithmetic mean of air temperature and mean radiant temperature. This approximation is appropriate for all normal indoor air velocities.

Assessment (1) above can be made quite simply from the above equation, in the following way: for a given room and time of day, a number of activities are to be expected. A judgment must be made as to what it is reasonable to be able to do without sweating or shivering and what are the reasonable limits for clothing for these activities. The equation will then give maximum and minimum temperatures, enabling the engineer to calculate frequency diagrams showing the number and duration of occasions during a typical year that given activities at given times of day will inevitably lead to discomfort. The consequences of refusing to adapt clothing can be calculated in terms of the increased frequency and duration of discomfort then predicted. Alternatively, the benefits of increased energy consumption can be expressed either as a reduction in the frequency and duration of periods of discomfort or in terms of a reduced need to adapt clothing or to reschedule activities to other rooms or other times of day.

Assessment (2) above can be made by considering the consequences of adjusting the body's heat balance to cope with unsuitable high or low temperatures. A general rule is that activities requiring mental concentration cannot be adequately performed when fully vasodilated, i.e., when at the limit of reduced thermal resistance of body tissues, 0.04 $m^{2}{}^{\circ}C/W$, close to the onset of sweating. For many activities, a lower limit of 0.065 $m^{2}{}^{\circ}C/W$ should be assumed. It is better to be slightly cool than slightly too hot. For social intercourse, too, it is unpleasant to be on the verge of sweating: it discourages animation and induces lethargy. On the other hand, it is not pleasant to be cold either. The region 0.075 to 0.055 is probably acceptable for pleasant, active relaxation. If skilled manual work is to be performed, it will be hampered by too great a degree of vasoconstriction. A maximum value of 0.07 should be assumed for writing, typing, dispensing, minor surgery, etc.

The above approach allows great flexibility in planning the hospital environment. Occupant behavior can be assumed to be adaptive or inflexible, as appropriate, and the consequences can be calculated quantitatively. Occupant decisions can be predicted. The effects of energy conservation on activities of various kinds can be estimated.

II. HOSPITAL WORKING CONDITIONS: PRACTICAL MEASURES

The practical process of dealing with indoor climate problems in the hospital environment should begin with two kinds of *in situ* investigation: quantification of strain, followed by quantification of stress. Strain in this context means assessing thermal and other discomfort, so as to clarify potentially conflicting opinions, e.g., those of patients and staff. Stress means here measurement of indoor climate parameters. Both assessments will be briefly discussed below.

A. Measurement of Thermal Comfort

Thermal comfort can be measured in a variety of ways. It is usually defined as the absence of discomfort. This negative definition defines a no-complaints zone that permits a certain latitude for variations in the thermal climate even for an individual. However, thermal comfort is sometimes defined as the state where a subject cannot decide whether he would like the temperature raised or lowered even if pressed. This point can be found fairly exactly by experiment. Unfortunately, it differs greatly between individuals even if they are all under the same conditions. The comfort zone is then defined as the region where a certain, quite arbitrary proportion of people are in exact thermal comfort. This provides very little information about the consequences, for an individual rather than a large group of people, of deviations from the ideal.

Complaints of thermal discomfort are caused in the first place by an unsuitable combination of the six factors determining the heat balance of the human body: temperature, thermal radiation, air velocity, humidity, clothing, and metabolic rate. They are also caused by a mismatch between what the occupant is trying to do and what he would have to do to be thermally comfortable — undress, stop work, or sweat if it is hot or dress up, work harder, or shiver if it is cold. Any study of thermal comfort must take account of what the subject is doing and would like to be doing. Thermal comfort cannot be measured as if it were some inevitable product of environmental factors alone. Occupants must be given an idea of what the environment they are being asked to assess is supposed to be comfortable for. Thermal comfort responses can be obtained by means of questionnaires with labeled categories of response. The seven-point scale below is used almost universally:

7 Much too hot
6 Too hot
5 Comfortably warm
4 Ideally comfortable
3 Comfortably cool
2 Too cold
1 Much too cold

If possible, the replies should be obtained by a verbal sequence of questions, starting by asking whether the temperature is comfortable. This places the response unequivocally in the 3, 4, 5 "comfort zone" or outside it. If the analysis later calls for an assessment of the proportion "too hot" or "too cold", there is then no doubt about where to draw the line: the subjects themselves have done so.

Mean skin temperature is closely related to thermal comfort in the cold, but hardly at all in the heat, where skin wetness is the best predictor. This is very difficult to measure. In most energy conservation work, it is a good idea merely to obtain a measure of finger temperature. This provides good information about the thermal state of the body in the region where vasodilation and constriction are sufficient to regulate the heat balance of the body. Finger temperature is itself of interest because of its influence on manual dexterity, usually important in dwellings as well as in workplaces. It is usually also worth supplementing the information obtained by the thermal questionnaire by asking for a simple three-category (too low/OK/too high) assessment of thermal radiation, air velocity, humidity, and floor temperature, if appropriate, for diagnostic purposes and remedial action. An assessment of air quality (stuffy/normal/fresh) is also informative and easy to obtain at the same time.

B. Instrumentation for Measuring the Indoor Climate in Hospitals

There are a number of instruments on the market for measuring the integrated effect of the indoor climate. In the present connection, where the problem is to diagnose which aspect of the indoor climate is causing complaints and what can be done about it, this superficially attractive simplification should be avoided. The absolute level indicated depends very strongly on the values assumed for clothing insulation and metabolic heat production, since these are not measured, and the "expected complaints level" is practically irrelevant, since the starting point is that complaints have been received and something must be done about them. Such instruments tend also to be expensive. They have value in teaching the relative importance of different situational factors, but for practical fault-finding it is better to invest in a sufficient number of sets of simple, reliable instruments for measuring each specific factor with good accuracy, avoiding known sources of systematic error.

Air temperature should be measured with an aspirated, radiation-shielded sensor, to eliminate radiation errors. Operative temperature can be measured with a globe thermometer, but this takes time and is insensitive to directional radiation that can be giving rise to complaints of "drafts" that are really due to an asymmetrical thermal radiation field causing certain parts of the body to feel colder than others. The globe can be exchanged for a "cube-thermometer" measuring directional operative temperature in six directions. If this indicates asymmetry, surface temperatures should be measured. Radiometers purporting to measure surface temperatures directly without contact should be avoided — they are expensive and their reading is critically dependent on an estimate of the emissivity of each surface. Instead, flat sensors such as the disposable thermistors now used for determining oral temperature (e.g., Craftemp®) should be taped onto each surface.[8]

Humidity can be measured with electronic sensors, but these are either unreliable, or very expensive, or both. Most of them can be permanently affected by the formation of condensation on the sensor element, which is almost impossible to avoid in practice. Instead, either Assman® thermometers, using aspirated wet- and dry-bulb sensors and direct calculation of humidity, or dew-point instruments are preferable. The latter tend again to be expensive, as a cooling system and programmed sequential measurement are necessary. The cooled mirror can become dirty, making detection of the dew point difficult. Assman® thermometers are cheap and reliable provided that the wet-bulb sensor is really wet and not just assumed to be wet. For this reason, the simplest kind, without a water reservoir, requiring manual wetting, is actually preferable. The dry-bulb value is available and fulfils the requirements of aspiration and radiation shielding mentioned above.

Air velocity must be measured or at least studied and assessed. Smoke visualization gives the most information — puffs of smoke reveal how air moves past the body of the complainant, which parts of the body are affected, and where the draft is coming from. If vents are then adjusted until all air movement over the body is upward, i.e., due most probably to self-convection, draft sensation has been eliminated with a high degree of reliability. If a figure, in meters per second, is required for air velocity, smoke puffs can be timed over a short distance. If this is found too difficult, a katathermometer is the next stage of instrumentation. This works by measuring the rate of cooling caused by a draft and thus integrates over time. Heated wire anemometers, though superficially attractive, are expensive, measure very badly in the interesting region below 0.25 m/s, due to excessive self-convection, are difficult to hold still, and must be very expensive indeed if they are to be capable of forming the required average over time. Their sole advantage is that they indicate variability in an airstream, if present, but then so does the much cheaper and more informative smoke visualization. Hot-wire and hot-bead anemometers are so much influenced by the direction of an air current that their reading is very suspect if turbulence is present. The recommendation is thus for a kit of rather simple, cheap instruments. Such a kit has become available in Sweden, following the author's recommendation, and is in widespread use for trouble-shooting the indoor climate in all categories of building (TCAK 1100).[9]

C. Thermal Aspects of Clothing in Hospitals

One of the main disadvantages of all uniform clothing, leaving psychological aspects aside, is that it removes almost completely the possibility of using clothing to adjust for individual differences in thermal requirements. As a general rule, the more uniform the clothing, the higher will be the proportion dissatisfied with the thermal environment. If activities are also uniform, so much the worse, for the same reason. It is hardly surprising that patients complain so much about the thermal environment, quite apart from their psychological need to direct their boredom, pain, and frustration toward some inanimate aspect of their unpleasant situation. The indoor climate in hospitals is usually as perfect as money can make it, as one feels disinclined to save energy and money at the expense of the sick, but it is still the object of many complaints, hence this chapter. Clothing could be used to alleviate the residual legitimate grounds for complaint, but is usually, for other reasons, a major part of the problem. Hospital clothing for patients and staff is chosen for other reasons than the thermally functional — to raise (or lower) status, identify patient from staff, to comply with tradition, to wash and wear well, to reduce bacterial dissemination, etc.

A laboratory study of hospital clothing was carried out at the National Swedish Institute for Building Research to investigate the possibility of reducing indoor temperatures in hospitals.[10] Different combinations of activity level and clothing, including bedclothing, were studied quantitatively, using a thermal manikin. The study revealed that the day clothing provided for patients in Swedish hospitals was far too thin for the relatively low levels of activity convalescent patients can reasonably be expected to maintain. It would otherwise have been perfectly possible to lower temperatures several degrees below the conventional 24°C, as bedclothing was adequate even for patients in nightgowns sitting up in bed. Staff clothing was far too hot for most nursing activities at 24°C, except for the minimal button-through shift Swedish nurses had for years been accustomed to wearing, but had recently rejected in favor of trouser suits. However, staff clothing was too thin to permit sitting still during rest periods or during long periods of passive supervision, at temperatrues below 24°C. Thus, hospital temperatures have been raised to a level permitting scantily clad patients to walk slowly and staff to sit still; this raises heating costs, causes serious discomfort when humidities are low in winter, as will be explained below, for both patients and staff, and makes it inevitable that staff should sweat and be uncomfortably hot when they are trying to work.

D. Energy Conservation and an Improved Indoor Climate?

The situation described above consists of a series of interlocking problems, each one incapable of solution if tackled separately. By means of an integrated approach, energy conservation — a major problem area — can be achieved at the same time as a series of diverse indoor climate problems are resolved. The following steps should all be taken:

1. Ward room temperatures should be reduced to no more than 21°C air temperature during the heating season. This will have the following consequences:

   1.1. Energy savings will be substantial — up to 5%/°C of the reduction, in cold climates.
   1.2. Complaints of low humidity will be dramatically reduced, making possible still greater energy conservation if humidification is no longer required. (The problems caused by humidification and its absence are documented in special sections below.)
   1.3 Hospital staff will be able to carry out most of their work without sweating, as shown above.
   1.4 The air will be judged "fresher" and of better quality. Energy losses due to window opening to "air out" will be reduced.

1.5 Hospital patients will be able to sleep more soundly, without having to wake up regularly to adjust the bedclothes.
1.6 Convalescent patients who are out of bed will feel cold.
1.7 Hospital staff will feel cold when sitting still.

2. Patients' day clothing should be chosen to give an insulation value of at least 1.5 clo, which can be as much as twice what is in use today, and bathrobes with good insulation should be available at the bedside for all patients. The costs should be set against the substantial savings from 1.1, 1.2, and 1.4 above.
3. Staff clothing should be supplemented with optional items, giving improved insulation, for use when sitting still. The additional cost of these items would be covered by the resulting energy conservation economies. The upper back requires specific protection.
4. Supplementary heating, most suitably at present from overhead panels controlled locally from the bedside, should be available for patients whose heat balance is a problem. Consideration should be given to providing all hospital beds with a means to regulate automatically the microclimate they provide to correspond to the requirements of each patient, awake or asleep, without affecting room climate.
5. Supplementary heating should be provided for work stations where staff must sit for long periods, e.g., on night duty, and in stand by rooms, e.g., those used for coffee breaks. Furniture that is well upholstered and designed to insulate the back when sitting is a measure that can reduce the need for raised air and radiation temperatures and the need to put on additional clothing items during short breaks.
6. Account should be taken of the positive advantages of Steps 2 to 5, as well as their costs: they are all means of achieving a better individual control of heat balance, adjusting for individual differences in skinfold thickness, age, sex, and preference, as well as for differences between individuals and occasions for the same individual, in heat production as affected by illness, activity, fatigue, sleep, physical handicap, or rate of working. These benefits are difficult to cost, as they are not available by altering room temperatures uniformly or by choosing different uniform clothing.

III. SPECIAL PROBLEMS ASSOCIATED WITH INDOOR CLIMATE

This section consists of a series of brief notes on special problem areas having some relation to certain aspects of the indoor climate.

A. Clothing — Conflicting Demands

A large number of quite different demands are placed on clothing function in hospitals: low bacteriological dissemination, washability, durability, cost, appearance, convention, convenience, and status denotation. These must be seen as additional to the thermal function of clothing, and no one set of considerations should be paramount. The final decision on these other demands must be subject to objective tests of whether the clothing works as clothing, i.e., thermally, preferably by measuring total heat loss and heat flow distribution over the body surface on a sectional, heated thermal manikin, such as TORE at Kansas State University, Manhattan (U.S.).

B. Vapor Diffusion Resistance

Thermal manikins available today measure dry heat loss: clo values based on these measurements quantify thermal insulation, but as the thermal manikins cannot sweat, these values are unaffected by the vapor diffusion resistance of clothing. Most types of clothing in use in hospitals have a sufficiently low vapor diffusion resistance; the material itself presents a low resistance, and there are enough openings at neck, wrist, and ankle for evaporation to take place at the low sweat rates that are usually necessary. However, if staff must work hard in a warm room; if textiles must be closely woven or plastic-treated, and garments made tightly fitting, to reduce bacterial dissemination; and particularly if surgical gloves must also be worn, vapor diffusion resistance may become a major problem. Realistic wearing trials are recommended. Soon thermal manikins that can sweat will permit quantification of this parameter of clothing. Already a sweating cylinder is available for testing simulated garments.[11] It was developed in collaboration with this institute.

C. Low Humidity

In cold climates, indoor humidities are low in winter. This causes widespread complaints of dryness, particularly if the relative humidity is below 20%. Field experiments carried out by this institute have shown that humidification to 40% does not increase satisfaction:[12] at temperatures that are usual in hospitals (23 to 24°C), humidification significantly reduced complaints of dryness, by about 50%, but caused a new set of complaints, equal in number, that the humidity was too high. At the temperatures recommended above for hospitals (21 to 22°C), over 80% were satisfied with the low humidity resulting from not humidifying outside air. Humidification actually reduced the percentage satisfied by increasing complaints of both low and high humidity. Thus, a decrease of 2°C in the air temperature, which increases relative humidity by a negligible amount, has been shown to have a disproportionately powerful effect on the sensation of dryness, probably due to a direct thermal effect on the rate of evaporation from the mucous membranes.

Women complained more than men of dryness when it was dry. There was no difference between the sexes when it was not dry. Older people similarly complained more. A very recent report of a similar study in Finland confirms the above thermal and sex effect, but reports the opposite effect of age.[13] Both apparent "age" effects may be caused by an association between age and sex, i.e., more older women in one study than in the other. While there was no difference between smoker and nonsmoker complaints of dryness in the Swedish study, the Finnish study found that passive smoking did increase complaints of dryness, as did any previous history of allergies. A further risk group has been identified by Nilsson and Andersson, who found that contact lens wearers had a reduced "break-up time" (BUT) of the corneal tear film in workplaces with low humidity.[14] This could be caused either by the low humidity itself, by increasing evaporation, or by a second factor, such as organic solvents, present at those workplaces. Humidities can be expected to be lower in workplaces with an increased air-change rate due to the presence of pollutants. A direct effect of humidity on BUT has not been proven either for contact lens wearers or for normal eyes.

D. High Humidity

Humidity is experienced as too high if sweating is necessary and vapor diffusion rates from the skin are too low. This can be due to clothing characteristics, as noted above, but is also experienced in normal indoor clothing during normal office work if temperatures are above 23°C and humidities are at 40% or above.[12] Many hospital activities such as bed-making, cleaning, physiotherapy, etc. imply a much higher rate of heat production. Hospital clothing should therefore have a lower vapor diffusion resistance than normal clothing. High humidity becomes a cause for complaint during sedentary, nonsweating activities above about 70%, when clothing fibers are noticeably damp. If it is necessary to move between areas of differing humidity, thermal transients in clothing can be a major cause of thermal imbalance; transit to lower humidity causes an immediate drop in the apparent temperature that can be equivalent to 4 to 6°C. The heat of evaporation during the 30 to 60 min it takes for clothing to achieve equilibrium with a lower humidity is taken from the body and experienced as a powerful drop in environmental temperature. The reverse effect occurs, equally powerfully and instantaneously, on transit from a dry to a humid enclosure. These effects are measurable on a thermal manikin and have been experienced by anyone entering a shower room fully clothed, although it is generally assumed that the effect is due to an actual difference in air temperature. Synthetic textiles reduce but do not eliminate the effect, as water vapor molecules can be trapped on the surfaces of the fibers even if they cannot be absorbed. If there must be differences in relative humidity between zones of a building, the transients described above can be reduced by maintaining the humid zones at a somewhat lower temperature (2 to 3°C). High humidity is often maintained at great expense as a means of reducing static electricity. Other measures are preferable (see below) for the above reasons.

E. Static Electricity

Static electricity builds up progressively as a function of the number of steps walked and is therefore much more of a problem in large buildings, such as offices, than in the home. However, the characteristics of the floor are normally more important than those of the shoe.[15] Most problems are experienced on carpeted floors, which for obvious reasons are not common in hospitals. It is the presence of explosive anesthetic gases, and of oxygen, that makes electrostatic sparks so much of a hazard in hospitals that general humidification is often installed for that reason alone. In view of the discomfort experienced as a result, all other means of reducing electrostatic charge should be used: Brundrett has reviewed this field for offices, and Guest et el.[16,17] and Bulgin[18] have reviewed reducing electrostatic charge with special reference to operating suites. Floors should be of conducting material, shoes should have a low electrical resistance, clothing should not contribute to the problem, and all room surfaces should regularly be treated with antistatic solutions as part of the normal cleaning routines. Only if these measures fail should humidity be raised, and then only to about 40%, which appears to be critical for assisting electrostatic discharge in practice. Particular attention should be paid to rooms in which explosive gases and oxygen are kept or used. They should have negative pressure difference to other rooms, to prevent the spread of gases, and modern industrial techniques for "point ventilation", to exhaust gases at source, should be used in every case as a first line of defense, especially in the operating room, which for bacteriological reasons must have positive pressure to its surroundings. Operating suites should be isolated from the rest of the hospital with efficient airlocks having negative pressure to both sides. The additional cost of these special measures to prevent the spread of gases must be set against the substantial investment and running cost savings achieved by not humidifying the whole hospital and the benefit thereby achieved in terms of improved comfort and heat balance.

F. Floors and Footwear

Floors in hospitals must be antistatic, electrically conducting, easy to clean, nonslip, and pleasant to walk or stand on. They should not emit air pollutants. Some of these requirements are difficult to reconcile. It appears to be difficult to make plastic flooring soft to walk on, without polluting the air with the softening agents. Plastic floors are often made conducting by letting in metal grids, thus making them harder to clean, and so on. These compromises are outside the scope of the present chapter. Footwear in hospitals should be chosen in conjunction with the floor, to compensate for its shortcomings, e.g., to make electrical contact, to provide good friction, to cushion the feet on hard floors, etc. Almost any footwear is thermally satisfactory if floor temperatures are maintained between 20 to 26°C.[19]

G. Noise

Noise in hospitals is a major problem for the patients rather than for the staff. Patients who are trying to rest and sleep are more easily disturbed than if they were alert and actively engaged in some normal work or activity.[20] Pain and apprehension lowers the threshold for subjective tolerance of unwanted sounds. Many sounds in hospitals are intrinsically frightening and disturbing for patients, but this is not the case for staff; on the contrary, it is part of their job to monitor sounds with this kind of significance, and they are therefore not noise, but one of many sources of information. However, background noise from ventilation, pipes, pumps, and motors, which conveys very little information even to the maintenance staff — the equipment is still in operation — does constitute noise. They are usually at such a low level of sound pressure that they are not regarded conventionally as constituting noise, but, if they irritate, distract, and disturb, they are sources of noise. Very little is known of the health effects of low-level background noise, but so many people experience such an intense feeling of relief if they suddenly stop, that there are *a priori* reasons for assuming that they are stressful. The sense of hearing is a ''watch dog'' function and cannot be switched off or ignored. It is possible that a continuously audible noise, even at low intensity, gives rise to an almost continual cycle of identification-interpretation-suppression that can be regarded as a secondary task and is therefore stressful and frustrating because it is meaningless — there is no new information, but the signal must be processed because it is there and the monitoring cycle is always operative. On this argument, such noise should be inaudible below a more meaningful background of sound: the murmur of voices, traffic, wind, trees, a river, and the sea are infinitely preferable, though similarly monotonous.

H. Windows

An extensive study of attitudes to windows among different categories of building occupants, in which half of the subjects were actually working in windowless rooms, revealed that windows had many important functions not related to admitting daylight. They were a source of information from outside, a potential escape route, an emergency source of fresh air, a source of diurnal variation (thermal, visual, and acoustic), and an aid to orientation in the building. Information on external weather conditions was most missed when absent. There was a significant tendency for those engaged in the most boring and montonous activities to appreciate windows the most. Windows were appreciated most by the blind — a striking demonstration of the nonvisual advantages of windows. There were no hospital patients in this study, but as their situation is often both boring and monotonous, they could be expected to appreciate windows for the above reasons.[21] In fact, Ulrich has shown that recovery from surgery is significantly shorter, and the amount of prescribed pain-killing drugs less, for patients with a stimulating view out through the window than for patients in the same department who had only a view of a brick wall.[22]

The above considerations lead to the recommendation that hospital wards should have windows that fulfil the specified functions — it is perhaps more important that they provide a view out, however limited, than that they provide useful amounts of daylight. If their size must be limited, it is then better to have low sills than to place them near the ceiling to increase the daylight. It should be possible to open them (emergency fresh air, escape). They should have clear glass, not frosted glass. Acoustic information from outside may well be such a positive feature of the environment that it should not be sacrified to reduce the general noise level from aircraft or traffic. The changing diurnal light from a window is far more important than its contribution to the lighting level. The thermal asymmetry introduced by a window — downdraft in winter or radiation gain or loss — may be a positive quality in a homogenous thermal environment, provided that it does not result in a direct feeling of draft.

I. SBS — Sick Building Syndrome

SBS is a complex set of subclinical symptoms relating to the eyes, nose, lips, mucous membranes, and skin, with irritation, itching, headache, fatigue, and a general malaise. When several of these otherwise rather common symptoms occur in a group of people sharing the same indoor environment, and appear to be aggravated by spending time in the building, SBS may be diagnosed. It is not yet possible to identify which air pollutants, if any, are responsible or at which concentrations, but SBS is often relieved by increasing the ventilation rate, lowering the temperature, or raising the humidity. The first measure reduces concentration; the second two enhance mucous clearance rates, so although the exact pollutants causing the problem have not been identified, measures tending to remove them from the air or from the mucous membranes do reduce the symptom intensity. Basic research on SBS continues to focus on chemical analysis and on epidemiological studies to identify buildings that can be said to cause SBS. Probably the best advice that can be given to hospital staff at this stage is to seek adaptive solutions by trying them out in limited parts of the building, using as objective tests of symptom intensity as can be devised as dependent variables. Conventional clinical tests are often not appropriate — SBS is a set of subclinical symptoms. Some progress has recently been made in using measurements of break up time (BUT — the time from a blink to the appearance of the first rupture of the precorneal film) as such a dependent measure: Franck showed that BUT was lower in those complaining of eye discomfort in SBS buildings;[23] Nilsson and Andersson showed that the BUT of contact lens wearers was lower in workplaces with low humidity, and Wyon and Wyon showed that BUT was reduced by even 10-min exposure to air movement typical for air-conditioned rooms and vehicles.[14,24] A new, much simpler method of self-reporting BUT was introduced by the last-named authors, suitable for repeated use on larger groups, in field studies of ameliorative measures as recommended above. Blink rate can be studied using video filming and subsequent observation, each blink being manually entered to a microcomputer for analysis of the blinking pattern. These measures, together with assessments of skin dryness, nasal blockage, and a standard saliva test, may be sufficient to reveal which of several alternative remedial measures has the best effect on SBS. They should naturally be supplemented by subjective assessments, but as it is often difficult to conceal from the subjects of field experiments the nature of the changes being made, subjective assessment can be an expression of preconceived ideas and objective measures, however seemingly variable, are to be preferred.

J. Radon

Radioactive risks in the hospital environment are discussed exhuastively in another chapter of this book, but it is appropriate to mention radon in the context of heating and ventilation. Radon is a radioactive gas that can be present in buildings, originating either from the use of inappropriate building materials, e.g., certain kinds of foam concrete, or from the ground below the building. It has been identified as a health hazard in many dwellings in Sweden. The immediate solution is to improve the ventilation and to ensure that there is a positive pressure in the house. Since hospitals are extremely well ventilated, to reduce the risk of cross-infection and to remove odors, chemicals, and anesthetic gases not present in dwellings, it is the opinion of the Swedish authorities responsible for health risks due to radon that such risks are not present in the hospital environment. At the time of writing, it has not been considered worthwhile even to make any field measurements of radon levels in hospitals, although a major survey of dwellings at risk has been made and considerable sums of public and private money have been spent on remedial measures. These usually consist either of improved ventilation or of measures to prevent radon entering the house from the ground. An interesting new example of the latter is the "radon well" developed at this institute, which removes radon from a whole area of ground and can protect a whole group of houses otherwise at risk for a low-investment and running cost. However, if a hospital is considered to be at risk, it would usually be more appropriate to increase the ventilation rate and thereby further reduce the risk of cross-infection at the same time.

IV. SPECIFIC HOSPITAL ENVIRONMENTS WITH THERMAL PROBLEMS

A. Hospital Wards

The special problem of the conflict between patient and staff requirements has been dealt with in the previous section on energy conservation.

B. Operating Theaters

The safety and best interests of the patient undergoing surgery are the first considerations in operating theater design. This has in the past led to a choice of thermal conditions that is believed to be correct for the patient regardless of the thermal requirements of the staff. However, it is not in the best interest of the patient if the surgeon is too hot, or the anesthetist too cold, for they will not then be able to perform to the best of their ability. The fact is that the most common situation in an operating theater is that the surgeon is too hot and the anesthetist is too cold, at the same time. If it is possible to adjust the temperature setting, as the anesthetist has more opportunity to do so, it will usually be too hot for the surgeon rather than too cold for the anesthetist. Thermal problems in the operating theater have been studied systematically.[6,25] It was widely believed of operating theaters that (1) the ventilation was inadequate, (2) the patient required a high temperature, and (3) the surgeon complained of the heat because of the psychological stress of being ultimately responsible. It was shown that temperatures could be lowered, that the patient's heat balance was a separate problem, better solved by direct application of cold or heat under the anesthetist's control, particularly during long operations, and that surgeons complained of heat only when it was hot and were only very marginally affected by the type of operation.

Thermal problems in operating theaters stem from the fact that there are four different categories of staff present in the same room and that while their activities differ, and therefore their rate of heat production, they are all expected to wear the same clothing. Such differences as exist — the extra gowns, aprons, and gloves of the scrubbed staff at the table — are in the wrong direction, as is the thermal effect of the operating lamp. The surgeon and his assistants are standing up and therefore producing about 10% more heat than the anesthetist, who is sitting down. One of the table staff, usually the surgeon, may be using his muscles more, either to perform some operation or to maintain an awkward posture. The operating lamp is hot and contributes radiant heat to the upper part of the bodies of those at the table. The extra gowns and aprons constitute extra thermal insulation and an increased barrier to vapor diffusion from the skin — they must be of tightly woven material to prevent shedding of bacteria, and the aprons are often waterproof and therefore an absolute vapor diffusion barrier. This is the case with the sterile gloves. Sweating, which is normally a powerful "safety valve" for losing excess heat, becomes ineffective. Surgeons sweat profusely from the forehead under conditions that are not felt to be hot by onlookers or by the anesthetist. It is natural for them to assume that this is nervous sweating, but the study cited above found that this was not so. Anesthetists have a passive, supervisory function for a large part of the time during a long operation. They must be relaxed so as to be vigilant for early signs of danger and unexpected failures of the life-support systems of the patient or theater. They sit still for long periods. They are not scrubbed and can easily walk over to a thermostat and turn it up. They do this rather often. If they were provided with extra items of clothing, and it became routine for them to wear them under a sterile gown, the conflicts described above could be resolved.

The situation of the "runner nurse" is somewhat different. They stand, go in and out, but must often be passive for long periods. They are not at the table. They tend to be too cold rather than too hot and should also have extra items of clothing under the gown. They can, like the anesthetist, lose heat by sweating, if necessary, much more easily than can the table staff. An increased awareness of the underlying causes of thermal discomfort in operating rooms is all that is necessary for the above simple remedies to be sufficient.

18. **Bulgin, D.,** Factors in the design of an operating theatre free from electrostatic risks, *Br. J. Appl. Phys.*, Suppl. 2, 87, 1953.
19. **Olesen, B. W.,** Termiske Komfortkrav til Gulve (Thermal Comfort Requirements for Floors), Doctoral thesis, Technical University of Denmark, Lyngby, 1975 (in Danish).
20. **MacDonald, D. G., Johnson, L. C., and Hord, D. J.,** Habituation of the orienting response in alert and drowsy subjects, *Psychophysiology,* 1, 163, 1964.
21. **Wyon, D. P. and Nilsson, I.,** Human experience of windowless environments in factories, offices, shops and colleges in Sweden, in *Proc. 8th CIB Congr.*, Norwegian Building Research Institute, Oslo, 1980,
22. **Ulrich, R. S.,** View through a window may influence recovery from surgery, *Science,* 224, 420, 1984.
23. **Franck, C.,** Eye symptoms and signs in buildings with indoor climate problems (''office eye syndrome''), *Acta Ophthalmol.*, 64, 306, 1986.
24. **Wyon, N. M. and Wyon, D. P.,** Measurement of acute response to draught in the eye, in *Biomechanics Seminar,* Lindgren, S., Ed., Centre for Biomechanics, CTH, Göteborg, Sweden, 1987, 59; *Acta Ophthalmol.*, 65, 385, 1987.
25. **Wyon, D. P., Lidwell, O. M., and Williams, R. E. O.,** Thermal comfort during surgical operations, *J. Inst. Heat. Vent. Eng.*, 37, 150, 1969 (republication of Reference 6).

Chapter 4

HUMAN IMMUNODEFICIENCY VIRUS (HIV)

Lars G. Burman

TABLE OF CONTENTS

I. INTRODUCTION

Just as hepatitis B has seemingly been brought under control, a new problem has dawned with the emergence of the acquired immune deficiency syndrome (AIDS) epidemic, posing a significant potential risk to health care workers (HCWs). AIDS is a serious clinical condition of different manifestations. It is characterized by various superinfections and an underlying cellular immune deficiency caused by a virus now designated HIV or HIV-1 (human immunodeficiency virus type 1, formerly human T-lymphotropic virus type III/lymphadenopathy associated virus or HTLV-III/LAV). When the viral origin of the disease became clear in 1983 to 1984, the potential risk of contracting HIV caused great concern both among the general public and HCWs.

During past years, numerous questions and concerns have been raised regarding AIDS and employee health issues, resulting in abundant literature consisting of major articles and letters in scientific journals as well as comments, reports, and formal statements from the World Health Organization (WHO), and national and local health authorities.[1-10] The data accumulated through these publications present an increasingly clear picture of the epidemiology of HIV in the health care arena. Even though the risk of contracting the virus now seems low for HCWs, sound guidelines based on scientific data are important both to minimize the occupational hazard and to reduce anxiety and thus to avoid too ambitious, elaborate, and costly control programs with negative effects on patient care.

II. THE HUMAN IMMUNODEFICIENCY VIRUS (HIV) INFECTION

A. The Causative Agent

HIV is a small, enveloped single-stranded RNA retrovirus capable of attacking certain human cells that have the so-called T4 receptor on their surface (e.g., helper/inducer lymphocytes, monocytes, and macrophages in various tissues, including Langerhans cells of the skin and microglia of the central nervous system [CNS]). These cells play important roles in our immune defense. HIV is now classified as a lentivirus and is characterized by an extreme genomic variation, with polymorphism in the envelope region complicating the development of effective vaccines.[11] This is illustrated, for example, by the observation that up to 30 isolations of HIV from the same patient may yield an equal number of variants of the virus.

In recent years, half a dozen HIV-like retroviruses have been isolated from humans. Although their modes of transmission appear similar to those for HIV, the HIV-related viruses are rarely occurring and seem to cause no or mild symptoms. However, they may be associated with immunodeficiency, leukemia, and lymphomas, and human T-cell lymphadenotropic virus type I (HTLV-I) associated with adult T-cell leukemia is currently disseminating among drug addicts in the U.S.[12] and also in Europe.

Through the viral enzyme reverse transcriptase, DNA copies of the viral RNA are made which are to be incorporated into the chromosomal DNA of the host cell which, thus, becomes a permanent carrier of HIV nucleic acid. The chronically infected cell continues to produce HIV particles capable of attacking new T-lymphocytes, monocytes, and macrophages until it succumbs. The spread of HIV infection in the host results in an increasingly weakened immune system with an extreme susceptibility to a variety of microbial pathogens including many of low virulence. The corresponding clinical conditions are collectively known as AIDS.

B. Epidemiology

According to one theory, HIV originates from one of a number of related viruses causing asymptomatic chronic infection in monkeys. In the 1960s, a variant of such a virus may have been transmitted and adapted to humans in central Africa. HIV now is widely disseminated in the human population of this region mainly by heterosexual contact, but also perinatally (from mother to infant), via blood transfusions, and possibly unsterile needles, surgical instruments, etc.

HIV was probably spread to the U.S. about 1975 with the emergence of the AIDS epidemic in New York City in 1977,[13] although it was not recognized until 1981. As of June 1, 1987, 36,000 Americans with AIDS had been reported to the Centers for Disease Control (CDC)[14] and half of them had died. The U.S. Public Health Service[15] estimates that 1 to 1.5 million U.S. citizens carry HIV and has projected that the number of AIDS cases will increase to 270,000 by the end of 1991. In 1981 AIDS was first described in western Europe where the epidemiological curve now parallels the U.S. development, but with a 2-year delay and about sixfold lower numbers of cases. Worldwide, 52,000 AIDS cases reported by 113 nations had accumulated by June 3, 1987.[16] Among the first 28,246 U.S. AIDS cases reviewed by the CDC in December 1986,[17] the main transmission category was homosexual/bisexual (HS/BS) males (65%), usually in the age group of 35 to 40 years, followed by intravenous (i.v.) drug abusers (17%), heterosexual cases (4%), recipients of blood and blood components (3%), and children of HIV-carrying mothers (perinatal transmission, 1%). A similar distribution of transmission categories has also been reported from western Europe.[18]

Transmission of HIV via blood or blood products now is largely eliminated in Western nations. The rate of increase of AIDS cases presently appears to be slowing down among HS/BS males and drug abusers, partly due to "saturation" of these risk groups (declining percentage of susceptible individuals). Therefore, heterosexual transmission has become the most rapidly increasing transmission category. It should be borne in mind, however, that a majority of heterosexually infected individuals have contracted HIV from drug abusers, from BS males, or in Africa, Haiti, etc. A nearly 1:1 male to female ratio of HIV infection in central Africa indicates that heterosexual contact is a major route of transmission in this region. Currently, there is a broad spectrum of opinions on the extent of further spread of HIV in the heterosexual population in the U.S. or western Europe.

For HCWs it is important to remember that the actual number of HIV carriers is thought to exceed the number of AIDS cases by up to 50-fold, which would mean 1.8 million carriers in the U.S. and 0.3 million in Europe. Also, despite easy access of HIV tests and various testing programs in Western countries in recent years, a majority of HIV carriers probably remain undiagnosed. According to current estimates in Sweden, only about 30 to 40% of HIV carriers have been identified and reported to the health authorities. Thus, the probability of encountering a patient or a blood/tissue specimen with undiagnosed HIV may still be comparable to that of handling material known to be contagious; therefore, each year a substantial number of HCWs will unwittingly be occupationally exposed to HIV.

The local incidence of HIV-positive patients or specimens can be assessed by blind screening of a random sample. In the Stockholm area (population of 1.5 million, 1020 known HIV carriers, and 29 live AIDS patients as of June 1, 1987), screening of 5000 blood specimens not known to contain HIV recently received at two hospital laboratories yielded three HIV-positive specimens.[135]

C. Symptoms and Clinical Course

The onset of infection is usually asymptomatic, although some patients show symptoms of an acute febrile illness resembling mononucleosis, the so-called acute HIV infection, a few weeks after inoculation. In adults the most common final stage of HIV disease, AIDS, takes several years, and occasionally perhaps as long as 10 years or more to develop, whereas infected infants show a more rapid clinical course.

Severe recurrent gingival or periodontal problems and oropharyngeal infections (*Herpes simplex* stomatitis, necrotizing tonsillitis, and candidiasis) may be early indicators of AIDS.

Another oral lesion probably caused by Epstein-Barr virus, "hairy" leukoplakia, typically involving the lateral margins of the tongue, is regarded to be nearly pathognomonic of HIV infection and predicting development of AIDS. CNS involvement is common, causing neuropsychological disturbances before the development of AIDS and resulting in dementia in about half of AIDS patients. AIDS is further characterized by periods with fever, diarrhea, weight loss, fatigue, and infections of the respiratory (*Pneumocystis carinii, Mycobacterium intracellulare-avium,* and other opportunists as well as the classic pathogens) and gastrointestinal (*Salmonella*, mycobacteria, parasites, etc.) tract, by generalized fungal (yeast, *Cryptococcus, Aspergillus, Histoplasma*), parasitic (*Toxoplasma, Cryptosporidium*), or viral (*Herpes simplex*, varicella-zoster, cytomegalovirus) infections, and by lymphomas and tumors in 10 to 20% of the cases. Kaposi's sarcoma is normally an extremely rare skin tumor, but may occur in AIDS patients. Due to the devastating courses of the superinfections, 80% of AIDS patients succumb within 3 years and nearly 100% have died within 5 years after diagnosis.

Earlier, only a small proportion of HIV-infected individuals were thought to develop AIDS, but the prospects are growing worse with the increasing time of observation of the earliest cohorts of HIV carriers followed up. The risk of clinical disease among HIV carriers was estimated to be 25 to 50% by an U.S. expert committee in December 1986,[19] and according to more recent data — nearly 40% of HIV-infected HS/BS men will develop AIDS within 7 years.[20] However, although extrapolations from small studies of groups with risk conduct should be done with caution, and the natural history of HIV infection still remains speculative; other experts fear that the time between contracting HIV and the occurrence of AIDS may occasionally be as long as 10 to 20 years[21] and that a majority of or perhaps all carriers will eventually develop the syndrome.

D. Laboratory Diagnosis

While diagnosing HIV by culture is expensive and has low sensitivity in carriers (see below), serological methods have proved to be reliable and economically feasible. For screening purposes (of blood donors, pregnant women, groups with risk conduct, etc.), detection of anti-HIV antibody in serum by enzyme linked immunosorbent assay (ELISA) techniques of various design and constantly improving performance is generally used. The sensitivity of these methods is deliberately adjusted to be very high (>99%), which in turn leads to lower specificity and a substantial risk of false positive results. Thus, positive screening tests need to be followed up by confirming serological tests (Western blot; ELISA; immunofluorescence; or radioimmunoprecipitation assay, RIPA), whereas isolation of HIV remains a supporting and research procedure. Recently also, ELISA tests to assay HIV antigen in serum have been developed to be used in special situations (see below).

In response to the HIV infection, nonprotective but diagnostic anti-HIV antibodies usually develop 2 to 8 weeks after inoculation. In occasional cases, latency periods for antibody response of up to 8 months have been observed, and also rare examples of isolation of HIV from consistently seronegative individuals have been published.[22]

HIV antigen assays can potentially be used to diagnose the virus in the window between contracting HIV and the production of serum antibodies. Regrettably, such tests currently detect only 60 to 70% of infected individuals during the first week of incubation followed by a declining positivity rate. Thus, this type of test has so far not solved the problem of reliable early diagnosis of HIV infection. On the other hand, the antigen tests may prove to be useful as an indicator of prognosis and contagiousness and to confirm the diagnosis in the occasional antibody-negative carriers of HIV. As the viral nucleic acid becomes an integral part of the host genome and infectious HIV particles can be isolated from the blood of 25% of asymptomatic individuals with HIV antibody and 100% of AIDS patients (see below), all seropositive individuals are to be regarded as contagious carriers of HIV.

E. Therapy and Vaccines

An increasing number of antiviral agents, usually directed against reverse trancriptase, are being developed. Several show anti-HIV effects *in vitro* and some of them also show anti-HIV effects *in vivo*, as indicated by fewer superinfections and a lowered mortality rate among the treated AIDS patients. Azidothymidine (Retrovir®) and phosphonoformate (Foscarnet®) are among agents showing some clinical effect in early trials. However, only suppression of HIV infection but not cure seems to be achieved. This indicates that lifelong treatment will be necessary, increasing the risk of adverse drug effects. Thus, no breakthrough in anti-HIV chemotherapy appears to be in sight and the immediate prospects for effective treatment of HIV-infected individuals remain poor. On the other hand, the current anti-HIV agents may prove to be of value for post-exposure prophylaxis among HCWs involved in accidents with, e.g., HIV-positive blood.

Despite the inherent problem of extreme antigenic variability in HIV (see above), several vaccines are being developed and the first clinical trials are under way. Although anti-HIV antibodies can be elicited, the protective effect of the immune response, if any, is not yet known.

III. SOURCES AND MODES OF HUMAN IMMUNODEFICIENCY VIRUS (HIV) TRANSMISSION

HIV can often be isolated from blood, serum, or plasma,[11,23,24] can be isolated less easily from semen,[11,24,25] cerebrospinal fluid,[11,26,27] and saliva,[11,28] can be isolated only sporadically from tears,[29,30] breast milk,[31] urine,[11] cervical secretions,[32] and feces,[33] and cannot be isolated from normal gastric juice.[34] There are no reports on isolation of HIV from sweat. Because of the difficulties in isolating HIV from the semen of carriers, despite clinical evidence of contagiousness, it has been maintained that transmission of HIV by semen and saliva is primarily through infected cells.[11]

The virus is also likely to be present in tissues containing human blood and in serous, bloody, or purulent discharge, e.g., from wounds. That certain inflammatory cells could be important vehicles for transmission of HIV is supported also by epidemiological evidence. For example, the risk of acquisition of HIV is greater for female sexual partners of HIV-carrying BS men (who often have urethral fluor due to various genital infections, unspecified urethritis, or other inflammatory states) than for partners of men infected by blood transfusion or blood products. Similarly, comparisons with hepatitis B virus (HBV) indicate a 100-fold lower transmission risk for HIV than for HBV upon parenteral exposure (see below), but only a 4-fold difference upon sexual exposure. This discrepancy could be due to the fact that cells in genital secretions may carry HIV, but not HBV. Thus, purulent secretions from sites of superinfection in AIDS patients may represent an underestimated HIV hazard to HCWs.

A large number of epidemiological studies have failed to show evidence for transmission of HIV via water, food, and close personal contact (other than sexual).[9] The following routes of transmission and nontransmission of HIV were emphasized by the WHO in December 1985:[35]

> Epidemiological evidence has only implicated direct contact with blood and semen in transmission. Thus, HIV is normally transmitted through sexual intercourse (including anal) or parenteral exposure to blood and blood products and from infected mother to child (*in utero* or in the postnatal period).
>
> Studies of nonsexual household contacts indicate that casual contact with saliva or tears does not result in transmission of infection. There is no evidence that HIV can be transmitted by close social contacts (family setting, schools, or other groups living or working together), blood-sucking insects, food, or water (fecal/oral or airborne routes).

In June 1987, the WHO reiterated these statements[36] and also excluded toilets, swimming pools, sweat, shared eating and drinking utensils, and telephones, in HIV transmission.

However, in addition to blood, semen, and vaginal/cervical fluid, the WHO now regards saliva as a possible risk. "Kissing has not been documented to pose a risk of HIV transmission. While unproven, some theoretical risk from vigorous "wet" kissing (deep kissing or tongue kissing) may exist."[36]

Thus far, only three cases of possible household transmission of HIV (without sexual or known parenteral exposures) have been reported and in none of these has the actual route of transmission been resolved. One case is a boy infected by his transfusion-infected little brother despite that no skin-penetrating bites, scratches, accidents involving bleeding, etc. are known to have occurred.[37] The second is a mother who nursed her transfusion-infected ill infant both in the hospital and at home (exposure to blood and secretions).[38] Similarly, a 45-year-old housewife without other risk factors may have contracted HIV during terminal home nursing care of a 33-year-old African man. He died from encephalitis of unknown etiology and was retrospectively diagnosed as a case of AIDS with cerebral toxoplasmosis.[39] In the two latter cases, nursing was performed without knowledge of the HIV risk and, thus, without any precautions, and in one case uncovered skin lesions were suggested as ports of entry.[39] However, an initial antibody-negative serum specimen to show an association between the period of exposure and seroconversion is lacking in all three cases.

Assuming that HIV can be transmitted via the same routes as HBV, although at a lower efficiency, the principal modes of HIV transmission within the health care setting probably include exposure to blood/body secretions by the following:[9]

1. Percutaneous exposure (e.g., injury by a needle, scalpel, or other sharp object that has been contaminated by HIV)
2. Mucous membrane splash (e.g., contamination of mouth, nose, or conjunctivae by HIV-containing blood)
3. Wound or skin exposure (contamination of fresh, open abrasions, incisions, or lacerations, chapped skin, and possibly acne or dermatitis lesions)

IV. OCCUPATIONAL TRANSMISSION OF HUMAN IMMUNODEFICIENCY VIRUS (HIV) TO HEALTH CARE WORKERS (HCWs)

A. Judgments by the World Health Organization (WHO)

Although the WHO concluded in 1985 that "the risk of transmission to HCWs seems to be very remote", countries were urged to issue "(i) guidelines for the care of HIV positive patients and the handling of their specimens in the hospitals and other settings and (ii) codes of good laboratory practice to protect staff against the risk of infection. Recommendations for the control of hepatitis B infection should, if followed, also effectively prevent spread of HIV infection."[35]

In 1987 the WHO issued similar statements.[36] "Reports of HIV infection of a small number of health workers have emphasized the need to adhere to existing guidelines for the prevention of blood-borne infections. Such existing guidelines refer to situations in which there is a possibility of exposure to blood or any body fluid regardless of their source. Available information indicates that health workers are normally at very low occupational risk of HIV infection. This very low risk can be further minimized if existing guidelines for avoiding any blood-borne infection are rigorously implemented and strictly enforced."

B. Evidence of Occupational Transmission

It is now established that there exists a possibility of occupational transmission of HIV to HCWs, but that the risk is very low. To confirm such transmission, the following criteria currently seem reasonable.[9]

1. Occupational exposure of mucous membranes or damaged skin to blood/serum/plasma, wound discharge, tissue, etc. from a patient infected with HIV
2. An initial serum specimen obtained on or soon after the date of exposure that is negative for HIV antibody and preferably also for HIV antigen
3. A follow-up specimen (ideally without interim exposure) that is confirmed to be positive for HIV antibody and/or HIV antigen
4. No other established risk factor (homosexual contacts, i.v. drug use, different heterosexual contacts, infected heterosexual partner, etc.) before or after the occupational exposure in question

Based on these criteria, seven documented cases of nosocomial transmission of HIV, all involving female HCWs (six nurses and one medical technologist), have been published to date (Fall of 1987; see also Appendix).

The first case reported involved needlestick and probably microinjection of patient blood,[40] and the second case involved a deep i.m. injury from a large-bore needle visibly contaminated with HIV-positive blood.[41] Two other cases have involved a superficial needlestick[42] and a superficial scrape with a needle not visibly contaminated.[43]

Recently, the CDC reported three additional documented cases among HCWs apparently not associated with percutaneous injuries.[44] One was an emergency room nurse who had chapped ungloved hands and applied pressure for 20 min with her index finger to a site where an arterial catheter had been removed from the patient. Another was a female phlebotomist filling a vacuum collection tube with blood from an outpatient when the top flew off and blood splattered on her face and on her mouth. She also had facial acne, but no open wounds. The seventh case involved a medical technologist who operated a blood separation device when blood spilled, covering her hands. She did not recall any open wounds or mucous membrane exposure, but may have touched one ear where she had a dermatitis.

The inherent limitation when documenting nosocomial or other occupational transmission of HIV is that existing alternative routes of transmission (notably sexual) may be difficult to disclose. On the other hand, there is also a risk that true occupational transmission of HIV is downplayed or dismissed because of incomplete documentation. For example, incidents involving contact with body secretions in the absence of percutaneous injuries may be difficult to recall and often lack initial HIV serology in patient or staff. Still, in addition to the seven published confirmed cases of occupational HIV transmission, seven less rigorously documented cases have been reported. Four were among HCWs,[45] another was the mother who unwittingly received "intense" exposure to her HIV-positive infant's blood and secretions both in hospital and at home,[38] and another was the housewife who without precautions provided terminal home nursing care for a man retrospectively diagnosed as a case of AIDS (see above).[39] Recently, a case of occupational HIV transmission to a dentist was also reported,[46] although BS conduct may be an alternative route of infection.

Several U.S. studies of various designs that estimate the risk of HIV infection among medical and dental HCWs after percutaneous injuries, mucous membrane splash accidents, etc., or no specific exposure during care of HIV-carrying patients have been published.[45-57] These studies have thus far collectively followed up 3995 individuals, and 1 documented and 3 apparent cases of transmission have been recorded. Disregarding about 50% of HCWs with no specific accident but included in these studies, about 0.2% of HCWs followed up after known exposures to HIV may have occupationally contracted the virus. Also a number of smaller studies from the U.K. and Canada indicate a low risk.[58-61]

In a recent update by the CDC on three ongoing prospective U.S. studies of HCWs exposed to blood of HIV-infected patients through needlestick wounds or contamination of open wound or mucous membrane, only one case of seroconversion after a total of 1492 occupational exposures to HIV (0.06%) has thus far been found.[44] The CDC Cooperative Needlestick Surveillance Study[50,57] presently represents the largest number of documented occupational exposures to HIV via percutaneous injuries. Thus far, one documented and one apparent case of transmission have been recorded after 856 parenteral exposures (0.2%), including one documented case of occupational transmission from 205 exposures involving patients who meet the surveillance definition of the CDC on AIDS (0.5%, 95% confidence interval, 0.00 to 2.30%).[57]

In summary, HCWs in the U.S. and western Europe have so far been exposed to more than 40,000 AIDS patients and, in addition, to a large number of patients with pre-AIDS states and to asymptomatic HIV carriers. A total of seven "documented" and a further seven possible cases of occupational acquisition of HIV among HCWs indeed supports the WHO view that the risk of such transmission is very low provided that traditional blood/body fluid precautions are adhered to.

C. Relation between Stage of HIV Disease and Contagiousness

As mentioned above, a majority of HIV-infected individuals probably still remain undiagnosed. Asymptomatic HIV carriers were earlier thought to represent a greater potential hazard to HCWs because they generally have higher levels of circulating T4-lymphocytes and, thus, possibly higher concentrations of circulating HIV particles than patients with AIDS.[23,28]

More recent work has, however, shown that along with declining immune defense and numbers of T4-cells, viral expression is increased in later stages of HIV infection as indicated, for example, by increasing positivity of HIV antigen assays[62-65] and by a rising HIV isolation rate (25% in asymptomatic carriers, 75 to 85% in patients with pre-AIDS stages, and 100% in AIDS cases).[66,67] Thus, appearance and persistence of HIV antigen and consistently low titers of neutralizing antibody[68,69] signal poor prognosis and full contagiousness, whereas absence of HIV antigen and high antibody titers correlate with better outcome and probably lower contagiousness. The rarely occurring antibody-negative carriers of HIV[22,70] may in fact belong to the most contagious patients.

The large group of healthy-appearing and usually undiagnosed HIV carriers should therefore generally represent a lesser HIV risk to HCWs than patients with pre-AIDS or AIDS. Of the seven documented instances of occupational transmission of HIV to HCWs described above, three involved patients with AIDS, one involved a patient with persistent generalized lymphadenopathy, and one involved a patient with undiagnosed HIV infection and *P. carinii* pneumonia, whereas two source patients were asymptomatic HIV carriers.

D. Risk Procedures in Hospitals and Staff at Risk

Studies of the prevalence of HBV infection among hospital workers show that frequency of blood contact and frequency of needle accidents are independently correlated with infection while degree of patient contact per se had no effect.[71,72] Thus, risk is best estimated by quantitating the degree of blood and needle contact during daily work.

Assuming that the major routes of occupational transmission of HIV are the same as for HBV, injuries from needles in various situations represent the most common event posing a potential risk of acquisition of HIV for HCWs (up to 70% of risk exposures, Table 1).[4,58] Major bleeding causing gross skin contamination and injuries from dirty sharp instruments in operating theaters are also relatively frequent, whereas blood splash in mouth or eye and bites and scratches from patients, etc. represent rare examples of potential exposure to HIV. However, studies with careful recording of minor splashing accidents also indicate that these may outnumber the needlestick accidents.[48] In the CDC Needlestick Study, 40% of accidents were judged potentially preventable (recapping of used needles, 17%; improperly disposing of used needles, 13%; contaminating open wounds, 10%).[57]

Table 1
BLOOD INOCULATION INJURIES TO HCWs INVOLVING POTENTIAL OR ACTUAL HBV OR HIV CARRIERS[a]

Type of injury	Number of staff
Needlestick, donor known (needle for collection of blood or injection)	93
Needlestick, donor unknown	
Sticking out of rubbish bag	25
Sticking out of box	10
Intravenous drip cannulae	13
Dirty sharp instrument in theater	24
Major bleeding causing gross skin contamination (in casualty or isolation ward)	25
Blood splash in eye	4
Razor blade accident	3
Bite from patient	3
Laboratory glass accident (pipette, test tube)	3
Scratch from patient	2
Cadaver (not in bag, blood-stained, discharging fluids)	1
Total	206

[a] St. Stephen's Hospital, London, 1982 to 1985.

From Shanson, D., *Infect. Control,* 7, 128, 1986. With permission.

Within the health care setting, those most likely to be at risk for occupational HIV infection include nurses, physicians, laboratory and patient care technicians, dentists, optometrists, podiatrists, chiropractors, morticians, housekeeping personnel, and any other employee whose duties require contact with patients, blood, body fluids, or tissues. These persons, termed HCWs both here and by the CDC, do not include persons such as ward clerks, administrative and medical records personnel, and dietary workers, who ordinarily are not at risk for exposure that might transmit infection.

Among HCWs involved with reported blood inoculation injuries, trained nurses are heavily dominating (about two thirds), followed by student nurses, laboratory technicians, doctors, and domestics (Table 2).[4,58,59] Regrettably, published studies lack information on the population size for each category of HCWs, preventing calculations of relative risks of exposure.

V. NONOCCUPATIONAL HUMAN IMMUNODEFICIENCY VIRUS (HIV) INFECTION AMONG HEALTH CARE WORKERS (HCWs)

As of May 1986, 922 (5.5%) of the 16,748 U.S. cases of AIDS (with occupational category identified) reported to the CDC were in HCWs. Out of these, 888 could be identified as members of known risk groups, including 27 of the 28 dentists and dental assistants/technicians reported.[50] In no case was specific occupational exposure to HIV infection documented or suspected. Thus, screening of HCWs for HIV antibody would disclose occasional individuals as carriers, but further investigation usually incriminated their private habits as the likely source of HIV infection. Nevertheless, there currently are substantial numbers of HCWs with HIV infection.

The risk of transmission of HIV from HCWs to patients is probably exceedingly low and has to date not been described (see Section IX). For this and a number of additional reasons (legal, psychological, ethical, financial, etc.), routine screening of HCWs for HIV antibody

Table 2
HCWs AND OTHER STAFF INVOLVED WITH BLOOD INOCULATION INJURIES DURING CARE OF POTENTIAL OR ACTUAL HBV OR HIV CARRIERS[a]

Professional	Number of staff	
	Male	Female
Trained nurse	4	112
Student nurse	0	26
Pathology technician	1	17
Physician	9	7
Domestic	3	7
Porter	2	0
Medical student	1	0
Laundry worker	0	1
Plumber	1	0
Others	4	11
Total	25	181

[a] St. Stephen's Hospital, London, 1982 to 1985.

From Shanson, D., *Infect. Control,* 7, 128, 1986. With permission.

is not recommended in current guidelines from the CDC[10,73-77] or other national or local authorities[1-3,78,79] nor is it suggested to exclude HCWs with anti-HIV antibody from employment. For HCWs performing invasive procedures, the question of screening (and restrictions for carriers) is currently being considered by U.S. health authorities (see Section IX).

Although nosocomial transmission of infections to HIV-infected HCWs has not been reported, such workers may be potentially immunocompromised and therefore extra susceptible to infections. Such employees should be well educated and consistent in their use of barrier precautions in appropriate situations in order to minimize both the possibility for acquisition of infections from the patients and transmission of HIV. Susceptibility to infection and a presence of CNS symptoms should further be considered in work assignments of HIV-infected HCWs. Direct patient contact should be avoided by HIV-infected HCWs who have exudative or weeping lesions (see Section IX).

VI. COMPARISON OF HUMAN IMMUNODEFICIENCY VIRUS (HIV) AND HEPATITIS B VIRUS (HBV) TRANSMISSION

The major modes of transmission are common to HBV and HIV (sexual, parenteral, perinatal contact). Besides percutaneous inoculation, splashing of contagious blood into the eye or mouth has caused transmission of HBV both in chimpanzee experiments[81,82] and occasionally in clincal practice, whereas classic airborne transmission (e.g., aerosols) of HBV has not been demonstrated. Thus, HCWs who do not have contact with blood, other body fluids, tissues, or needles/sharp objects contaminated with blood are not at risk for acquiring HBV or HIV infection in the workplace.

Despite epidemiological similarities of HBV and HIV infection, the risk of nosocomial transmission is much lower for HIV than for HBV. This may, in part, be due to the marked difference in concentrations of infectious virus particles in the blood of patients (10^7 to more than 10^8 per milliliter for HBV[83] and none to 10^4 per milliliter, for HIV,[11] depending on the phase of the respective infection). Whereas the risk of acquiring HIV within the health care setting is currently estimated to be about 0.5% following percutaneous exposure (see above), the same risk for HBV has been reported to be 6 to 30%,[84,85] with the higher rates if the source carried the virulence marker HBV e-antigen. In fact, an HBV transmission rate of 19% from e-antigen-positive patients has been recorded despite postexposure administration of specific anti-HBV immune globulin in various dosages.[86]

That the risk of acquiring HIV infection is low has been further indicated by two case reports where needlestick exposures enough to result in transmission of HBV[87] and *Cryptococcus neoformans*[88] infection, respectively, did not lead to transmission of HIV concomitantly carried by the source patients. HBV thus represents a "worst case" condition with regard to nosocomial transmission and, as also emphasized by the WHO,[35] recommendations for the control of HBV infection should, if followed, also effectively prevent the spread of HIV. Thus, despite the difference in transmission risk, the model of nosocomial HBV transmission and available data regarding HIV provide the basis for generating precautions to HCWs.

VII. PREVENTION OF TRANSMISSION OF HUMAN IMMUNODEFICIENCY VIRUS (HIV) TO HEALTH CARE WORKERS (HCWs)

Several major, formal consensus statements have appeared in the U.S.,[1,10,73-76] U.K.,[2] and most other West European countries (e.g., Sweden[78]) primarily addressed at prevention of nosocomial HIV transmission. Updated comprehensive recommendations by the U.S. Department of Health and Human Services were published in a nursing journal[77] and the subject has been reviewed also by several authors, focusing on different aspects of prevention and different clinical disciplines.[3-9,89-95] Since microorganisms know no frontier, these and similar publications have provided a general basis for locally modified guidelines now effective in most U.S. and European hospitals.

Sound guidelines in writing, combined with education of HCWs regarding the epidemiology, modes of transmission, and prevention of HIV infection, are fundamental to minimize occupational acquisition of HIV in the health-care arena. This is of particular importance for staffs frequently exposed to AIDS patients, both to reduce anxiety and risk.[96-98] In the future, possibilities to assess individual HCWs with respect to their genetic susceptibility to HIV infection may influence work assignment and, thus, become a further method to reduce the transmission risk.[99]

It is universally agreed (see above) that existing approaches to reduce occupational exposure to blood and other body fluids will also protect HCWs from acquiring HIV. New infection-control guidelines for patients with HIV infection are therefore not required, but rather a stricter application of current guidelines.[79] Thus, awareness of the potential for nosocomial transmission of HIV has resulted in renewed respect for the present principles and practice of infection control.

As complete information on which patients, laboratory specimens, etc. represent an HIV risk will probably never become available to HCWs, the logical and safest policy is to use "blood/body fluid precautions" for all patients and laboratory specimens.[79,100] This strategy will become of increasing importance as long as the AIDS epidemic continues to grow. In the following, the most important aspects of preventing transmission of HIV, HBV, and other less common blood-borne infections in the workplace are summarized.

Sharp objects — Accidental skin punctures represent the leading type of blood inoculation injuries in hospitals (Table 1). Also, four out of the seven documented instances of occupational transmission of HIV to HCWs involved needle accidents (see Section IV). As about half of such skin punctures are potentially preventable,[57,100,101] proper equipment and work routines and taking the maximum caution possible when handling sharp objects, notably needles, are the most important areas for preventing occupational transmission of HIV in the health care setting besides education of HCWs. Consequently, only trained and experienced personnel should perform venepunctures, injections, operations, laboratory procedures involving patient blood, etc., particularly in connection with HIV-positive patients. Particular care should be taken when removing and disposing of needles, scalpels, etc. into puncture-proof containers, and when ejecting blood into specimen bottles.[2] New products are available that reduce needlestick accidents in venepunctures.[102]

Nursing — Routine handwashing before and after patient care using soap and water is generally sufficient. Gloves should be used when contact with blood, body fluids, and tissues is anticipated[75] (plus impervious gowns or plastic aprons for gross contamination). Based on experience with HBV (occasional transmission due to contact between contagious blood and oral mucosa or conjunctiva), masks and protective eyewear or a face shield are recommended when splashing of blood is anticipated.[75] A mask should be used also when pulmonary tuberculosis is suspected. Special isolation units or single rooms are not necessary for care of cooperating HIV patients unless they have superinfections due to classical pathogens such as *Mycobacterium tuberculosis* or salmonellae.

Surgery, obstetrics, post-mortem examinations — "All HCWs who perform or assist in operative, obstetric or dental invasive procedures must wear gloves when touching mucous membranes or nonintact skin of all patients and use other appropriate barrier precautions when indicated (e.g., masks, eye coverings, and gowns, if aerosolization or splashes are likely to occur)."[76] Since arterial squirting must always be anticipated, e.g., in surgery, eye protection should also be routinely used. Probably even more important is the CDC recommendation to "use extraordinary care to prevent injuries to hands caused by needles, scalpels, and other sharp instruments or devices during procedures."[76]

Specimens — Blood-contaminated tubes/containers should always be disinfected before transport to the laboratory. In many countries HBV or HIV risk is indicated by a biohazard label (e.g., "blood precautions"). All such tubes/containers should be put into a protective sleeve or an outer safety container for transport.

Laboratory — All blood specimens should be handled as being contagious.[100] Whenever possible open/manual laboratory methods should be replaced by mechanical pipetting devices, closed automatic procedures, etc. Gloves should always be used during handling of blood specimens and during cleaning and repair of analysis equipment. Glass pipettes should be avoided and mouth pipetting should not be allowed. "Biological safety cabinets (Class I or II) and other primary containment devices (e.g., centrifuge safety caps) are advised whenever procedures are conducted that have a high potential for creating aerosols or infectious droplets. These include centrifuging, blending, sonicating, vigorous mixing, and harvesting infected tissues from animals or embryonated eggs." [3] However, as there is no evidence of HIV transmission via aerosolized blood, masks and eye coverings, a face shield, or a protective perspex or glass shield attached to the workbench are budget alternatives that offer as good protection as specially ventilated safety cabinets. Potentially contaminated materials should be decontaminated, preferably by autoclaving, before disposal or reprocessing. For chemical disinfection of equipment, work surfaces, and spill of potentially infectious material, see the section entitled "Sterilization and Disinfection".

Dentistry — Although dentists and to some extent dental assistants show excess morbidity for HBV infection[103] and one doubtful case of occupational HIV in a Manhattan dentist has recently been claimed (see above),[46] there is no other evidence to date of HIV transmission in dental/oral surgery clinics.[46,54,56,104,105] Because HBV- or HIV-infected patients may be particularly difficult to disclose in this setting, gloves must always be worn when touching blood, saliva, or mucous membranes and surgical masks and protective eyewear (or face shield) must be worn when splashing of blood or saliva is likely according to the CDC.[106]

occupational transmission of HIV to one Swedish HCW was recently estimated to be about 0.5 billion.[132] Also, the WHO takes a cautious attitude towards the screening of patients. "Routine HIV screening of patients to protect health workers should not be implemented without careful and detailed consideration of all of the HIV screening criteria developed by the World Health Organization."[35]

IX. PREVENTION OF TRANSMISSION OF HUMAN IMMUNODEFICIENCY VIRUS (HIV) FROM HEALTH CARE WORKERS (HCWs) TO PATIENTS

In case of HBV, there exists, to this author's knowledge, only rare examples of transmission of infection from a HCW carrier to a patient. Since HIV appears to be about 100-fold less contagious than HBV in the hospital setting (see Section VI), the risk of nosocomial HCW to patient transmission of HIV must be nearly nil. Thus, despite efforts to evaluate the possible risk due to HIV-positive HCWs, including follow-up of 400 patients operated on by a HIV-positive surgeon,[133] there is currently no evidence of HCW to patient transmission of HIV. Such transmission could, however, theoretically occur in situations where there is both (1) a high degree of trauma to the patient that would provide a portal of entry for HIV and (2) access of blood from the infected HCW to the open tissue of the patient (e.g., scalpel injury to the HCW during an invasive procedure).

Current guidelines do not include routine HIV testing of HCWs or restricting HCWs from work because of known carriage of HIV (see Section V). Whether HIV testing of HCWs who perform invasive procedures (and restrictions for carriers) is indicated is currently being considered by the U.S. Department of Health and Human Services.[75,77] HCWs known to be HIV carriers should wear gloves for direct contact with mucous membranes or nonintact skin of all patients. HCWs who have exudative lesions or weeping dermatitis should refrain from direct patient care and from handling patient care equipment until the condition resolves.[75,77]

X. CONCLUSION

Blood/body fluid precautions in combination with availability of diagnostic tests (not used for routine screening of patients) have proven to be highly effective in preventing HBV transmission in hospitals and have already reduced the occupational risk among HCWs by about tenfold before the advent of anti-HBV immune globulin and hepatitis B vaccine.[103] Because of the many similarities between HBV and HIV with regard to routes of transmission, although with about 100-fold lower apparent efficiency for HIV, current guidelines to prevent occupational HBV infection are also potentially effective against occupational HIV infection. A stricter application of these precautions is still possible, as suggested by the further decline of occupational HBV infections in Sweden since the appearance of HIV.[134] Similarly, in a recent study from San Francisco, compliance with recommended (CDC) infection-control precautions in HCWs performing procedures with potential exposure to HIV on patients with AIDS was still only 56% in 1984 to 1986.[48] Sound guidelines, rigorously implemented and strictly enforced, and well-educated HCWs together ensure a very low risk of occupational HIV infection and represent the only preventive approach currently available.

APPENDIX

During the processing of this volume a large number of scientific reports and even journals entirely dedicated to AIDS have been published. More sensitive diagnostic methods (e.g., DNA polymerase chain reaction, PCR) are being evaluated and differentiation between types of anti-HIV antibodies yields better prognostic information.

The number of reported cases of AIDS has increased by more than 100% over the last 18 months and about 20 cases of occupational HIV infection in HCWs have been documented. A majority of these have been associated with needlesticks and in one case AIDS has developed.

In recent statements from the CDC, preventing exposures to patient blood is emphasized, whereas other body fluids are downplayed as risks.[136-138] Detailed guidelines for prevention of transmission of blood-borne infection during clinical and anatomical laboratory procedures have been issued by the U.S. National Committee for Laboratory Standards (NCCLS).[134]

REFERENCES

1. **Rahme, F. S.,** Prevention of in-hospital transmission of the acquired immune deficiency syndrome virus (HTLV III): current USA policy, *J. Hosp. Infect.,* 6(Suppl. C), 53, 1985.
2. U.K. Hospital Infection Society, Acquired immune deficiency syndrome: recommendations of a HIS Working Party, *J. Hosp. Infect.,* 6(Suppl. C), 67, 1985.
3. **Ayliffe, G., Jeffries, D., Geddes, A., Casewell, M., Keane, C. T., Craske, J., Longson, M., Percival, A., Gazzard, B., Shanson, D. C., Reeves, D. S., Ridgway, G., Hofbrand, V., Kernoff, P., and Brumfitt, W.,** AIDS and the health professions, *Br. Med. J.,* 290, 853, 1985.
4. **Shanson, D.,** The risks of transmission of the HTLV-III and hepatitis B virus in the hospital, *Infect. Control,* 7, 128, 1986.
5. **Garibaldi, R. A.,** Transmission of hepatitis B and AIDS, *Infect. Control,* 7, 132, 1986.
6. **Castro, K. G., Hardy, A. M., and Curran, J. W.,** The acquired immunodeficiency syndrome: epidemiology and risk factors for transmission, *Med. Clin. North Am.,* 70, 635, 1986.
7. **Valenti, W. M.,** AIDS update: HTLV-III testing, immune globulins and employees with AIDS, *Infect. Control,* 7, 427, 1986.
8. **Healey, M. J.,** AIDS: definition and guidelines for the health professional, *Radiol. Technol.,* 57, 233, 1986.
9. **Vlahov, D. and Polk, B. F.,** Transmission of human immunodeficiency virus within the health care setting, in *Occupational Medicine, State of the Art Reviews,* Hanley & Belfus, Philadelphia, 1987.
10. CDC, AIDS — Recommendations and Guidelines, November 1982-November 1986, Centers for Disease Control, Public Health Service, U.S. Department of Health and Human Services, Atlanta, 1987, 1.
11. **Levy, J. A., Kaminsky, L. S., Morrow, W. J. W., Steimer, K., Luciw, P., Dina, D., Hoxie, J., and Oshiro, L.,** Infection by the retrovirus associated with the acquired immunodeficiency syndrome. Clinical, biological, and molecular features, *Ann. Intern. Med.,* 103, 694, 1985.
12. **Weiss, S. H., Ginzburg, H. M., Saxinger, W. C., Cantor, K. P., Mundon, F. K., Zimmerman, D. H., and Blattner, W. A.,** Emerging high rates of human T-cell lymphotrophic virus type I (HTLV-I) and HIV infection among U.S. drug abusers, in *Proc. 3rd Int. Conf. on AIDS,* Washington, D.C., June 1 to 5, 1987, 211 (Abstr.).
13. **Biggar, R. J., Nasca, P. C., and Brunett, W. S.,** AIDS-related Kaposi's sarcoma in New York City in 1977, in *Proc. 3rd Int. Conf. on AIDS,* Washington, D.C., June 1 to 5, 1987, 121 (Abstr.).
14. CDC, U.S. AIDS Program, AIDS Weekly Surveillance Report, Centers for Disease Control, Public Health Service, U.S. Department of Health and Human Services, Atlanta, June 1, 1987.
15. **Cassens, B. J.,** Social consequences of the acquired immunodeficiency syndrome, *Ann. Intern. Med.,* 103, 768, 1985.
16. WHO, AIDS update, *WHO Wkly. Epidemiol. Rec.,* 62, 202, 1987.
17. Leading article, CDC update, Acquired Immunodeficiency Syndrome — United States, Morbidity and Mortality Weekly Rep. No. 35, Centers for Disease Control, Public Health Service, U.S. Department of Health and Human Services, Atlanta, 1986, 757.
18. WHO Collaborating Center on AIDS, Paris, AIDS Surveillance in Europe, Rep. 12, World Health Organization, Geneva, 1983.
19. **Church, G. J.,** Call to battle, U.S. National Academy of Sciences report on AIDS, *TIME,* November 10, 1986.
20. **Hessol, N. A., Rutherford, G. W., O'Malley, P. M., Doll, L. S., Darrow, W. W., and Jaffe, H. W.,** The natural history of human immunodeficiency virus infection in a cohort of homosexual and bisexual men: a 7-year prospective study, in *Proc. 3rd Int. Conf. on AIDS,* Washington, D. C., June 1 to 5, 1987, 1 (Abstr.).

21. **Rees, M.,** The sombre view of AIDS, *Nature (London),* 326, 343, 1987.
22. **Farber, C. M., Sprecher-Goldberger, S., Liesnard, C., Huyghen, K., Coginaux, J., and Thiry, L.,** Persistent human immunodeficiency virus (HIV) detection in seronegative asymptomatic carriers, in *Proc. 3rd Int. Conf. on AIDS,* Washington, D.C., June 1 to 5, 1987, 30 (Abstr.).
23. **Gallo, R. C., Salahuddin, S. Z., Popovic, M., Shearer, G. M., Kaplan, M., Haynes, B. F., Palker, T. J., Redfield, R., Oleske, J., Safai, B., White, G., Foster, P., and Markham, P. D.,** Frequent detection and isolation of cytopathic retroviruses (HTLV-III) from patients with AIDS and risk for AIDS, *Science,* 224, 500, 1984.
24. **Ho, D. D., Schooley, R. T., Rota, T. R., Kaplan, J. C., Flynn, T., Salahuddin, S. Z., Gonda, M. A., and Hirsch, M. S.,** HTLV-III in the semen and blood of a healthy homosexual man, *Science,* 226, 451, 1984.
25. **Zagury, D., Bernard, J., Leibowitch, J., Safai, B., Groopman, J. E., Feldman, M., Sarngadharan, M. G., and Gallo, R. C.,** HTLV-III in cells cultured from semen of two patients with AIDS, *Science,* 226, 448, 1984.
26. **Levy, J. A., Shimabukuro, J., Hollander, H., Mills, S., and Kaminski, L. S.,** Isolation of AIDS associated retroviruses (ARV) from cerebrospinal fluid and brain from patients with neurologic symptoms, *Lancet,* 2, 586, 1985.
27. **Shaw, G. M., Harper, M. E., Hahn, B. H., Epstein, L. G., Gajdusek, C., Price, R. W., Navia, B. A., Petito, C. K., O'Hara, C. J., Groopman, J. E., Cho, E. S., Oleske, J. M., Wong-Staal, F., and Gallo, R. C.,** HTLV-III infection in brains of children and adults with AIDS encephalopathy, *Science,* 227, 177, 1985.
28. **Groopman, J. E., Salahuddin, S. Z., Sarngadharan, M. G., Markham, P. D., Gonda, M., Sliski, A., and Gallo, R. C.,** HTLV/III in saliva in people with AIDS-related complex and healthy homosexual men at risk for AIDS, *Science,* 226, 447, 1984.
29. **Fujikawa, L. S., Salahuddin, S. Z., Palestine, A. G., Masur, H., Nussenblatt, R. B., and Gallo, R. C.,** Isolation of human T-lymphotrophic virus type. III from the tears of a patient with the acquired immunodeficiency syndrome, *Lancet,* 2, 529, 1985.
30. **Tervo, T., Lahdevirta, J., Vaheri, A., Valle, S. L., and Suni, J.,** Recovery of HTLV-III from contact lenses, *Lancet,* 1, 379, 1986 (letter).
31. **Thiry, L., Sprecher-Goldberger, S., Jonckheer, T., Levy, J., van de Perre, P., Henrivaux, P., Cogniaux-LeClerc, J., and Clumeck, N.,** Isolation of AIDS virus from cell-free breast milk of three healthy virus carriers, *Lancet,* 2, 891, 1985 (letter).
32. **Wofsy, C. B., Cohen, J. B., Hauer, L. B., Padian, N. S., Michaelis, B. A., Evans, L. A., and Levy, J. A.,** Isolation of AIDS-associated retrovirus from genital secretions of women with antibodies to the virus, *Lancet,* 1, 527, 1986.
33. **Lycke, E.,** personal communication, 1986.
34. **Levy, J. A., Hoffman, A. D., Kramer, S. M., Landis, J. A., and Shimabukuro, J. M.,** Isolation of lymphocytopathic retroviruses from San Francisco patients with AIDS, *Science,* 225, 840, 1984.
35. WHO, In point of fact, AIDS/SIDA, *WHO Media Service,* Geneva, December, 1985.
36. **Mann, J.,** Concensus statements from the Third Meeting of the WHO Collaborative Centres on AIDS, Washington, D.C., June 6, 1987, *WHO Facsimile,* No. 1363, 1, 1987.
37. **Wahn, V., Kramer, H. V., Voit, T., Brüster, H. T., Scrampical, B., and Scheid, A.,** Horizontal transmission of HIV infection between two siblings, *Lancet,* 2, 694, 1986 (letter).
38. Leading article, CDC, Apparent Transmission of Human T-Lymphotrophic Virus Type III/Lymphadenopathy Associated Virus from a Child to a Mother Providing Health Care, Morbidity and Mortality Weekly Rep. No. 35, Centers for Disease Control, Public Health Service, U.S. Department of Health and Human Services, Atlanta, 1986, 76.
39. **Grint, P., Rademaker, M., and McEvoy, M. B.,** A case of acquired immune deficiency syndrome without the recognised risk factors, in *Proc. 3rd Int. Conf. on AIDS,* Washington, D.C., June 1 to 5, 1987, 21 (Abstr.).
40. **Anon.,** Needlestick transmission of HTLV-III from a patient infected in Africa, *Lancet,* 2, 1376, 1984.
41. **Stricof, R. L. and Morse, D. L.,** HTLV-III/LAV seroconversion following a deep intramuscular needlestick injury, *N. Engl. J. Med.,* 314, 1115, 1986 (letter).
42. **Oksenhendler, E., Harzic, M., Le Roux, J. M., Rabian, C., and Clauvel, J. P.,** HIV infection with seroconversion after a superficial needlestick injury to the finger, *N. Engl. J. Med.,* 315, 582, 1986 (letter).
43. **Neisson-Vernant, C., Mathez, S., Arfi, S., Leibowitch, J., and Monplaisir, N.,** Needlestick HIV seroconversion in a nurse, *Lancet,* 2, 814, 1986.
44. CDC update, Human Immunodeficiency Virus Infections in Health-Care Workers Exposed to Blood of Infected Patients, Morbidity and Mortality Weekly Rep. No. 36, Centers for Disease Control, Public Health Service, U.S. Department of Health and Human Services, Atlanta, 1987, 285.

45. **Weiss, S. H., Saxinger, W. C., Rechtman, D., Grieco, M. H., Nadler, J., Holman, S., Ginzburg, H. M., Groopman, J. E., Goedert, J. J., Markham, P. D., Gallo, R. C., Blattner, W. A., and Landesman, S.,** HTLV-III infection among health care workers: association with needle-stick injuries, *JAMA,* 254, 2089, 1985.
46. **Klein, R. S., Phelan, J., Friedland, G. H., Freeman, K., Schable, C., and Steigbigel, N. G.,** Low occupational risk of HIV infection for dental professionals, in *Proc. 3rd Int. Conf. on AIDS,* Washington, D.C., June 1 to 5, 1987, 155 (Abstr.)
47. **Weiss, S. H., Goedert, J. J., Sarngadharan, M. G., Bodner, A. J., Gallo, R. C., and Blattner, W. A.,** Screening test for HTLV-III (AIDS agent) antibodies, specificity, sensitivity and applications, *JAMA,* 253, 221, 1985.
48. **Gerberding, J. L., Bryant-LeBlanc, C. E., Moss, A. R., Nelson, K., Osmond, D., Chambers, H. F., Carlson, J. C., Drew, W. L., Levy, J. A., and Sande, M. A.,** Risk of transmitting the human immunodeficiency virus, cytomegalovirus, and hepatitis B virus to health care workers exposed to patients with AIDS and AIDS-related conditions, *J. Infect. Dis.,* 156, 1, 1987.
49. **Hirsch, M. S., Wormser, G. P., Schooley, R. T., Ho, D. D., Felsenstein, D., Hopkins, C. C., Joline, C., Duncanson, F., Sarngadharan, M. G., Saxinger, C., and Gallo, R. C.,** Risk of nosocomial infection with human T-cell lymphotropic virus III (HTLV-III), *N. Engl. J. Med.,* 312, 1, 1985.
50. **McCray, E.,** The Cooperative Needlestick Surveillance Group, occupational risk of the acquired immunodeficiency syndrome among health care workers, *N. Engl. J. Med.,* 314, 1127, 1986.
51. **Henderson, D. K., Saah, A. J., Zak, B. J., Kaslow, R. A., Lane, H. C., Folks, T., Blackwelder, W. C., Schmitt, J., LaCamera, D. J., Masur, H., and Fauci, A. S.,** Risk of nosocomial infection with human T-cell lymphotropic virus type III/lymphadenopathy associated virus in a large cohort of intensively exposed health care workers, *Ann. Intern. Med.,* 104, 644, 1986.
52. **Henderson, D. K., Saah, A. J., Zak, B. J., Kaslow, R. A., Lane, C., Folks, T., Blackwelder, W. C., Schmitt, J., LaCamera, D. J., Masur, H., and Fauci, A. S.,** Risk of nosocomial infection with LAV/HTLV-III in a large cohort of extensively exposed health care workers, in *Proc. 2nd Int. Conf. on AIDS,* Paris, France, June 23 to 25, 1986, 124 (Abstr.).
53. **Kuhls, T. L., Viker, S., Parris, N. B., Garakian, A., Sullivan-Bolyai, J., and Cherry, J. D.,** A prospective cohort study of the occupational risk of AIDS and AIDS-related infections in health care personnel, in *Proc. 2nd Int. Conf. on AIDS,* Paris, France, June 23 to 25, 1986, 117 (Abstr.).
54. **Gerberding, J. L., Bryant-LeBlanc, C. E., Greenspan, J., Greenspan, D., Greene, J., Carlson, J., and Sande, M. A.,** Risk to dentists from exposure to patients with AIDS virus, in *Proc. 26th Interscience Conf. on Antimicrobial Agents and Chemotherapy,* American Society for Microbiology, Washington, D.C., 1986, 283.
55. **Moss, A., Osmond, D., Bacchetti, P., Gerberding, J., Levy, J., Carlson, J., and Casavant, C.,** Risk of seroconversion for acquired immunodeficiency syndrome (AIDS) in San Fransisco health workers, *J. Occup. Med.* 28, 821, 1986.
56. **Klein, R. S., Phelan, J., Friedland, G. H., Schable, C., Trieger, N., and Steigbigel, N. G.,** Prevalence of antibodies to HTLV-III/LAV among dental professionals, in *Proc. 26th Interscience Conf. on Antimicrobial Agents and Chemotherapy,* American Society for Microbiology, Washington, D.C., 1986, 283.
57. **Marcus, R.,** The Cooperative Needlestick Surveillance Group, Update, prospective evaluation of health care workers parenterally exposed to blood of patients infected with human immunodeficiency virus, in *Proc. 3rd Int. Conf. on AIDS,* Washington, D.C., June 1 to 5, 1987, 200 (Abstr.)
58. **Shanson, D. C., Evans, R., and Lai, L.,** Incidence and risk of transmission of HTLV-III infections to staff at a London hospital 1982-85, *J. Hosp. Infect.,* 6(Suppl. C), 15, 1985.
59. **Jones, P. and Hamilton, P.,** HTLV-III antibodies in haematology staff, *Lancet,* 1, 217, 1985.
60. **Gilmore, N., Ballachey, M. L., O'Shaughnessey, M., Jothy, S., McDonald, J. C., and Gill, P.,** HTLV-III/LAV serologic survey of health care workers in a Canadian teaching hospital, in *Proc. 2nd Int. Conf. on AIDS,* Paris, France, June 23 to 25, 1986, 124 (Abstr.).
61. **McEvoy, M., Mortimer, P., Simmons, N., and Shanson, D.,** A prospective study of clinical laboratory and ancillary staff with parenteral or mucosal exposures to blood or body fluids infected with HTLV-III/LAV, in *Proc. 2nd Int. Conf. on AIDS,* Paris, France, June 23 to 25, 1986, 124 (Abstr.).
62. **Lange, J. M. A., Paul, D. A., Huisman, H. G., de Wolf, F., van den Berg, H., Coutinho, R. A., Danner, S. A., van der Noordaa, J., and Goudsmit, J.,** Persistent HIV antigenemia and decline of HIV core antibodies associated with transition to AIDS, *Br. Med. J.,* 293, 1459, 1986.
63. **Lange, J. M. A., Paul, D. A., de Wolf, F., Coutinho, R. A., and Goudsmit, J.,** Viral gene expression, antibody production and immune complex formation in human immunodeficiency infection, *AIDS,* 1, 15, 1987.
64. **de Wolf, F., Goudsmit, J., Paul, J. A., Lange, J. M. A., Hooykaas, C., and Coutinho, R. A.,** HIV antigenemia: association with decreased numbers of T4-cells and increased risk for AIDS, in *Proc. 3rd Int. Conf. on AIDS,* Washington, D.C., June 1 to 5, 1987, 19 (Abstr.).

65. **Osmond, D., Chaisson, R., Leuther, M., Allain, J. P., and Moss, A. R.,** Serum HIV antigen (HIV-Ag) as a predictor of progression to AIDS and ARC in homosexual men, in *Proc. 3rd Int. Conf. on AIDS,* Washington, D.C., June 1 to 5, 1987, 24.
66. **Albert, J. Jr., Gaines, H., Sönnerborg, A. Nyström, G., Pehrson, P. O., Chiodi, F., von Sydow, M., Moberg, L., Lidman, K., Christensson, B., Asjö, B., and Fenyö, E. M.,** Isolation of human immunodeficiency virus (HIV) from plasma during primary HIV infection, *J. Med. Virol.,* 23, 67, 1987.
67. **Coombs, R. W., Collier, A., Nikora, B., Chase, M., Gjerset, G., and Corey, L.,** Relationship between recovery of HIV from plasma and stage of disease, in *Proc. 3rd Int. Conf. on AIDS,* Washington, D.C., June 1 to 5, 1987, 119 (Abstr.).
68. **Arendrup, M., Ulrich, K., Nielsen, J. O., Lindhardt, B. O., Pedersen, C., and Krogsgaard, K.,** Neutralizing antibodies against HIV in relation to AIDS related diseases, in *Proc. 3rd Int. Conf. on AIDS,* Washington, D.C., June 1 to 5, 1987, 76 (Abstr.).
69. **Robert-Guroff, M., Goedert, J. J., Jennings, A., Blattner, W. A., and Gallo, R.,** High HTLV-III/LAV neutralizing antibody titers correlate with better clinical outcome, in *Proc. 3rd Int. Conf. on AIDS,* Washington, D.C., June 1 to 5, 1987, 53 (Abstr.).
70. **Salahuddin, S. Z., Groopman, J. E., Markhan, P. D., Sarngadharan, M. G., Redfield, R. R., McLane, M. F., Essex, M., Sliski, A., and Gallo, R. C.,** HTLV-III in symptomfree seronegative persons, *Lancet,* 2, 1418, 1984.
71. **Dienstag, J. L. and Ryan, D. M.,** Occupational exposure to hepatitis B virus in hospital personnel: infection on immunization?, *Am. J. Epidemiol.,* 115, 26, 1982.
72. **Hadler, S. C., Doto, I. L., Maynard, J. E., Smith, J., Clark, B., Mosley, J., Eickhoff, T., Himmelsbach, C. K., and Cole, W. R.,** Occupational risk of hepatitis B infection in hospital workers, *Infect. Control,* 6, 24, 1985.
73. Leading article, CDC, Acquired Immunodeficiency Syndrome (AIDS), Precautions for Clinical and Laboratory Staffs, Morbidity and Mortality Weekly Report No. 31, Centers for Disease Control, Public Health Service, U.S. Department of Health and Human Services, Atlanta, 1982, 577.
74. Leading article, CDC, Acquired Immunodeficiency Syndrome (AIDS) Precautions for Health Care Workers and Allied Professionals, Morbidity and Mortality Weekly Report No. 32, Centers for Disease Control, Public Health Service, U.S. Department of Health and Human Services, Atlanta, 1983, 450.
75. Leading article, CDC, Recommendations for Preventing Transmission of Infection with Human T-Lymphotropic Virus Type III/Lymphadenopathy-Associated Virus in the Workplace, Morbidity and Mortality Weekly Report No. 34, Centers for Disease Control, Public Health Service, U.S. Department of Health and Human Services, Atlanta, 1985, 681,691.
76. Leading article, CDC, Recommendations for Preventing Transmission of Infection with Human T-Lymphotropic Virus Type III/Lymphadenopathy Associated Virus During Invasive Procedures, Morbidity and Mortality Weekly Report No. 35, Centers for Disease Control, Public Health Service, U.S. Department of Health and Human Services, Atlanta, 1986, 221.
77. Leading article, U.S. Department of Health and Human Services, AIDS in the workplace; how to prevent the transmission of infection, *Int. Nurs. Rev.,* 33, 117, 1986.
78. The National Board for Occupational Safety and Health, Blood-borne Infection in the Workplace (Blodsmitta), Stockholm, 1987 (in Swedish).
79. **Gerberding, J. L.,** California task force on AIDS update: recommended infection — control policies for patients with human immunodeficiency virus infection, *N. Engl. J. Med.,* 315, 1562, 1986.
80. **Melbye, M., Biggar, R. J., Ebbesen, P., Weiss, S. M., Blattner, W. A., Sarngadharan, M. G., and Gallo, R. C.,** Seroepidemiology of HTLV-III antibody in Danish homosexual men, prevalence, transmission, and disease outcome, *Br. Med. J.,* 289, 573, 1984.
81. **Maynard, J. E.,** Modes of hepatitis B virus transmission, in Japan Medical Research Foundation, *Hepatitis Viruses, Proc. 1976 Int. Symp. Hepatitis Viruses,* University of Tokyo Press, Tokyo, 1978, 125.
82. **Francis, D. P., Favero, M. S., and Maynard, J. E.,** Transmission of hepatitis B virus, *Semin. Liver Dis.,* 1, 27, 1981.
83. **Tabor, E., Prucell, R. H., and Gerety, R. J.,** Primate animal models and titered inocula for the study of human hepatitis A, hepatitis B and non-A, non-B hepatitis, *J. Med. Primatol.,* 12, 305, 1983.
84. **Grady, G. F., Lee, V. A., Prince, A. M., Gitnick, G. L., Fawaz, K. A., Vyas, G. N., Levitt, M. D., Senior, J. R., Galambos, J. T., Bynum, T. E., Singleton, J. W., Clowdus, B. F., Akdamar, K., Aach, R. D., Winkelman, E. I., Schiff, G. M., and Hersh, T.,** Hepatitis B immune globulin for accidental exposures among medical personnel: final report of a multicenter controlled trial, *J. Infect. Dis.,* 138, 625, 1978.
85. **Seeff, L. B., Wright, E. C., Zimmerman, H. J., Alter, H. J., Dietz, A. A., Felsher, B. F., Finkelstein, J. D., Garcia-Pont, P., Gerin, J. L., Grenlee, H. B., Hamilton, J., Holland, P. V., Kaplan, P. M., Kiernan, T., Koff, R. S., Leevy, C. M., McAuliffe, V. J., Nath, N., Purcell, R. M., Schif, E. R., Schwarz, C. C., Tamburro, C. H., Vlahcevic, Z., Zemel, R., and Zimmon, D. S.,** Type B hepatitis after needlestick exposure: prevention with hepatitis B immune globulin, final report of the Veterans Administration Cooperative Study, *Ann. Intern. Med.,* 88, 285, 1978.

86. **Werner, B. G. and Grady, G. F.,** Accidental hepatitis B surface antigen positive inoculations: use of e-antigen to estimate infectivity, *Ann. Intern. Med.*, 97, 367, 1982.
87. **Gerberding, J. L., Hopewell, P. C., Kaminsky, L. W., and Sande, M. S.,** Transmission of hepatitis B without transmission of AIDS by accidental needlestick, *N. Engl. J. Med.*, 31, 56, 1985 (letter).
88. **Glaser, J. B. and Garden, A.,** Inoculation of cryptococcosis without transmission of the acquired immunodeficiency syndrome, *N. Engl. J. Med.*, 313, 266, 1985 (letter).
89. **Ungvarski, P. J.,** Infection control in the patient with AIDS, *J. Hosp. Infect.*, 5(Suppl. A), 111, 1984.
90. **Schaffner, W.,** Editorial, the response of hospital infection control programs to the AIDS epidemic, *Infection,* 13, 201, 1985.
91. **Bennett, J.,** AIDS: what precautions do you take in the hospital, *Am. J. Nurs.*, August, 952, 1986.
92. **Geddes, A. M.,** Editorial, hepatitis B and HTLV-III infection in injured patients, *Injury,* 17, 293, 1986.
93. **Tomasi, T. J.,** Editorial, AIDS and the occupational physician, *J. Occup. Med.*, 28, 517, 1986.
94. **Conte, J. E., Jr.,** Infection with human immunodeficiency virus in the hospital, epidemiology, infection control and biosafety considerations, *Ann. Intern. Med.*, 105, 730, 1986.
95. **Jenner, E., Levi, A., and Houghton, D.,** Nursing, in *The Management of AIDS Patients,* Miller, D., Weber, J., and Green, J., Eds., Macmillan, London, 1986, 93.
96. **Seidl, O. and Goebel, F. D.,** Psychological problems in nursing patients with AIDS, in *Proc. 3rd Int. Conf. on AIDS,* Washington, D.C., June 1 to 5, 1987, 98 (Abstr.).
97. **O'Dowd, M. A., Adachi, N., Razin, A. M., and Klein, R. S.,** Nursing staff knowledge, attitudes and self-reported behaviour with respect to patients with AIDS, in *Proc. 3rd Int. Conf. on AIDS,* Washington, D.C., June 1 to 5, 1987, 145 (Abstr.).
98. **Cooke, M. and Koenig, B.,** Housestaff attitudes towards the acquired immunodeficiency syndrome, in *Proc. 3rd Int. Conf. on AIDS,* Washington, D.C., June 1 to 5, 1987, 154 (Abstr.).
99. **Eales, L. J., Nye, K. E., Parkin, J. M., Weber, J. N., Forster, S. M., and Pinching, A. J.,** A genetic factor affecting susceptibility to HIV infection and to disease progression, in *Proc. 3rd Int. Conf. on AIDS,* Washington, D.C., June 1 to 5, 1987, 81 (Abstr.).
100. Leading article, CDC update, Prospective Evaluation of Health Care Workers Exposed Via the Parenteral or Mucous-Membrane Route to Blood or Body Fluid from Patients with Acquired Immunodeficiency Syndrome — United States, Morbidity and Mortality Weekly Report No. 34, Centers for Disease Control, Public Health Service, U.S. Department of Health and Human Services, Atlanta, 1985, 101.
101. Leading article, CDC, Recommendations for Preventing Possible Transmission of Human T-Lymphotropic Virus Type III/Lymphadenopathy-Associated Virus from Tears, Morbidity and Mortality Weekly Report No. 34, Centers for Disease Control, Public Health Service, U.S. Department of Health and Human Services, Atlanta, 1985, 533.
102. **Goldwater, P. N., Nixon, A. D., Law, R., Officer, J., and Cleland, J. F.,** Use of the "needle guard" in the prevention of needle-stick injuries, in *Proc. 3rd Int. Conf. on AIDS,* Washington, D.C., June 1 to 5, USA, 1987, 44 (Abstr.).
103. **Christenson, B.,** Acute infections with hepatitis B virus in medical personnel during a 15-year follow-up, *Am. J. Epidemiol.*, 122, 411, 1985.
104. **Ebbesen, P., Scheutz, F., Bodner, A. J., and Biggar, R. J.,** Lack of antibodies to HTLV-III/LAV in Danish dentists, *JAMA,* 256, 2199, 1986 (letter).
105. **Lubick, H. A., Schaeffer, L. D., and Kleinman, S. H.,** Occupational risk of dental personnel survey, *J. Am. Dent. Assoc.*, 113, 10, 1986 (letter).
106. Leading article, CDC, Recommended Infection-Control Practices for Dentistry, Morbidity and Mortality Weekly Report No. 35, Centers for Disease Control, Public Health Service, U.S. Department of Health and Human Services, Atlanta, 1986, 237.
107. **Barré-Sinoussi, F., Nugeyre, M. T., and Chermann, J. C.,** Resistance of AIDS virus at room temperature, *Lancet,* 2, 721, 1985.
108. **Resnick, L., Veren, K., Salahuddin, S. Z., Tondreau, S., and Markham, P. D.,** Stability and inactivation of HTLV-III/LAV under clinical and laboratory environments, *JAMA,* 255, 1887, 1986.
109. **Martin, L. S. and Loskoski, S. L.,** Effects of drying on the human immunodeficiency virus (HIV) diluted in heparinized blood, serum or media, in *Proc. 3rd Int. Conf. on AIDS,* Washington, D.C., June 1 to 5, 1987, 50 (Abstr.).
110. **Spire, B., Barré-Sinoussi, F., Montagnier, L., and Chermann, J. C.,** Inactivation of lymphadenopathy-associated virus by chemical disinfectants, *Lancet,* 2, 899, 1984.
111. **McDougal, J. S., Martin, L. S., Cort, S. P., Mozen, M., Heldebrant, C. M., and Evatt, B. L.,** Thermal inactivation of the acquired immunodeficiency syndrome virus, human T-lymphotropic virus-III/lymphadenopathy-associated virus with special reference to antihemophilic factor, *J. Clin. Invest.*, 76, 875, 1985.
112. **Harada, S., Yoshiyama, H., and Yamamoto, N.,** Effect of heat and fresh human serum on the infectivity of human T-Cell lymphotropic virus type III evaluated with new bioassay systems, *J. Clin. Microbiol.*, 22, 908, 1985.

113. **Martin, L. S., McDougal, J. S., and Loskoski, S. L.,** Disinfection of the human T lymphotropic virus type III/lymphadenopathy-associated virus, *J. Infect. Dis.*, 152, 400, 1985.
114. **Spire, B., Dormont, D., Barré-Sinoussi, F., Montagnier, L., and Chermann, J. C.,** Inactivation of lymphadenopathy-associated virus by heat, gamma rays and ultraviolet light, *Lancet*, 1, 188, 1985.
115. **Wallbank, A. M.,** Disinfectant inactivation of AIDS virus in blood or serum, *Lancet*, 1, 642, 1985.
116. **Shikata, T., Karasawa, T., Abe, K., Takahashi, T., Mayami, M., and Oda, T.,** Incomplete inactivation of hepatitis B virus after heat treatment at 60°C for 10 hours, *J. Infect. Dis.*, 138, 242, 1978.
117. **Bond, W. W., Favero, M. S., Petersen, N. J., and Ebert, J. W.,** Inactivation of hepatitis B virus by intermediate-to-high level disinfectant chemicals, *J. Clin. Microbiol.*, 18, 535, 1983.
118. **Howard, C. R., Dixon, J., Young, P., van Eerd, P., and Shellekens, H.,** Chemical inactivation of hepatitis B virus: the effect of disinfectants on virus-associated DNA polymerase activity, morphology and infectivity, *J. Virol. Methods*, 7, 135, 1983.
119. **Tabor, E., Buynak, E., Smallwood, L. A., Snoy, P., Hilleman, M., and Gerety, R.,** Inactivation of hepatitis B virus by three methods: treatment with pepsin, urea or formalin, *J. Med. Virol.*, 11, 1, 1983.
120. **Kobayashi, H. and Tsuzuki, M.,** The effect of disinfectants and heat on hepatitis B virus, *J. Hosp. Infect.*, 5(Suppl. A), 93, 1984.
121. **Kobayashi, H., Tsuzuki, M., Koshimizu, K., Toyama, H., Yoshihara, N., Shikata, T., Abe, K., Mizuno, K., Otomo, N., and Oda, T.,** Susceptibility of hepatitis B virus to disinfectants or heat, *J. Clin. Microbiol.*, 20, 214, 1984.
122. **Kobayashi, H., Tsuzuki, M., Koshimizu, K., Oda, T., Toyama, H., Yoshihara, N., Shikata, T., Karasawa, T., and Abe, K.,** Inactivation of hepatitis B virus, *Jpn. J. Med. Instrum.*, 50, 524, 1980.
123. **Thraenhart, O., Kuwert, E. K., Scheiermann, N., Dermietzel, R., Paar, D., Maruhn, D., Alberti, A., Richter, H.-J., and Hotz, J.,** Comparison of the morpological alteration and disintegration test (MADT) and the chimpanzee infectivity test for determination of hepatitis B virucidal activity of chemical disinfectants, *Zentralbl. Bakteriol. Parasitenkd. Infektionskr. Hyg. Abt. 1 Orig. Reihe B*, 176, 472, 1982.
124. **Kuwert, E., Thraenhart, O., Dermietzel, R., and Scheiermann, N.,** The morphological alteration and disintegration test (MADT) for quantitative and kinetic determination of hepato-virucidal effect of chemical disinfectants, *Hyg. Med.*, 9, 379, 1984.
125. **Thraenhart, O. and Kuwert, E.,** Zur Wirksamkeitsprüfung von Desinfektionsmitteln gegenüber dem Hepatitis B-Virus unter besonderer Berücksichtigung des MADT, *Hyg. Med.*, 9, 385, 1984.
126. **Kuwert, E.,** MADT im Vergleich zu anderen Nachweisverfahren zur Hepatitis B-Wirksamkeit von Desinfektionsmitteln, *Hyg. Med.*, 9, 391, 1984.
127. **Steinmann, J., Böse, A., and Arnold, W.,** HBV-Wirksamkeit von chemischen Desinfektionsmitteln im DNS-Polymerase-Test, *Hyg. Med.*, 10, 255, 1985.
128. **Valenti, W. M.,** Infection control and the pregnant health care worker, *Am. J. Infect. Control*, 14, 20, 1986.
129. Leading article, CDC; Recommendations for Assisting in the Prevention of Perinatal Transmission of Human T-Lymphotropic Virus Type III/Lymphadenopathy-Associated Virus and Acquired Immunodeficiency Syndrome, Morbidity and Mortality Weekly Report No. 34, Centers for Disease Control, Public Health Service, U.S. Department of Health and Human Services, Atlanta, 1985, 721, 731.
130. Leading article, CDC; Safety of Therapeutic Immune Globulin Preparations with Respect to Transmission of Human T-Lymphotropic Virus Type III/Lymphadenopathy-Associated Virus Infection, Morbidity and Mortality Weekly Report No. 35, Centers for Disease Control, Public Health Service, U.S. Department of Health and Human Services, Atlanta, 1986, 231.
131. **Fujikawa, L. S., Salahuddin, S. Z., Ablashi, D., Palestine, A. G., Masur, H., Nussenblatt, R. B., and Gallo, R. C.,** Human T-cell leukemia/lymphotropic virus type III in the conjunctival epithelium of a patient with AIDS, *Am. J. Opthalmol.*, 100, 507, 1985.
132. **Burman, L. G.,** Routine HIV testing of patients yields very poor cost-effectiveness, *Lakartidningen*, 36, 2764, 1987, (letter, in Swedish).
133. **Sacks, J. J.,** AIDS in a surgeon, *N. Engl. J. Med.*, 313, 1017, 1985 (letter).
134. **Christenson, B.,** personal communication, 1987.
135. The National Bacteriological Laboratory, Stockholm, Sweden, 1987, unpublished data.
136. Centers for Disease Control, Recommendations for prevention of HIV transmission in health-care settings, *Morbid. Mort. Wkly. Rep.*, 36(2), 1, 1987.
137. Centers for Disease Congrol, Agent summary statement for human immunodeficiency viruses (HIVs) including HTLV-III, LAV, HIV-1 and HIV-2, *Morbid. Mort. Wkly. Rep.*, 37(4), 1, 1988.
138. Centers for Disease Control, Universal precautions for prevention of transmission of human immunodeficiency virus, hepatitis B virus and other bloodborne pathogens in health care settings, *Morbid. Mort. Wkly Rep.*, 37, 1, 1988.
139. National Committee for Clinical Laboratory Standards, Protection of laboratory workers from infectious diseases transmitted by blood and tissue, Proposed guidelines, NCCLS Document M29-P, National Committee for Clinical Laboratory Standards, Villanova, PA, 1987.

Chapter 5

MICROBIOLOGICAL HAZARDS

Egil Lingaas and Tore Kalager

TABLE OF CONTENTS

I. INTRODUCTION

It must have been an observation older than the medical profession that many diseases spread from person to person and that taking care of patients with such contagious diseases imposed a risk on the caretaker. Throughout medical history, there are numbers of anecdotal reports on medical personnel succumbing from infectious diseases contracted during work with patients or specimens from patients. One of the most well known of these is the fate of Professor Jacob Kolletscka, a forensic pathologist and a close friend of Ignaz Philipp Semmelweis. He was stuck in the finger by a student's knife during an autopsy and soon developed an acute infection and died. His autopsy report made an important contribution to Semmelweis' puzzle on the etiology of childbed fever.[1]

Professor Kolletscka obviously shared his fate with a large number of health care personnel. In a review of the lives of 1276 Norwegian doctors in the time period from 1800 to 1886, a large number were noted to have contracted diseases like tuberculosis, typhoid fever, cholera, diptheria, erysipelas, and other infections from their patients.[2] However, even though infectious diseases acquired from patients must have been rather common, there are few historical data on the incidence of such infections compared to that of the general population before the 20th century. In Florence Nightingale's book, *Notes on Hospitals,* she refers to data collected and analyzed by William Farr on mortality among hospital personnel.[3] These data showed that the mortality from contagious diseases among female hospital employees was significantly higher than that of the female population of London not employed by hospitals.

Correspondingly, in a paper read before the Royal Medical and Chirurgical Society on January 26, 1886, Ogle presented a comparison of the mortality in the medical profession and the general population during the years from 1880 to 1882 (Table 1).[4] For most infectious diseases against which no prophylactic remedy was known, such as scarlet fever, diphtheria, typhus, enteric fever, and erysipelas, the mortality among medical men was found to be considerably in excess of the population average.

During recent years, along with the increasing interest in occupational medicine in general, attention has also focused on health care facilities, and it has turned out that infectious diseases constitute a significant proportion of occupational medicine in this setting. As reported by Sherertz and Hampton, out of the total visits to the employee health service at Shands Hospital (a 450-bed university hospital in Gainesville, FL. U.S.), 57% was related to infection prevention or surveillance efforts.[5] Of the remaining 43% of personnel visits, 17% were due to infectious diseases and 6% to infectious disease exposures.

II. CLASSIFICATION OF MICROORGANISMS

Classification of microorganisms according to their levels of hazard is used to establish safe, practical, and economic ways of dealing with the pathogens. The purpose is to reduce the risk for spread of the organisms, both within hospitals and laboratories and in the community at large. Usually, lists of pathogens with the same levels of hazards have been generated, and the laboratory conditions under which the microorganisms should be handled have also been specified. The U.S. Public Health Service classification and the U.K. classification differ in the respect that the U.K. classification has three categories and uses a descending order of risk, compared to the U.S. classification with four categories and an ascending order of risk.[6,7] Both differ from the classification of the World Health Organization (WHO) which has four risk groups, but does not list individual microorganisms.[8] By this, the WHO seeks to avoid the difficulties encountered because of varying opinions on the levels of hazard of some pathogens. It is expected that each country generates its own lists within the framework provided. The WHO classification can be summarized as follows:

Table 1
MORTALITY AMONG THE MEDICAL PROFESSION IN ENGLAND AND WALES DURING THE YEARS 1880 TO 1882

	Annual deaths per million living males over 20 years of age	
Causes of death	**Medical men**	**General population**
Smallpox	13	73
Scarlet fever	59	16
Typhus	79	38
Diphtheria	59	14
Simple or ill-defined continued fever	33	40
Enteric fever	311	238
Diarrhea, cholera	205	274
Malarial fever	46	11
Erysipelas	172	136
Alcoholism	178	130
Gout	291	78
Rheumatic affections	251	215
Malignant diseases	879	790
Phtisis	1,738	3,145
Diabetes	284	108
Diseases of nervous system	4,565	4,268
Diseases of circulatory system	4,142	2,934
Diseases of respiratory system	3,237	4,408
Liver diseases	1,744	744
Other diseases of digestive system	973	632
Calculus	86	30
Diseases of bladder and prostate	634	287
Other diseases of urinary system	1,520	665
Hernia	13	88
Accident	793	1,105
Suicide	363	238
All other causes	2,869	2,124
Total from all causes	25,535	22,829

Adapted from Ogle, W., *Med. Chir. Trans.*, 69, 217, 1886.

- Risk Group I — Agents of low risk; can be handled in an ordinary laboratory. (If supervised, the general public may enter the laboratory.)
- Risk Group II — Agents of moderate individual and low community risk. (Conditions for work and access are the same as for Risk Group I.)
- Risk Group III — Agents of high individual but low community risk; should only be handled in specially designed laboratories. (Staff should be specially trained and identified, and only authorized personnel should have access to the laboratory.)
- Risk Group IV — Agents of high individual and community risk. (These should be handled only in highly sophisticated laboratories, and entry should be restricted to only a very few authorized staff.)

Arguments in favor of the WHO classification are that conditions and specific situations which may influence the hazard considerably will be taken into consideration in addition to the infectivity and virulence of the organism. Such considerations may be the climate, the prevalence of the pathogens in the country, the presence of vectors and reservoirs, and the amount and concentration of the microorganisms. Also, available resources for treatment of an outbreak may have to be considered.

III. TRANSMISSION OF INFECTIOUS AGENTS

Knowledge of the mode of transmission of the various infectious agents is essential for the establishment of proper safety precautions for health care personnel. There are four main routes of transmission: contact, airborne, vehicleborne, or vectorborne. Each organism may be spread by one or more of these routes and the relative signficance of a certain mode of spread in health care facilities may sometimes differ from that in the community. The dominating mode of spread for a specific organism may also vary in different types of health care facilities. Furthermore, diagnostic and therapeutic procedures, including various types of laboratory work, will often influence both the mode and potential for spread. General knowledge of how infections are transmitted should be part of the basic training of all categories of health care personnel. In order to reduce the risk of occupationally acquired infections, knowledge of the way diagnostic and therapeutic procedures may contribute to dissemination of microorganisms is also important. Whether or not transmission and infectious disease will occur depends in each case on a number of factors such as the type and duration of contact, the type and dose of infectious agent, and the susceptibility of the host.

A. Contact Transmission

This is the most important and frequent mode of spread both within and outside hospitals. There are three types of contact transmission: direct contact, indirect contact, and droplet contact.

Direct contact — Direct contact spread involves direct physical contact between the infected source and a susceptible host. The source may be patients or personnel with an acute symptomatic or asymptomatic infection, but persons in the incubation phase of an infection, reconvalescents, and healthy carriers may also transmit the infectious agent.

Indirect contact — Indirect contact spread involves the contamination of an intermediate object, usually inanimate, followed by contact between this object and the susceptible host. The hands of health care personnel play a pivotal role in both direct and indirect contact transmission and are considered to be the most important factor in the transmission of infections in health care facilities.

Droplet contact — Droplet contact occurs when droplets are dispersed from an infected source, such as, for example, occurs during talking, coughing, and sneezing, suctioning from infected airways, or splashing of infected body fluids. Depending on the size of the droplets, they will either settle within approximately 1 m of their origin or they will evaporate and turn into droplet nuclei (see below). Microorganisms carried by droplets may come in contact with the conjunctivae, nose, mouth, or skin of subjects in the close vicinity of the source, and this form of transmission is therefore classified as a form of contact transmission. The droplets may also contaminate horizontal surfaces or other inanimate objects close to the patient (within approximately 1 m) and be the origin of indirect contact transmission.

B. Airborne Transmission

Airborne transmission occurs by droplet nuclei or dust particles and involves a true airborne phase of the organism. Among health care personnel, airborne infection usually occurs through inhalation. Infectious droplet nuclei arise from evaporation of small aerosolized droplets containing microorganisms. Droplets with a diameter of 0.1 mm have been reported to evaporate in 1.7, s, and those with a diameter of 0.05 mm evaporate in 0.4 s.[9] However, in the presence of protein, e.g., in sputum, mucus, and serum, evaporation will be much slower, and the droplets will settle more rapidly, resulting in fewer airborne droplet nuclei.

The smaller the number of organisms and amount of dried material in the droplet nuclei, the longer they will remain airborne, enhancing their potential for traveling long distances.[10] However, due to the scarcity of water in the droplet nuclei, many microorganisms will not survive very long in this state. This is especially true for many Gram-negative bacteria. Transmission by droplet nuclei is therefore first of all of importance for dessication-resistant microorganisms using the airways of the host as an entry port. Infected airborne dust particles may arise from dried materials like exudates and secretions, pus, skin squamae, dressings, bed linen, etc. Dust particles are usually larger than droplet nuclei and to large to reach the alveoli by inhalation.

C. Vehicleborne Transmission

In vehicleborne transmission, disease is transmitted through a contaminated vehicle. Common examples are food- and waterborne transmission of enteric pathogens, transmission of several viral agents by blood transfusion (non-A and non-B hepatitis, cytomegalovirus, and human immunodeficiency virus), and transmission by contaminated medications. Hospital personnel may occasionally be infected by vehicle transmission, most often by the foodborne route, but generally this form of transmission is less important for health care workers than contact spread and airborne spread.

D. Vectorborne Transmission

Vectorborne agents are transmitted by arthropods, either passively by simple external or internal carriage or biologically, whereby the organism undergoes physiological changes in the vector. In developed countries, vectorborne transmission is for all practical purposes not seen in hospitals and other health care facilities. In underdeveloped countries, however, the opportunity for vector transmission may also be present in this setting.

IV. PREVENTION OF INFECTIOUS DISEASES — GENERAL MEASURES

Infection control in health care workers has recently been outlined by the Centers for Disease Control (CDC).[11] The objectives include

1. Maintenance of sound habits in personal hygiene and individual responsibility in infection control
2. Monitoring and investigating infectious diseases and outbreaks of infections among personnel
3. Providing care to personnel for work-related illnesses or exposures
4. Identifying infection risks related to employment and instituting appropriate preventive measures
5. Containing costs by eliminating unnecessary procedures and preventing infectious diseases resulting in absenteeism and disability

The main factors needed to establish an infectious disease are the infectious agent, the route of transmission, and the susceptible host.

A. The Infecting Organism

Elimination of the infecting organism prevents further spread of the disease. Thus, efficient diagnosis and treatment of patients with communicable diseases are of major importance. Antimicrobial agents and surgery are the main therapeutic modalities. Usage of antibiotics must take into consideration the development of microbial resistance and in some cases also the prolongation of the carrier state and an increased tendency to relapse.

B. Transmission

Transmission describes the movement of the infecting organism from the source to the host. Detailed knowledge of the route of transmission makes it possible to stop the spread of the organisms. The control measures will vary according to how transmission occurs, whether it is by contact, air, common vehicle, or vector. It is important to acknowledge that some organisms may be transmitted by various routes. For many infectious diseases, interruption of the route of transmission is the most important control measure. In particular, interruption of contact spread by proper handwashing may still be the most important single preventive procedure, both within health institutions and in the community.[12] Other means of interrupting the transmission is by judicious use of gloves, masks, and gowns. The CDC, Atlanta, GA has published detailed guidelines for isolation precautions in hospitals.[13]

C. The Host

An individual exposed to an infectious agent may or may not be at risk of developing disease, depending on the host's defense against infections. For the health care worker, both nonspecific mechanisms such as the skin and specific mechanisms such as the natural or artificial immunity are of great importance. Thus, proper attention to both defense systems (e.g., care of the skin and use of immunoglobulins and vaccine) is necessary to obtain and maintain optimal resistance against development of various infectious diseases. Comprehensive programs aiming at the maintenance and promotion of good health for health care personnel should be formulated and implemented.

D. Handwashing

The hands play a pivotal role for the spread of infection in health care facilities as well as in the community at large. The hands of health care workers are regularly contaminated with large numbers of microorganisms during patient care. For example, Jensen reported that common nursing procedures resulted in contamination of the hands with 10^5 to 10^{10} organisms.[14] To cause infection, these transient microorganisms have several different portals of entry at hand. First of all, they may penetrate the skin directly, resulting in localized or generalized infection (*Staphylococcus aureus, Streptococcus pyogenes, Herpes simplex,* etc.) This usually requires a break in the integrity of the skin, since most microorganisms are unable to penetrate unbroken skin on their own. However, health care personnel frequently have minor wounds, scratches, erosions, and eczematous skin on their hands, often unnoticed. Microorganisms contaminating the hands may also be transferred to other skin areas, usually uncovered skin of the face, neck, and forearms, and to the mucous membranes of the eyes, nose, and mouth. These areas are frequently, and often unnoticed, touched by the hands.

Prompt removal of contaminating microroganisms from the hands is therefore important for the protection of both health care workers and their patients, and handwashing is the single most important procedure for the prevention of nosocomial infections.[12] However, more than a century after Semmelweis the practice of handwashing among health professionals is far from optimal. Several studies have shown inferior quality, frequency, and timing of handwashing.[15-22] Taylor found that 89% of 129 ward staff missed some part of the hand surface during handwashing. She also reported that only 38% of dirty nursing procedures was followed by handwashing.[15,16] Several studies, including different categories of health professionals from different countries, have shown handwashing frequencies between 7 and 55% after dirty procedures.[18-22] Obviously, there is much to improve with regard to this seemingly simple protective procedure.

Handwashing with plain unmedicated soap removes the majority of transient organisms on the hands.[23-27] In general, less than 1% of contaminating bacteria persist after cleansing with soap and water.[27] However, the use of antiseptic-containing handwashing preparations or alchohol disinfection of the hands has in several laboratory studies been found to be even more efficient for the elimination of transient bacteria.[25-28] Washing with soap and water is also an efficient method for removing viruses.[29,30] Recently, Faix reported that ordinary soap and water were as efficient as chlorhexidin gluconate and povidone iodine solution for the removal of cytomegalovirus from the fingertips.[30] Many infection-control experts advocate hand disinfection as a supplement to or partly as an alternative to handwashing, but the recommendations for the use of regular hand disinfection varies from country to country.[27] In the guidelines issued by the CDC, they state as a Category II measure: "Plain soap should be used for handwashing unless otherwise indicated."[12] (Category II means that the statement is supported by highly suggestive clinical studies in general hospitals or by definite studies in speciality hospitals that might not be representative of general hospitals. Measures that have not been adequately studied, but have a logical or strong theoretical rationale indicating probable effectiveness, are included in this category.) Reference 27 should be consulted for a more thorough review of handwashing and hand disinfection.

E. Research and Education

The informed and well-educated employee in less likely to become ill.[31] Information, education, and training of the health care workers will prevent them from most infectious diseases. A broad and profound understanding of the workplace achieved by good data collection and research is necessary to provide such education.

V. BACTERIAL INFECTIONS

A. *Bacillus anthracis*

Human anthrax is nowadays very uncommon in developed countries. However, animal anthrax remains endemic in parts of Asia, Africa, and South America.[32] Human-to-human transmission of antrax has not been documented. Several cases have, however, been reported in laboratories handling specimens and cultures with *B. anthracis*.[33-35] Pike reported a total of 45 accumulated laboratory-acquired cases (40 from the U.S.) with 5 deaths.[35] The organism may be present in wound exudates, vesicular fluid, and blood or biopsy specimens and is rarely present in urine and feces. In the laboratory, transmission occurs most commonly through direct or indirect contact, including accidental inoculation, but transmission through inhalation of aerosol has also been described.[36]

B. *Bordetella pertussis*

Pertussis (whooping cough) is highly communicable by droplet contact. There is no evidence of spread by indirect transmission, and asymptomatic carriers probably do not exist.[37] Pertussis is primarily a disease of children, but the disease may also occur in adults. During the last decades, the epidemiology of pertussis seems to have changed in many countries. After the introduction of childhood immunization, young adults with waning immunity and mild illness may often be the source.[38-42]

Exposure to pertussis is frequent among pediatric health care personnel as compared to the general population, and nosocomial spread of *Bordetella pertussis* to pediatric staff has been reported on several occasions.[41,43-45] Transmission from health care personnel to their co-workers and from staff to patients may also occur. Linnemann et al.[44] also found that antibody titers among hospital employees decreased with increasing age and that the level of antibody correlated with degree of patient contact.

During an outbreak, removal of personnel with cough or upper respiratory tract symptoms may be important in preventing further spread. Infected contacts may be indentified by fluorescent antibody technique, but more reliably by culture. Erythromycin used within a few days after exposure appears to prevent the aquisition of clinical disease.[44,46] Routine pertussis vaccination of hospital personnel is not recommended, but a booster dose of adsorbed pertussis vaccine may be considered for health care personnel exposed during nosocomial or community outbreaks.[44,47] A reduced dose is used for adults.

C. *Brucella*

Brucellosis is a zoonosis transmitted mainly from cattle, swine, goats, and sheep through ingestion of contaminated milk or through direct exposure to their excreta, products of abortion, or carcasses. Transmission by ingestion requires a large dose and a virulent strain.[48] Infection is more readily produced by inoculation on abraded skin, injection through skin, or splashing on mucous membranes. Brucellosis therefore represents an occupational hazard to meat packers, farmers, and veterinarians. Brucellosis in man is generally not considered to be contagious, usually not representing any risk for personnel providing care for the patient. However, a few patients are reported to develop abcesses, requiring drainage/secretion precautions.[13,49] The organism may also be shed in the urine in some patients, but the disease has apparently not spread to their contacts.[50] On the other hand, laboratory workers handling *Brucella* cultures have a high risk of aquiring brucellosis through accidents, aerosols, or inadequate sanitary precautions. In the surveys of laboratory-acquired infection by Sulkin and Pike, brucellosis was the most commonly reported infection, at least in the U.S.[34,35,51-53] A total of 426 cases with 5 deaths were reported.[52] Included in their figures are 45 cases of *Brucella* infections resulting from a common exposure, probably an aerosol.[54] Despite the declining incidence of brucellosis in the general population, brucellosis still represents a problem at least in some laboratories. Fox and Kaufman reported that in the decade between 1965 and 1974, 1.7% of brucellosis cases in the U.S. resulted from accidents with cultures of *Brucella*.[55] However, they did not state how many of these were acquired in medical laboratories.

D. *Campylobacter*

Campylobacter jejuni has become established as one of the most common causes of bacterial enteritis in humans. The infective dose has been demonstrated to be quite small.[56] Like *Salmonella*, one would expect this organism to represent a certain risk to health care personnel. However, at present there is very limited information to confirm this supposition.[57,58] On the contrary, despite a pattern of person-to-person spread during an outbreak of *C. jejuni* infection in a neonatal intensive care unit with seven infected neonates, Hershkowici et al.[58] could not identify any transmission to staff members. Among 25 cases of laboratory-acquired bacterial bowel infections in British clinical laboratories during 1980 to 1985, Grist reported only 1 case of campylobacteriosis.[59-61]

E. *Chlamydia*

1. Chlamydia psittaci

The predominant source of human psittacosis is infected avian species. The route of infection is airborne. Human-to-human transmission of *C. psittacii* can occur from a heavily infected pneumonia case, but this is rare. Broholm et al.[62] reported an outbreak of nosocomial ornithosis affecting eight employees of an infectious disease clinic in Sweden. A nurse accompanying the patient to the hospital was infected as well. The son of the index case and a patient who had been treated in the same room as the index case also fell ill. All had a favorable outcome after treatment with doxycycline, but the index case, a 53-year-old man, who had been healthy during all his life, finally died from the infection.

In the laboratory, the organism presents a definite risk. The organism may be present in blood, sputum, and tissues of infected humans. Exposure to infectious aerosols and droplets created during the handling of infected tissues is the primary hazard to laboratory personnel working with psittacosis. Pike reported 116 cases of laboratory-associated psittacosis with 10 deaths, one of the highest case fatality rates among more than 4000 laboratory infections recorded in his review.[52] However, only one of the ten fatal cases of laboratory-acquired psittacosis occurred since 1949, a fact possibly related to the effectiveness of antibiotics.[53]

2. Chlamydia trachomatis

The risk of taking care of patients infected with *C. trachomatis* (trachoma, oculogenital disease, or lymphogranuloma venerum) seems to be almost negligible. Depending on the clinical presentation, the organism may be present in ocular, genital, or respiratory secretions which should be handled as potentially infectious. The use of gloves is recommended when touching infective material, and handwashing is as always important. Airborne transmission has occasionally been reported in laboratory personnel working with large quantities of the organism.[63] One case of infection of a dentist resulting from conjunctival exposure to aerosol from a patient with oral secretions positive for *C. trachomatis* has also been reported.[64]

F. *Clostridium difficile*

C. difficile, the most important etiologic agent in antibiotic-associated colitis, has recently turned out to be an important nosocomial pathogen.[65] The most important risk factor is antibiotic treatment. Hospital personnel are not reported to be at increased risk for *C. difficile* disease. Kim et al.[66] and Fekety et al.[67] isolated the organism from the hands or stools of several asymptomatic staff members working in wards with patients with *C. difficile* colitis. Scherertz and Sarubbi[68] recovered the organism from the hand of 1 of 11 nursing personnel during an outbreak in a nursery. None of 24 stool samples from physicians and nurses revealed *C. difficile,* but neutralizable toxin was present in a stool sample from a nurse. Mulligan et al.[69] and Malamou-Ladas et al.[70] also isolated the organism from hospital staff. However, several other studies have failed to isolate the organism from medical personnel during outbreaks of *C. difficile* colitis in hospitals.[71-74] However, it has been claimed that this might be due to failure to use selective media for low numbers of *C. difficile.*[65]

G. *Corynebacterium diphtheriae*

C. diphtheriae is mainly transmitted by droplets from the upper respiratory tract. Shedding appears to increase with clinical disease and pseudomembrane formation, but asymptomatic carriers are probably a more important reservoir. The nasal form of the disease may also favor spread due to mild symptomatology and the presence of the organism in the nose as well as the pharynx. Patients with the cutaneous form of the disease may also constitute an important reservoir.[75] They have been shown to frequently harbor *C. diphtheriae* in the respiratory tract and to have a greater spread of infection to other household contacts than patients with respiratory forms of the infection.[76]

Since the introduction of large-scale immunization against diptheria, this disease has become very uncommon in many countries and almost forgotten in the medical community. As a consequence, delay in diagnosis and treatment of sporadic cases may easily occur, resulting in enhanced exposure of medical personnel. The majority of such cases are seen in adults of lower socioeconomic status.[77,78]

In spite of this, hospital personnel are not at a substantially higher risk than the general adult population of acquiring diptheria.[79] During an epidemic of diphtheria in Gothenburg, Sweden in 1984 and 1985, involving a total of 12 patients with clinical disease and 64 carriers, Larson et al.[80] failed to detect *C. diphtheriae* in 533 cultures from 328 members of the hospital staff; 192 of these samples were from the ear, nose, and throat (ENT) and intensive care unit (ICU) departments where 2 fatal cases of diphtheria were treated before the diagnosis was suspected and before special precautions were taken. They also found that environmental contamination around patients was rare except from a carrier with a chronic skin disorder. In an epidemic of occult diphtheria in a hospital for the mentally subnormal, Gray and James identified carriers of *C. diphtheriae* var. *mitis* in 35 of 484 patients and in only 2 of 341 staff.[81] The majority had not been immunized against diphtheria. Despite the treatment of carriers, immunization of patients and staff, and other efforts to eradicate the organism, this proved to be difficult; 4 months later, screening identified 12 carriers, 4 old cases, and 8 new ones, and after another 5 months, swabbing revealed 5 carriers, including 2 members of the nursing staff.

C. diphtheriae may represent a certain risk to laboratory personnel. Pike reported 33 cases, none of them fatal, in his survey of the world literature.[35]

Hospital personnel closely exposed to patients with diphtheria should immediately receive a booster dose of diphtheria toxoid, if more than 5 years have elapsed since the last dose.[47] Inadequately immunized personnel, i.e., those with less than three previous doses of diphtheria toxoid, should complete the series according to the recommended schedule for their country. Chemoprophylaxis, preferrably with erythromycin, 1 g/d for 7 to 10 d, should also be given to close contacts not properly immunized.[30] Alternatively, intramuscular benzathine penicillin, 1,200,000 U, should be given.

H. *Coxiella burneti* (Q Fever)

Human infection with *C. burnetii* is usually acquired by inhalation and the primary source of infection is domestic animals that shed the organism in urine, feces, milk, and especially in birth products. *C. burnetii* is considerably resistant to dessication and to other detrimental environmental conditions, and the organism has been shown to survive for months in the environment.[82,83] The disease is mainly found in agricultural workers or persons with similar occupational exposure to animals or animal products. Inhalation experiments with human volunteers have suggested that a single organism is capable of initiating the disease.[84] Person-to-person transmission has been described, but seems to be rare. Marmion and Stoker reported the apparent infection of a nurse, two pathologists, and a mortuary attendant during the terminal care and the necropsy of an unrecognized case of Q fever.[85] In 1950 Siegert et al. reported 38 cases that were traced to an infected patient in a hospital in Frankfurt.[86]

Laboratory aquisition of the disease has been a significant feature since inception of laboratory work with *C. burnetii*.[87] Johnson and Kadull reported 50 cases from 1 laboratory during a 15-year period;[88] 37 of the subjects were laboratory workers, while the 13 others included clerical workers, engineering personnel, animal handlers, steam fitters, carpenters, repairmen, and janitors. Pike reported 278 cases with 1 death that had come to his attention in 1974.[35] Q fever was the second most common laboratory infection in his survey.

Laboratory workers need not work with *C. burnetii* to be at risk for Q fever. In recent years it has also been reported from laboratories not directly working with the organism, but rather using sheep as research animals. The mere exposure to infected laboratory animals has been documented to be an occupational hazard.[89,90] Schachter et al. found that 16% of employees who had contact with sheep in a university hospital environment were seropositive, compred to a prevalence rate of only 0.1 to 0.3% among the general population in the same area.[89] Simor and associates noted that an outbreak of Q fever among research personnel in Toronto was unrecognized despite considerable morbidity.[90] Among 331 staff members tested, 59 (18%) were found to be seropositive. The highest rate of seropositivity was found among the animal attendants, but 63% of the seropositive staff members had no direct contact with sheep. Bernard et al.[91] have published detailed recommendations for reducing the risk of exposure to Q fever in persons not working with sheep in research facilities that use sheep as well as in persons working with sheep in such institutions.

I. *Enterobacteriaceae* (Except Salmonella/Shigella)

The Gram-negative enteric bacteria belonging to the *Enterobacteriaceae* family (*Escherichia coli, Klebsiella* spp., *Enterobacter* spp., *Proteus* spp., *Citrobacter* spp., *Serratia* spp., etc.) are the etiologic agents of a large proportion of nosocomial infections. They are also a common cause of many community-acquired infections. Despite this, most infectious disease specialists would probably deny that this group of organisms poses a significant risk to health care personnel. There are, however, few studies on this subject, and most of these deal with colonization and not clinical disease. Rahal et al.[92] reported that upper respiratory tract carriage of Gram-negative enteric bacilli by hospital personnel was not different from that among personnel without patient contact. Several workers have reported carriage rates of Gram-negative bacilli around 20% on the hands of hospital personnel.[93-96] Maki found 44.5% of 348 hand cultures from 25 medical personnel to be positive for Gram-negative bacilli, all 25 being positive at least once during a 4-month period.[97] However, studies comparing hand carriage of Gram-negative bacteria have in fact shown significantly lower carriage rates among hospital personnel than among controls without patient contact.[96,98,99] The reason for this seems to be that medical personnel wash their hands more frequently. The use of handwashing agents with antiseptics was also very common among the hospital personnel in these studies. In general, because of their low virulence, carriage of Gram-negative enteric bacilli on the hands of health care workers is a much larger hazard to their patients than to themselves.

However, some strains may have enhanced virulence, and occasionally transmission to hospital personnel may result in clinical disease. Carter et al.[100a] recently reported a severe outbreak of *E. coli* 0157:H7-associated hemorrhagic colitis in a nursing home, affecting 55 of 169 residents and 18 of 137 staff members. The origin was a common source, but 11 of the 18 staff members probably acquired the infection by person-to-person spread from the residents. Likewise, Karmali et al.[100b] reported a case of hospital-aquired *E. coli* 0157:H7-associated hemolytic uremic syndrome in a nurse. The transmission probably occurred because she failed to use enteric precautions when cleaning an infant extensively soiled with feces.

Enterobacteriaceae (with the exception of *Salmonella/Shigella*) also seems to be a minor risk for laboratory workers. Pike reported only 7 cases (5, *Serratia marcescens*; 2, *E. coli*; and 1, *Klebsiella pneumoniae*) among 3921 laboratory-associated infections.[35]

J. *Francicella tularensis*

F. tularensis is a highly infectious agent for man. The disease is usually acquired from animals, especially cottontail rabbits, voles, muskrats, and lemmings. Transmission occurs by direct contact or inhalation or by bites of insects that infect such animals. Studies with volunteers have shown that less than 50 organisms administered intradermally or by the respiratory route are able to cause disease.[101,102] Despite its reputation as a highly infectious agent, however, no person-to-person transmission has been reported. For example, large numbers of viable organisms can be grown from cough plates obtained from patients with pneumonia, but no disease in hospital personnel has been reported following this method of aerosolization.[103]

In the laboratory setting, however, *F. tularensis* is a definite occupational hazard. Tularemia is the third most common bacterial laboratory-associated infection reported by Pike.[35] Direct contact of skin or mucous membranes with infectious materials, accidental parenteral inoculation, and ingestion and exposure to aerosols and infectious droplets have resulted in infection. For example, Barbeito et al.[104] reported tularemia in five persons after petri dishes containing the organism had been dropped. Administration of a live attenuated tularemia vaccine has been proven effective against tularemia and should be considered for laboratory workers exposed to *F. tularensis*.[105]

K. *Haemophilus influenzae*

Infection with *H. influenzae* is predominantly a disease of children.[106] In adults there is a nearly uniform presence of bactericidal humoral antibodies against the organism, and infection with *H. influenzae* is seldomly seen. When it occurs, it is usually in patients with some underlying disease.[107] Among children with close contact of patients with *H. influenzae* disease, like siblings or residents of the same day care center, there is also a definite risk of secondary cases.[108-110] There are, however, no data to suggest any occupational risk of *H. influenzae* disease for health care workers. The risk for colonization also seems to be negligible. During an outbreak with 5 cases of meningitis in an enclosed hospital population of chronically ill children, Glode and co-workers were not able to find any carrier among 160 adult employees and volunteers working with these children.[11] Ginsburg et al.[112] detected 1 carrier among 11 adult personnel during an outbreak in a day care center. In two other outbreaks in day care centers, the nasopharyngeal colonization rate for adult staff members was 0%.[113-115]

L. *Legionella*

L. pneumophilia has emerged as a frequently recognized pathogen in either community-acquired or nosocomial pneumonia.[116-118] A number of nosocomial outbreaks have been reported involving hospitalized patients as well as personnel.[116,119,120] Legionnaire's disease is believed to be transmitted predominantly by inhalation of infectious aerosols from environmental water sources like water towers, air-conditioning apparatus, and potable water. When hospital employees are infected, they seem to acquire the disease from the same environmental source as their patients. Immunosuppressed patients, especially renal transplant recipients, seem to be particularly susceptible. There has been no documented person-to-person spread of Legionnaire's disease.[116-118,121,122]

Saravolatz et al.[123] reported a higher prevalence of seropositives among hospital staffs with patient contact than among staffs not in a position of contact with patients diagnosed with Legionnaires' disease. However, this study has been disputed, and others have not been able to confirm their observations.[124-126] Morgan et al.[125] prospectively tested the sera of 122 staff members exposed to two patients with Legionnaires' disease. In one of the patients, the diagnosis was made post-mortem after 11 d of standard nonisolation nursing. All tests were negative and no case of clinical or subclinical disease was identified among the ward staff or laboratory staff. Correspondingly, Rashed and co-workers were not able to identify any secondary case among members of the staff at Stafford District General Hospital during an extensive epidemic despite very limited formal barrier nursing and the fact that at the height of the epidemic as many as 40 patients were hospitalized at the same time.[126] The prevailing view, therefore, is that medical personnel do not appear to be at increased risk from exposure to patients with Legionnaires' disease and no isolation precautions are considered necessary.[13,116-118]

Several studies have shown an increased prevalence of *L. pneumophilia* antibodies among dental personnel.[127,128] Reinthaler et al.[128] recently reported the results of a seroepidemiological study among dentists, dental assistants, and dental technicians: 9 of 18 (50%) dentists, 20 of 53 (38%) dental assistants, and 7 of 36 (20%) the technicians were found to be seropositive. The 36 positive persons were found among 57 employees in 13 different dental offices, whereas 50 employees from 18 other offices were all negative. The source was thought to be stagnant water in the dental equipment aerosolized during use and inhaled by the personnel (as well as their patients).

M. *Leptospira interrogans*

Indirect transmission of *leptospirae* from animal urine to man is the predominant source of human leptospirosis. The organisms enter the body through abrasions of the skin or through mucous membranes, but may also penetrate intact skin. Direct human-to-human transmission of *leptospirae* occurs only rarely. However, blood and urine from patients with leptospirosis may be infective and should be handled with blood and body fluid precautions. Laboratory personnel are included among the occupational groups with enhanced risk of leptospirosis.[35,53,129] Laboratory workers may acquire the infection either during the handling of specimens from patients or from experimentally or naturally infected laboratory animals.[53,130] Pike reported 67 laboratory-associated infections and 10 deaths.[35]

N. *Mycobacterium leprae*

Historically, leprosy has a bad reputation for being highly contagious, leading to prolonged segregation of the victims of the disease. *M. leprae* is generally considered to be the etiologic agent of leprosy, but more than 110 years since Hansen's discovery of the *M. leprae* bacillus, the epidemiology of this disease still is not well understood.[131,132] Although it is considered a communicable disease transmitted by person-to-person contact, the mechanism of transmission of *M. leprae* has not been firmly determined. In fact, even though the risk of acquiring leprosy among household contacts of a patient is severalfold higher than in households without a case of leprosy, in 50 to 70% of all sporadic cases of leprosy no history of contact with another known patient can be documented.[133-135] Strong arguments exist for the possibility that leprosy may also be spread by indirect transmission via enviromental nonhuman sources.[136]

In the case of direct person-to-person spread, the prevailing theory for a long time was that transmission resulted from prolonged close skin-to-skin contact. However, the number of acid-fast bacilli shed from the intact skin of lepromatous patients appear to be very small.[137] On the other hand, untreated patients with lepromatous leprosy may have extensive involvement of the nasal mucous membranes with consequent shedding of millions of viable *M. leprae* in sputum and nasal secretions.[138] Furthermore, primary leprosy lesions have been found in the nose of apparently healthy contacts of leprosy patients.[139] Bacilli have been shown to remain viable for at least 7 d in dried nasal secretions.[140,141] Therefore, at present most authorities hold that transmission by the respiratory route is the most likely possibility in the majority of cases, even though entrance through broken skin or spread by insect vectors may also occur.[142-144] When transmission from a patient to a contact occurs, the great majority seem to be able to resist the infection and subclinical infection seems to be the most common outcome.[145]

There are few studies on the occupational risk for health care personnel having contact with leprosy patients. Myrvang et al.[145] tested the immune responsiveness to *M. leprae* in 52 medical attendants to leprosy patients and found them to respond significantly more strongly than individuals not exposed to patients. By classifying responses of strengths found in noncontacts as negative, 71.2% of medical attendants were responders by the leukocyte migration inhibition test, 44.2% responders by the lymphocyte transformation assay, and 50.0% responders by the early lepromin reaction.

Regarding clinical disease, however, Mathai and co-workers found that the prevalence rate of leprosy among hospital employees in Vellore, India was less than one fourth of that found among the general population of the area, and they concluded that the medical staff serving leprosy patients for whom no isolation is practiced do not run an additional risk of infection.[146] Gray and Dreisbach found a prevalence rate of 13.2/1000 persons among foreign missionaires in northern Nigeria compared to a prevalence of 35/1000 in the indigenous population.[147] Missionaires doing leprosy work had higher attack rates than those doing general medical work.

Inadvertent parenteral human-to-human transmission of leprosy following an accidental needlestick in a surgeon has been reported.[148] Despite the historically bad reputation of leprosy, in modern care for leprosy patients, no isolation is required.[13]

O. *Mycobacterium tuberculosis*

For many years, including the first 3 decades of the 20th century, the view prevailed that tuberculosis was less common among hospital personnel than among the general population. In 1878 Williams, quoted by Mücke, stated:[149] "In the care of 15,662 tuberculous patients at the Brompton Hospital for Consumption in London, over a period of twenty years not a single case of pulmonary disease developed among the physicians and the hospital personnel in spite of the worst hygienic conditions." In 1930 the same conclusions were drawn by

Baldwin at Trudeau Sanatorium, where no case of pulmonary tuberculosis developed in forty-five years among hundreds of employees.[150]

In 1928, however, Heimbeck reported the development of tuberculosis in 47 of 337 Norwegian nursing students during the years from 1924 to 1927.[151] Later, he reported a moribidity of 152.9/1000 observation years among 387 nurses that were tuberculin skin test negative on entrance to nursing school.[152] In the following decades, a large number of reports appeared comparing the incidence of tuberculosis among health care personnel with that of the general population. Most of these studies demonstrated an increased incidence among medical students, students of nursing, physicians, nurses, and other groups of health care personnel, and tuberculosis became recognized as an occupational hazard for health care workers.[153-158]

Two main factors determine the level of hazard for occupationally acquired tuberculosis among health care personnel: the degree of exposure to patients with tuberculosis and the immunity of the personnel exposed. The degree of exposure varies widely around the world. In many countries, tuberculosis has become a rare disease after many decades of continuous decline, and the reported annual incidence rate has fallen below 10/100,000 inhabitants in several developed countries. This has been accompanied by a decrease of the absolute risk of occupationally acquired infection among health care personnel. For example, in a study performed at Philadelphia General Hospital, Philadelphia, PA (U.S.), Israel and associates reported that 100% of tuberculin-negative nursing students became tuberculin reactive during their 3-year study in the years from 1935 to 1939.[155] They also found that approximately one of every ten students developed active disease. Between 1959 and 1971, however, Weiss reported that only 37 out of 1891 students at the same hospital converted, and only 5 cases of active tuberculosis were identified.[159] Correspondingly, Levine, in a study from New York, reported conversion to a positive tuberculin test reaction in 66% of students of nursing entering in 1948 through 1951, 21% in those entering in 1952 through 1958, and 3.5% in those students who entered study in 1959 through 1964.[160]

However, despite this dramatic decline in conversion rates, the incidence among health care personnel still seems to be higher than that among the general population. For example, in British Columbia, Burhill et al.[161] reported that between 1969 and 1979 the rate of active tuberculosis among female nurses born in Canada was almost twice that of other Canadian-born women adjusted for age. Interestingly, this was about the same relative risk as that reported by Daniels et al.[157] among over 5000 nurses in London between 1934 and 1943.

Recent studies in the U.S. also showed enhanced risks among physicians. Barrett-Connor found that the cumulative percentage of tuberculin-positive physicians in California remains at least twice the age-specific infection rate for the U.S. population.[162] Tuberculosis developed in nearly one in ten physicians infected after medical school entry. She found highest rates in medicine, pediatrics, and surgery, intermediate rates in obstetrics and gynecology and orthopedics, and lowest rates in radiology and psychiatry. In Illinois Geiseler et al.[163] found that even among physicians graduating since 1970, the incidence of active tuberculosis was higher than in the general population. They did not find, however, that any medical speciality or type of clinical practice was more frequently associated with development of active tuberculosis. This was in contrast to the findings in a study of British doctors carried out in 1977, where respiratory tuberculosis had occurred almost three times more often in hospital doctors than in general practitioners.[164] All of these studies showed that the majority of cases among physicians occurred during medical school or in the first years after graduation. For example, in the study of Geiseler et al.,[163] the age distribution revealed that 89% of the physicians with tuberculosis were between 25 and 44 years of age at the time of diagnosis. More than two thirds of all cases of tuberculosis occurred during medical school or within 6 years of graduation.

Other health care professionals may be at even higher risk. Among medical laboratory workers in England and Wales, the risk of pulmonary tuberculosis in 1971 was five times higher than that of the general population.[165] The same observation was made by Grist in a survey of infections in British clinical laboratories during 1980 to 85.[59-61] Out of 33 cases of tuberculosis, 20 were considered to be of occupational origin, an "occupational" attack rate of 22.9/100,000 person years. Technicians were at greatest risk, especially if they worked in morbid anatomy departments. This latter point corroborates that of Meade from 1948, where the chief source of primary infection in medical students was found to be the autopsy.[166] Dentists also run a higher risk than the general population.[167]

Some recent studies challenge the view that health care personnel are infected with *M. tuberculosis* more frequently than the general population in communities with a low prevalence of the disease. Ruben et al.[168] and Lowenthal and Keys found no correlation between degree of patient contact and tuberculin skin test conversion rates in Pittsburgh and Cleveland, respectively.[169] On the contrary, the latter found highest conversion rates among groups with no significant exposure to patients. These groups were made up predominantly of inner-city minority persons. A similar observation was made by Berman et al.[170] in a 5-year study from 1971 to 1976 in Baltimore. In North Carolina, Price et al.[171] found that the incidence of tuberculosis among approximately 100,000 hospital employees in 1983 and 1984 was less than the incidence in the general population. The same conclusion was drawn by Weinstein et al.[172] for medical students at Mount Sinai School of Medicine, New York, in the period from 1974 to 1982.

The studies referred to above were all performed in communities with a low prevalence of active tuberculosis. In a global perspective, however, tuberculosis is still a major problem, and in many impoverished and densely populated communities, the annual incidence of pulmonary tuberculosis is many tenfold higher.[173] Even though data are scarce, it is reasonable to assume that health care personnel in such areas run a significantly higher risk of occupationally acquired disease.

The immunity to *M. tuberculosis* is the other main determinant of tuberculosis hazard to health care workers. In the study of Heimbeck, even though the distribution of tuberculin-positive and -negative nursing students was about fifty-fifty on entrance to nursing school, 44 of the 47 students later aquiring tuberculosis were originally tuberculin negative.[151] A number of studies later confirmed these observations.[155,157,158,174,175] However, this view has recently been challenged by two studies from Canada where the tuberculosis rates among nurses whose tuberculin tests were initially positive were much higher than the rates among their tuberculin-negative colleagues.[161,176]

Immunity to *M. tuberculosis* may also be induced through vaccination with Bacille-Calmette-Guerin (BCG). Actually, one of the first reports ever on the protective effect of injection (as opposed to oral administration) of BCG was a study involving student nurses.[151] Later quite variable results have been reported from studies on the effect of BCG vaccination, but most experts agree that the vaccine induces protective immunity against tuberculosis.[177-180] There is no consensus, however, on the degree of protection it may offer, figures from nil to 80% protection have been reported.

A large number of studies have been published on the protective effect of BCG against occupationally acquired tuberculosis among health care personnel, and the degree of protection reported varies between 40 and almost 90%.[161-163,181-183] In recent studies from communities with low prevalences of tuberculosis, recipients of the vaccine still are reported to have 40 to 80% less tuberculosis than nonvaccinated colleagues.

1. Transmission

Aerosolized droplet nuclei less than 10 μm in diameter containing only one or a few bacilli are the primary means of transmission of tuberculosis. The risk of infection is related to the density of bacilli in the respired air.[184,185] Current evidence indicates that only droplet nuclei that reach the alveoli are effective in infecting a susceptible host. Other particles are too voluminous to reach the pulmonary alveoli. Therefore, bacilli present on clothing, bed linen, and other objects contaminated by the patient do not play a significant part in infection since they cannot be dispersed into aerosols of this small size.[186,187] Thin watery sputum is more easily dispersed into infective particles than viscous material. The presence of cough increases the probability of aerosol generation, as do other forceful exhalational maneuvers such as sneezing, singing, or shouting.[188,189]

Tubercle bacilli are considered to have unusual resistance to conditions destructive to other vegetative bacteria, but are also quite susceptible to specific physical and chemical agents. They are among the most resistant to moist heat of the nonspore-forming pathogenic bacteria, but are killed by boiling for 2 min. They are also highly resistant to dessication. Droplet nuclei may remain infectious for 8 to 10 d. Smith found that the bacilli retained their viability for up to 3.5 to 5 months in dried sputum if kept in the dark.[190] When exposed to sunlight, they are killed within several hours, and the bacilli are very susceptible to UV light irradiation of a wavelength around 2537 Å.

From the point of view of transmission, patients with tuberculosis represent very different levels of hazard. Several factors are important in this respect. Subjects that present sputum which is positive on direct microscopical examination infect many more of their contacts than do patients with merely culture-positive sputum and patients with culture- (and microscopy) negative sputum. Furthermore, transmission from smear-positive patients more often results in disease than transmission from other categories of patients with tuberculosis. The closeness of contact also plays a considerable role. Intimate contacts run a much higher risk than occasional contacts, i.e., friends and colleagues at work.[191] For the latter group, only patients with microscopically positive sputum smear constitute a risk, while patients who are only culture positive and those who are culture negative seem hardly to be dangerous at all.[192]

For practical purposes, therefore, the only source of tuberculous infection is a person with pulmonary tuberculosis in whose sputum tubercle bacilli are present in sufficient numbers to be seen on direct examination of sputum smears. Patients with pulmonary tuberculosis in whom three or more sputum smears give negative results on direct smear examination should be regarded as noninfectious.[193]

The unrecognized patient with tuberculosis often represents the greatest risk. Along with the decline of tuberculosis, health care personnel have become increasingly unfamiliar with the disease and often fail to consider the diagnosis. Consequently, diagnosis and treatment are often delayed, thereby enhancing exposure of personnel. Edlin reported that 12 of 24 cases of active tuberculosis which came to necropsy in Dundee hospitals from 1968 to 1975 were diagnosed after death.[194] Correspondingly, in two separate studies, MacGregor in Pennsylvania and Vogeler and Burke in Utah found that almost half the cases hospitalized with tuberculosis in the 1970s were misdiagnosed, exposing an average of 35 and 25.6 hospital personnel to each unisolated patient with positive smear.[195,196] Craven and associates found that 15 out of 59 inpatients with tuberculosis were not promptly placed in isolation on admission.[197] Delay in making the diagnosis in these patients was associated with failure to apply a tuberculin skin test on admission in all but 1 patient and radiologic misinterpretation in all 15. The median time interval between admission and isolation was 5 d. Personnel exposed to these patients had a greater than sixfold increased incidence of infection compared to nonexposed personnel.

A new group of patients have recently been included among those with enhanced incidence of tuberculosis. Mycobacterial diseases, including tuberculosis, are common among patients with the acquired immunodeficiency syndrome (AIDS) and human immunodeficiency virus (HIV) infection.[198] Among 15,181 patients with AIDS in the U.S., 645 (4.2%) also have had tuberculosis.[199] Intravenous drug abusers and Haitians seem to be particularly at risk.[200] Tuberculosis among patients with AIDS is believed to be partly responsible for the failure of tuberculosis morbidity in the U.S. to decline as expected in 1985 and for the increase in 1986.[199] The fact that tuberculosis often presents with unusual manifestations in these patients may result in delayed diagnosis and enhance the exposure of health care workers.[201,202]

Modern medical diagnostic and therapeutic procedures may sometimes enhance the risk of transmission from unrecognized cases, despite negative sputum smear for acid-fast bacilli. Catanzaro reported an outbreak in an intensive care unit where 14 of 45 (31%) of susceptible hospital staff who were exposed over a 5-d period to a single patient with undiagnosed smear-negative tuerculosis became infected.[203] The patient required bronchoscopy, intubation, and assisted ventilation; 10 of 13 (77%) susceptible hospital staff members present at the time of bronchoscopy converted. In this context, the potential hazard of the increasing use of bronchoscopy for diagnostic procedures among immunosuppressed patients is also worth mentioning.

In another ICU outbreak, a single patient with undiagnosed tuberculosis infected 43.5% of tuberculin-negative medical students and physician contacts.[204] This outbreak also illustrated the importance of adequate ventilation for prevention of secondary cases. On a ward with faulty ventilation where the patient spent 57 h, there were 21 converters in 60 tuberculin-negative personnel (35%), including 10 who had little or no direct contact with the patient. On a better ventilated ward, where the patient spent 67 h, conversions occurred in only 2 of 19 personnel at risk (10.6%), and both were in close contact.

2. Containment Procedures

Patients with active pulmonary tuberculosis with sputum smear positive for acid-fast bacilli should be accommodated in a private room with special ventilation. Patients with extra-pulmonary tuberculosis do not require segregation; neither do infants and young children because they seldom cough and their bronchial secretions contain few TB organisms. Patients should be instructed to cover their mouth and nose during coughing and sneezing. Masks are usually not required, but may be worn if the patient coughs and does not reliably cover the mouth.[13,193] Alternatively, patients unable or unwilling to control coughs should wear a mask during nursing procedures. Gowns are not necessary except if gross contamination of clothing is likely. Gloves need not be used. Hand washing is efficient to remove organisms possibly picked up from fomites or direct contact with infectious sputum or other discharges.

Patients become noninfectious very soon after starting chemotherapy with includes rifampicin and long before the disappearance of acid-fast bacilli from sputum smears. The number of live tubercle bacilli in sputum (as shown by growth in culture) falls by 99% after the first 2 weeks of chemotherapy.[205] The risk to contacts occurs before chemotherapy is started and is abolished rapidly thereafter.[192,206]

3. Surveillance and Control

Prompt diagnosis and treatment of patients with tuberculosis are the primary means of reducing the occupational hazard to health care personnel. The other elements of surveillance and control of tuberculosis in hospitals include BCG vaccination, tuberculin skin testing, chest roentgenograms, and preventive isoniazid treatment of tuberculin skin test converters. The design of the control program will depend on the prevalence of tuberculosis in the community and the hospital as well as on the financial resources available.

BCG vaccination policies for the general population as well as for health care personnel vary widely around the world. In 1980 BCG vaccination for the general population was reported to be compulsory in 64 countries and officially recommended in a further 118 countries and territories. Almost all developing countries were included among these.[180] BCG vaccination is inexpensive, its administration is simple, and it is the preventive method of choice in health care personnel as well as in the general population in communities with high prevalences of the disease. In areas with a low risk of infection, however, the BCG vaccination policies differ, as for example across the Atlantic Ocean. In Europe, the prevailing strategy for prevention of tuberculosis is based on BCG vaccination. In the U.S., however, vaccination is generally not recommended, and control programs are mainly based on regular tuberculin skin testing. One of the arguments for the latter approach is to preserve the usefulness of the tuberculin test in the detection of naturally acquired infection.

4. Preemployment Examination of Personnel

All personnel should receive a tuberculin test before employment unless they have a documented medical record of a significant tuberculin skin test reaction. There are several tests in current use. The Mantoux test is considered to give the most precise measurement of tuberculin sensitivity and is the method recommended for routine use in the U.S. In the U.K., the Heaf multiple puncture test is usually preferred. Several other types of tests are used in other countries.

In Britain, a preemployment chest X-ray is also recommended.[193] In the U.S. and in many other countries, this policy has now been abandoned, and a chest radiograph on employment is only recommended for those with a significant tuberculin test reaction.[11] As demonstrated by Belfield and co-workers, the efficiency of a preemployment program depends on how it is practiced.[207] Among six junior doctors with tuberculosis, only three were offered a preemployment screening; two accepted the offer, and both had a Heaf skin test reaction of grade 3 to 4 which was erroneously ascribed to prior BCG vaccination.

5. Follow-up of Tuberculin-Negative Individuals

In many western European countries, BCG vaccination of tuberculin-negative personnel is recommended. For employees who refuse vaccination, regular tuberculin tests are usually ordered. In the U.S., most medical centers recommend annual tuberculin testing of nonreactors. However, along with the falling prevalence of the disease, regular skin testing is no longer considered necessary in hospitals rarely admitting patients with tuberculosis and serving a community with an extremely low prevalence of infection.[208] The CDC guidelines for infection control in hospital personnel recommend that the need for repeat testing should be determined individually in each hospital.[11] Different policies may be established for different groups of employees at the same hospital.[169]

The efficiency of a screening program is also dependent on compliance. As noted by Barrett-Connor, the value of preserving the tuberculin tests as a detection device for recent tuberculosis infection may not be relevant in medical students and physicians.[162] The majority of them fail to adhere to established public health standards of annual tuberculin tests in tuberculin-negative health workers exposed to patients with active tuberculosis. Correspondingly, in the study of Geiseler et al.,[163] only 30% of the tuberculin-negative physicians had a tuberculin skin test every 1 to 2 years, and 35% did not have a tuberculin skin test as often as every 10 years.

One of the problems with interpreting the results of repeated tuberculin tests is the so-called booster phenomenon whereby the first in a series of tests has a stimulating effect on the reaction to subsequent tests.[209] As a result, falsely elevated conversion rates may be recorded.[210] Infection with nontuberculous mycobacteria is considered to be most frequently responsible for the boosting. The frequency of the booster effect therefore varies from area to area. Among ten hospitals in different parts of the U.S., Snider and Cauthen found boosting rates ranging from 0 to almost 10%.[211] The necessity of the use of the two-step procedure, whereby personnel with tuberculin reactions of less than 10 mm are retested within 1 to 3 weeks, will therefore depend on the boosting rate in the area in question.[212] Some countries also recommend annual chest X-ray examination for high-risk personnel. This latter approach has, however, been shown to have no significant effect on the occurrence of subsequent smear-positive cases, as these develop so rapidly that they appear between roentgenograms.[213] For the same reason, annual X-ray screening of tuberculin-positive employees should also largely be abandoned.[214-216]

6. Preventive Treatment with Isoniazid

Recent tuberculin skin test converters run a substantial risk of clinical tuberculosis, ranging from 4 to 23% during several years of follow-up.[158-160,217,218] In a number of studies, chemoprophylaxis with isoniazid has been documented to prevent tuberculous infection to progress into clinical disease.[219-223] Among groups with a high degree of compliance, the level of protection exceeded 90%.

The generally accepted regiment for isoniazid prophylaxis is a single daily oral dosage of 300 mg in adults for 9 months to 1 year. The major concern regarding side effects from isoniazid is hepatitis, the risk increasing with age.[224] After the age of 35, the risk of hepatitis has been considered to be more important than the risk of tuberculosis, providing that additional risk factors for tuberculosis other than a positive tuberculin reaction are absent.[225] However, in the presence of additional risk factors, such as household or other close association with persons who have infectious tuberculosis, abnormal chest roentgenogram, or special clinical conditions (diabetes, renal insufficiency, steroid treatment, hematological malignancy, etc.), age ceases to be a factor in the risk-benefit analysis.[226] Subjects over the age of 35 who are receiving isoniazid chemoprophylaxis should have the early symptoms of hepatitis carefully explained to them and should have hepatic function monitored with determinations of levels of transaminases at 1, 3, 6, and 9 months.[226]

The use of chemoprophylaxis in individuals under the age of 35, when a positive tuberculin skin test is the sole risk factor, has, however, been disputed.[227] Not surprisingly, therefore, the compliance with isoniazid prophylaxis among physicians may be poor. Barrett-Connor reported that only 41.3% of physicians with documented tuberculin conversion initiated prophylaxis, and prophylaxis was not completed in 29% of those who began it.[162] In the study of Ruben and associates, 28 of 46 converters among hospital employees refused to begin isoniazid prophylaxis despite great efforts to educate them regarding their risk of developing tuberculous disease.[168]

7. Contact Tracing

The extent of contact tracing among health care personnel after exposure to an undiagnosed case of tuberculosis should be guided by the probability of transmission from the patient. Factors to be considered are closeness and duration of contact, results of sputum smears, the presence of cough, a cavitary infiltrate, or tuberculous laryngitis, ventilation in the area where the patient resided, and the use of diagnostic or therapeutic procedures (bronchoscopy, endotracheal intubation, etc.) known to enhance dissemination of airborne bacilli. For patients with sputum smear negative for acid-fast bacilli, investigation should be limited only to personnel with close contact, i.e., persons caring directly for the patients. For patients with positive sputum smear, however, causual contacts should also be included, especially if the patient has a cough. A two-step procedure may also be chosen, whereby investigation of casual contacts will depend on the degree of infection among close contacts.

Personnel elected for investigation should receive a tuberculin test, preferably by the Mantoux technique, as soon as possible after exposure and 8 to 15 weeks later. Those with significant reactions also need a chest roentgenogram to exclude the possibility of tuberculous pulmonary disease.[11]

The interpretation of the tuberculin test after exposure may be difficult in BCG-vaccinated personnel. However, using the Mantoux technique, the results of skin tests in persons who have had a prior BCG vaccination should be interpreted and acted on in the same manner as those in personnel who have not been vaccinated with BCG.[11] Personnel compliance may also prove to be a problem in contact tracing. After an outbreak of tuberculosis in a pediatric department, Stewart reported that whereas nearly all the nonmedical adult contacts were traced and examined, fewer than half the medical contacts attended for chest radiography.[228]

P. *Mycoplasma pneumoniae*

Infections with *M. pneumoniae* are most frequently seen in children 5 to 14 years old. The mode of transmission of *M. pneumoniae* is not entirely clarified, but prolonged close contact is usually required for secondary cases to occur. However, airborne transmission may occasionally be involved. The organism has only rarely been isolated from sites other than the respiratory tract. As noted by Foy, the disease is frequently observed among physicians and other hospital personnel.[229] However, comparative data on incidence by occupation are scarce. Gerth and co-workers studied the occupational risk of acquiring *M. pneumoniae* infection for pediatric student nurses over a 3-year period.[230] They were not able to demonstrate a higher incidence among the student nurses compared to a group of medical technology students.

Only a few reports on the occurrence among health care workers have been published. Fisher et al.[231] reported an outbreak involving 14 employees of a large hospital. In the absence of documented cases among the inpatient population, it was not possible to identify any single area of the hospital frequented by enough of the involved personnel to explain the spread of the infection among the employees. The personnel worked in widely separated parts of the medical center, including unconnected buildings covering two city blocks. However, since only personnel with severe respiratory symptoms were included in the study, the possibility of tracing person-to-person spread via personnel with minor symptoms was not possible. Sande and co-workers described an outbreak involving ten persons in a large prosthodontics laboratory. The organism may have been spread by a fine-particle aerosol generated by the abrasive drilling of contaminated false teeth.[232] Except for handwashing and careful handling of respiratory secretions, no special precautions are necessary in the care of patients with mycoplasma infections.[13]

Q. *Neisseria meningitidis*

Nasopharyngeal carriage of *N. meningitidis* is frequently found in healthy individuals. The carriage rate is, however, quite variable, depending on serotype and the population studied.[233-235] The relationship between meningococcal disease and carriage is still poorly defined. Meningococci are transmitted mainly by direct contact with droplets and secretions from the respiratory tract.[11] Close contacts of patients with meningococcal disease run a much higher risk than the general population. In this context, as defined by Kaiser et al.,[236] a close contact is an individual who frequently sleeps and eats in the same dwelling with the index case. By this definition, therefore, health care personnel will seldom be regarded as close contacts. However, procedures such as mouth-to-mouth resuscitation, nasotracheal intubation, suctioning, and other instrumentation of the respiratory tract prior to treatment are likely to increase the risk of transmission to health care personnel. Several cases of meningococcal disease have occurred among physicians who gave mouth-to-mouth resuscitation to patients with severe meningococcal infection.[237] In general, however, the risk of colonization from exposure to a patient with meningococcal meningitis seems to be quite low, and the risk of secondary cases of clinical infection seems to be almost negligible.[238,239] In a study of 370 patients in Finland, no member of the hospital staff developed the disease.[240] Although one family member of a nurse did develop meningococcal disease, no relation between the two infections was documented. The low risk also seems to account for meningococcal pneumonia. In a study of 349 hospital personnel working in or near an area where a patient with meningococcal pneumonia resided, Cohen et al.[241] only found 1 positive throat swab for the organism.

In the laboratory setting, however, a few cases of meningococcal infections have been reported. At least two fatal cases, in 1918 and 1935, respectively, as well as six nonfatal cases have been reported in the literature.[35,53] In Norwegian laboratories, at least four unpublished cases have occurred during the last 20 years.[242] Proper precautions against transmission of meningococci include handwashing supplemented by protection of the mucosal surfaces of eyes, nose, and mouth from exposure to droplets and respiratory tract secretions during close contact. Persons having suffered mucosal splash by infected secretions or oral-to-oral contact should receive antimicrobial treatment with penicillin or prophylaxis with rifampin.

R. *Pseudomonas aeruginosa*

Pseudomonas aeruginosa is an opportunistic pathogen causing disease only in persons with impaired host defense against infection. Under normal circumstances, the organism represents no risk for health care personnel. However, *P. aeruginosa* is a frequent cause of microbial keratitis in contact lens wearers.[243-246] Bowden and Sutpin recently reported a case of nosocomial pseudomonas keratitis in a critical care nurse wearing extended-wear contact lens.[247] The infection was probably acquired from a patient with pneumonia and septicemia.

S. *Pseudomonas pseudomallei*

Person-to-person transmission of melioidosis is extremely rare.[248] Health care personnel are not at risk when taking care of patients with the disease. However, transmission has been reported to occur in the laboratory during work with cultures of the organism.[249,250]

T. Rickettsiae (Tick-Borne)

Rickettsiae are not transmitted from person to person and there is no occupational risk for health care personnel taking care of patients suffering from tick-borne rickettsioses. Laboratory work with these agents is, however, a documented hazard. In his survey of laboratory-acquired infections, excluding Q fever, Pike reported a total of 295 cases of rickettsial infections with 22 deaths.[35]

U. *Salmonella*

Hospital outbreaks of *Salmonella* infection are quite common and have been reported to be second in frequency only to those caused by staphylococci.[251] In England and Wales, 522 such outbreaks were reported in the 10-year period from 1968 to 1977, comprising 32 to 50% of all general outbreaks of salmonellosis each year.[252] In the U.S., 28% of outbreaks reported to the CDC between 1963 and 1972 occurred in institutions.[253] For the year 1979, this figure was 13%.

The epidemiology of nosocomial salmonellosis differs in many aspects from that of salmonellosis in the community. In the community, *Salmonella* outbreaks are often explosive following the consumption of food or drink in which the organisms have multiplied. They are characteristically associated with a cluster of cases, with onset 6 to 48 h after exposure. Within hospitals this common source type of outbreak is much less common. In a 2-year prospective survey of outbreaks of *Salmonella* infection in hospitals in England and Wales between 1980 and 1982, food-borne infection probably accounted for only 6 of 55 identified outbreaks.[254] Correspondingly, only 30 of 112 outbreaks in U.S. hospitals between 1963 and 1972 were classified as common source outbreaks.[253] Nosocomial salmonellosis more often follows a pattern of propagation by cross-infection, tending to spread among patients in a ward, apparently by person-to-person contact or by fomites. Nosocomial outbreaks are most commonly seen in geriatric units, nurseries, pediatric units, psychiatric units, and institutions for the mentally subnormal.

As noted by Taylor, "The typical history of a hospital outbreak is that quite suddenly it is appreciated that several patients in a ward have complained of diarrhoea during the previous few days, the patients having been admitted for conditions in which diarrhoea was not a symptom or in which diarrhoea would occur as the result of the primary disease. At this stage specimens of faeces are investigated and a *Salmonella* serotype isolated, this will probably take about 2 days, during which time more patients will develop symptoms. All contacts, patients and staff, are then investigated and a number of excretors of the same salmonella will be found among the patients, usually a few among the nurses and rarely among the medical staff and other categories.''[255]

The number of persons infected in hospital outbreaks of *Salmonella* infection is usually smaller than the number infected in community outbreaks. On average, eight patients and staff were affected per hospital outbreak in the study of Palmer and Rowe in England and Wales.[254] In the U.S., the average institutional outbreak involved 31 persons between 1963 and 1972, but involved only 6 persons in 1979, that year contrasting with an average of 69 cases in noninstitutional outbreaks.

Hospitalized patients appear to be unusually susceptible to *Salmonella* infection and are much more frequently involved in nosocomial outbreaks than healthy personnel. A ratio of patients to staff with illnesses of 5.5:1 was reported in the 55 nosocomial outbreaks studied by Palmer and Rowe.[254] However, infected personnel are more commonly asymptomatic than their patients, and the full extent of an outbreak will not be revealed until screening of all contacts has been performed. Palmer and Rowe reported that 26 (79%) of 33 infected employees in 9 nosocomial outbreaks were asymptomatic compared to only 36 (29%) of 125 patients in 15 outbreaks. They also reported that 3% of asymptomatic patients and 5% of asymptomatic staff were found to be infected on fecal screening.[254]

The ratio of patients to employees infected has been reported from several studies where screening of asymptomatic contacts has been done. Some of these are summarized in Table 2. The most common pattern of transmission during *Salmonella* outbreaks in hospitals is that of apparent person-to-person spread or cross-infection by fomites. For most outbreaks, this conclusion has mainly been based on the shape of the epidemic curve and the lack of identification of a common source. Definite proof of the mode of transmission is seldom provided, even though environmental factors such as air, dust, a thermometer, food-warming water bath, delivery room resuscitator, suctioning tubing, bedside tables, cribs, and fiberoptic endoscopes have been included in some epidemics.[256,258,267,270-275] However, in several outbreaks, extensive culturing has failed to identify environmental sources of the *Salmonella* strain involved.[262,263,267,268]

The role of the hands of personnel and patients is probably at least as important for the spread of *Salmonella* as for other organisms involved in nosocomial infections. Interestingly, despite negative fecal cultures among medical and nursing personnel, Robins-Browne and associates obtained positive cultures from the hands of a nurse and a physiotherapist, each of whom had recently attended an infected patient.[265] Both claimed to have disinfected their hands. Pether and Gilbert demonstrated that *Salmonella anatum* could be recovered from the fingertips 3 h after artificial contamination with between 500 and 2000 organisms.[276] A 15-s handwash was not sufficient to remove this organism completely. Similarly, the survivors from minimal inocula of less than 100 *S. anatum* per finger tip were, after 10 min, still capable of infecting samples of corned beef and ham. On the other hand, MacGregor and Reinhart argued that person-to-person spread of *Salmonella* in hospitals is difficult to accomplish without an intermediary common vehicle.[277] They were not able to detect a single secondary case among 265 patients and staff exposed to 8 patients admitted to hospital with acute diarrhea due to different *Salmonella* species, despite no use of enteric precautions for at least 72 h and the fact that extensive fecal contamination occurred.

Salmonella may also be a threat to the laboratory worker. Typhoid fever has been reported to account for more fatalities than any other laboratory-acquired infection.[53] By 1978, Pike reported a total of 258 recorded cases of laboratory-acquired typhoid fever with 20 deaths in this century.[52] However, a significant number of cases are not reported. Blaser et al.[278] retrospectively detected 24 cases of laboratory-acquired typhoid fever in the U.S. during a

Table 2
ISOLATION OF *Salmonella* SPP. FROM HEALTH CARE PERSONNEL DURING NOSOCOMIAL OUTBREAKS OF SALMONELLA INFECTIONS

Number of patients	Number of staff: Screened	Number of staff: Infected	Organism	Hospital type	Ref.
21	n.s.	1 Nurse 2 Orderlies 2 Porters	*S. typhimurium*	Pediatric	256
178	n.s.	1 Doctor 67 Nurses 10 Domestic Staff	*S. typhimurium*	General	257
526	n.s.	161 (Staff category not stated)	*S. derby*	12 hospitals or chronic care institutions	258
22	110	1 Nurse	*S. indiana*	Pediatric ward	259
22	n.s.	4 Nurses/aides 4 Food Service 1 Ward clerk 1 Laundry worker	*S. typhimurium*	General	260
55	75	5 Pediatric staff	*S. heidelberg*	Pediatric wards	261
12	351	1 Nurse 1 Kitchen worker	*S. heidelberg*	General	262
5	493	0	*S. typhi*	Maternity	263
38	n.s.	5 Staff members	*S. newport*	Nursing home	264
26	n.s.	0	*S. typhimurium*	Pediatric surgery	265
10	n.s.	0 (Son of a ward clerk infected)	*S. nienstedten*	Nursery	266
19	173	0	*S. oranienburg*	Nursery	267
8	n.s.	0	*S. bareilly*	Nursery	267
5	51	0	*S. typhimurium*	Nursery	268
11	n.s.	7 Nurses 2 Porters 1 Cook 1 Domestic staff 1 Handicrafts instructor	*S. typhimurium*	Psychiatric	269

Note: n.s., not stated.

33-month period from 1977 to 1979. Later, Blaser and Lofgren extended the number to 32 cases in 3.5 years.[279] They noted that the number of individuals in the U.S. exposed to *S. typhi* from voluntary sources (proficiency testing, educational purposes) probably far exceeded the number of individuals exposed to a strain isolated from a clinical specimen. They also reported several cases of typhoid fever passively transmitted from laboratory workers to family members.

Feces is the predominating infective material from patients with *Salmonella* infection. However, the organism may occasionally also be isolated from urine, respiratory secretions, wound secretions, and other sources.[262] Transmission is prevented by the application of enteric precautions including the use of gloves when touching infected material and careful handwashing after touching the patient or potentially contaminated articles. However, since a significant proportion of infectious patients are asymptomatic, stringent handwashing routines should not be restricted to the care of patients with diarrhea or positive cultures for *Salmonella*. Patients must also be instructed to practice handwashing properly.

Personnel suffering from an acute diarrheal illness should be excluded from food handling and direct patient contact until the diarrhea has resolved. Employees from whom fecal specimens grow *Salmonella* should not handle food or care for high-risk patients (newborns, the elderly, immunocompromised patients, and patients in intensive care units) until fecal cultures are negative. At least two negative specimens collected no less than 24 h apart are required.[11] However, intermittent excretion of detectable *Salmonella* is common, a fact to be remembered when propagation of a nosocomial outbreak is difficult to explain.[280] Asymptomatic *Salmonella* carriers are not considered to be a hazard for low-risk patients and may continue their duties provided there is careful attention to thorough handwashing after the use of toilet facilities.

V. Shigella

Reports of occupationally acquired shigellosis among health care personnel are few. One of the reasons for this may be that asymptomatic carriage of *Shigella* is less common than for *Salmonella*.[281,282] Furthermore, compared to *Salmonella*, *Shigella* generally survives poorly outside the human body. Not surprisingly therefore, the group at highest risk appears to be laboratory workers.[35,165,283,284] Pike's worldwide review of laboratory-acquired infections includes 56 cases of shigellosis.[35] However, Harrington and Shannon recorded as many as 37 cases in the U.K. alone during the year 1971.[165] Furthermore, as reported by Grist, 18 out of 25 cases of laboratory-acquired bowel infections in British clinical laboratories during 1980 to 1985 were caused by *Shigella* spp., a fact possibly reflecting a higher degree of virulence for this organism than for other enteric pathogens during laboratory work.[59-61,284] As few as 10 to 200 viable cells of *Shigella dysenteriae* have been shown to be able to establish an infection.[285,286] The most common mode of acquisition is by the oral route, often indirectly via contaminated hands, but mucosal splashing is also a hazard, and protection of the eyes, nose, and mouth is necessary. The CDC recommends Biosafety Level 2 practices and containment of equipment and facilities for all activities utilizing known or potentially infectious clinical material or cultures.[287]

In 1979 an outbreak of shigellosis was reported among hospital employees in a children's hospital in Pennsylvania (U.S.).[288] A total of 32% of employees reported being ill, but no hospitalized patients became culture positive for *Shigella* as the result of the outbreak. The source was proven not to be a patient, but a kitchen worker preparing salad in the hospital cafeteria.

W. *Staphylococcus aureus*

In most studies on the prevalence of colonization with *S. aureus*, the prevalence rate has been found to be higher among hospital employees than among the general population. Nasal carrier rates usually range from 10 to 40% outside hospitals and from 20 to 90% among hospital employees.[289-292] Staphylococci among hospital staff have also been reported to be more resistant to antibiotics than those of the nonhospital population.[293-295] On the other hand, when it comes to the risk of clinically significant staphylococcal infections, comparative data are obviously harder to obtain and are therefore more scarce.

Davies performed a 12-month study of staphylococcal infections among nurses at The London Hospital from 1958 to 1959.[293] Among approximately 725 nurses, there were 146 cases of staphylococcal infections with an average of 8.5 d of lost working time each. The author noted that staphylococcal infections in nurses have a nuisance value out of proportion to their severity since even trivial lesions resulted in removal from contact with patients. The most common site of infection was the finger (29%), and 83% of all lesions were on the face, neck, axilla, or upper limb. Whether or not these figures were higher than in the general population was, however, not examined in this study.

In a study of septic lesions of the skin among hospital personnel in Aarhus, Denmark, 700 cases with a mean of 8.7 d lost from work per case were identified during the 7-year period from 1959 to 1966.[295] About 90% of these were caused by *S. aureus*. The average incidence rate was found to be slightly higher than that of the general population of Denmark

(5.7% as compared to 4.5 to 5%). This excess morbidity could be related exclusively to infections following skin injuries, chiefly septic fingers, among the personnel of the general wards. The causes of the finger infections were predominantly minor stab lesions and manicure lesions. The risk of infection varied greatly in different departments. In the large general wards, the orderlies section, the pathological institute, and the laundry, the prevalence rate was twice as high as that of the general population. The frequency of infections within the various staff groups also differed greatly. Student nurses and hospital orderlies had the highest relative risks. Other groups with an increased risk were the domestic staff, nurses aides, the laundry staff, and to a somewhat lesser degree, nurses. Doctors and technical staff appeared to be much less affected than the general population. Phage typing and antibiogram comparison disclosed that three fourths of the infections following skin injuries and one fourth of the spontaneously occurring boils of the staff might have been due to bacterial transmission from in-patients with suppurating infections. Among the staff, 36% were nasal carriers, and about 40% of these harbored strains identical to the ones recovered from patient infections during the previous 5 months, whereas about 25% had strain identity with nosocomial infections that broke out during the following 7 months. The results of this study also indicated that efficient hand protection would greatly reduce the extra risk of the exposed categories of staff.

The epidemiology of methicillin-resistant staphylococci (MRSA) does not appear to differ much from that of methicillin-sensitive strains. However, nasal acquisition of MRSA seems to occur less commonly.[296] Repeated screening of 321 staff nursing cohorted patients with MRSA during an outbreak at The London Hospital revealed a persistent carriage rate of 5.6%, but more than half of the staff took between 1 and 4 months to acquire the outbreak strain.[297] There seems, however, to be a consensus that the methicillin-resistant staphylococci of the 1950s and 1960s did possess enhanced pathogenicity, often infecting intact skin, including that of hospital personnel which may not be shared by the phage group III methicillin-resistant staphylococci causing recent outbreaks.

It is well known that preoperative nasal carriage of *S. aureus* enhances the risk of postoperative infections with this organism.[298,299] When hospital employees turn into surgical patients, one would therefore expect that they run a somewhat greater risk of postoperative infections than the general population due to their higher carriage rate.

The main mode of transmission of *S. aureus* is by direct or indirect contact and the most important preventive measure is frequent and thorough handwashing.[300] Gloves should be used when handling patients with purulent infections. General skin care, especially of the hands, avoiding soreness, cuts, and abrasions, also seems to be an important preventive measure.[293,295] Among health care personnel, especially those having frequent contact with water, occupational hand dermatoses seem to be common, enhancing the risk of secondary infections.[301]

X. Streptococci

1. Streptococcus pyogenes (Group A)

From a historical point of view, *Streptococcus pyogenes* is an important cause of occupational infections among medical personnel, first of all through infection of finger pricks and scratches during attendance to septic patients.[302] In the preantibiotic era, a very small inoculum was often sufficient to induce devastating effects after these trivial surgical accidents, due to the enhanced virulence acquired by the *Streptococcus* in a series of hosts successively infected at short intervals.

Today, the majority of patients with streptococcal infections have skin or respiratory tract infections that are handled by health care personnel in the out-patient setting, and acute local infections can so easily be arrested that the stage of septicemia is very seldom reached. These infections first of all spread by close contact and droplets.[300,303,304] It is not clear to what degree exposure to such patients results in sore throat or skin infections among health care personnel, but the duration and closeness of contact in the out-patient setting will probably in most cases not be sufficient for transmission of infection. However, occasional reports clearly show that *S. pyogenes* still represents a certain risk to health care personnel.[305,306]

2. Group B Streptococci

The Group B *Streptococcus* is an important pathogen for neonates, young infants, and post-partum women. These organisms are frequently carried in the GI and genitourinary tracts of adults.[307] The prevalence of Group B streptococcal colonization among adults has been reported to vary between 2.2 and 45%.[308] Several investigators have studied group B streptococcal carriage rates in hospital workers closely associated with mothers and neonates and have concluded that the carriage rate is not different from that of the general population.[309,310]

3. Streptococcus pneumoniae

Nosocomial epidemics of pneumococcal pneumonia definitely occurred in the preantibiotic period, involving patients as well as health care personnel.[311] Transmission to laboratory personnel was also described, but Pike lists only five cases in his survey.[35,311] However, no data are available on the relative occupational risk before the antibiotic era. After the introduction of effective antibiotic therapy, the risk of secondary cases of pneumococcal pneumonia among health care workers is probably neglible. Respiratory secretions should, however, be considered potentially infectious the first 24 h after start of therapy.

Y. *Treponema pallidum* (Syphilis)

The overwhelming majority of cases of syphilis are transmitted venerally. *T. pallidum* enters the body by penetrating intact mucous membranes, via minute abrasions of the epithelium or possibly through unbroken skin by way of hair follicles.[312] Survival of *T. pallidum* outside the body is brief, and transmission by fomites is extremely rare. Infection therefore virtually requires direct contact with infected lesions. Both primary and secondary syphilitic lesions contain large numbers of treponemes and the number of organisms sufficient to cause infection has been found very small. In inoculation experiments in volunteers, Magnuson et al. found that a 50% infectious inoculum was about 57 organisms.[313] In health care personnel, accidental direct inoculation may occur by needle prick, when touching infectious lesions, or when handling infected clinical material. Occupationally acquired lesions in dentists, physicians, and nurses are usually seen as a chancre on a finger, usually masquerading as a paronychia.[314] Indeed, syphilis of the finger is most commonly seen in medical personnel.[315]

Syphilis is also a documented hazard to laboratory personnel who handle or collect clinical material from infected lesions. The survey of Pike includes 20 cases of laboratory-associated infection.[35]

Z. *Vibrio cholerae*

Historically, cholera has a bad reputation for attacking health care personnel taking care of patients with the disease.[316] However, doctors were reported to be affected less frequently than other health care workers.[316,317] Snow explained this by the fact that doctors had much less close contact with the patients than nurses and other personnel.[317] From modern time, only a few reports deal with the occurrence of cholera among health care workers.[318,319] Ryder et al.[319] recently reported an outbreak of nosocomial cholera in a rural Bangladesh hospital. During an 8-week period, 170 patients had admission stool cultures positive for *V. cholerae*, and 48 patients with admission stool cultures negative for *V. Cholerae* were studied prospectively and nosocomial transmission was detected in 14 (29% of these). *V.*

cholerae was, however, not detected from any of the rectal swabs taken from hospital personnel or from attendants of patients who had acquired nosocomial cholera infection. All handwashing cultures from the hospital staff were also negative.

In his survey of laboratory-acquired infections, Pike reports only 12 cases of cholera with 4 deaths.[35] Grist reported one case of laboratory-acquired *V. cholerae* (non-01) infection in British clinical laboratories in 1985.[61] As noted by Collins, laboratory workers seem to have been able to protect themselves against cholera since the early days of laboratory work with the organism.[320]

Cholera is traditionally a disease associated with poverty, nutritional deficiency, and poor sanitary conditions. In addition, ingestion of a large infective dose is necessary to induce disease.[321] All these factors are probably explanatory for the seemingly low risk for health care personnel. Needless to note, an adequate handwashing routine is the dominating protective measure.

AA. *Yersinia enterocolitica*

The primary clinical sydromes caused by *Y. enterocolitica* are acute enteritis, abdominal pain and appendicitis-like symptomes, arthritis, and erythema nodosum, but a variety of other clinical forms including pharyngitis and pneumonia have been reported.[322] The epidemiology of *Y. enterocolitica* infection is not fully clarified, but oral ingestion of the organism seems to be the final link of the chain.[322,323] The infective dose was in one experimental study shown to be rather large in healthy volunteers, 3.5×10^9 organisms.[324]

Only one report on the occurrence of *Yersinia* infection among health care personnel has been published. Toivanen et al.[325] described a hospital outbreak of *Y. enterocolitica* involving one cleaner and five nurses with possible secondary spread to four relatives of two nurses. The source was a patient hospitalized for suspected appendicitis with abdominal pain, diarrhea, and fever. One of the nurses became infected after the patient left the ward, probably from her colleagues already infected.

BB. *Yersinia pestis* (Plague)

Plague is a zoonotic infection cased by *Y. pestis* and is transmitted in the natural rodent reservoir, predominantly by flea bites or ingestion of contaminated animal tissues by the host.[322] The disease is widespread throughout the world with the highest incidence in Southeast Asia, particularly in Vietnam.[326] The dominating route of human infection is from infected rodent fleas. However, human-to-human transmission can occur through infected exudate from skin or mucous membrane lesions or through airborne droplets from patients with plague pneumonia. Asymptomatic carriage can also occur. Marshall et al.[327] demonstrated pharyngeal carriage in 16 of 212 patients with bubonic plague and in 15 of 114 healthy contacts of patients. None of these pharyngeal carriers developed plague, however, and there was no evidence that they were contagious.

Primary plague pneumonia is a potential threat following exposure to a patient with plague who has a cough. It occurs primarily after close and prolonged contact with another person with pneumonic plague. Hence, respiratory transmission occurs most frequently to household contacts or medical personnel who are directly involved with the care of the patient.[327] It can be so rapidly fatal that persons reportedly have been exposed, became ill, and died on the same day.[328]

On hospitalization of patients with plague, the potential risk of transmission by fleas carried by the patient or by his or her clothing or baggage should be recognized and eliminated through the use of an effective insecticide. Patients with bubonic plague, negative chest X-ray, and no cough should be isolated with drainage/secretion precautions for 3 d after start of effective therapy. For patients with pneumonic plague, strict isolation precautions are required until 3 full days of appropriate therapy have been completed and there has been a favorable clinical response.[329] Close contacts of confirmed or suspected plague pneumonia cases (including medical personnel) should be provided chemoprophylaxis with tetracycline (15 to 30 mg/kg) or sulfonamides (40 mg/kg) daily in four divided doses for 1 week.[329]

VI. VIRAL INFECTIONS

A. Adenovirus

Several outbreaks of adenovirus conjunctivitis and pharyngitis involving medical personnel have been reported from ophthalmic units and ICUs.[330-334] The source may be patients with adenovirus keratoconjunctivitis or pneumonia, and those at highest risk are personnel that have the closest contact. In most cases, full recovery is reported, but corneal infiltrates may take up to 1 year to resolve spontaneously.[335] Levandowski and Rubenis described one nurse who developed a persistent visual disturbance with subepithelial deposits and required prolonged opthalmic care.[333] Pike reported ten cases of laboratory-acquired infections with adenovirus.[35]

Drainage/secretion precautions, the mainstay of which is the use of gloves and handwashing, are recommended for patientts with adenovirus infections.[13] It is important to be aware of the possible presence of the virus in feces. The CDC does not recommend the use of masks for patients with adenovirus pneumonia. Protection of eyes, nose, and mouth against droplet contact might, however, be advisable.[334]

B. Cytomegalovirus

Cytomegalovirus (CMV) is associated with congenital malformations. It has therefore been of concern to female health care workers of childbearing age. However, the epidemiology and mode of transmission of the virus are not entirely known. Conflicting studies as to the prevalence of CMV antibodies among nurses working in high risk areas have been reported.[336,337] Rigid adherence to careful handwashing techniques and other means to prevent cross-infection seems to reduce the serocoversion rates, and a recent study did not find the virus to be an occupational hazard for nurses in renal transplant and neonatal units.[338,339] In spite of this, reassignment of pregnant nurses or other female health care workers to low-risk areas has been recommended.[340]

C. Epstein-Barr Virus

Ho and co-workers reported that the prevalence of seropositivity against Epstein-Barr virus was significantly higher among staff members treating patients with nasopharyngeal carcinoma than among controls in Hong Kong.[341] The geometric mean titers of serum IgG antibodies against the virus capsid antigens correlated with the degree of patient contact, and physicians and nurses had the highest titers.

D. Hepatitis A

Hepatitis A virus is mainly spread through the fecal-oral route. The concentration of the virus in feces decreases sharply after the debut of jaundice, and the risk of transmission to hospital personnel is therefore low. However, nosocomial outbreaks have been reported, and transmission from asymptomatic infected infants has recently been proven.[342,343] Transmission to health care workers can effectively be prevented by the use of gloves and meticulous handwashing. Isolation of the patients in single rooms may be of value if they have severe diarrhea, are incontinent, or do not cooperate. Gowns should be used to avoid fecal contamination of clothes. With a high prevalence of the disease, immunglobulin may be offered to personnel who have close contacts with the patients. Given early after the exposure, immunglobulin reduces the attack rate by approximately 90%, and little or no protection is observed when it is used 2 weeks or more after exposure.[344]

E. Hepatitis B

The increased risk for acquiring hepatitis B virus (HBV) among health personnel has been known for about 35 years.[345,346] During the last decade, studies of serologic markers have shown that health care workers have a higher prevalence of antibodies to HBV than the general population and that the antibody findings are correlated to the extent of blood handling and the frequency of patient contact.[347,348] The hepatitis B surface antigen (HB_sAg) implies infectivity; anti-HB_s shows immunity or recent exposure to the virus. In a small number of patients with hepatitis B, both markers may be absent, but serologic evidence of infection may be detected by the presence of antibody to the core antigen (anti-HB_c). The acute case fatality rate of hepatitis B is usually less than 1%. However, the sequelae (such as chronic active hepatitis) and the strong association to hepatocellular carcinoma are of major concern to the health care workers. Although hepatitis B is the most commonly reported occupational infectious disease in health care personnel, adherence to well-established procedures in the handling of blood and blood products and in the care of patients with the disease has been effective in preventing the spread of hepatitis B.[349] Blood and blood products are the major vehicles of transmission, but infectivity has also been proven for saliva and semen.[350] Frank inoculation accounts for the majority of reported cases among hospital employees. Spillage of blood into fresh wounds may result in transmission, while splashing of blood or saliva onto intact skin or mucous membranes represents an even less risk of transmission.

To prevent the spread of hepatitis B virus (and other blood-borne diseases), blood and blood products should — independently of what is known about the infectivity of the material — always be handled with great care. Gloves should be used when direct contact with infectious material is expected, and gowns should be used when splattering may occur. Immediate handwashing is imperative if one becomes contaminated. Masks and goggles may be used if splattering of blood is unavoidable. Recapping or bending of used needles by hand should never be attempted. Puncture-proof sharp boxes should be available in all patient care areas and where needles and other sharp objects are used. Blood specimens, accompanying papers, and charts should be marked with a biohazard label if Hb_sAg has been detected.

The recently developed vaccines against hepatitis B offers a high degree of protection against infection, and it should be given to personnel at high risk of acquiring the disease.[351] The risk of hepatitis B is mainly determined by the prevalence of infectivity in the patient population and the frequency of contact with blood, blood products, and with patients. Vaccination programs must therefore be tailored to each individual institution. If inoculation accidents occur, hepatitis B immune globulin (HBIG) should be given as soon as possible and always within 7 d. HBIG should only be given to previously vaccinated employees if the antibody response to the vaccine has been insufficient (Table 3). In unvaccinated health care workers, a combined active-passive immunization provides a long-term protection against hepatitis B virus and reduces the need for a second injection of HBIG. A schedule for immunization following needlestick or other percutaneous exposure has been proposed by the CDC and is shown in Table 3.[352,353] (References 352 and 353 should be consulted for further details.)

The second generation hepatitis B vaccine developed through recombinant DNA technique or fully synthetic vaccines may reduce the cost of vaccination programs. Hopefully, all health care workers at risk may in the future be offered the vaccine, and a further reduction in the incidence of hepatitis B among health personnel may be expected.

F. Hepatitis Non-A, Non-B (HNANB)

The prevalence of HNANB among health care workers is unknown. Epidemiologic evidence indicates more than one etiologic agent, at least one of which is transmitted through blood and blood products and one of which is spread by the fecal-oral route.[354] Only the

Table 3
POSTEXPOSURE PROPHYLAXIS FOR HEBATITIS B AFTER PERCUTANEOUS EXPOSURE

Source	HBIG	HB vaccine
	Exposed Person Unvaccinated	
HB_sAg positive	0.06ml/kg i.m. within 24 h	1.0 ml (20 μg) i.m. within 7 d, at 1 and 6 months
Possibly HB_sAg positive	Source tested positive,0.06 ml/kg i.m. once within 7 d	As above
Unlikely HB_sAg positive	Testing source not necessary; HBIG not required	As above
Unknown	Nothing required	As above
	Exposed Person Vaccinated	
HB_sAg positive	Test of exposed shows anti-HB_s <10 SRU (RIA) or negative (EIA), 0.06 ml/kg i.m. within 24 h and after 1 month	Nothing required[a]
Possibly HB_sAg positive	If exposed person is nonresponder and source is tested positive, 0.06 ml/kg i.m. within 24 h and after 1 month	Nothing required
Unlikely HB_sAg positive	Nothing required	Nothing required

[a] If the exposed person has not completed a full series of vaccine, the series should be completed. If a previously adequate antibody level is found inadequate, a booster dose of vaccine should be given.

blood-borne virus seems to represent any risk to health personnel, and preventive measures are identical to those described for hepatitis B. Vaccine is not available but ordinary immunglobulin has sometimes been used after exposure.[355]

G. Herpes Simplex Virus

Herpes simplex virus is spread by direct contact, and caring for patients with active lesions poses a threat to the health care worker. The most common herpetic lesion among health care personnel is probably the herpetic paronychia (whitlow), occurring on the thumb or index finger.[356] The disease usually incapacitates employees for 1 to 3 weeks. Routine handwashing and use of gloves when dealing with respiratory secretions effectively prevent the acquisition and spread of the disease.

H. Influenza

Morbidity of influenza among healthy hospital personnel is usually low. However, during epidemics the attack rate may be high. Vaccination is recommended for health care workers caring for elderly and critically ill patients and for personnel in essential services. Chemoprophylaxis by amantadin or rimantadin effectively prevents influenza A and may be used for unvaccinated employees providing essential services.

I. Parotitis Virus (Mumps)

Parotitis virus (a paromyxovirus) spreads through close contact and droplets and probably also through fomites. Clinical meningitis occurs in 1 to 10% of patients with mumps parotitis.[357] More feared is epidydimo-orchitis and oophoritis seen in males and females in 20 to 30 and 5%, respectively.[358] Pediatric personnel are at highest risk for acquiring the disease, and in the absence of a previous history of clinical illness, vaccine should be offered.[359]

The vaccine, being of the live, attenuated type, should not be given to pregnant personnel or to personnel with congenital or acquired immunodeficiency.

J. Poliomyelitis

Introduction of the polio virus vaccines has dramatically reduced the occurrence of poliomyelitis. However, outbreaks of the disease have occurred in subpopulations not being vaccinated, emphasizing the need for continued immunization of the population.[360]

Certain health care workers may be at an increased risk of exposure, particularly to the vaccine strain of the virus, and should be fully immunized. This includes laboratory personnel handling potentially infected feces and personnel on pediatric wards and in emergency rooms. Alkan et al.[361] studied the immunity of 275 employees of three hospitals in Israel and found that only 77% had detectable antibodies against all three types of polio virus. Employees who were directly exposed to patients had a significantly higher seroprevalence rate than other groups. Nurses were more immune than doctors, and nurses in pediatrics showed especially high immunity.

Inactivated polio virus vaccine is recommended for routine use because of the slightly lesser risk of vaccine-associated paralysis compared to the oral poliovirus vaccine. However, immediate protection is best achieved with oral vaccine, and this vaccine should also be used to boost the response in already vaccinated personnel in the case of an emergency situation.

K. Rabies

Inoculation is the major mode of transmission of rabies and rabies-related viruses, but transmission by inhalation may also occur if the virus is aerosolized. Human-to-human spread of the virus has not been documented, but it is considered possible.[362] Vaccine is recommended for use in persons at high risk for acquiring the disease. Among these are veterinarians, animal handlers, including those in medical research laboratories, and laboratory workers who come in contact with the virus.[363] Postexposure immunization of health care workers with rabies immune globulin and vaccine has been considered necessary after treating patients with rabies.[364] In the laboratory, Biosafety Level 2 should be used, and Biosafety Level 3 is recommended when aerosols or droplets may be generated.

L. Respiratory Syncytial Virus

Respiratory syncytial virus is the most common cause of lower respiratory disease in infants and children below school age, and personnel caring for infected children have a high risk of acquiring the disease.[365a] Even though the disease is not serious, the employee frequently has to refrain from work for some days. The virus is transmitted by close contact with droplets from the respiratory tract, directly infecting the eyes or nose, and is transmitted indirectly from contaminated fomites. Agah et al.[365b] recently reported that the use of masks and goggles significantly reduced the risk of RSV infection in personnel caring for children with such infections. The RSV illness rate in health-care workers using masks and goggles was 5%, but the rate for those not using masks and goggles was 61%. Similar results were reported by Gala et al.[365c] However, the degree of protection provided by stringent handwashing alone was not examined in these studies. Since the virus is often inoculated into the mucous membrane from contaminated hands, frequent handwashing is important. However, due to suboptimal compliance with handwashing, the use of masks and goggles or similar protection of the mucous membranes seems to be required for maximal protection against RSV infection. An effective vaccine is not available.

M. Rotavirus

Rotavirus is a major cause of gastroenteritis in infants and young children. Transmission occurs by the fecal-oral route. Staff members attending children with diarrhea frequently have rotavirus on their hands.[366] Seriologic studies indicate a high rate of infection among the nursing staff on the infant/toddler wards.[367] Characteristically, the symptoms are either mild or absent. Prevention is difficult because viral shedding can occur in asymptomatic individuals. Handwashing and cohorting of both patients and personnel are probably the best method of controlling an outbreak.

N. Rubella

Rubella virus was first isolated in 1962, but the link between maternal rubella and congenital defects has been recognized since 1941 and has caused great concern among health care workers.[368,369] The virus is spread through droplets shed from the respiratory tract of infected persons. Since vaccine against rubella became available (1962), immunization of health personnel is the optimal control strategy. Both male and female personnel should be vaccinated, because transmission from males in the incubation period to pregnant, susceptible females may occur. If patients with rubella are admitted to the hospital, they should have private rooms, and only personnel immune to the disease should be allowed to care for them.[370] Prevention by the use of masks is probably of little or no value.

O. Varicella-Zoster Virus

Varicella-zoster virus is the etiologic agent of chicken pox and herpes zoster. In chicken pox, the virus is transmitted primarily via airborne spread by small-particle aerosols and by large particles.[371] However, spread by person-to-person contact occurs and is probably the main mode of transmission of the virus during zoster.

Because of widespread immunity in the adult population, epidemics of chicken pox are uncommon. However, the disease may be severe in adults and is often life-threatening in immunocompromized patients. Therefore, employees without a history of chicken pox should not care for infected patients. If exposure has occurred, the employee should refrain from work during the period of transmissability (days 10 to 21) and if overt disease develops. Varicella-zoster immunglobulin may be administered, but it should probably be restricted to immunosuppressed employees and to personnel providing essential services. Health care workers with herpes zoster should not care for susceptible patients unless their lesion can be completely covered. Uncomplicated varicella or zoster does not need specific therapy. However, various recently developed antiviral drugs have been shown to modify the course of the disease and may be considered for use on key personnel.

Live, attenuated varicella-zoster vaccine may be considered for susceptible health-care workers. It has been shown to be safe and effective for normal and immunocompetent hosts.[372] Used rationally, the vaccine may contribute to decreased transmission of the virus within the hospital.

P. Kuru and Creutzfeldt-Jacob Disease

The agents of kuru and Creutzfeldt-Jacob disease are normally referred to as slow virus. Most probably, they are transmissible to humans only by inoculation. Kuru has not been reported during the last 25 years, and Creutzfeldt-Jacob disease occurs with a frequency of about one per million.[373] No excess cases among health care workers have been reported.[374] However, the viruses are of concern to medical personnel because of their unusual resistance to inactivation. For sterilization of reusable instruments or materials, treatment in a 131°C saturated steam autoclave for 1 h is recommended. If heat cannot be used, 1 *N* sodium hydroxide or 5.25% sodium hypochlorite is recommended.[375] To avoid laboratory accidents with the viruses, hepatitis B precautions should be observed.

Q. Other Viral Infections

An extended list of viruses covering all reported laboratory infections as well as proposed biosafety measures is available and should be consulted for details.[376] Lassa fever, first described in 1961, has had a reputation for causing very severe nosocomial infections with a case fatality rate of about 50%. Recent investigations have shown Lassa fever to be an extremely common infection in West Africa (Sierra Leone) and a major cause of death.[377] How the virus is spread from person to person is still not clear, but accidental inoculation as well as airborne spread seem to play a role. Vaccine is not available. Strict isolation of patients should be applied, and handling of infectious materials in the laboratory should only be undertaken in a maximum containment facility.[13,287] These recommendations are also valid for the other arena virus infections (lymphocytic choriomeningitis virus and the tacaribe group of viruses) with the exception that lymphocytic choriomeningitis virus can be handled in laboratories classified as Biosafety Level 3.

Laboratory infections with the virus of Rift Valley fever have been known since the mid 1970s. The most likely route of transmission is by aerosol. Control measures are difficult. Procedures generating aerosols should be avoided, and only Biosafety Level 3 laboratories should handle the virus. An unlicensed vaccine is available.[378]

Infections caused by the Marburg and Ebola viruses were described in laboratory workers in 1967. The mode of transmission is probably through blood, and hepatitis B precautions should be used. Maximum containment facilities are recommended for laboratory investigations, and strict isolation is indicated for hospitalized patients.[13]

VII. FUNGAL INFECTIONS

A. *Blastomyces dermatitidis*

The natural infection with *B. dermatitidis* is in most cases thought to be acquired by the respiratory route.[379] However, the saprophytic source of the organism in nature is largely unknown.[380] There are no reports on the occurrence of occupationally acquired blastomycosis among personnel taking care of patients with blastomycosis. Among laboratory personnel, however, a small number of cases have been reported. The most commonly reported manifestation is primary cutaneous blastomycosis acquired by accidental inoculation during laboratory work with the organism or during autopsy. Larson et al.[381] reported two cases and reviewed another seven cases from the literature. Infection acquired by inhalation has also been reported.[382]

B. *Coccidioides immitis*

Coccidioidomycosis is a disease acquired by the inhalation of the arthrospores of *C. immitis* of environmental origin. The natural infection is largely confined to the southwest part of the U.S., Central America, Venezuela, Colombia, Paraguay, and Argentina.[383] Coccidioidomycosis does not spread from person to person, and there is for all practical purposes no risk of secondary infection of health care personnel providing bedside care for patients with the disease. One exception to this rule was reported by Eckman et al.,[384] although strictly speaking, this was a case of transmission via growth on a fomite. The dressings and plaster cast of a patient with draining coccidial osteomyelitis was contaminated with spherules of *C. immitis*. These germinated on the plaster cast forming arthrospores, resulting in six cases of primary pulmonary infection among personnel involved in changing dressings. In addition, there were considerably more reactors to the coccidioidin skin test among 78 possibly exposed personnel (19.4% positive) than among 38 nonexposed individuals (5.3% positive).

Coccidioidomycosis has been associated with a variety of occupations involving exposure to dust and soil, like agricultural work, military service, archeology work, etc.[384] Laboratory work with the organism is also a significant hazard.[384-391] Among 353 laboratory-associated fungal infections reported by Pike in 1976, 93 were caused by *C. immitis*.[35] Unlike other occupations, laboratory workers may also acquire coccidioidomycosis outside the endemic area.[389-391] The infection is usually acquired by the respiratory route through the inhalation of arthrospores that are easily dispersed from mold cultures of *C. immitis*. Smith et al. reported that the frequency of clinical infection as opposed to a subclinical course was higher after inhalation in the laboratory than in the community.[392] On the other hand, Johnson et al.[387] reported that only a small fraction of laboratory workers seropositive against *C. immitis* have experienced clinical manifestations. Whether symptomatic disease will occur or not depends on the magnitude of the infective dose, which in laboratories often will be large. The CDC recommends Biosafety Level 3 practices and facilities for all activities with sporulating mold from cultures of *C. immitis*.[287]

The spherule-endospore or tissue phase of the organism poses no risk for inhalation. However, like the arthrospores, they may cause infection by inoculation. Primary cutaneous coccidioidomycosis resulting from inoculation directly into the skin has been reported among laboratory workers, although much less frequently than the pulmonary form.[393,394] The CDC recommends Biosafety Level 2 practices for handling and processing clinical specimens.

C. Dermatophytes

Superficial fungal infections of the skin and its appendages caused by members of the genera *Trichophyton, Epidermophyton,* and *Microsporum* have a worldwide distribution. Although the prevalence of the different forms of dermatomycosis varies, health care personnel all around the world are frequently in contact with patients with such infections. Despite this, the occupational hazard to personnel getting in touch with such patients seems to be negligible. However, infections have been reported in laboratory workers. In Pike's survey of laboratory-associated infections, dermatomycoses were the most numerous mycotic infections.[35] They were usually reported to be acquired from laboratory animals, and most of them gave only minor clinical symptoms and little time was lost from work.

D. *Histoplasma Capsulatum*

Natural infection with *H. capsulatum* is acquired by the respiratory route through inhalation of spores from environmental sources. The infection is worldwide in distribution. Direct human-to-human transmission does not occur and no isolation precautions are considered necessary.[13] However, *H. capsulatum* is a hazard to laboratory workers much in the same manner as *C. immitis*. Histoplasmosis was the third most common laboratory-associated infection reported by Pike; a total of 71 cases with 1 death had come to his attention through 1974.[35]

Furcolow et al.[395] prospectively studied 56 employees at the Mycoses Laboratory in Kansas City, KS (U.S.), which is an endemic area for histoplasmosis. They found a histoplasmin skin test conversion rate of 13.23/100 susceptible person months. This was 26 times higher than the conversion rate for school children in the same area. In most cases, laboratory-associated histoplasmosis is acquired by the inhalation of microconidia from cultures of the organism. The infective *Histoplasma* spores are resistant to dessication, and the majority are less than 4.8 μm in size, which is ideal for intrapulmonary retention.[396] The CDC recommends Biosafety Level 3 practices and facilities for processing mold cultuers of *H. capsulatum*.[287]

Like coccidioidomycosis, primary cutaneous histoplasmosis has occasionally been reported after inoculation accidents during autopsy or laboratory work with the organism.[397,398] For the handling of clinical specimens from patients with histoplasmosis, Biosafety Level 2 practices are recommended.[287]

E. *Sporothrix schenckii*

Natural infection with *S. schenkii* has a worldwide distribution and is usually acquired from environmental sources by accidental inoculation. Such inoculation may also occur in the laboratory, and laboratory-acquired infections have been reported after recognized inoculations as well as in personnel unaware of trauma during work with cultures of *S. schenkii.*[399-401] Conjunctival infections have also been reported.[402,403] Pike records only 12 cases.[35]

VIII. PARASITIC INFECTIONS

Toxoplasma gondii infections have been reported in the laboratory and in autopsy workers.[35] Fecal-oral transmission or accidental inoculation are the main modes of transmission (besides vertical transmission), and human-to-human transmission has not been documented. Seronegative females should not work with this organism. Biological safety cabinet, gloves, and mask are usually recommended. Handwashing is important.

Malaria may be transmitted to the laboratory worker by an infected vector, through infected primate hosts and by inoculation of infective blood. Prevention consists of control of vectors in laboratories where mosquito colonies are maintained and blood precautions.

The orally transmitted parasites *Entamoeba histolytica, Giardia lamblia, Ascaris, Trichuris, Hymenolepis nana,* and *Taenia solium* are all preventable by proper handwashing. Several nosocomial outbreaks of scabies involving a large number of hospital personnel and their family members have been reported.[404-407] The source is usually a patient with unrecognized disease. When first recognized, scabies is easily controlled by contact isolation precautions. In particular, the use of gloves and handwashing will protect the personnel from becoming infested.

Recently, evidence for person-to-person spread of *Cryptosporidium* involving hospital personnel caring for patients with cryptosporidiosis has been reported.[408] Fecal-oral spread of oocysts seems to be the predominant mode of transmission, and enteric precautions should be followed.

REFERENCES

1. **Semmelweis, I. F.,** Die Aetiologie, der Begriff und die Prophylaxis der Kindbettfiebers, C. A. Hartleben's Verlags-Expedition, Pest, 1861.
2. **Kiær, F. C.,** Norges Læger (1800-1886), Alb. Cammermeyer, Christiania (Oslo), 1890.
3. **Nightingale, F.,** *Notes on Hospitals,* 3rd ed., London, 1863.
4. **Ogle, W.,** Statistics of mortality in the medical profession, *Med. Chir. Trans.,* 69, 217, 1886.
5. **Sherertz, R. J. and Hampton, A. L.,** Infection control aspects of hospital employee health, in *Prevention and Control of Nosocomial Infections,* Wenzel, R. P., Ed., Williams & Wilkins, Baltimore, 1987, chap. 13.
6. **USPHS,** Classification of etiological agents on the basis of hazard, *Fed. Regis.,* 46, 59,379, 1981.
7. Department of Health and Social Security, Code of Practice for the Prevention of Infection in Clinical Laboratories and Post-Mortem Rooms, Her Majesty's Stationary Office, London, 1978.
8. World Health Organization, Safety measures in microbiology. Minimum standards of laboratory safety, *W.H.O. Wkly. Epidemiol. Rec.,* 28, 154, 1979.
9. **Wells, W.F.,** Airbone infection. II. Droplets and droplet nuclei, *Am. J. Hyg.,* 20, 611, 1934.
10. **Green, H. L. and Lane, W. R.,** *Particulate Clouds: Dusts, Smokes and Mists,* E. & F.N. SPON. LTD, London, 1967.
11. **Williams, W. W.,** CDC guidelines for infection control in hospital personnel, *Infect. Control,* Suppl. 4, 326, 1983.
12. **Garner, J. S. and Favero, M. S.,** CDC guidelines for the prevention and control of nosocomial infections. Guideline for handwashing and hospital environmental control, 1985, *Am. J. Infect. Control,* 14, 110, 1986.
13. **Garner, J. S. and Simmons, B. P.,** CDC guidelines for isolation precautions in hospitals, *Infect. Control,* 4, 245, 1983.
14. **Jensen, K.,** Disinfection of the hands. A practical trial, *Ugeskr. Læg.,* 136, 579, 1974.

15. **Taylor, L. J.,** An evaluation of hand washing techniques. I, *Nurs. Times,* 74, 54, 1978.
16. **Taylor, L. J.,** An evaluation of hand washing techniques. II, *Nurs. Times,* 74, 108, 1978.
17. **Quraishi, Z. A., McGuckin, M., and Blais, F. X.,** Duration of handwashing in intensive care units: a descriptive study, *Am. J. Infect. Control,* 11, 83, 1984.
18. **Preston, G. A., Larsson, E. L., and Stamm, W. E.,** The effect of private isolation rooms on patient care practices, colonization and infection in an intensive care unit, *Am. J. Med.,* 70, 641, 1981.
19. **Albert, R. K. and Condie, F.,** Hand-washing patterns in medical intensive care units, *N. Engl. J. Med.,* 304, 1465, 1981.
20. **Larson, E.,** Compliance with isolation technique, *Am. J. Infect. Control,* 11, 221, 1983.
21. **Zimakoff, J. D. A.,** The hand hygienic behaviour of the staff in intensive care units, *Ugeskr. Læg.,* 147, 3946, 1985.
22. **Fox, M. K., Langner, S. B., and Wells, R. W.,** How good are handwashing practices?, *Am. J. Nurs.,* 74, 1676, 1974.
23. **Lowbury, E. J. L., Lilly, H. A., and Bull, J. P.,** Disinfection of hands: removal of transient organisms, *Br. Med. J.,* 2, 230, 1964.
24. **Ayliffe, G. A. J., Babb, J. R., and Quoraishi, A. H.,** A test for hygienic hand disinfection, *J. Clin. Pathol.,* 31, 923, 1978.
25. **Ojajärvi, J.,** Effectiveness of hand washing and disinfection methods in removing transient bacteria after patient nursing, *J. Hyg.,* 85, 193, 1980.
26. **Mackintosh, C. A. and Hoffman, P. N.,** An extended model for transfer of microorganisms via the hands: differences between organisms and the effect of alcohol disinfection, *J. Hyg.,* 92, 345, 1984.
27. **Reybrouck, G.,** Handwashing and hand disinfection, *J. Hosp. Infect.,* 8, 5, 1986.
28. **Bartzokas, C. A., Corkill, J. E., and Makin, T.,** Evaluation of the skin disinfecting activity and cumulative effect of chlorhexidine and triclosan handwash preparations on hands artificially contaminated with *Serratia marcescens, Infect. Control,* 8, 163, 1987.
29. **Hendley, J. O.,** Epidemic keratoconjunctivitis and hand washing, *N. Engl. J. Med.,* 289, 1368, 1973.
30. **Faix, R. G.,** Comparative efficacy of handwashing agents against cytomegalovirus, *Infect. Control,* 8, 158, 1987.
31. World Health Organization, Occupational Hazards in hospitals, EURO Reports and Studies 80, Report on a WHO meeting, World Health Organization, Copenhagen, 1983.
32. World Health Organization, *World Health Statistics Annual, 1973-1976,* Vol. 2, Infectious Diseases: Cases, Deaths and Vaccinations, World Health Organization, Geneva, 1976.
33. **Ellingson, H. V., Kadull, P. J., Bookwalter, H. L., and Howe, C.,** Cutaneous anthrax: report of twenty-five cases, *JAMA,* 131, 1105, 1946.
34. **Sulkin, S. E. and Pike, R. M.,** Laboratory-acquired infections, *JAMA,* 147, 1740, 1951.
35. **Pike, R. M.,** Laboratory-associated infections. Summary and analysis of 3921 cases, *Health Lab. Sci.,* 13, 105, 1976.
36. **Brachman, P. S.,** Inhalation anthrax, *Ann. N.Y. Acad. Sci.,* 353, 83, 1980.
37. **Linnemann, C. C., Jr., Bass, J. W., and Smith, M. H. D.,** The carrier state in pertussis, *Am. J. Epidemiol.,* 88, 422, 1968.
38. **Phillips, J.,** Whooping-cough contracted at the time of birth, with report of 2 cases, *Am. J. Med. Sci.,* 161, 163, 1921.
39. **Linnemann, C. C., Jr. and Nasenbeny, J.,** Pertussis in the adult, *Annu. Rev. Med.,* 28, 179, 1977.
40. **Nelson, J. D.,** The changing epidemiology of pertussis in young infants, *Am. J. Dis. Child.,* 132, 371, 1978.
41. **Trollfors, B. and Rabo, E.,** Whooping cough in adults, *Br. Med. J.,* 283, 696, 1981.
42. **MacLean, D. W.,** Adults with pertussis, *J. R. Coll. Gen. Pract.,* 32, 298, 1982.
43. **Kurt, T. L., Yeager, A. S., Guenette, S., and Dunlop, S.,** Spread of pertussis by hospital staff, *JAMA,* 221, 264, 1972.
44. **Linnemann, C. C., Jr., Ramundo, N., Perlstein, P. H., Minton, S. D., Englender, G. S., McCormick, J. B., and Hayes, P. S.,** Use of pertussis vaccine in an epidemic involving hospital staff, *Lancet,* 2, 540, 1975.
45. **Valenti, W. M., Pincus, P. H., and Messner, M. K.,** Nosocomial pertussis: possible spread by a hospital visitor, *Am. J. Dis. Child.,* 134, 520, 1980.
46. **Benenson, A. S., Ed.,** *Control of Communicable Diseases in Man,* American Public Health Association, Washington, D.C., 1985, 280.
47. Centers for Disease Control, Recommendation of the Immunization Practices Advisory Committee (ACIP): Diphtheria, Tetanus, Pertussis: Guidelines for Vaccine Prophylaxis and Other Preventive Measures, *Morbid. Mortal. Wkly. Rep.,* 34, 405, 1985.
48. **Morales-Otero, P.,** Further attempts at experimental infections of man with a bovine strain of *Brucella abortus, J. Infect. Dis.,* 52, 54, 1933.
49. **Smith, I. M.,** *Brucella* species (brucellosis), in *Principles and Practice of Infectious Diseases,* Mandell, G. L., Douglas, R. G., and Bennet, J. E., Eds., John Wiley & Sons, New York, 1979, chap. 183.

50. **Hall, W. H.,** Brucellosis, in *Bacterial Infections of Humans. Epidemiology and Control,* Evans, A. S. and Feldman, H. A., Eds., Plenum Press, New York, 1982, chap. 6.
51. **Pike, R. M., Sulkin, S. E., and Schulze, M. L.,** Continuing importance of laboratory-acquired infections, *Am. J. Publ. Health,* 55, 190, 1965.
52. **Pike, R. M.,** Past and present hazards of working with infectious agents, *Arch. Pathol. Lab. Med.,* 102, 333, 1978.
53. **Pike, R. M.,** Laboratory-associated infections: incidence, fatalities, causes and prevention, *Annu. Rev. Microbiol,* 33, 41, 1979.
54. **Huddleson, I. F. and Munger, M.,** A study of an epidemic of brucellosis due to *Brucella melitensis, Am. J. Publ. Health,* 30, 944, 1940.
55. **Fox, M. D. and Kaufmann, A. F.,** Brucellosis in the United States, 1965-1974, *J. Infect. Dis.,,* 136, 312, 1977.
56. **Robinson, D. A.,** Infective dose of *Campylobacter jejuni* in milk, *Br. Med. J.,* 282, 1584, 1981.
57. **Cadranel, S., Rodesch, P., Butzler, J.-P., and Dekeyser, P.,** Enteritis due to ''related *Vibrio*'' in children, *Am. J. Dis. Child.,* 126, 152, 1973.
58. **Hershkowici, S., Barak, M., Cohen, A., and Montag, J.,** An outbreak of *Campylobacter jejuni* infection in a neonatal intensive care unit, *J. Hosp. Infect.,* 9, 54, 1987.
59. **Grist, N. R.,** Infections in British clinical laboratories 1980-81, *J. Clin. Pathol.,* 36, 121, 1983.
60. **Grist, N. R.,** Infections in British clinical laboratories, 1982-83, *J. Clin. Pathol.,* 38, 721, 1985.
61. **Grist, N. R.,** Infections in British clinical laboratories, 1984-85, *J. Clin. Pathol.,* 40, 826, 1987.
62. **Broholm, K.-A., Böttiger, M., Jernelius, H., Johansson, Grandien, M., and Sölver, K.,** Ornithosis as a nosocomial infection, *Scand. J. Infect. Dis.,* 9, 263, 1977.
63. **Bernstein, D. I., Hubbard, T., Wenman, W. M., Johnson, B. L., Holmes, K. K., Liebhaber, H., Schechter, J., Barnes, R., and Lovett, M. A.,** Mediastinal and supraclavicular lymphadenitis and pneumonitis due to *Chlamydia trachomatis* serovars L_1 and L_2, *N. Engl. J. Med.,* 311, 1543, 1984.
64. **Midulla, M., Sollecito, D., Feleppa, F., Assensio, A. M., and Ilari, S.,** Infection by airborne *Chlamydia trachomatis* in a dentist cured with rifampicin after failure with tetracycline and doxycycline, *Br. Med. J.,* 294, 742, 1987.
65. **McFarland, L. V. and Stamm, W. E.,** Review of *Clostridium difficile*-associated diseases, *Am. J. Infect. Control,* 14, 99, 1986.
66. **Kim, K.-H., Fekety, R., Batts, D. H., Brown, D., Cudmore, M., Silva, J., Jr., and Waters, D.,** Isolation of *Clostridium difficile* from the environment and contacts of patients with antibiotic-associated colitis, *J. Infect. Dis.,* 143, 42, 1981.
67. **Fekety, R., Kim, K.-H., Brown, D., Batts, D. H., Cudmore, M., and Silva, J., Jr.,** Epidemiology of antibiotic-associated colitis. Isolation of *Clostridium difficile* from the hospital environment, *Am. Med. J.,* 70, 906, 1981.
68. **Scherertz, R. J. and Sarubbi, F. A.,** The prevalence of *Clostridium difficile* and toxin in a nursery population: a comparison between patients with necrotizing enterocolitis and an asymptomatic group, *J. Pediatr.,* 100, 435, 1982.
69. **Mulligan, M. E., George, W. L., Rolfe, R. D., and Finegold, S. M.,** Epidemiological aspects of *Clostridium difficile* induced diarrhoea and colitis, *Am. J. Clin. Nutr.,* 33, 2533, 1980.
70. **Malamou-Ladas, H., O'Farrel, S., Nash, J. Q., and Tabaqchali, S.,** Isolation of *Clostridium difficile* from patients and the environment of hospital wards, *J. Clin. Pathol.,* 36, 88, 1983.
71. **Keighley, M. R. B., Burdon, D. W., Mogg, G. A. G., George, R. H., Alexander-Williams, J., and Thompson, H.,** Pseudomembranous colitis, *Lancet,* 1, 559, 1979.
72. **Al-Jumaili, I. J., Shibley, M., Lishman, A. H., and Record, C. O.,** Incidence and origin of *Clostridium difficile* in neonates, *J. Clin. Microbiol.,* 19, 77, 1984.
73. **Pierce, P. F., Jr., Wilson, R., Silva, J., Jr., Garagusi, V., F., Rifkin, G. D., Fekety, R., Nunez-Montiel, O., Dowell, V. R., Jr., and Hughes, J. M.,** Antibiotic-associated pseudomembranous colitis: an epidemiological investigation of a cluster of cases, *J. Infect. Dis.,* 145, 269, 1982.
74. **Walters, B. A. J., Stafford, R., Roberts, R. K., and Seneviratne, E.,** Contamination and cross infection with *C. difficile* in an intensive care unit, *Aust. N.Z. J. Med.,* 12, 255, 1982.
75. **Bray, J. P., Burt, E. G., Potter, E. V., Poon-King, T., and Earle, D. P.,** Epidemic diphtheria and skin infections in Trinidad, *J. Infect. Dis.,* 126, 34, 1972.
76. **Belsey, M. A. and LeBlanc, D. R.,** Skin infections and the epidemiology of diphtheria: acquisition and persistence of *C. diphtheriae* infections, *Am. J. Epidemiol.,* 102, 179, 1975.
77. **Dixon, J. M. S.,** Diphtheria in North America, *J. Hyg.,* 93, 419, 1984.
78. **Björkholm, B., Böttiger, M., Christenson, B., and Hagberg, L.,** Antitoxin and antibody levels and the outcome of illness during an outbreak of diphtheria among alcoholics, *Scand. J. Infect, Dis.,* 18, 325, 1986.
79. **Williams, W. W. and Garner, J. S.,** Personnel health services, in *Hospital Infections,* Bennett, J. V. and Brachman, P.S., Eds., Little, Brown, Boston, 1986, 17.

140. **Davey, T. F. and Rees, R. J. W.,** The nasal discharge in leprosy: clinical and bacteriologic aspects, *Lepr. Rev.*, 45, 121, 1974.
141. **Desikan, K. B.,** Viability of *Mycobacterium leprae* outside the human body, *Lepr. Rev.*, 48, 231, 1977.
142. **Jacobson, R. R.,** Leprosy, in *Bacterial Infections of Humans. Epidemiology and Control,* Evans, A. S. and Feldman, H. A., Eds., Plenum Press, New York, 1982, 293.
143. **Benenson, A. S., Ed.,** *Control of Communicable Diseases in Man,* American Public Health Association, Washington, D.C., 1985, 210.
144. **Levy, L.,** Leprosy, in *Infectious Diseases,* Hoeprich, P. D., Ed., Harper & Row, Philadelphia, 1983, 945.
145. **Myrvang, B., Negassi, K., Løfgren, M., and Godal, T.,** Immune responsiveness to *Mycobacterium leprae* of healthy humans, *Acta Pathol. Microbiol. Scand. Sect. C.*, 83, 43, 1975.
146. **Mathai, R., Rao, P. S. S., and Job, C. K.,** Risk of treating leprosy in a general hospital, *Int. J. Lepr.*, 47, 322, 1979.
147. **Gray, H. H. and Dreisbach, J. A.,** Leprosy among foreign missionaires in northern Nigeria, *Int. J. Lepr.*, 29, 279, 1961.
148. **Marchoux, P. E.,** Un cas d'inoculation accidentelle du bacille de Hansen en pays non lepreux, *Int. J. Lepr.*, 2, 1, 1934.
149. **Mücke, H.,** Die tuberculose des Pflegepersonals, *Beitr. Klin. Tuberk. Spezifischen. Tuberk. Forsch.*, 64, 115, 1926.
150. **Baldwin, E. R.,** Danger of tuberculous infection in hospitals and sanatoria, *U.S. Veterans Bur. Med. Bull.*, 6, 1, 1930.
151. **Heimbeck, J.,** Immunity to tuberculosis, *Arch. Intern. Med.*, 41, 336, 1928.
152. **Heimbeck, J.,** Tuberculosis in hospital nurses, *Tubercle,* 18, 97, 1936.
153. **Hetherington, H. W., McPhedran, F. M., Landis, H. R. M., and Opie, E. L.,** Tuberculosis in medical and college students, *Arch. Intern. Med.*, 48, 734, 1931.
154. **Hetherington, H. W. and Israel, H. L.,** Tuberculosis in medical students and young physicians, *Am. J. Hyg.*, 31, 35, 1940.
155. **Israel, H. L., Hetherington, H. W., and Ord, J. G.,** A study of tuberculosis among students of nursing, *JAMA,* 117, 839, 1941.
156. **Morris, S. I.,** Tuberculosis as occupational hazard during medical training, *Am. Rev. Tuberc.*, 54, 140, 1946.
157. **Daniels, M., Ridehalgh, F., Springett, V. H., and Hall, I. M.,** *Tuberculosis in Young Adults: report of the Prophit Tuberculosis Survey,* 1935-1944, H. K. Lewis, London, 1948.
158. **Abruzzi, W. A. and Hummel, R. J.,** Tuberculosis: incidence among American medical students, prevention and control and the use of BCG, *N. Engl. J. Med.*, 248, 722, 1953.
159. **Weiss, W.,** Tuberculosis in student nurses at Philadelphia General Hospital, *Am. Rev. Respir. Dis.*, 107, 136, 1973.
160. **Levine, I.,** Tuberculosis risk in students of nursing, *Arch. Intern. Med.*, 121, 545, 1968.
161. **Burhill, D., Enarson, D. A., Allen, E. A., and Grzybowski, S.,** Tuberculosis in female nurses in British Columbia: implications for control programs, *Can. Med. Assoc. J.*, 132, 137, 1985.
162. **Barrett-Connor, E.,** The epidemiology of tuberculosis in physicians, *JAMA,* 241, 33, 1979.
163. **Geiseler, P. J., Nelson, K. E., Crispen, R. G., and Moses, V. K.,** Tuberculosis in physicians: a continuing problem, *Am. Rev. Respir. Dis.*, 133, 773, 1986.
164. **Allibone, A., Oakes, D., and Shannon, H. S.,** The health and health care of doctors, *J. R. Coll. Gen. Pract.*, 31, 728, 1981.
165. **Harrington, J. M. and Shannon, H. S.,** Incidence of tuberculosis, hepatitis, brucellosis and shigellosis in British medical laboratory workers, *Br. Med. J.*, 1, 759, 1976.
166. **Meade, G. M.,** The prevention of primary tuberculosis infection in medical students, *Am. Rev. Tuberc.*, 58, 675, 1948.
167. **Cuthbertson, W. C.,** Causes of death among dentists, a comparison with the general male population, *J. Calif. State Dent. Assoc. Nev. State Dent. Soc.*, 30, 159, 1954.
168. **Ruben, F. L., Norden, C. W., and Schuster, N.,** Analysis of a community hospital employee tuberculosis screening program 31 months after its inception, *Am. Rev. Respir. Dis.*, 115, 23, 1977.
169. **Lowenthal, G. and Keys, T.,** Tuberculosis surveillance in hospital employees: are we doing to much?, *Infect. Control,* 7, 209, 1986.
170. **Berman, J., Levin, M. L., Tangerose, S., and Desi, L. D.,** Tuberculosis risk for hospital employees: analysis of a five-year tuberculin skin testing program, *Am. J. Public Health,* 71, 1217, 1981.
171. **Price, L. E., Rutala, W. A., and Samsa, G. P.,** Tuberculosis in hospital personnel, *Infect. Control,* 8, 97, 1987.
172. **Weinstein, R. S., Oshins, J., and Sacks, H. R.,** Tuberculosis infection in Mount Sinai medical students: 1974-1982, *Mt. Sinai J. Med.*, 51, 283, 1984.
173. **Lowell, A. M.,** Tuberculosis in the World, Publ. No. CDC 76-8317, Centers for Disease Control, Public Health Service, U.S. Department of Health, Education and Welfare, Atlanta, 1976.

174. **Brahdy, L.,** Tuberculosis in hospital personnel, *JAMA,* 114, 102, 1940.
175. **Dickie, H. A.,** Tuberculosis in student nurses and medical students at University of Wisconsin, *Ann. Intern. Med.,* 33, 941, 1950.
176. **Ashley, M. J. and Wigle, W. D.,** The epidemiology of active tuberculosis in hospital employees in Ontario, *Am. Rev. Respir. Dis.,* 104, 851, 1971.
177. **Clemens, J. D., Choung, J. J. H., and Feinstein, A. R.,** The BCG controversy. A methodological and statistical reappraisal, *JAMA,* 249, 2362, 1983.
178. Editorial, BCG vaccination after the Madras study, *Lancet,* 1, 1981.
179. **Barclay, W. R.,** BCG: an effective immunizing agent, *JAMA,* 249, 2376, 1983.
180. BCG Vaccination Policies, Report of a WHO study group, WHO Tech. Rep. Ser. 652, World Health Organization, Geneva, 1980.
181. **Ferguson, R. G.,** BCG vaccination in hospitals and sanatoria of Sascatchewan, *Am. Rev. Tuberc.,* 54, 325, 1946.
182. **Heimbeck, J.,** BCG vaccination of nurses, *Tubercle,* 29, 84, 1948.
183. **Rosenthal, S. R., Afremov, M. L., Nikurs, L., Loewinsohn, E., Leppmann, M., Katele, E., Liveright, D., and Thorne, M.,** BCG vaccination and tuberculosis in students of nursing, *Am. J. Nurs.* 63, 88, 1963.
184. **Riley, R. L.,** The hazard is relative, *Am. Rev. Respir. Dis.,* 96, 623, 1967.
185. **Riley, R. L., Mills, C. C., O'Grady, F., Sultan, L. U., Wittstadt, F., and Shivpuri, D. N.,** Infectiousness of air from a tuberculosis ward, *Am. Rev. Respir. Dis.,* 85, 511, 1962.
186. American Thoracic Society, Infectiousness of tuberculosis, a statement of the ad hoc committee on treatment of tuberculosis patients in general hospitals, *Am. Rev. Respir. Dis.,* 96, 836, 1967.
187. Centers for Disease Control, Guidelines for Prevention of TB Transmission in Hospitals, HHS Publ. No. ICDCI 82-8371 Centers for Disease Control, Public Health Service, U.S. Department of Health and Human Services, Atlanta, 1982.
188. **Loudon, R. G. and Roberts, R. M.,** Singing and the dissemination of tuberculosis, *Am. Rev. Respir. Dis.,* 98, 297, 1968.
189. **Loudon, R. G. and Spohn, S. K.,** Cough frequency and infectivity in patients with pulmonary tuberculosis, *Am. Rev. Respir. Dis.,* 99, 109, 1969.
190. **Smith, C. R.,** Survival of tubercle bacilli, *Am. Rev. Tuberc.,* 45, 334, 1942.
191. **Grzybowsky, S., Barnet, G. D., and Styblo, K.,** Contacts of cases of active pulmonary tuberculosis. Report No. 3 of TSRU, *Bull. Int. Union Tuberc.,* 60, 90, 1975.
192. **van Geuns, H. A., Meijer, J., and Styblo, K.,** Results of contact examinations in Rotterdam 1967-1969. Report No. 3 of TSRU, *Bull. Int. Union Tuberc.,* 50, 107, 1975.
193. Joint Tuberculosis Committee of the British Thoracic Society, Control and prevention of tuberculosis: a code of practice, *Br. Med. J.,* 287, 1118, 1983.
194. **Edlin, G. P.,** Active tuberculosis unrecognised until necropsy, *Lancet,* 1, 650, 1978.
195. **MacGregor, R. R.,** A year's experience with tuberculosis in a private urban teaching hospital in the postsanatorium era, *Am. J. Med.,* 58, 221, 1975.
196. **Vogeler, T. M. and Burke, J. P.,** Tuberculosis screening of hospital employees. A five year experience in a large community hospital, *Am. Rev. Respir. Dis.,* 117, 227, 1978.
197. **Craven, R. B., Wenzel, R. P., and Atuk, N. O.,** Minimizing tuberculosis risk to hospital personnel and students exposed to unsuspected disease, *Ann. Intern. Med.,* 82, 628, 1975.
198. CDC, U.S. DHHS, Diagnosis and management of mycobacterial infection and disease in persons with human immunodeficiency virus infection, *Ann. Intern. Med.,* 106, 254, 1987.
199. **Anon.**, Tuberculosis provisional data — United States, 1986, *Morbid. Mortal. Wkly. Rep.,* 36, 254, 1987.
200. **Sunderam, G., McDonald, R. J., Maniatis, T., Oleske, J., Kapila, R., and Reichman, L. B.,** Tuberculosis as a manifestation of the acquired immunodeficiency syndrome (AIDS), *JAMA,* 256, 362, 1986.
201. **Sunderam, G., Maniatis, T., Kapila, R., McDonald, R., and Reichman, L. B.,** Mycobacterium tuberculosis disease with unusual manifestations is relatively common in acquired immuno-deficiency syndrome (AIDS), *Am. Rev. Respir. Dis.,* 129, A191, 1984.
202. **Pitchenik, A. E. and Rubinson, H. A.,** The radiographic appearance of tuberculosis in patients with the acquired immunodeficiency syndrome (AIDS) and pre-AIDS, *Am. Rev. Respir. Dis.,* 131, 393, 1985.
203. **Catanzaro, A.,** Nosocomial tuberculosis, *Am. Rev. Respir. Dis.,* 125, 559, 1982.
204. **Ehrenkranz, N. J. and Kicklighter, J. L.,** Tuberculosis outbreak in a general hospital: evidence for airborne spread of infection, *Ann. Intern. Med.,* 77, 377, 1972.
205. **Jindani, A., Aber, V. R., Edwards, A., and Mitchison, D. A.,** The early bactericidal activity of drugs in patients with pulmonary tuberculosis, *Am. Rev. Respir. Dis.,* 121, 939, 1980.
206. **Pouillon, A., Perdrizet, S., and Parrot, R.,** Transmission of tubercle bacilli: the effects of chemotherapy, *Tubercle,* 57, 275, 1976.
207. **Belfield, P. W., Arnold, A. G., Williams, S. E., Bostock, A. D., and Cooke, N. J.,** Recent experience of tuberculosis in junior hospital doctors in Leeds and Bradford, *Br. J. Dis. Chest,* 78, 13, 1984.
208. American Thoracic Society, Control of tuberculosis, *Am. Rev. Respir. Dis.,* 128, 336, 1983.

209. **McGowan, J. E.,** The booster effect — a problem for surveillance of tuberculosis in hospital employees, *Infect. Control,* 1, 14, 1980.
210. **Bass, J. B. and Serio, R. A.,** The use of repeat skin tests to eliminate the booster phenomenon in serial tuberculin testing, *Am. Rev. Respir. Dis.,* 123, 394, 1981.
211. **Snider, D. E. and Cauthen, G. M.,** Tuberculin testing of hospital employees: infection "boosting" and two-step testing, *Am. J. Infect. Control.,* 12, 305, 1984.
212. **Valenti, W. M., Andrews, B. A., Presley, B. A., and Reifler, C. B.,** Absence of the booster phenomenon in serial tuberculin skin testing, *Am. Rev. Respir. Dis.,* 125, 323, 1982.
213. WHO Expert Committee on Tuberculosis, 9th Report, WHO Tech. Rep. Ser. 1974, No. 552, World Health Organization, Geneva, 1974.
214. **Barrett-Connor, E.,** The periodic chest roentgenogram for the control of tuberculosis in health care personnel, *Am. Rev. Respir. Dis.,* 122, 153, 1980.
215. **Cope, R. and Harstein, A. I.,** The annual chest roentgenogram for the control of tuberculosis in hospital employees: recent changes and their implications, *Am. Rev. Respir. Dis.,* 125, 106, 1982.
216. **Reeves, S. A. and Noble, R. C.,** Ineffectiveness of annual chest roentgenograms in tuberculin skin test positive hospital employees, *Am. J. Infect. Control,* 11, 212, 1983.
217. **Bailey, W. C., Albert, R. K., Davidson, P. T., Farer, L. S., Glassroth, J., Kendig, E., Loudon, R. G., and Inselman, L. S.,** Treatment of tuberculosis and other mycobacterial diseases: an official statement of the American Thoracic Society, *Am. Rev. Respir. Dis.,* 127, 790, 1983.
218. **Egsmosse, T., Ang'awa, J. W. C., and Porti, S. J.,** The use of isoniazid among household contacts of open cases of pulmonary tuberculosis, *Bull. W. H. O.* 33, 419, 1965.
219. **Comstock, G. W., Ferebee, S. H., and Hammes, L. M.,** A controlled trial of community-wide isoniazid prophylaxis in Alaska, *Am. Rev. Respir. Dis.,* 95, 935, 1967.
220. **Debre, R., Perdrizet, S., Lotte, A., Naveau, M., and Lert, F.,** Isoniazid chemoprophylaxis of latent primary tuberculosis in five trial centers in France from 1959 to 1969, *Int. J. Epidemiol.,* 2, 153, 1973.
221. **Ferebee, S. H.,** Controlled chemoprophylaxis trials in tuberculosis: a general review, *Adv. Tuberc. Res.,* 17, 28, 1970.
222. **Falk, A. and Fuchs, G. F.,** Prophylaxis with isoniazid in inactive tuberculosis, *Chest,* 73, 44, 1978.
223. **Comstock, G. W., Baum, C., and Snider, D. E.,** Isoniazid prophylaxis among Alaskan Eskimos: a final report of the Bethel isoniazid studies, *Am. Rev. Respir. Dis.,* 119, 827, 1979.
224. **Mitchell, J. R., Zimmerman, H. J., Ishak, K. G., Thorgeirsson, U. P., Timbrell, J. A., Snodgrass, W. R., and Nelson, S. D.,** Isoniazid liver injury: clinical spectrum, pathology and probable pathogenesis, *Ann. Intern. Med.,* 84, 181, 1976.
225. **Comstock, G. W. and Edwards, P. Q.,** The competing risk of tuberculosis and hepatitis for adult tuberculin reactors, *Am. Rev. Respir. Dis.,* 111, 573, 1975.
226. **Bailey, W. C., Byrd, R. B., Glassroth, J. L., Hopewell, P. C., and Reichman, L. B.,** Preventative treatment of tuberculosis, *Chest,* 87, 129S, 1985.
227. **Tailor, W. C., Aronson, M. D., and Delbanco, T. L.,** Should young adults with a positive tuberculin test take isoniazid?, *Ann. Intern. Med.,* 94, 808, 1981.
228. **Stewart, C. J.,** Tuberculosis infection in a pediatric department, *Br. Med. J.,* 1, 30, 1976.
229. **Foy, H. M.,** *Mycoplasma pneumoniae,* in *Bacterial Infections of Humans. Epidemiology and Control,* Evans, A. S. and Feldman, H. A., Eds., Plenum Press, New York, 1982, chap. 19.
230. **Gerth, H. J., Grüner, C., Müller, R., and Dietz, K.,** Seroepidemiological studies on the occurrence of common respiratory infections in pediatric student nurses and medical technology students, *Epidemiol. Infect.,* 98, 47, 1987.
231. **Fisher, B., Yu, B., Armstrong, D., and Magill, J.,** Outbreak of *Mycoplasma pneumoniae* infection among hospital personnel, *Am. J. Med. Sci.* 276, 205, 1978.
232. **Sande, M. A., Gadot, F., and Wenzel, R. P.,** Point source of *Mycoplasma pneumoniae* infection in a prostodontics laboratory, *Am. Rev. Respir. Dis.,* 112, 213, 1975.
233. **Greenfield, S., Sheehe, P. R., and Feldman, H. A.,** Meningococcal carriage in a population of "normal" families, *J. Infect. Dis.,* 123, 67, 1971.
234. **Holten, E., Bratlid, D., and Bøvre, K.,** Carriage of *Neisseria meningitidis* in a semi-isolated arctic community, *Scand. J. Infect. Dis.,* 10, 36, 1978.
235. **Peltola, H.,** Meningococcal disease: still with us, *Rev. Infect. Dis.,* 5, 71, 1983.
236. **Kaiser, A. B., Hennekens, C. H., Saslaw, M. S., Hayes, P. S., and Bennett, J. V.,** Seroepidemiology and chemoprophylaxis of disease due to sulfonamide-resistant *Neisseria meningitidis* in a civilian population, *J. Infect. Dis.,* 130, 217, 1974.
237. **Feldman, H. A.,** Meningococcal infections, in *Bacterial Infections of Humans. Epidemiology and Control,* Evans, A. S. and Feldman, H. A., Eds., Plenum Press, New York, 1982, 327.
238. **Arkwright, J. A.,** Cerebrospinal meningitis: the interpretation of epidemiological observations by the light of bacteriological knowledge, *Br. Med. J.,* 1, 494, 1915.

239. **Artenstein, M. S. and Ellis, R. E.,** The risk of exposure to a patient with meningococcal meningitis, *Mil. Med.*, 133, 474, 1968.
240. **Salmi, I., Pettay, O., Simula, I., Kallio, A.-K., and Waltimo, O.,** An epidemic due to sulphonamide-resistant group. A meningococci in the Helsinki area (Finland). Epidemiological and clinical observations, *Scand. J. Infect. Dis.*, 8, 249, 1976.,
241. **Cohen, M. S., Steere, A. C., Baltimore, R., von Graevenitz, A., Pantelick, E., Camp, B., and Root, R. K.,** Possible nosocomial transmission of group Y *Neisseria meningitidis* among oncology patients. *Ann. Intern. Med.*, 91, 7, 1979.
242. **Bøvre, K.,** personal communication.
243. **Wilson, L. A., Schlitzer, R. L., and Ahearn, D. G.,** *Pseudomonas* corneal ulcers associated with soft contact-lens wear, *Am. J. Ophthalmol.*, 92, 546, 1981.
244. **Hassman, G. and Sugar, J.,** *Pseudomonas* corneal ulcer with extended-wear soft contact lenses for myopia, *Arch. Ophthalmol.*, 101, 1549, 1983.
245. **Weissman, B. A., Mondino, B. J., Pettit, T. H., and Hofbauer, J. D.,** Corneal ulcers associated with extended-wear soft contact lenses, *Am. J. Ophthalmol.*, 97, 476, 1984.
246. **Galentine, P. G., Cohen, E. J., Laibson, P. R., Adams, C. P., Michaud, R., and Arentson, J. J.,** Corneal ulcers associated with contact lens wear, *Arch. Ophthalmol.*, 102, 891, 1984.
247. **Bowden, H. H. and Sutpin, J. E.,** Nosocomial *Pseudomonas* keratitis in a critical care nurse, *Am. J. Ophthalmol.*, 101, 612, 1986.
248. **Benenson, A. S., Ed.**, *Control of Communicable Diseases in Man,* American Public Health Association, Washington, D.C., 1985, 237.
249. **Green, R. N. and Tuffnell, P. G.,** Laboratory-acquired melioidosis, *Am. J. Med.*, 44, 599, 1968.
250. **Schlech, W. F., Turchik, J. B., Westlake, R. E., Klein, G. C., Band, J. D., and Wever, R. E.,** Laboratory-acquired infection with *Pseudomonas pseudomallei, N. Engl. J. Med.*, 305, 1133, 1981.
251. **Stamm, W. E., Weinstein, R. A., and Dixon, R. E.,** Comparison of endemic and epidemic nosocomial infections, *Am. J. Med.*, 70, 393, 1981.
252. **Abbot, J. D., Hepner, E. D., and Clifford, C.,** Salmonella infections in hospital. A report from the Public Health Laboratory Service Salmonella Subcommittee, *J. Hosp. Infect.*, 1, 307, 1980.
253. **Baine, W. B., Gangarosa, E. J., Bennett, J. V., and Barker, W. H.,** Institutional salmonellosis, *J. Infect. Dis.*, 128, 357, 1973.
254. **Palmer, S. R. and Rowe, B.,** Investigation of outbreaks of salmonella in hospitals, *Br. Med. J.*, 287, 891, 1983.
255. **Taylor, J.,** Salmonella infections in hospitals, in *Infection in Hospitals. Epidemiology and Control,* Williams, R. E. O., and Shooter, R. A., Eds., Blackwell Scientific, Oxford, 1963, 145.
256. **Bate, J. G. and James, U.,** *Salmonella typhimurium* infection dust-borne in a childrens ward, *Lancet,* 2, 713, 1958.
257. **Datta, N. and Pridie, R. B.,** An outbreak of infection with *Salmonella typhimurium* in a general hospital, *J. Hyg.*, 58, 229, 1960.
258. **Sanders, E., Sweeney, F. J., Friedman, E. A., Boring, J. R., Randall, E. L., and Polk, L. D.,** An outbreak of hospital-associated infections due to *Salmonella derby, JAMA,* 186, 110, 1963.
259. **Adler, J. L., Anderson, R. L., Boring, J. R., III, and Nahmias, A. J.,** A protracted hospital-associated outbreak of salmonellosis due to a multiple-antibiotic-resistant strain of *Salmonella indiana, J. Pediatr.*, 77, 970, 1970.
260. **Steere, A. C., Craven, P. J., Hall, W. J., III, Leotsakis, N., Wells, J. G., Farmer, J. J., III, and Gangarosa, E. J.,** Person-to-person spread of *Salmonella typhimurium* after a hospital common-source outbreak, *Lancet,* 1, 319, 1975.
261. **Rice, P. A., Craven, P. C., and Wells, J. G.,** *Salmonella heidelberg* enteritis and bacteremia. An epidemic on two pediatric wards, *Am. J. Med.*, 60, 509, 1976.
262. **Lintz, D., Kapila, R., Pilgrim, E., Tecson, F., Dorn, R., and Louria, D.,** Nosocomial *Salmonella* epidemic, *Arch. Intern. Med.*, 136, 968, 1976.
263. **Ayliffe, G. A., Geddes, A. M., Pearson, J. E., and Williams, T. C.,** Spread of *Salmonella typhi* in a maternity hospital, *J. Hyg.*, 82, 353, 1979.
264. **Anand, C. M., Finlayson, M. C., Garson, J. Z., and Larson, M. L.,** An institutional outbreak of salmonellosis due to a lactose-fermenting *Salmonella newport, Am. J. Clin. Pathol.*, 74, 657, 1980.
265. **Robins-Browne, R. M., Rowe, B., Ramsaroop, R., Naran, A. D., Threlfall, E. J., Ward, L. R., Lloyd, D. A., and Mickel, R. E.,** A hospital outbreak of multiresistant *Salmonella typhimurium* belonging to phage type 193, *J. Infect. Dis.*, 147, 210, 1983.
266. **Seals, J. E., Parrott, P. L., McGowan, J. E., and Feldman, R. A.,** Nursery salmonellosis: delayed recognition due to unusually long incubation period, *Infect. Control,* 4, 205, 1983.
267. **Epstein, H. C., Hochwald, A., and Ashe, R.,** Salmonella infections of the newborn infant, *J. Pediatr.* 38, 723, 1951.

268. **Lamb, V. A., Mayhall, C. G., Spadora, A. C., Markowtiz, S. M., Farmer, J. J., III, and Dalton, H. P.,** Outbreak of *Salmonella typhimurium* gastroenteritis due to an imported strain resistant to ampicillin, chloramphenicol, and trimethoprim-sulfamethoxazole in a nursery, *J. Clin. Microbiol.,* 20, 1076, 1984.
269. **Galloway, A., Roberts, C., and Hunt, E. J.,** An outbreak of *Salmonella typhimurium* gastroenteritis in a psychiatric hospital, *J. Hosp. Infect.,* 10, 248, 1987.
270. **Mushin, R.,** An outbreak of gastro-enteritis due to *Salmonella derby, J. Hyg.,* 46, 151, 1948.
271. **Ip, H. M. H., Sin, W. K., Chau, P. Y., Tse, D., and Teoh-Chan, C. H.,** Neonatal infection due to *Salmonella worthington* transmitted by a delivery room suction apparatus, *J. Hyg.,* 77, 307, 1976.
272. **Rubenstein, A. D. and Fowler, R. N.,** Salmonellosis of the newborn with transmission by delivery room resuscitators, *Am. J. Public Health,* 45, 1109, 1955.
273. **Szmuness, W., Sikorska, J., Szymanek, E., Mikosz, L., and Cechowicz, A.,** The microbiological and epidemiological properties of infections caused by *Salmonella enteritidis, J. Hyg.,* 49, 9, 1966.
274. **Watt, J., Wegman, M. E., Brown, O. W., Schliessmann, M. S., Maupin, E., and Hephill, E. C.,** Salmonellosis in a premature nursery unaccompanied by diarrheal disease, *Pediatrics,* 22, 689, 1958.
275. **Beecham, H. J., Cohen, M. L., and Parkin, W. E.,** *Salmonella typhimurium:* transmission by fiberoptic upper gastrointestinal endoscopy, *JAMA,* 241, 1013, 1979.
276. **Pether, J. V. S. and Gilbert, R. J.,** The survival of salmonellas on finger-tips and transfer of the organisms to food, *J. Hyg.,* 69, 673, 1971.
277. **MacGregor, R. R. and Reinhart, J.,** Person-to-person spread of salmonella: a problem in hospitals?, *Lancet,* 2, 1001, 1973.
278. **Blaser, M. J., Hickman, F. W., Farmer, J. J., III, Brenner, D. J., and Feldman, R. A.,** *Salmonella typhi:* the laboratory as a reservoir of infection, *J. Infect. Dis.,* 142, 934, 1980.
279. **Blaser, M. J. and Lofgren, J. P.,** Fatal salmonellosis originating in a clinical microbiology laboratory, *J. Clin. Microbiol.,* 13, 855, 1981.
280. **Linnemann, C. C., Jr., Cannon, C. G., Staneck, J. L., and McNeely, B. L.,** Prolonged hospital epidemic of salmonellosis: use of trimethoprim-sulfamethoxazole for control, *Infect. Control,* 6, 221, 1985.
281. **DuPont, H. L., Gangarosa, E. J., Reller, L. B., Woodward, W. E., Armstrong, R. W., Hammond, J., and Morris, G. K.,** Shigellosis in custodial institutions, *Am. J. Epidemiol.,* 92, 172, 1970.
282. **Levine, M. M., DuPont, H. L., Khodabandelou, M., and Hornick, R. B.,** Long-term shigella-carrier state, *N. Engl. J. Med.,* 288, 1169, 1973.
283. **Aleksic, S., Bochemühl, J., and Degner, I.,** Imported shigellosis: aerogenic *Shigella boydii* 14 (Sachs A 12) in a traveller followed by two cases of laboratory-associated infections, *Tropenmed. Parasitol.,* 32, 61, 1981.
284. **Ghosh, H. K.,** Laboratory-acquired shigellosis, *Br. Med. J.,* 285, 695, 1982.
285. **Levine, M. M., DuPont, H. L., Formal, S. B., Hornick, R. B., Takeuchi, A., Gangarosa, E. J., Snyder, M. J., and Libonati, J. P.,** Pathogenesis af *Shigella dysenteriae* 1 (Shiga) dysentery, *J. Infect. Dis.,* 127, 261, 1973.
286. **Wedum, A. G., Barkley, W. E., and Hellman, A.,** Handling of infectious agents, *J. Am. Vet. Med. Assoc.,* 161, 1557, 1972.
287. U.S. DHHS, Biosafety in Microbiological and Biomedical Laboratories, HHS Publ. No. (CDC) 84-8395760, U.S. Department of Health and Human Services, U.S. Government and Printing Office, Washington, D.C., 1984.
288. **Lampert, J., Plotkin, S., Campos, J., Trendler, M., and Schlagel, D.,** Shigellosis in a Childrens' Hospital — Pennsylvania, *Morbid. Mortal. Wkly. Rep.,* 28(No. 42), 498, 1979.
289. **Godfrey, M. E. and Smith, I. M.,** Hospital hazards of staphylococcic sepsis, *JAMA,* 166, 1197, 1958.
290. **Wheat, J. L., Kohler, R. B., and White, A.,** Treatment of nasal carriers of coagulase-positive staphylococci, in *Skin Microbiology,* Maibach H. and Aly, R., Eds., Springer-Verlag, New York, 1981, 50.
291. **Bruun, J. N.,** Post-operative wound infection. Predisposing factors and the effect of a reduction in the dissemination of staphylococci. Thesis, *Acta Med. Scand.,* Suppl. 514, 1970.
292. **Fekety, R. F.,** The epidemiology and prevention of staphylococcal infection, *Medicine,* 43, 593, 1964.
293. **Davies, D. M.,** Staphylococcal infection in nurses, *Lancet,* 1, 644, 1960.
294. **Gedebou, M.,** Nasal carrier rates and antibiotic resistance of *Staphylococcus aureus* isolates from hospital and nonhospital populations, Addis Ababa, *Trans. R. Soc. Trop. Med. Hyg.,* 78, 314, 1984.
295. **Behrendt, F. B.,** Septic lesions of the skin among hospital personnel. An epidemiological investigation of 700 cases, registered during the period 1959-1966, *Danish Med. Bull.,* 16, 120, 1969.
296. **Tewodros, W., Thompson, R. L., Cabezudo, I., and Wenzel, R. P.,** Epidemiology of nosocomial infections caused by methicillin-resistant *Staphylococcus aureus, Ann. Intern. Med.,* 97, 309, 1982.
297. **Casewell, M. W.,** Epidemiology and control of the "modern" methicillin-resistant *Staphylococcus aureus, J. Hosp. Infect.,* 7(Suppl. A), 1, 1986.
298. **Bassett, H. F. M., Ferguson, W. G., Hoffmann, E., Walton, M., Blowers, R., and Conn, C. A.,** Sources of staphylococcal infection in surgical wound sepsis, *J. Hyg.,* 61, 83, 1963.
299. **Cruse, P.,** Surgical infection: incisional wounds, in *Hospital Infections,* Bennett, J. V. and Brachman, P. S., Eds., Little, Brown, Boston, 1986, 423.

300. **Rammelkamp, C. H., Jr., Mortimer, E. A., Jr., and Wolinsky, E.,** Transmission of streptococcal and staphylococcal infections, *Ann. Inter. Med.,* 60, 753, 1964.
301. **Singgih, S. I. R., Lantinga, H., Nater, J. P., Woest, T. E., and Kruyt-Gaspersz, J. A.,** Occupational hand dermatoses in hospital cleaning personnel, *Contact Dermatitis,* 14, 14, 1986.
302. **Garrod, L. P.,** The eclipse of the hemolytic streptococcus, *Br. Med. J.,* 1, 1607, 1979.
303. **Perry, W. D., Siegel, A. C., Rammelkamp, C. H., Jr., Wannamaker, L. W., and Marple, E. C.,** Transmission of group-A streptococci. I. The role of contaminated bedding, *Am. J. Hyg.,* 66, 85, 1957.
304. **Perry, W. D., Siegel, A. C., and Rammelkamp, C. H., Jr.,** Transmission of group-A streptococci. II. The role of contaminated dust, *Am. J. Hyg.,* 66, 96, 1957.
305. **Hawkey, M., Pedler, S. J., and Southall, P. J.,** *Streptococcus pyogenes:* a forgotten occupational hazard in the mortuary, *Br. Med. J.,* 281, 1058, 1980.
306. **Webster, A., Scott, G. M. S., Ridgeway, G. L., and Grüneberg, R. N.,** An outbreak of group A streptococcal skin infection: control by source isolation and teichoplanin therapy, *Scand. J. Infect. Dis.,* 19, 205, 1987.
307. **Easmon, C. S. F.,** The carrier state: Group B streptococcus, *J. Antimicrob. Chemother.,* 18(Suppl. A), 59, 1986.
308. **Green, S. L., Nodell, C. C., and Porter, C. W.,** The prevalence and persistence of Group B streptococcal colonization among hospital personnel, *Int. J. Gynaecol. Obstet.,* 16, 99, 1978.
309. **Franciosi, R. A., Knostman, J. D., and Zimmerman, R. A.,** Group B streptococcal neonatal and infant infections, *J. Pediatr.,* 82, 707, 1973.
310. **Steere, A. C., Aber, R. C., Warford, L. R., Murphy, K. E., Feeley, J., Hayes, P. S., Wilkinson, H. W., and Facklam, R. R.,** Possible nosocomial transmission of group B streptococci in a newborn nursery, *J. Pediatr.,* 87, 784, 1985.
311. **Benjamin, J. E., Ruegsegger, J. M., and Senior, F. A.,** Cross infection in pneumococcic pneumonia, *JAMA,* 112, 1127, 1939.
312. **Rudolph, A. H.,** Syphilis, in *Infectious Diseases,* Hoeprich, P. D., Ed., Harper & Row, Philadelphia, 1983, 611.
313. **Magnuson, H. J., Thomas, E. W., Olansky, S., Kaplan, B., deMello, L., and Cutler, J. C.,** Inoculation syphilis in human volunteers, *Medicine,* 35, 33, 1956.
314. **Kampmeier, R. H.,** Syphilis, in *Bacterial Infections of Humans. Epidemiology and Control,* Evans, A. S. and Feldman, H. A., Eds., Plenum Press, New York, 1982.
315. **Tramont, E. C.,** *Treponema Pallidum* (syphilis), in *Principles and Practice of Infectious Diseases,* Mandell, G. L., Douglas, R. G., and Bennett, J. E., Eds., John Wiley & Sons, New York, 1979.
316. **Knarberg Hansen, L. I.,** Koleraen i Christiania i 1853, Thesis, University of Oslo, 1986.
317. **Snow, J.,** Cholera and the water supply in the south districts of London, in 1854, *J. Public Health Sanit. Rev.,* 2, 239, 1856.
318. **Mhalu, F. S., Mtango, F. D., and Msengi, A. E.,** Hospital outbreaks of cholera transmitted through close person-to-person contact, *Lancet,* 2, 82, 1984.
319. **Ryder, R. W., Mizanur Rahman, A. S. M., Alim, A. R. M. A., Yunis, M. D., and Houda, B. S.,** An outbreak of nosocomial cholera in a rural Bangladesh hospital, *J. Hosp. Infect.,* 8, 275, 1986.
320. **Collins, C. H.,** *Laboratory-Acquired Infections,* Butterworths, London, 1983, 14.
321. **Hornick, R. B., Music, S. I., Wenzel, R., Cash, R., Libonati, J. P., Snyder, M. J., and Woodward, T. E.,** The Broad Street pump revisited: response of volunteers to ingested cholera vibrios, *Bull. N.Y. Acad. Med.,* 47, 1181, 1971.
322. **Butler, T.,** Plague and other *Yersinia* infections, in *Current Topics in Infectious Disease,* Greenough, W. B., III and Merigan, T. C., Eds., Plenum Press, New York, 1983,
323. **Tauxe, R. V., Wauters, G., Goossens, V., van Noyden, R., Vanderpitte, J., Martin, S. M., DeMol, P., and Thiers, G.,** *Yersinia enterocolitica* infections and pork: the missing link, *Lancet,* 1, 1129, 1987.
324. **Szita, J., Kali, M., and Redey, B.,** Incidence of *Yersinia enterocolitica* infection in Hungary: observations on the laboratory diagnosis of yersinosis, in *Contributions to Microbiology and Immunology,* Vol. 2, Winblad, S., Ed., S. Karger, Basel, 1973, 106.
325. **Toivanen, P., Olkkonen, L., Toivanen, A., and Aantaa, S.,** Hospital outbreak of *Yersinia enterocolitica* infection, *Lancet,* 1, 801, 1973.
326. **Anon.,** Human plague in 1981, *Wkly. Epidemiol. Rec.,* 57, 289, 1982.
327. **Marshall, J. D., Quy, D. V. and Gibson, F. L.,** Asymptomatic pharyngeal plague infection in Vietnam, *Am. J. Trop. Med. Hyg.,* 16, 175, 1967.
328. **Meyer, K. F.,** Pneumonic plague, *Bacteriol. Rev.,* 25, 249, 1961.
329. **Benenson, A. S., Ed.,** *Control of Communicable Diseases in Man,* American Public Health Association, Washington, D.C., 1985, 285.
330. **Dawson, C. and Darrell, R.,** Infections due to adenovirus type 8 in the United States. I. An outbreak of epidemic keratoconjuncivitis originating in a physician's office, *N. Engl. J. Med.,* 268, 1031, 1963.
331. **Laibson, P. R., Ortolan, G., and Dupré-Strachan, S.,** Community and hospital outbreak of epidemic keratoconjunctivitis, *Arch. Ophthal.,* 80, 467, 1968.

332. **Faden, H., Gallager, M., Ogra, P., and McLaughlin, S.,** Nosocomial Outbreak of Pharyngoconjunctival Fever due to Adenovirus Type 4 — New York, *Morbid. Mort. Wkly. Rep.*, 27, 49, 1978.
333. **Levandowski, R. A. and Rubenis, M.,** Nosocomial conjunctivitis caused by adenovirus type 4, *J. Infect. Dis.*, 143, 28, 1981.
334. **Larsen, R. A., Jacobson, J. T., Jacobson, J. A., Strikas, R. A., and Hierholzer, J. C.,** Hospital-associated epidemic of pharyngitis and conjunctivitis caused by adenovirus (21/H21 + 35), *J. Infect. Dis.*, 154, 706, 1986.
335. **Pavan-Langston, D.,** Ocular viral infections, *Med. Clin. North Am.*, 67, 973, 1983.
336. **Hatherley, L. I.,** Is primary cytomegalovirus infection an occupational hazard for obstetric nurses? A serologic study, *Infect. Control,* 7, 452, 1986.
337. **Adler, S.,** Nosocomial transmission of cytomegalovirus, *Pediatr. Infect. Dis.*, 5, 239, 1986.
338. **Haneberg, B., Bertnes, E., and Haukenes, G.,** Antibodies to cytomegalovirus among personnel at a children's hospital, *Acta Paediatr. Scand.*, 69, 407, 1980.
339. **Balfour, C. L. and Balfour, H. H.,** Cytomegalovirus is not an occupational risk for nurses in renal transplant and neonatal units, *JAMA,* 256, 1909, 1986.
340. **Votra, E. M., Rutala, W. A., and Sarubbi, F. A.,** Recommendations for pregnant employees interaction with patients having communicable infectious diseases, *Am. J. Infect. Control,* 11, 10, 1983.
341. **Ho, H. C., Kwan, H. C., Poon, Y. F., Tse, K. C., and No, M. H.,** Epstein-Barr virus infection in staff treating patients with nasopharyngeal carcinoma, *Lancet,* 1, 710, 1978.
342. **Goodman, R. A., Carter, C. C., Allen, J. R., Orenstein, W. A., and Finton, R. F.,** Nosocomial hepatitis A transmission by an adult patient with diarrhea, *Am. J. Med.*, 73, 220, 1982.
343. **Klein, B. S., Michaels, J. A., Rytel, M. W., Berg, K. G., and Davis, P. D.,** Nosocomial hepatitis A. A multinursery outbreak in Wisconsin, *JAMA,* 252, 2716, 1984.
344. **Mosley, J. W., Reiser, D. M., Brakoll, D., et al.,** Comparison of two lots of immun serum globulins for prophylaxis of infectious hepatitis, *Am. J. Epidemiol.*, 85, 539, 1968.
345. **Kuh, C. and Ward, W. E.,** Occupational virus hepatitis; an apparent hazard for medical personnel, *JAMA,* 143, 631, 1950.
346. **Trumbull, M. L. and Greiner, D. J.,** Homologous serum jaundice: an occupational hazard to medical personnel, *JAMA,* 145, 965, 1951.
347. **Dienstag, J. L. and Ryan, D. M.,** Occupational exposures to hepatitis B virus in hospital personnel: infection or immunization, *Am. J. Epidemiol.*, 115, 26, 1982.
348. **Janzen, J., Tripatzis, I., Wagner, U., Schlieter, M., Muller-Dethard, E., and Wolters, E.,** Epidemiology of hepatitis B surface antigen (HB_sAg) and antibody to HB_sAg in hospital personnel, *J. Infect. Dis.*, 137, 261, 1978.
349. **Zoulek, G.,** Vaccination against hepatitis B. An overview with consideration of the epidemiology of hepatitis B among medical personnel, *J. Oslo City Hosp.*, 36, 25, 1986.
350. **Villarejos, V. M., Visona, K. A., Gutiervez, A., and Rodriguez, A.,** Role of saliva, urine and feces in the transmission of type B hepatitis, *N. Engl. J. Med.*, 291, 1375, 1974.
351. **Dienstag, J. L., Werner, B. G., Polk, B. F., Snydmann, D. R., Craven, D. E., Platt, R., Crumpacker, C. S., Quellet-Hellstrom, R., and Grady, G. F.,** Hepatitis B vaccine in health care personnel: safety, immunogenicity, and indicators of efficacy, *Ann. Intern. Med.*, 101, 34, 1984.
352. Advisory Committee on Immunization Practices, Postexposure Prophylaxis of Hepatitis B, *Morbid. Mortal. Wkly. Rep.*, 33, 285, 1984.
353. CDC, Recommendation of the Immunization Practices Advisory Committee: Recommendation for Protection Against Viral Hepatitis, *Morbid. Mortal. Wkly. Rep.*, 34, 313, 1985.
354. **Maynard, J. E.,** Epidemic non-A, non-B hepatitis, *Semin. Liver Dis.*, 4, 336, 1984.
355. **Patterson, W. B., Craven, D. E., Schwartz, D. A., Nardell, E. A., Kasmer, J., and Noble, J.,** Occupational hazards to hospital personnel, *Ann. Intern. Med.*, 102, 658, 1985.
356. **Rosato, F. E. and Plotkin, S. A.,** Herpetic paronychia — an occupational hazard of medical personnel, *N. Engl. J. Med.*, 283, 804, 1970.
357. **McLean, D. M., Bach, R. D., Larke, R. P. B., and McNaughton, G. A.,** Mump meningo-encephalitis, Toronto, 1963, *Can. Med. Assoc. J.*, 90, 458, 1964.
358. **Candel, S.,** Epididymitis in mumps, including orchitis: further clinical studies and comments, *Ann. Intern. Med.*, 34, 20, 1951.
359. **Valenti, W. M., Hruska, J. F., Menegus, M. A., and Freeburn, M. J.,** Nosocomial virus infections. III. Guidelines for prevention and control of exanthematous viruses, gastrointesinal viruses, picornaviruses and uncommonly seen viruses, *Infect. Control,* 2, 38, 1981.
360. Centers for Disease Control, Poliomyelitis — Finland, *Morbid. Mortal Wkly. Rep.*, 34, 5, 1985.,
361. **Alkan, M., Morag, R., Rubinstein, E., and Derazne, E.,** Immunity of hospital personnel against polio virus, *Infect. Control,* 7, 27, 1986.
362. **Anderson, L. J., Winkler, W. G., Vernon, A. A., Helmick, C. G., and Roberts, M. R.,** Prophylaxis for persons in contact with patients who have rabies, *N. Engl. J. Med.*, 302, 967, 1980.

363. Centers for Disease Control, Recommendation of the Immunization Practices Advisory Committee (ACIP): Rabies Prevention — United States, 1984, *Morbid. Mortal. Wkly. Rep.*, 33, 393, 1984.
364. **Remmington, P. L., Shope, T., and Andrews, J.,** A recommended approach to the evaluation of human rabies exposure in an acute-care hospital, *JAMA*, 254, 67, 1985.
365a. **Hall, C. B. and Douglas, R. G., Jr.,** Nosocomial respiratory syncytial virus infections. The roles of gowns and masks in prevention, *Am. J. Dis. Child.*, 135, 512, 1981.
365b. **Agah, R., Cherry, J. D., Garakian, A. J., and Chapin, M.,** Respiratory syncytial virus (RSV) infection rate in personnel caring for children with RSV infections, *Am. J. Dis. Child.*, 141, 695, 1987.
365c. **Gala, C. L., Hall, C. B., and Schnabel, K. C.,** The use of eye-nose goggles to control nosocomial respiratory syncytial virus infection, *JAMA*, 256, 2706, 1986.
366. **Samadi, A. R., Huq, M. I., and Ahmed, Q. S.,** Detection of rotavirus in handwashings of attendants of children with diarrhoea, *Br. Med. J.*, 286, 188, 1983.
367. **Hjelt, K., Grauballe, P. C., Henriksen, L., and Krasilnikoff, P. A.,** Rotavirus infections among the staff of a general paediatric department, *Acta Paediatr. Scand.*, 74, 617, 1985.
368. **Parkman, P. D., Buescher, E. C., and Artenstein, M. S.,** Recovery of rubella virus from army recruits, *Proc. Soc. Exp. Biol. Med.*, 111, 225, 1962.
369. **Gregg, N. M.,** Congenital cataract following German Measles in the mother, *Trans. Opthalmol. Soc. Aust.*, 3, 35, 1941.,
370. Advisory Committee on Immunization Practices, Rubella Prevention, *Morbid. Mortal. Wkly. Rep.*, 33, 301, 1984.
371. **LeClair, J. M., Zaia, J. A., Levin, M. J., Congdon, D. E., and Goldman, D. A.,** Airborne transmission of chickenpox in a hospital, *N. Engl. J. Med.*, 302, 450, 1980.
372. **Gershon, A. A.,** The success of varicella vaccine, *Pediatr. Infect. Dis.*, 3, 500, 1984.
373. **Gajudusek, D. C.,** Unconventional viruses and the origin and disappearance of kuru, *Science*, 197, 943, 1977.
374. **Masters, C. L., Harris, J. O., Gajudusek, D. C., Gibbs, C. J., Jr., Bernoulli, C., and Asher, M.,** Creutz-Feldt Jacob disease: patterns of world wide occurrence and the significance of familial and sporadic clustering, *Ann. Neurol.*, 5, 177, 1979.
375. **Asher, D. M., Gibbs, C. J., Jr., and Gajudusek, D. C.,** Slow viral infections: safe handling of the agents of subacute spongiform encephalopathies, *Laboratory Safety: Principles and Practices*, Miller, B. M., Ed., American Society for Microbiology, Washington, D.C., 1986, 59.
376. **Miller, B. M., Ed**., *Laboratory Safety: Principles and Practices*, American Society for Microbiology, Washington, D.C., 1986.
377. **Monath, T. P.,** Lassa fever — new issues raised by field studies in West Africa, *J. Infect. Dis.*, 155, 433, 1987.
378. **Scherer, W. F.,** Laboratory safety for arboviruses and certain other viruses of vertebrates (The Subcommittee on Arbovirus Laboratory Safety of the American Committee on Arthropod-Borne Viruses), *Am. J. Trop. Med. Hyg.*, 29, 1359, 1980.
379. **Schwarz, J. and Baum, G. L.,** Blastomycosis, *Am. J. Clin. Pathol.*, 11, 999, 1951.
380. **Utz, J. P.,** Blastomycosis, in *Infectious Diseases*, Hoeprich, P. D., Ed., Harper & Row, Philadelphia, 1983, 479.
381. **Larson, D. M., Eckman, M. R., Alber, R. L., and Goldschmidt, V. G.,** Primary cutaneous blastomycosis: an occupational hazard to pathologists, *Am. J. Clin. Pathol.*, 79, 253, 1983.
382. **Baum, G. L. and Lerner, P. I.,** Primary pulmonary blastomycosis: a laboratory acquired infection, *Ann. Intern. Med.*, 73, 263, 1970.
383. **Hoeprich, P. D.,** Coccidioidomycosis, in *Infectious Diseases*, Hoeprich, P. D., Ed., Harper & Row, Philadelphia, 1983, 457.
384. **Eckmann, B. H., Schaefer, G. L., and Huppert, M.,** Bedside interhuman transmission of coccidioidomycosis via growth on fomites, *Am. Rev. Respir. Dis.*, 89, 175, 1964.
385. **Johnson, W. M.,** Occupational factors in coccidioidomycosis, *J. Occup. Med.* 23, 369, 1981.
386. **Fiese, M. J.,** *Coccidioidomycosis*, Charles C. Thomas, Springfield, Ill., 1958, 77.
387. **Johnson, J. E., Perry, J. E., Fekety, F. R., Kadull, P. J., and Cluff, L. E.,** Laboratory-acquired coccidioidomycosis. A report of 210 cases, *Ann. Intern. Med.*, 60, 941, 1964.
388. **Hanel, E. and Kruse, R. H.,** Laboratory-Acquired Mycoses, Fort Detrick Misc. Publ. No. 28, Department of the Army, Frederick, Md., 1967.
389. **Nabarro, J. D. N.,** Primary pulmonary coccidioidomycosis; a case of a laboratory infection in England, *Lancet*, 1, 982, 1948.
390. **Drouchet, E., Segretain, G., and Mariat, F.,** Coccidioidomycosis infection of a laboratory worker, *Bull. Soc. Mycol. France*, 3, 163, 1974.
391. **Wegmann, T. and Plempel, M.,** Das Krankenheitsbild der Coccidioidomykose, *Dtsch. Med. Wochenschr.*, 99, 1653, 1974.
392. **Smith, C. E., Pappagianis, D., Levine, H. B., and Saito, M.,** Human coccidioidomycosis, *Bacteriol. Rev.*, 25, 310, 1961.

393. **Overholt, C. E. L. and Hornick, R. B.,** Primary cutaneous coccidioidomycosis, *Arch. Intern. Med.*, 114, 149, 1964.
394. **Carrol, G. F., Haley, L. D., and Brown, J. M.,** Primary cutaneous coccidioidomycosis. A review of the literature and a report of a new case, *Arch. Dermatol.*, 13, 933, 1977.
395. **Furcolow, M. L., Guntheroth, W. G., and Willis, M. J.,** The frequency of laboratory infections with *Histoplasma capsulatum:* their clinical and X-ray characteristics, *J. Lab. Clin. Med.*, 40, 182, 1952.
396. **Furcolow, M. L.,** Airborne histoplasmosis, *Bacteriol. Rev.*, 25, 301, 1961.
397. **Tosh, F. E., Balhuizen, J., Yates, J. L., and Brasher, C. A.,** Primary cutaneous histoplasmosis: report of a case, *Arch. Intern. Med.*, 114, 118, 1964.
398. **Tesh, R. B. and Schneidau, J. D., Jr.,** Primary cutaneous histoplasmosis, *N. Engl. J. Med.*, 275, 597, 1966.
399. **Norden, A.,** Sporotrichosis: clinical and laboratory features and a serological study in experimental animals and humans, *Acta Pathol. Microbiol. Scand.*, Suppl. 89, 3, 1951.
400. **Mikkelsen, W. M., Brandt, R. L., and Harrell, E. R.,** Sporotrichosis: a report of 12 cases, including two with skeletal involvement, *Ann. Intern. Med.*, 47, 435, 1957.
401. **Thompson, D. W. and Kaplan, W.,** Laboratory-acquired sporotricosis, *Sabouraudia,* 15, 167, 1977.
402. **Fava, A.,** Un cas de sporotrichose conjunctivale et palpebrale primitives, *Ann. Ocul.*, 141, 338, 1909.
403. **Wilder, W. H. and McCullough, C. P.,** Sporotrichosis of the eye, *JAMA,* 62, 1156, 1914.
404. **Bernstein, B. and Mihan, R.,** Hospital epidemic of scabies, *J. Pediatr.*, 83, 1086, 1973.
405. **Gooch, J. J., Strasius, S. R., Beamer, B., Reiter, M. D., and Correll, G. W.,** Nosocomial outbreak of scabies, *Arch. Dermatol.*, 114, 897, 1978.
406. **Belle, E. A., D'Souza, T. J., Zarzour, J. Y., Lemieux, M., and Wong, C. C.,** Hospital epidemic of scabies: diagnosis and control, *Can. J. Publ. Health,* 70, 133, 1979.
407. **Pancoast, S. J. and Kishel, J. J.,** Patient-source scabies among hospital personnel — Pennsylvania, *JAMA,* 250, 1817, 1983.
408. **Koch, K. L., Phillips, D. J., Aber, R. C., and Current, W. L.,** Cryptosporidiosis in hospital personnel. Evidence for person-to-person transmission, *Ann. Intern. Med.*, 10, 2593, 1985.

Chapter 6

ANESTHETIC GASES

Christer Edling

TABLE OF CONTENTS

I. INTRODUCTION

The first report that indicated that health professionals could fall ill as a consequence of exposure to anesthetic gases in operating rooms came at the end of the 19th century. However, it was the physical properties of anesthetic agents that attracted the most interest. Particularly, questions concerning volatility, flammability, and dangers of explosion came to the fore and systems for elimination of anesthetic gases from operating rooms were described at the beginning of the 20th century. Since the mid 1950s, the anesthetic compounds have become less explosive and flammable through the introduction of halogenated anesthetic agents, and during the last decade, the interest has been focused on the hazards for health professionals with regard to pregnancy outcome, cancer risk, and impaired central nervous system (CNS) function.

II. MUTAGENIC AND CYTOGENIC EFFECTS

There are a few and contradictory studies on the mutagenicity of anesthetic agents in bacterial systems. Halothane, methoxyflurane, enflurane, trichlorethylene, and isoflurane were tested without any evidence of mutagenicity.[1-3] Fluroxene was mutagenic in the same test system[2] and another study has reported trichloroethylene as a mutagen. In the conventional Ames assay, the urine of anesthesiologists exposed to halothane induced mutations in one study,[4] whereas another reported negative findings.[5]

Sister chromatid exchanges (SCE) and sister exchange points (SCE points) in lymphocytes in peripheral blood can also be used as indicators of mutagenicity. No indication of such a mutagen effect after long-term exposure to waste anesthetic agents like halothane and nitrous oxide was found among persons exposed to anesthetics when compared to persons nonexposed.[6] In another study, no increase of chromosome gaps or SCE was reported among nurses exposed to anesthetic gases such as halothane, enflurane, and nitrous oxide.[7]

III. REPRODUCTIVE AND TERATOGENIC EFFECTS

A. Animal Tests

Several animal studies have indicated reproductive and teratogenic effects of nitrous oxide and halogenated anesthetic agents.[8,9] Since different rat and mice strains (with various susceptibility to chemicals) have been used and in most studies very high exposure levels were also used, it is difficult to compare the results of these animal experiments with what could be expected from the far lower exposure levels for occupationally exposed personnel. However, some recent studies have shown that exposure to trace levels of anesthetic gases might be hazardous. After continuous exposure of pregnant Wistar rats to nitrous oxide in air mixtures, the threshold value for an effect of fetal development, like reduced litter size, skeletal anomalies, and prenatal growth retardation, is reported to be somewhere between 500 and 1000 ppm.[10-12]

Following intermittent exposure, the threshold of nitrous oxide in air for reduction in litter size has been claimed to lie between 1000 and 5000 ppm[13] and the effect is dose dependent. Pregnant rats exposed to 10 ppm of halothane gave birth to fetuses with CNS damage.[13] The results of several other studies, on the other hand, do not support the hypothesis that trace levels of anesthetic agents adversely affect reproductive processes.[15-22] However, in most of these studies, mice strains have been used as test animals.

B. Epidemiology

In 1967, Vaisman[23] reported a Russian study where among 31 pregnant anesthesiologists, 1 pregnancy in 3 had ended in spontaneous abortion, 2 women gave birth prematurely, and in 1 case, the child had congenital malformations. The anesthetic gas in use during the period of exposure was mostly ether, but nitrous oxide and halothane had also been used.

This report was followed by a Danish study[24] where the spontaneous abortion frequency was significantly higher among exposed women (20%) when compared with nonexposed women (10%). The authors even reported a higher frequency of spontaneous abortions among nonexposed women whose husbands worked as anesthesiologists. Similar results were reported in an American study[25] in which wives of exposed dentists had a higher frequency of spontaneous abortion in comparison with nonexposed referents. In two English questionnaire surveys,[26,27] the results indicated a higher rate of spontaneous abortion among exposed women, whereas the husband's exposure did not seem to influence the outcome of pregnancy. The findings of a higher frequency of spontaneous abortion among exposed women have been reported in several studies from many countries, e.g., Czechoslovakia,[8] Finland,[28] the U.K.[29] Sweden,[30] West Germany,[31] and the U.S.[32]

However, two recent case control studies have reported negative findings. A Swedish study[33] claimed that work in anesthesiology or operating rooms had no effect on the incidence of hospitalization for miscarriage, perinatal deaths, or malformation detected in the neonatal period. A Finnish study[34] found no significant increase in risk of spontaneous abortion or malformations after exposure to anesthetic gases. Neither the results of another Finnish[35] nor a Belgian[36] study assess any stastically significant effects on pregnancy attributable to work in the operating theater.

In some of the aforementioned studies, a risk of malformations among children born by women exposed to anesthetic gases during pregnancy was present. In his study of English anesthesiologists, Tomlin[29] reported an increased frequency of malformations, especially in the CNS and musculoskeletal system. Pharoah et al.[37] surveyed the obstretic history of women physicians in England and Wales and reported that when conceptions had occurred when the mother was in an anesthetic appointment, this resulted in smaller babies, higher stillbirth rates, and more congenital malformations of the cardiovascular system than those of other women physicians. There was, however, no significant difference in the spontaneous abortion rate between the two groups. A Swedish survey[38] followed the pregnancy outcome, i.e., perinatal mortality and congenital malformations, for women working in medical occupations before giving birth. The outcome of pregnancies was checked with the aid of existing central registers and compared with the outcome of pregnancies for the whole country. There was no difference in the observed and expected number of perinatal mortality and congenital malformations among women working in anesthesia and operation departments.

Although there are methodlogical weaknesses in many of the epidemiological studies, i.e., low response rate, response bias, a lack of or poor definition of reference groups, or no information on the anesthetic agents used or the environmental concentrations of the gases, many studies indicate a higher risk of spontaneous abortion among women occupationally exposed to anesthetic gases during pregnancy. This is particularly true for most of the earlier studies (1967 to 1978) where the risk ratios are more than doubled in most studies. In the later studies (1979 and onward), there is a marked decrease in the risk and several studies are negative. This might be due to a better epidemiological methodology in the more recent studies, indicating that the earlier findings have been biased. Another interpretation could be that the observed decrease in risk reflects a better occupational hygienic standard with low exposure to anesthetic gases.

IV. CARCINOGENIC EFFECTS

The possibility of a carcinogenic effect of anesthetic gases also has attracted some attention. The anxiety about this effect is partly due to the structural similarities between known human carcinogens (i.e., dibromoethane, dichloroethane, bischloromethyleter, and chloromethyl

methyl ether) and several of the inhalation anesthetic agents now in use (i.e., isoflurane, methoxyflurane, and enflurane). Besides, anesthetic compounds can be transformed into reactive metabolites which can combine with tissue macromolecules[39] and turn into chemical carcinogens.[40]

A. Animal Tests

Studies on rats and mice have demonstrated that some volatile anesthetic agents are capable of inducing cancer. In an early experiment, Eschenbrenner[41] induced hepatoma by giving mice very high doses of chloroform perorally. In 1976 the National Cancer Institute (NCI) reported[42] animal tests in which rats have been given chloroform perorally in doses of one fifth to one tenth of those of Eschenbrenner. A significant increase in kidney tumors and hepatocellular carcinomas was observed. The methods of exposure were quite unusual when compared with occupational exposure and no other conclusion can be drawn on the basis of these studies than that the results indicate that the tested substance might have some carcinogenic potential.

Repeated exposure of the mother during pregnancy and of the offspring after delivery to subanesthetic concentrations (0.1 to 0.5%) of isoflurane induced pulmonary and hepatic neoplasms in the offspring of mice.[43] In another study,[44] mice were exposed to low doses (1/2 to 1/32 of the minimum alveolar anesthetic concentration, MAC) of either enflurane, isoflurane, halothane, methoxyflurane, or nitrous oxide, both in utero during the last half of pregnancy and after delivery. The result did not confirm the suggestion that isoflurane is a hepatocarcinogen, nor did the data suggest that the inhaled anesthetics currently in use pose a significant threat of carcinogenicity. Coate et al.[45] studied the effects on tumor incidence of prolonged and simultaneous exposure to halothane and nitrous oxide. Rats were exposed to filtered air, 1 ppm of halothane, and 50 ppm of nitrous oxide or 10 ppm of halothane and 500 ppm of nitrous oxide. Histologic evaluation of the reticuloendothelial system and of other major organs did not reveal any enhancement of the spontaneous tumor rate or any unusual neoplasms.

B. Epidemology

The mortality among U.S. anesthesiologists was studied in a retrospective cohort study.[46] An increased frequency of malignant disease in the lymphoid and reticuloendothelial systems was reported among 441 persons deceased in 1947 to 1966. A later follow-up of the mortality during 1967 to 1971 did not indicate any increase in deaths from tumors.[47] Another large American study followed the morbidity through a questionnaire survey of 73,496 persons and found that women, but not men, exposed to anesthetic gases had a higher frequency of cancer, especially leukemia and lymphoma.[48] In other U.S. studies.[49,50] the incidence of malignant diseases among anesthetic nurses was followed. Excluding skin cancer, a higher incidence of unusual tumors, such as malignant lymphoma, leiomyosarcoma, hepatocellular carcinoma, and cancer of the pancreas, was found.

Furthermore, the occurrence of cancer among children born by anesthetic nurses was studied, and for them, also, a higher cancer incidence was found. The same findings, cancer among children, were observed by Tomlin.[29] These data could indicate that anesthetic gases can act as a transplacental chemical carcinogen, as has been reported for some other substances in man, i.e., barbiturates,[51] diethylbestrol,[52] chlorinated pesticides,[53] and smoking.[54] Some U.K. studies[26,37,55] did not disclose any increase in cancer among anesthesiologists, nor did a study by Cohen et al.[32] regarding occupational diseases in dentistry.

Taken together, the results of a short-term test, genototoxic studies, animal tests, and epidemiology do not indicate any stronger risk for cancer after occupational exposure to anesthetic agents. However, better epidemiological studies are needed for a definite conclusion and to rule out a slightly increased risk of cancer in the lympoid and reticuloendothelial systems.

V. OTHER ORGAN EFFECTS

A. Central Nervous System (CNS)

Some industrial solvents now in use have been used earlier as anesthetic agents, i.e., trichloroethylene and chloroform. Solvents, as well as anesthetic agents, are fat-soluble, which is one of the conditions for the narcotic effects. Furthermore, anesthetic gases can form toxic metabolites[39] in the same manner as many solvents; these metabolites react on the cellular level and can induce cell damage.[56] Many studies have shown that long-term exposure to solvents can cause chronic damage effects on the CNS.[57] Today there are no studies on the chronic nervous effects on anesthetic personnel exposed for many years to anesthetic gases. However, animal as well as human experiments indicate that exposure to anesthetic gases does affect the CNS.

Animal experiments have shown CNS damage in rats after exposure to halothane in concentrations of 8 to 12 ppm for 8 h a day 5 d a week during 8 weeks.[58] Controversy exists over the effects of trace levels of anesthetic gases on performance. High levels of nitrous oxide, 20,000 ppm and above, cause a deterioration on tests of reaction time.[59,60] After exposing volunteers to a range of concentrations, Allison et al.[61] concluded that the threshold at which nitrous oxide started to affect performance was between 8000 and 12,000 ppm. Bruce et al.[62] and Bruce and Bach,[63] on the other hand, reported that exposure to 500 ppm of nitrous oxide for 4 h affected performance on the digit span test for short-term memory. In a later study, the same authors[64] claimed that exposure to 50 ppm of nitrous oxide for 4 h caused a deterioration in the performance of an audiovisual task and that 500 ppm also caused a deterioration in the performance of vigilance and tachistoscopic tasks and the digit span test. The effects were enhanced by simultaneous exposure to 1 ppm of halothane. Other studies[65-67] have failed to replicate the results of Bruce et al.,[62] even when using a design and a test battery which included an audiovisual task very similar or similar to that used by Bruce and Bach.[63]

Gamberale and Svensson[68] and Snyder et al.[69] found no acute behavioral effects of a trace level of anesthetic gases on operating theater personnel. In Finland, Korttila et al.[70] looked into the driving skills of operating room nurses after occupational exposure to halothane (0 to 43.7 ppm) and nitrous oxide (100 to 1200 ppm). The driving skills of 19 nurses were measured and compared to those of 11 nurses working in other wards at the same hospital, but with no exposure to anesthetic agents. The authors found no differences in driving skill between the two groups and stated that no impairment of driving skills should be expected after daily exposure to halothane and nitrous oxide among long-term employees working in operating rooms.

B. Peripheral Nervous System

Layzer et al.[71] reported a disabling peripheral neuropathy in three health workers who habitually abused nitrous oxide. Nerve conduction studies suggested an axonal rather than demyelinative neuropathy. The neurologic disorder improved slowly when the patients abstained from further nitrous oxide abuse. In another study, Layzer[72] reported a neurologic disorder in 15 patients after prolonged exposure to nitrous oxide — 13 of the patients had abused nitrous oxide to some extent for periods ranging from 3 months to several years, but 2 patients were exposed to nitrous oxide only professionally, as dentists, during work in poorly ventilated offices. Neurologic examinations showed sensorimotor polyneuropathy and a picture similar to that of subacute combined degeneration of the spinal cord. A dental survey in America suggested that long-term exposure to anesthetics among male dentists and female chairside assistants might result in an increased incidence of numbness, tingling, and muscle weakness.[32]

Interestingly, nitrous oxide neuropathy closely resembles the neuropathy of pernicious anemia and it has been shown that a few hours of exposure to concentrations of nitrous oxide greater than 1000 ppm result in oxidation of vitamin B_{12}, causing inactivation of methionine synthase, of which B_{12} is the cofactor.[73] This inactivation also affects folate metabolism and taken together these changes might explain not only findings of neuropathy, but also the megaloblastic changes and fetotoxicity.

C. Liver

In many animal experiments, liver effects have been reported after exposure to high doses of halothane and methoxyflurance[8] and also after long-term exposure to subanesthetic concentrations of halothane, isoflurance, and diethyl ether.[74,75] There are very few reports available on the effects on health professionals. A temporary increase of liver transaminases is reported among operating personnel working with halothane.[76] There are also case reports on icterus and liver cirrhosis among anesthetists exposed to halothane.[77,78] In some of the larger epidemiological studies in the U.S.[25,32,48] and Czechoslovakia,[8] an increased frequency of liver diseases among anesthesiologists or dental personnel was observed. Nunn et al.[73] reported that the hepatic function enzyme activities (measured as aspartate transaminase and gammaglutamyl transpeptidase) showed no significant difference between a group of ten members of an operating theater staff exposed to mean concentration of nitrous oxide ranging from 150 to 400 ppm, compared to a nonexposed group. In a Danish study,[79] no change was found in the hepatic microsomal function as assessed by the antipurine test in two groups with occupational exposure to less than 10 ppm of halothane and less than 100 ppm of nitrous oxide. A Finnish and an Italian study could not reveal any hepatic dysfunction among exposed anesthetists either.[80,81]

D. Kidneys

In animal experiments, kidney impairment has been reported after exposure to levels of halothane in the range of 100 to 500 ppm.[82] An effect on man has been reported on several occasions, where side effects have been noticed in patients anesthetized with methoxyflurane.[83,84] In 1968 a retrospective cohort study of anesthetic personnel reported an increase in chronic kidney diseases.[46] In the aforementioned American and Czechoslovakian studies, there was a higher frequency of kidney diseases among exposed persons.[8,25,32,48] In a Swedish report,[85] a moderate and temporary effect on kidney function was shown among personnel (and patients) exposed to methoxyflurane.

E. Bone Marrow

Animal studies and clinical experience on the effect of prolonged exposure to high concentrations of nitrous oxide have shown that leukopenia may be induced.[86,87] This effect has been used in the treatment of chronic and acute myeloid leukemia.[88] However, because of the rapid recovery of marrow, it has been ineffective as a choice of treatment for leukemia. There are no reports concerning bone marrow depression in health professionals.

VI. EXPOSURE AND SCAVENGING

When no scavenging is used, the levels of anesthetic gases in operation rooms are very variable and might be rather high.[8,70,76,81] Usually the levels of halothane are reported to be about 1 to 70 ppm and those of nitrous oxide are reported to be 400 to 3000 ppm. The highest levels are found in older operating rooms with poor general ventilation or when performing pediatric anesthesia. However, scavenging systems have been in use for a long time and better systems for close or local scavenging have been introduced[89-91] as well as a system with double mask.[92] The double mask can in comparison with a conventional mask reduce the amount of escaped anesthetic gases by 90%. When using these local scavenging systems, it is possible to reduce the exposure to levels well below 5 ppm of halothane and 100 ppm of nitrous oxide.

VII. THRESHOLD LIMIT VALUES AND MONITORING

Although adverse health effects of anesthetic gases have been discussed in many countries; quite a few have settled on threshold limit values (TLVs). In 1977, the National Institute for Occupational Safety and Health (NIOSH)[8] concluded (based on their compilation) that a safe level of exposure to the halogenated agents could not be defined, but recommended that exposure should be controlled to levels around 2 ppm and that the permissible level of exposure to nitrous oxide alone should be 25 ppm. Today (1988), however, there is still no TLV settled in the U.S., but a time-weighted average (TWA) of 50 ppm (400 mg/m^3) is proposed for halothane.

In Sweden, Denmark, and Norway, a TWA of 100 ppm (180 mg/m^3) for nitrous oxide has been in use since about 1981. These countries togehter with Holland, Switzerland, and West Germany also have a TWA of 5 ppm (40 mg/m^3) for halothane. In Sweden the TWA for enflurane is 10ppm (80 mg/m^3). In Australia a TWA of 0.5 ppm (4.1 mg/m^3) for halothane, 0.5 ppm (3.8 mg/m^3) for enflurane, and 25 ppm (45 mg/m^3) for nitrous oxide has been proposed (August, 1988).

The establishment of TLVs results in the requirement of simple methods for monitoring. Environmental monitoring is still the matter of choice, i.e., by pump-bag sampling and analyzing by an I.R. spectrophotometer or a gas chromatograph. Methods for diffuse or "passive" sampling are a rapidly developing technology and diffuse samples are now available for nitrous oxide,[93] halothane,[94] and enflurane.[95] These might prove to be an important and easy tool for environmental monitoring.

Some attempts have been made to use biological monitoring as an estimate of exposure. Sonander et al.[96] found a rather high correlation between N_2O in air and urine and stated that monitoring of urinary nitrous oxide with headspace extraction and gas chromatographic analysis might be a simple and accurate method for monitoring of exposure to nitrous oxide.

REFERENCES

1. **Baden, J. M., Brickenhoff, M., Wharton, R. S., Hitt, B. A., Simmon, V. F., and Mazze, R.,** Mutagenicity of volatile anesthetics: halothane, *Anesthesiology*, 45, 311, 1976.
2. **Baden, J. M., Kelly, M., Wharton, R. F., Hitt, B., Simmon, V. F., and Mazze, R.,** Mutagenicity of halogenated ether anesthetics, *Anesthesiology*, 46, 346, 1977.
3. **Waskell, L. A.,** A study of the mutagenicity of anesthetics and their metabolities, *Mutat. Res.*, 57, 141, 1978.
4. **McCoy, E. C., Hankel, R., Robbins, K., Rosenkrantz, H. S., Ginffrida, I. Y., and Bizarri, D. V.,** Presence of mutagenic substances in the urine of anesthesiologists, *Mutat. Res.*, 53, 71, 1978.
5. **Baden, J. M., Kelley, M., Cheung, A., and Mortelmans, K.,** Lack of mutagens in urines of operating room personnel, *Anesthesiology*, 53, 195, 1980.
6. **Husum, B. and Wulf, H. C.,** Sister chromatid exchanges in lymphocytes in operating room personnel, *Acta Anaesthesiol. Scand.*, 24, 22, 1980.
7. **Waksvik, H., Klepp, O., and Brogger, A.,** Chrosome analyses of nurses handling cytostatic agents, *Cancer Treat. Rep.*, 65, 607, 1981.
8. National Institute for Occupational Safety and Health, Occupational Exposure to Waste Anesthetic Gases and Vapors, U.S. Department of Health, Education and Welfare, Washington, D.C., 1977.
9. **Mazze, R. I., Fujinaga, M., Ricc, S. A., Harris, S. B., and Baden, J. M.,** Reproductive and teratogenic effects of nitrous oxide, halothane, isoflurane and enflurane in Sprague-Dawley rats, *Anesthesiology*, 64, 339, 1986.
10. **Vieira, E.,** Effect of the chronic administration of nitrous oxide 0.5% to gravid rats, *Br. J. Anaesth.*, 51, 283, 1979.

11. **Vieira, E., Cleaton-Jones, P., Austin, J., and Fatti, L.,** Intermittent exposure of gravid rats to 1% nitrous oxide and the effect on the post-natal growth of their offspring, *S. Afr. Med. J.*, 53, 106, 1978.
12. **Vieira, E., Cleaton-Jones, P., Austin, J. C., Moyes, D. G., and Shaw, R.,** Effects of low concentrations of nitrous oxide on rat fetuses, *Anesth. Analg.*, 59, 175, 1980.
13. **Vieira, E., Cleaton-Jones, P., and Moyes, D.,** Effects of low intermittent concentrations of nitrous oxide on the developing rat fetus, *Br. J. Anaesth.*, 55, 67, 1983.
14. **Chang, L., Dudley, A. W., Jr., Katz, J., and Martin, A. H.,** Nervous system development following in utero exposure to trace amounts of halothane, *Teratology,* 9, A15, 1974.
15. **Mazze, R. I., Wilson, A. I., Rice, S. A., and Baden, J. M.,** Reproduction and fetal development in mice chronically exposed to nitrous oxide, *Teratology*, 26, 11, 1982.
16. **Pope, W. D., Halsey, N. J., Landsdown, A. B., Simmonds, A., and Bateman, P. E.,** Fetotoxicity in rats following chronic exposure to halothane, nitrous oxide or methoxyflurane, *Anesthesiology*, 48, 11, 1978.
17. **Wharton, R. S., Mazze, R. I., Baden, K. M., Hitt, B. A., and Dooley, J. R.,** Fertility, reproduction and postnatal survival in mice chronically exposed to halothane, *Anesthesiology*, 48, 167, 1978.
18. **Wharton, R. S., Wilson, A. I., Mazze, R. I., Baden, J. M., and Rice, S. A.,** Fetal morphology in mice exposed to halothane, *Anesthesiology*, 51, 532, 1979.
19. **Wharton, R. S., Sievenpiper, T. S., and Mazze, R. I.,** Developmental toxicity of methoxyflurane in mice, *Anesth. Analg.*, 59, 421, 1980.
20. **Wharton, R. S., Mazze, R. I., and Wilson, A. I.,** Reproduction and fetal development in mice chronically exposed to enflurane, *Anesthesiology*, 54, 505, 1981.
21. **Richard, I. and Mazze, M. D.,** Fertility, reproduction, and postnatal survival in mice chronically exposed to isoflurane, *Anesthesiology*, 63, 663, 1985.
22. **Landsdown, A. B. G., Pope, W. D. B., Halsey, M. J., and Bateman, P. E.,** Analysis of fetal development in rats following maternal exposure to subanesthetic concentrations of halothane, *Teratology*, 13, 299, 1976.
23. **Vaisman, A. I.,** Working conditions in surgery and their effect on the effect on the health of anesthesiologists, *Eksp. Khir. Anesteziol.*, 12, 44, 1967 (in Russian with English summary).
24. **Askrog, V. and Harvald, B.,** Teratogenic effects of inhalation anaestetics, *Nord. Med.*, 83, 498 1970 (in Danish).
25. **Cohen, E. N., Brown, B. W., Bruce, D. L., Cascorbi, H. F., Corbett, T. H., Jones, T. W., and Whicher, C. H.,** A survey of anesthetic health hazards among dentists, *J. Am. Dent. Assoc.*, 90, 1291, 1975.
26. **Knill-Jones, R. P., Newman, B. J., and Spence, A. A.,** Anaesthetic practice and pregnancy, controlled survey of male anaesthetists in United Kingdom, *Lancet*, 2, 807, 1975.
27. **Knill-Jones, R. P., Rodriques, L. V., Moir, D. D., and Spence, A. A.,** Anaesthetic practice and pregnancy, controlled survey of women anaesthetists in the United Kingdom, *Lancet*, 1, 1326, 1972.
28. **Rosenberg, R. and Kirves, A.,** Miscarriages among operating theater staff, *Acta Anaesthesiol. Scand.*, 53, 37, 1973.
29. **Tomlin, P. J.,** Health problems of anaesthetists and their families in the West Midland, *Br. Med. J.*, 1, 779, 1979.
30. **Axelsson, G. and Rylander, R.,** Exposure to anaesthetic gases and spontaneous abortion: response bias in a postal questionnaire study, *Int. J. Epidemiol.*, 11, 250, 1982.
31. **Garstka, K., Wagner, K. L., and Hamacher, M.,** Pregnancy complications in anesthesiologists, paper presented at the 159th Meeting of the Lower Rhine, Westphalia, Gynecology and Obstetric Association, Bonn, West Germany, June 23, 1974.
32. **Cohen, E. N., Brown, B. W., Whittcher, C. E., Brodsky, J. B., Gift, H. C., Greenfield, W., Jones, T. W., and Driscoll, E. J.,** Occupational disease in dentistry and chronic exposure to trace anesthetic gases, *J. Am. Dent. Assoc.*, 101, 21, 1980.
33. **Ericson, A. and Källén, B., Hospitalization for miscarriage and delivery outcome Swedish nurses**
34. **Hemminki, K., Kyyrönen, P., and Lindbohm, M. L.,** Spontaneous abortions and malformations in the offspring of nurses exposed to anaesthetic gases, cytostatic drugs, and other potential hazards in hospitals, based on registered information of outcome, *J. Epidemiol. Community Health*, 39, 141, 1985.
35. **Rosenberg, P. M. and Vänttinen, H.,** Occupational hazards to reproduction and health in anaesthetists and paediatricians, *Acta Anaesthesiol. Scand.*, 22, 202, 1978.
36. **Lauwerys, R., Siddons, M., Mison, C. B., Borlee, I., Bouckaert, A., Lechat, M. F., and De Temmerman, O.,** Anaesthetic health hazards among Belgian nurses and physicians, *Int. Arch. Occup. Environ. Health*, 48, 195, 1981.
37. **Pharaoh, P. O. D., Alberman, E., Doyle, P., and Chamberlain, G.,** Outcome of pregnancy among women in anaesthetic practice, *Lancet*, 1, 34, 1977.
38. **Baltzar, B., Ericson, A., and Källén, B.,** Delivery outcome in women employed in medical occupations in Sweden, *J. Occup. Med.*, 21, 543, 1979.

39. **Cohen, E. N.,** Toxicity of inhalation anaesthetic agents, *Br. J. Anaesth.*, 50, 665, 1978.
40. **Boyland, E.,** Biochemistry of occupational cancer, *J. Soc. Occup. Med.*, 27, 97, 1977.
41. **Eschenbrenner, A. B.,** Induction of hepatomas in mice by repeated oral administration of chloroform with observations on sex differences, *J. Natl. Cancer Inst.*, 5, 251, 1945.
42. National Cancer Institute, Report on Carcinogenesis Bioassay of Chloroform, Carcinogenesis Program, Division of Cancer Cause and Prevention, Bethesda, MD, 1976.
43. **Corbett, T. H.,** Cancer and congenital anomalies associated with anesthetics, *Ann. N.Y. Acad. Sci.*, 271, 58, 1976.
44. **Eger, E. I., II, White, A. E., Brown, C. L., Biava, C. G., Corbett, T. H., and Stevens, W. C.,** A test of the carcinogenicity of enflurane, isoflurane, halothane, methoxyflurane and nitrous oxide in mice, *Anesth. Analg.*, 57, 678, 1978.
45. **Coate, W. B., Ulland, B. M., and Lewis, T. R.,** Chronic exposure to low concentrations of halothane, nitrous oxide: lack of carcinogenic effect in the rat, *Anesthesiology*, 50, 306, 1979.
46. **Bruce, D. L., Eide, K. A., Linde, H. W., and Eckenhoff, J. E.,** Causes of death among anesthesiologists — a 20-year survey, *Anesthesiology*, 29, 565, 1968.
47. **Bruce, D. L., Eide, K. A., Smith, M. J., Seltzer, F., and Dykes, M. H. M.,** A prospective survey of anesthesiologists mortality 1967—1971, *Anesthesiology*, 41, 71, 1974.
48. **Cohen, E. N., Brown, B. W., Bruce, D. L., Cascorbi, H. F., Corbett, T. H., Jones, T. W., and Whitcher, C. H.,** Occupational disease among operating room personnel: a national study, *Anesthesiology*, 41, 321, 1974.
49. **Corbett, T. H., Cornell, R. G., Endres, J. L., and Lieding K.,** Birth defects among children of nurse-anesthetists, *Anesthesiology*, 41, 341, 1974.
50. **Corbett, T. H., Cornell, R. G., Lieding, K., and Endres, J. L.,** Incidence of cancer among Michigan nurse-anesthetists, *Anesthesiology*, 38, 260, 1973.
51. **Gold, E., Gordis, L., Tonascia, I., and Szklo, M.,** Increased risk of brain tumors in children exposed to barbiturates, *J. Natl. Cancer Inst.*, 60, 1031, 1978.
52. **Herbst, A. L., Ulfelder, H., and Poskanzer, D. C.,** Adenocarcinoma of the vagina, association of maternal stilbestrol therapy with appearance in young women, *N. Engl. J. Med.*, 284, 878, 1971.
53. **Infante, P. F., Epstein, S. S., and Newton, W. A.,** Blood dyscrasia and childhood tumours in exposure to chlorodane and heptachlor, *Scand. J. Work Environ. Health*, 4, 137, 1978.
54. **Neutel, C. J. and Buck, C.,** Effect of smoking during pregnancy on the risk of cancer in children, *J. Natl. Cancer Inst.*, 47, 59, 1971.
55. **Doll, R. and Peto, R.,** Mortality among doctors in different occupations, *Br. Med. J.*, 1, 1433, 1977.
56. **Savolainen, H.,** Some aspects of the mechanism by which industrial solvents produce neurotoxic effects, *Chem. Biol. Interact.*, 18, 1, 1977.
57. Proceedings of the Workshop on Neurobehavioral Effects of Solvents, *Neurotoxicology*, 7, 1, 1986.
58. **Chang, L. W., Dudley, A. W., Jr., Lee, Y. K., and Katz, J.,** Ultrastructural changes in the nervous system after chronic exposure to halothane, *Exp. Neurol.*, 45, 209, 1974.
59. **Cook, T. L., Smith, M., Starkweather, J. A., Winter, P. M., and Eger, E. L.,** Behavioural effects of trace and subanesthetic halothane and nitrous oxide in man, *Anesthesiology*, 49, 419, 1978.
60. **Garfield, J. M., Garfield, F. B., and Sampson, J.,** Effects of nitrous oxide on decision-strategy and sustained attention, *Psychopharmacology (Berlin)*, 42, 5, 1975.
61. **Allison, R. H., Shirley, A. W., and Smith, G.,** Threshold concentration of nitrous oxide affecting psychomotor performance, *Br. J. Anaesth.*, 51, 177, 1979.
62. **Bruce, D. L., Bach, M. J., and Arbit, J.,** Trace anesthetic effects on perceptual cognitive and motor skills, *Anesthesiology*, 40, 453, 1974.
63. **Bruce, D. L. and Bach, M. J.,** Psychological studies of human performance as affected by traces of enflurane and nitrous oxide, *Anesthesiology*, 42, 194, 1975.
64. **Bruce, D. L. and Bach, M. J.,** Effects of trace anaesthetic gases on behavioural performance of volunteers, *Br. J. Anaesth.*, 48, 871, 1976.
65. **Frankhuizen, J. L., Vlek, C. A. J., Burm, A. G. L., and Rejger, V.,** Failure to replicate negative effects of trace anesthetics on mental performance, *Br. J. Anaesth.*, 50, 229, 1978.
66. **Smith, G. and Shirley, A. W.,** Failure to demonstrate effect of trace concentrations of nitrous oxide and halothane on psychomotor performance, *Br. J. Anaesth.*, 49, 65, 1977.
67. **Venables, H., Cherry, N., Waldron, H. A., Buck, L., Edling, C., and Wilson, H. K.,** Effects of trace levels of nitrous oxide on psychomotor performance, *Scand. J. Work Environ. Health*, 9, 391, 1983.
68. **Gamberale, F. and Svensson, G.,** The effect of anesthetic gases on the psychomotor and perceptual functions of anesthetic nurses, *Work Environ. Health*, 11, 108, 1974.
69. **Snyder, B. D., Thomas, R. S., and Gyorky, Z.,** Behavioural toxicity of anesthetic gases, *Ann. Neurol.*, 3, 67, 1978.
70. **Korttila, K., Pfäffli, P., Linnoila, M., Blomgren, E., Hänninen, H., and Häkkinen, S.,** Operating room nurses psychomotor and driving skills after occupational exposure to halothane and nitrous oxide, *Acta Anaesthesiol. Scand.*, 22, 33, 1978.

71. **Layzer, R. B., Fishman, R. A., and Schafer, J. A.,** Neuropathy following abuse of nitrous oxide, *Neurology,* 28, 504, 1978.
72. **Layzer, R. B.,** Myeloneuropathy after prolonged exposure to nitrous oxide, *Lancet,* 2, 1227, 1978.
73. **Nunn, J. F., Sharper, N., Royston, D., Watts, R. W. E., Purkiss, P., and Worth, H. G.,** Serum methionine and hepatic enzyme activity in anaesthetists exposed to nitrous oxide, *Br. J. Anaesth.,* 54, 593, 1982.
74. **Stevens, W. C., Eger, E. I., II, White, A., Halsey, M. J., Munger, W., Gibbons, R. D., Dolan, W., and Shargol, R.,** Comparative toxicities of halothane, isoflurane and diethyl ether at subanesthetic concentrations in laboratory animals, *Anesthesiology,* 42, 408, 1975.
75. **Plummer, J. L., Hall, P., Jenner, M. A., Ilsley, A. H., and Cousins, M. J.,** Effects of chronic inhalation of halothane, enflurane or isoflurane in rats, *Br. J. Anaesth.,* 58, 517, 1986.
76. **Götell, P. and Ståhl, R. R.,** Exposure of nurse anesthesists to halothane, *Lakartidningen,* 69, 6179, 1972 (in Swedish with English summary).
77. **Belfrage, S., Ahlgren, J., and Axelson, S.,** Halothane hepatitis in an anesthetist, *Lancet,* 2, 1466, 1966.
78. **Klatsking, G. and Kimberg, D. W.,** Recurrent hepatistis attributable to halothane sensitization in an anesthetist, *N. Engl. J. Med.,* 280, 515, 1969.
79. **Dossing, M. and Weihe, P.,** Hepatic microsmal enzyme function in technicians and anesthesiologists exposed to halothane and nitrous oxide, *Int. Arch. Occup. Environ. Health,* 51, 91, 1982.
80. **Rosenberg, P. H. and Oikkonen, M.,** Effects of working environment on the liver in 10 anaesthetists, *Acta Anaesthesiol. Scand.,* 27, 131, 1983.
81. **De Zotti, R., Negro, C., and Gobbato, F.,** Results of hepatic and hemopoietic controls in hospital personnel exposed to waste anesthetic gases, *Int. Arch. Occup. Environ. Health,* 52, 33, 1983.
82. **Chang, L. W., Dudley, A. W., Jr., Lee, Y. K., and Katz, J.,** Ultrastructural changes in the kidney following chronic exposure to low levels of halothane, *Am. J. Patol.,* 78, 225, 1975.
83. **Cousins, M. J., Greenstein, L. R., Hitt, B. A., and Mazze, R. I.,** Metabolism and renal effects of enflurane in man, *Anesthesiology,* 44, 44, 1974.
84. **Cousins, M. J. and Mazze, R. I.,** Methoxyflurane nephrotoxicity. A study of dose-response in man, *JAMA,* 225, 1611, 1973.
85. **Dahlgren, B.-E.,** Fluoride concentration in urine of delivery ward personnel following exposures to low concentrations of methoxyflurane, *J. Occup. Med.,* 21, 624, 1979.
86. **Green, C. D. and Eastwood, D. W.,** Effects of nitrous oxide inhalation on hemopoiesis in rats, *Anesthesiology,* 24, 341, 1963.
87. **Lassen, H. S. A., Henriksen, E., Neukirch, R., and Kristensen, H. S.,** Treatment of tetanus: severe bone marrow depression after prolonged nitrous oxide anaesthesia, *Lancet,* 1, 527, 1956.
88. **Eastwood, D. W., Green, C. D., Lambdin, M. A., and Gardner, R.,** Effect of nitrous oxide on the white cell count in leukemia, *N. Engl. J. Med.,* 268, 297, 1963.
89. **Bernow, J., Björdal, J., and Wiklund, K. E.,** Pollution of delivery ward air by nitrous oxide. Effects of various modes of room ventilation, excess and close scavenging, *Acta Anaesthesiol. Scand.,* 28, 229, 1984.
90. **Carlsson, P., Ljungqvist, B., and Hall, B.,** The effect of local scavenging on occupational exposure to nitrous oxide, *Acta Anaesthesiol. Scand.,* 27, 470, 1983.
91. **Nilsson, K., Sonander, H., and Stenqvist, O.,** Close scavenging of anaesthetic during mask anaesthesia. Further experimental and clinical studies of a method of reducing anaesthetic gas contamination, *Acta Anaesthesiol. Scand.,* 25, 421, 1981.
92. **Reiz, S., Gustavsson, A.-S., Häggmark, S., Lindkvist, A., Lindkvist, R., Norman, M., and Strömberg, B.,** The double mask — a new local scavenging system for anaesthetic gases and volatile agents, *Acta Anaesthesiol. Scand.,* 30, 260, 1986.
93. **Cox, P. C. and Brown, R. H.,** A personnel sampling method for the determination of nitrous oxide exposure, *Am. Ind. Hyg. Assoc. J.,* 45, 345, 1984.
94. **Purnell, C. J., Wright, M. D., and Brown, R. H.,** Performance of the Porton Down charcoal cloth diffusive sampler, *Analyst,* 106, 590, 1981.
95. **Norbäck, D. and Michel, J.,** Diffuse sampling of halothane and enflurane, in *Diffuse Sampling. An Alternative Approach to Workplace Air Monitoring,* Berlin, A., Brown, R. H., and Saunders, K. J., Eds., Royal Society of Chemistry, London, 1987, 256.
96. **Sonander, H., Stenqvist, O., and Nilsson, K.,** Exposure to trace amounts of nitrous oxide. Evaluation of urinary gas content monitoring in anaesthetic practice, *Br. J. Anaesth.,* 55, 1225, 1983.

Chapter 7

ANTINEOPLASTIC AGENTS

John Widström and Christer Edling

TABLE OF CONTENTS

I. INTRODUCTION

The use of antineoplastic agents in the treatment of malignant diseases has increased substantially during the last 20 years. Antineoplastic drugs may be administered alone, together with other antineoplastic agents, or in combination with other treatments. Antineoplastic drugs consist of many chemically unrelated substances with different modes of action.

The common feature of antineoplastic agents is the ability to disturb growth and kill tumor cells by blocking or interfering with different biochemical pathways. These pathways are present not only in cancer cells, but also in other cells. The selectivity of the drug between cancer and normal cells depends on differences in activity of these pathways. Rapidly dividing cells, e.g., cells from bone marrow, hair follicles, testes, and fetus, are therefore usually more affected by antineoplastic drugs than other cells are.

All antineoplastic agents cause adverse effects of some kind when therapeutically administered to patients. These effects range from nausea to the induction of secondary cancers. Mutagenic, carcinogenic, and teratogenic properties are also documented for many of these substances through studies *in vitro* and *in vivo*. A summary of mutagenic, carcinogenic, and teratogenic properties of some antineoplastic drugs is presented in Table 1.

Is there a health risk for health care personnel handling these potent drugs and if that is the case, how can this risk be eliminated? Research on exposure and uptake of antineoplastic agents and on the effects in occupationally exposed health professionals has been reviewed in several articles.[10-17] Guidelines and recommendations for safe handling of antineoplastic agents have also been proposed by both organizations and researchers.[18-28]

II. ROUTES OF EXPOSURE

Antineoplastic agents are prepared and used in many hospitals, clinics, and pharmacies. Personnel handling these agents are, for example, nurses, pharmacists, and physicians. Antineoplastic drugs are usually dissolved in a liquid for intravenous administration to the patient. It is the preparation of these solutions that possesses the greatest hazard to health care personnel. Another important source of exposure is the administration of solubilized drugs and antineoplastic drugs in tablet form to the patient. The person preparing or administering drugs may be exposed through direct skin contact or through inhalation of an aerosolized drug. Direct skin contact may occur after spills or from spraying and spattering during preparation. Use of contaminated work clothes or penetration of protective gloves[29] may also cause contact with the skin.

Aerosols may be created in many ways, e.g., by dissolving and diluting drugs, insertion or withdrawal of an injection syringe from drug vials with rubber seals, transferring a drug with a syringe, expelling air and excess liquid from a syringe, and by opening glass ampules with snap-off tops. Drugs can also be released to the environmental air from trash receptacles with discarded drug vials and syringes or from other contaminated objects and areas.

Another source of exposure to health care staff is contact with vomitus, urine, and feces from patients undergoing chemotherapy that may be contaminated with unmetabolized or metabolized rests of the given drug. Uptake due to ingestion of contaminated foodstuff may in some cases be of importance. In a survey of drug handling practices at ten oncology clinics in the Chicago area in 1981, de Werk Neal et al.[30] found that refrigerators were used for both food and drugs and that eating and drinking occurred in preparation areas.

III. LEVELS OF EXPOSURE AND UPTAKE

There are a few studies where exposure levels and uptake of antineoplastic drugs have been measured directly in air and/or in the urine of exposed personnel. These studies show

Table 1
TOXIC PROPERTIES OF SOME ANTINEOPLASTIC AGENTS — A SUMMARY OF POSITIVE FINDINGS

Agent	Mutagenic[a]	Chromosomal effects[b]	Carcinogenic[c]	Teratogenic[d]	Ref.
Actinomycin D	—	Chr ab	+ r, m; (+) hum	+ sev sp	7
Adriamycin	+	Chr ab, SCE	+ r	—	1,3-5,9
Azacytidine	+	—	(+) m	+ m	8
Azathioprine	+	—	(+) m, r; + hum	+ sev sp	8
Bleomycin	—	SCE	—	—	9
Busulfan	+	Chr ab, SCE	(+) m; + hum	—	9
Carmustine (BCNU)	+	Chr ab	+ r	+ r	3,8,9
Chlorambucil	+	Chr ab	(+) m, r; (+) hum	—	8
Cisplatin	+	Chr ab	—	—	8
Cyclophosphamide	+	Chr ab, SCE	+ m, r; + hum	+ sev sp	1,3-5,8
Dacarbazine	+	—	+ m, r	+ sev sp	3,8
Daunorubicin	+	Chr ab	—	—	1,3-5,7
Fluorouracil	—	—	—	+ sev sp	8
Isophosphamide	+	Chr ab	(+) m, r	+ m	1,8
Lomustine (CCNU)	+	SCE	+ r	+ r	2,8
Melphalan	+	Chr ab, SCE	+ m, r; + hum	—	1,3,9
Mercaptopurine	+	Chr ab	—	+ sev sp	4,5,8
Methotrexate	+	Chr ab	—	+ sev sp, + hum	9
Mytomycin C	+	Chr ab	—	—	5, 7
Prednisone	—	—	—	+ rod	9
Procarbazine	+	—	+ m, r	+ r	8
Streptozotocin	+	—	—	—	9
Thiotepa	+	Chr ab	+ m, r	+ m, r	1,4,8
Treosulfan	—	Chr ab	+ hum	—	8
Uracil mustard	+	—	+ m, r	+ r	1,6,9
Vinblastine sulfate	—	—	—	+ sev sp	8
Vincristine sulfate	—	—	—	+ sev sp	8

[a] +, mutagenic to bacterial or mammalian cells in culture.
[b] Chr ab, increased incidence of chromosomal aberrations; SCE, increased incidence of sister-chromatid exchange.
[c] +, sufficient evidence; (+), limited evidence for carcinogenicity to mice (m), rats (r), or humans (hum) according to the International Agency for Research on Cancer (IARC).
[d] +, teratogenic to mice (m); rats (r); rodents (rod); several animal species (sev sp); or humans (h).

that measurable levels of antineoplastic agents may be present both in environmental air during drug preparation and in the urine of the preparer.

In a study by de Werk Neal and co-workers,[30] air levels of fluorouracil and cyclophosphamide were measured in the medication-preparation room of three oncology clinics. The samples were collected during 40- or 80-h periods and the total time of monitoring was 320 h. Fluorouracil was detected in 200 of 320 h in concentrations ranging from 0.1 to 82 ng/m^3. Cyclophosphamide was present in air in 80 h at 370 ng/m^3. Measured levels of the two substances are relatively low, but this shows the possibility of air contamination and subsequent exposure of the personnel.

In another study, Kleinberg and Quinn[31] measured the concentration of fluorouracil inside a horizontal laminar flow hood during normal mixing procedures. No measurements of breathing air concentrations were made. Detectable amounts of fluorouracil were found in five out of nine trials (0.00 to 0.07 μg/l). In this type of hood, filtered air is blown toward the operator to hinder contamination of drugs being prepared. Aerosolized drugs from inside the hood may thus reach and be inhaled by the operator. The authors suggest that the differences in air levels inside the hood depended on handling techniques and on the type of vial and rubber seal used.

Absorption of anticancer drugs in health care personnel has been studied by Hirst et al.[32] They studied uptake in nurses during regular working conditions and cutaneous uptake in volunteers. From two nurses handling anticancer drugs in an oncology clinic, urine was sampled during a 6-week period and the samples were analyzed for cyclophosphamide. Samples were collected in the morning and after preparation of cyclophosphamide mixtures. The nurses prepared the drugs in an open room with no special ventiliation. Cyclophosphamide was found in 1 (2%) of the 57 morning samples and in 7 (23%) of the 30 samples collected after preparation of the drug. The urine concentrations were not related to the amount of drug handled.

Cutaneous uptake of cyclophosphamide was studied in five volunteers by topical application to the cubital fossa area. Urine samples were collected over a 24-h period at 2-h intervals for the first 6 h and when possible they were collected during the next 18 h. Cyclophosphamide was in most cases present only in samples collected 6 h or more after application. The authors suggest that the faster appearance of cyclophosphamide in urine of the nurses than in urine of the volunteers is due to different routes of absorption. Inhalation of aerosolized drug might be a possible way of uptake in the nurses.

IV. BIOLOGICAL MARKERS AS INDICATORS OF EXPOSURE AND UPTAKE

The uptake of antineoplastics has been indirectly indicated in studies of chromosomal aberrations, sister-chromatid exchange (SCE) rates, urinary thioether excretion, and mutagenic activity in urine. Sorsa and co-authors[33] report sampling of lymphocytes from exposed nurses for testing of SCE rates. Patients undergoing chemotherapy were used as positive controls and nonexposed health care and office personnel were used as negative controls. Exposed nurses had significantly raised levels of SCE compared to office workers. The difference was not significant between exposed and nonexposed health care personnel. The patients constituting the positive control group had exchange rates four to five times higher than the other groups.

In a French investigation,[34] several biological markers were studied in a group of nurses occupationally exposed to antineoplastic drugs. The markers studied were urine mutagenicity, chromosomal abnormalities, and SCE in lymphocytes. A statistically significant increase in urine mutagenicity of the 17 exposed nurses was found when compared to a nonexposed control group. Comparisons of chromosomal abnormalities and SCE rates revealed no differences between the groups. However, the occurrence of the three studied phenomena in smoking nurses was significantly increased when compared with nonsmokers.

Waksvik et al.[35] found a significant increase in the number of chromatid gaps and SCEs in a group of ten nurses with long-term handling of anticancer drugs when compared with a control group of female clerks. This difference was not found between a group of nurses with shorter period of handling of anticancer drugs and the control group.

Thioethers in urine have been proposed as an indicator of exposure to alkylating agents. Jagun and co-workers[36] studied thioether excretion in 15 nurses who regularly handled anticancer drugs. Preexposure samples were collected after at least 3 work-free days and postexposure samples were collected after the nurse had worked 5 or more days in a spell. Twenty nurses without contact with anticancer drugs were used as a reference group. The thioether levels were significantly increased in postexposure samples when compared with both preexposure and reference samples. No significant difference was found between the two latter groups of samples. Smoking habits were not found to correlate with levels of thioether in urine.

Mutagenic urine of nurses handling antineoplastic substances was first reported by Falck and co-authors in 1979.[37] They compared the mutagenicity in urine of patients undergoing chemotherapy, nurses administering anticancer drugs, and nonexposed controls. Mutagenic activity, measured by the bacterial fluctuation test, was found in the urine of all patients and most nurses, but was significantly higher among the former. No activity was found in the urine of nonsmoking controls. Other authors also described increased mutagenicity in urine of exposed personnel[34,38-41] while others find no difference between exposed and nonexposed groups.[33,43-44] This inconsistency may have several explanations, e.g., differences in exposure, test systems with unalike sensitivity, and high background mutagenicity rates due to flaws in the test system.

Several researchers have shown that it is possible to eliminate urinary mutagenicity by improvement of working conditions. Andersson and co-authors[38] (also reported by Nguyen et al.[41]) compared the urine mutagenicity in urine of six persons preparing mixtures of anticancer drugs in either horizontal laminar-flow hoods or vertical-flow biological safety cabinets. Mutagenicity was seen in the urine of all six when preparing drugs both with and without gloves or masks in the horizontal-flow hood. No mutagenicity was observed when preparations were made in the vertical-flow cabinet using gloves. The same group also found an increase in mutagenic activity toward the end of the week. After a duty-free weekend, the activity returned to the spontaneous level. Kolmodin-Hedman et al.[40] observed that urinary mutagenicity in pharmacy prescriptionists was eliminated by changing glove material and improvement of ventilation in the safety cabinet used.

The Finnish researchers,[33] who first reported on mutagenicity in urine of nurses exposed to antineoplastic drugs, measured urinary mutagenicity in the same group after introducing new handling instructions and the use of a laminar-flow chamber, gloves, and protective clothing during preparation. Mutagenicity was considerably lowered after these improvements.

V. EFFECTS IN HEALTH-CARE PERSONNEL

A. Acute Effects

Several pharmacists complained of lightheadedness, dizziness, and facial flushing while preparing admixtures of antineoplastic drugs dacarbazine and cisplatin. The problems started when the pharmacy department of an obstetrical-gynecological hospital became responsible for preparations of antineoplastic drugs. No complaints were reported after the installation and use of a glovebox.[45]

Reynolds et al.[46] report of different symptoms in pharmacy and nursing personnel mixing the anticancer drug AMSA (Amsacrine). One 21-year-old pharmacy technician developed an urticarial rash on the arms and trunk in direct connection with mixing AMSA in a horizontally ventilated hood. The rash reappeared on two occasions in the following week while he was preparing the drug. He used surgical gloves at the first of these occasions and wore gloves, surgical mask, and surgical gown the second time. A female pharmacist felt nauseated and vomited 4 h after having spilled AMSA on her hands and clothing. She washed her hands directly after the spillage, but did not change clothing.

Reynolds et al.[46] also report of seven pharmacy technicians and nurses responsible for AMSA preparations who developed specific symptoms 5 min to 3 h after having handled the drug or been close to the mixing. The symptoms included nausea, lightheadedness, headache, epigastric pain, and malaise. After these episodes, it was recommended that all mixing of AMSA be done in vertically ventilated hoods. No symptoms were observed in personnel handling AMSA during the next 3 months.

B. Effects in Pregnancy

In 1985 a statistically significant association between fetal loss and occupational exposure to antineoplastic agents in the first trimester was observed by Selevan et al.[47] in a matched case control study. The studied pregnancies were identified through Finnish national registers of health care personnel and hospital discharges and also through multiclinic data. Self-administered questionnaires were mailed to nurses at 17 Finnish hospitals with fetal losses. A response rate of 87% was obtained after three mailings. Drugs associated with fetal loss were cyclophosphamide, doxorubicin (adriamycin), and vincristine. No particular drug could be connected with specific effects since most of the nurses reported having handled several drugs.

C. Liver Toxicity

Sotaniemi et al.[48] reported that three nurses from an oncology department developed liver injuries. The nurses had prepared anticancer drugs daily for 6, 8, and 16 years. All three presented neurological symptoms and increased serum levels of alanine aminotransferase (ALAT) and alkaline phosphatase (ALP). Liver histology showed portal hepatitis with piecemeal necrosis in one case and hepatic fibrosis with accumulation of fat in the other two. According to the authors, these findings suggest that the handling of anticancer drugs may lead to irreversible fibrosis of the liver.

VI. SAFE HANDLING OF ANTINEOPLASTIC AGENTS

Occupational exposure of hospital professionals to antineoplastic agents is most prominent during preparation and administration. Studies of both acute effects and of urinary mutagenicity show that the observed effects can be eliminated or decreased, e.g., by improved routines for drug handling, the use of vertical flow safety cabinets or hoods in drug preparation, and by use of gloves with good resistance to drug penetration. The routines and safety precautions used for limiting exposure to antineoplastic agents vary from one workplace to another.

Stellman et al.[49] studied handling practices at five American hospitals and found that of all preparations of antineoplastic drugs, 80% were made under a vertical laminar-flow hood, 3% under a horizontal flow hood, and 17% in an open area. Many of the vertical laminar-flow hoods were installed in busy walk-through areas which affects the capacity of the hoods in a negative way. Gloves were used for protection during preparation by 75% of those surveyed, but none used gloves while administering drugs. In a survey of ten oncology clinics in Chicago by de Werk Neal et al.,[30] a ventilated hood was used in only one clinic and gloves were used routinely for drug preparations at four.

VII. GUIDELINES AND RECOMMENDATIONS

The following recommendations are a summary of several guidelines published with the aim of minimizing occupational exposure to antineoplastic agents and at the same time maintaining drug sterility. Guidelines have been formulated by the National Institutes of Health and also by several private organizations and researchers.[18-28]

1. Handling of antineoplastic drugs should only be performed by personnel trained in the proper use of safety equipment and the handling of antineoplastic agents.
2. During pregnancy and breast-feeding, personnel should be given duties without contact with antineoplastic agents.
3. Both personnel preparing and administering drugs should wear protective clothing and gloves. Contaminated clothes and gloves should be changed.
4. Hands should be washed before and after drugs have been handled. After direct contact with drugs, the affected skin areas should be washed thoroughly with soap and water.
5. All drug manipulations should be done inside a ventilated safety cabinet where incoming air is filtered through a high-efficiency particulate filter (HEPA), blown away from the manipulator, and exhausted to the outside.
6. The work area should be covered by plastic-backed paper that should be changed after larger spills and also at the end of each workshift.
7. Vials should be ventilated before use to lower internal pressure. A sterile, alcohol-dampened pledget should be wrapped around the top of the vial and the needle during withdrawal from the septum. The pledget should be held against the needle tip when ejecting air bubbles from the syringe. The neck of glass ampules with snap-off tops should be wrapped with an alcohol-dampened pledget when opening.
8. Needles and syringes should be disposed of intact to minimize formation of aerosols. Contaminated materials should be placed in airtight bags and together with waste antineoplastic drugs should be disposed of in accordance with federal and state requirements applicable to toxic chemical waste.

REFERENCES

1. **Benedict, W. F., Baker, M. S., Haroun, L., Choi, E., and Ames, B. N.,** Mutagenicity of cancer chemotherapeutic agents in the *Salmonella*/microsome test, *Cancer Res.,* 37, 2209, 1977.
2. **Franza, B. R., Oeschger, N. S., Oeschger, M. P., and Schein, P. S.,** Mutagenic activity of nitrosurea antitumour agents, *J. Natl. Cancer Inst.,* 65, 149, 1980.
3. **Matheson, D., Brusick, D., and Carrano, R.,** Comparison of the relative mutagenic activity for eight antineoplastic drugs in the Ames *Salmonella*/microsome and TK+/mouse lymphoma assays, *Drug Chem. Toxicol.,* 3, 277, 1978.
4. **Pak, K., Iwasaki, T., Miyakawa, M., and Yoshida, O.,** The mutagenic activity of anti-cancer drugs and the urine of rats given these drugs, *Urol. Res.,* 7, 119, 1979.
5. **Seino, Y., Nagao, M., Yahagi, T., Hoshi, A., Kawachi, T., and Sugimura, T.,** Mutagenicity of several classes of antitumour agents to *Salmonella typhimurium* TA98, TA100 and TA92, *Cancer Res.,* 38, 2148, 1978.
6. International Agency for Research on Cancer, Evaluation of carcinogenic risk of chemicals to man: some naturally occurring substances, *IARC (Int. Agency Res. Cancer) Monogr.* 9, 1975.
7. International Agency for Research on Cancer, Evaluation of carcinogenic risk of chemicals to man: some aziridines, N-, S- & O-mustards and selenium, *IARC (Int. Agency Res. Cancer) Monogr.,* 10, 1976.
8. International Agency for Research on Cancer, Evaluation of carcinogenic risk of chemicals to man: some antineoplastic and immunosuppressive agents, *IARC (Int. Agency Res. Cancer) monogr.,* 26, 1981.
9. International Agency for Research on Cancer, Evaluation of carcinogenic risk of chemicals to man: chemicals, industrial processes and industries associated with cancer in humans, *IARC (Int. Agency Res. Cancer) Monogr. Suppl.* 4, 1982.
10. **D'Arcy, P. F.,** Reactions and interactions in handling anticancer drugs, *Drug Intell. Clin. Pharm.,* 17, 532, 1983.
11. **Hillcoat, B. L., Levi, J., and Snyder, R.,** Preparation and administration of antineoplastic agents. Risks and recommendations, *Med. J. Aust.,* 1, 424, 1983.

12. **Knowles, R. S. and Virden, J. E.,** Handling of injectable antineoplastic agents, *Br. Med. J.*, 281, 589, 1980.
13. **Mattia, M. A. and Blake, S. L.,** Hospital hazards: cancer drugs, *Am. J. Nurs.*, 83, 759, 1983.
14. **Reich, S. D.,** Antineoplastic agents as potential carcinogens: are nurses and pharmacists at risk?, *Cancer Nurs.*, 4, 500, 1981.
15. **Stellman, J. M. and Zoloth, S. R.,** Cancer chemotherapeutic agents as occupational hazards: a literature review, *Cancer Invest.*, 4, 127, 1986.
16. **Vainio, H.,** Inhalation anesthetics, anticancer drugs and sterilants as chemical hazards in hospitals, *Scand. J. Work Environ. Health,* 8, 94, 1982.
17. **Wilson, J. P. and Solimando, D. A.,** Antineoplastics: a safety hazard, *Am. J. Hosp. Pharm.*, 38, 624, 1981.
18. **Harrison, B. R.,** Developing guidelines for working with antineoplastic drugs, *Am. J. Hosp. Pharm.*, 38, 1686, 1981.
19. Canadian Society of Hospital Pharmacists, Guidelines for the handling of hazardous pharmaceuticals, *Can. J. Hosp. Pharm.*, 34, 126, 1981.
20. Council on Scientific Affairs, Division of Drugs and Technology, American Medical Association, Guidelines for handling parenteral antineoplastics, *JAMA,* 253, 1590, 1985.
21. **Jones, B. R., Frank, R., and Mass, T.,** Safe handling of chemotherapeutic agents: a report from the Mount Sinai Medical Center, *Ca.*, 33, 258, 1983.
22. **Hillcoat, B. L., Levi, J., and Snyder, R.,** Guidelines and recommendations for handling of antineoplastic agents, *Med. J. Aust.*, 1, 426, 1983.
23. **Hoffman, D. M.,** The handling of antineoplastic drugs in a major cancer center, *Hosp. Pharm.*, 15, 302, 1980.
24. National Board of Occupational Safety and Health, Regulations Concerning Cytostatica, Stockholm, Sweden, 1984, 8 (in Swedish).
25. Division of Safety, National Institutes of Health, Recommendations for the Safe Handling of Parenteral Antineoplastic Drugs, U.S. Government Printing Office, Washington, D.C., 1982.
26. **Stolar, M. H. and Power, L. A.,** Safe handling of cytotoxic drugs in hospitals, *Qual. Rev. Bull.*, 209, 1984.
27. The Society of Hospital Pharmacists of Australia's Specialty Practice Committee on Parenteral Service, Guidelines for safe handling of cytotoxic drugs in pharmacy departments and hospital wards, *Hosp. Pharm.*, 16, 17, 1981.
28. **Zimmerman, P. F., Larsen, R. K., Barkley, E. W., and Gallelli, J. F.,** Recommendations for the safe handling of injectable antineoplastic drug products, *Am. J. Hosp. Pharm.*, 38, 1693, 1981.
29. **Laidlaw, J. L., Connor, T. H., Theiss, J. C., Anderson, R. W., and Matney, T. S.,** Permeability of latex and polyvinyl chloride gloves to 20 antineoplastic drugs, *Am. J. Hosp. Pharm.*, 41, 2618, 1984.
30. **de Werk Neal, A., Wadden, R. A., and Chiou, W. L.,** Exposure of hospital workers to airborne antineoplastic agents, *Am. J. Hosp. Pharm.*, 40, 597, 1983.
31. **Kleinberg, M. L. and Quinn, M. J.,** Airborne drug levels in a laminar-flow hood, *Am. J. Hosp. Pharm.*, 38, 1301, 1981.
32. **Hirst, M., Tse, S., Mills, D. G., Levin, L., and White, D. F.,** Occupational exposure to cyclophosphamide, *Lancet,* 1, 186, 1984.
33. **Sorsa, M., Falck, K., Norppa, H., and Vainio, H.,** Monitoring genotoxicity in the occupational environment, *Scand. J. Work Environ. Health,* 7(4), 61, 1981.
34. **Stucker, I., Hirsch, A., Doloy, T., Bastie-Sigeac, I., and Hemon, D.,** Urine mutagenicity, chromosomal abnormalities and sister chromatid exchanges in lymphocytes of nurses handling cytostatic drugs, *Int. Arch. Occup. Environ. Health,* 57, 195, 1986.
35. **Waksvik, I., Klepp, O., and Brøgger, A.,** Chromosome analyses of nurses handling cytostatic agents, *Cancer Treat. Rep.*, 65, 607, 1981.
36. **Jagun, O., Ryan, M., and Waldron, H. A.,** Urinary thioether excretion in nurses handling cytotoxic drugs, *Lancet,* 2, 443, 1982.
37. **Falck, K., Gröhn, P., Sorsa, M., Vainio, H., Heinonen, E., and Holsti, L. R.,** Mutagenicity in urine of nurses handling cytostatic drugs, *Lancet,* 1, 1250, 1979.
38. **Anderson, R. W., Puckett, W. H., Dana, W. J., Nguyen, T. V., Theiss, J. C., and Matney, T. S.,** Risk of handling injectable antineoplastic agents, *Am. J. Hosp. Pharm.*, 39, 1881, 1982.
39. **Bos, R. P., Leenaars, A. O., Theuws, J. L. G., and Henderson, P. T.,** Mutagenicity of urine from nurses handling cytostatic drugs, influence of smoking, *Int. Arch. Occup. Environ. Health,* 50, 359, 1982.
40. **Kolmodin-Hedman, B., Hartvig, P., Sorsa, M., and Falck, K.,** Occupational handling of cytostatic drugs, *Arch. Toxicol.*, 54, 25, 1983.
41. **Nguyen, T. V., Theiss, J. C., and Matney, T. S.,** Exposure of pharmacy personnel to mutagenic antineoplastic drugs, *Cancer Res.*, 42, 4792, 1982.

42. **Gibson, J. F., Baxter, P. F., Hedworth-Whitty, R. B., and Gomperz, D.,** Urinary mutagenicity assays: a problem arising from presence of histidine associated growth factors in XAD-2 prepared urine concentrates, with particular relevance to assays carried out using the bacterial fluctuation test, *Carcinogenesis,* 4, 1471, 1983.
43. **Staiano, N., Gallelli, J. F., Adamson, R. H., and Thorgeirsson, S. S.,** Lack of mutagenic activity in urine from hospital pharmacists admixing antitumour drugs, *Lancet,* 1, 615, 1981.
44. **Venitt, S., Crofton-Sleigh, C., Hunt, J., Speechley, V., and Briggs, K.,** Monitoring exposure of nursing and pharmacy personnel to cytotoxic drugs: urinary mutation assays and urinary platinum as markers of absorption, *Lancet,* 1, 74, 1984.
45. **Ladik, C. F., Stoehr, G. P., and Maurer, M. A.,** Precautionary measures in the preparation of antineoplastics, *Am. J. Hosp. Pharm.,* 37, 1184, 1980.
46. **Reynolds, R. D., Ignoffo, R., Lawrence, J., Torti, F. M., Koretz, M., Anson, N., and Meier, A.,** Adverse reactions to AMSA in medical personal, *Cancer Treat. Rep.,* 66, 1885, 1982.
47. **Selevan, S. G., Lindbohm, M.-L., Hornung, R. W., and Hemminki, K.,** A study of occupational exposure to antineoplastic drugs and fetal loss in nurses, *N. Engl. J. Med.,* 313, 1173, 1985.
48. **Sotaniemi, E. A., Sutinen, S., Arranto, A. J., Sutinen, S., Sotaniemi, K. A., Lehtola, J., and Pelkonen, R. O.,** Liver damage in nurses handling cytostatic agents, *Acta Med. Scand.,* 214, 181, 1983.
49. **Stellman, J. M., Aufiero, B. M., and Taub, R. N.,** Assessment of potential exposure to antineoplastic agents in the health care setting, *Prev. Med.,* 13, 245, 1984.

Chapter 8

METHYL METHACRYLATE

John Widström

TABLE OF CONTENTS

I. INTRODUCTION

Methyl methacrylate is an ester of methacrylic acid and methanol and is used in the manufacturing of polymethyl methacrylate. The polymer is a clear and strong plastic material that has both industrial and other applications.

The American Conference of Governmental Industrial Hygienists (ACGIH) time-weighted threshold limit value for monomethyl methacrylate in air is 100 ppm (410 mg/m^3) for 8-h daily exposure.[1] When used by health care professionals, polymethyl methacrylate is usually prepared by mixing liquid monomethyl methacrylate with polymethyl methacrylate powder. The mixture may either be formed to the required shape by hand or be molded. Liquid monoacrylic compounds are often used for repairing dental prostheses. The plastic is cured at different temperatures, depending on the concentration ratios between monomer and polymer and on the type and amounts of additives. Both methyl methacrylate cured at room temperature and heat-cured methyl methacrylate are used in medical applications.

The following groups of health care personnel may come in contact with methyl methacrylate: technicians and technical aides making and mending acrylic dentures and hearing aid apparatus, orthopedic operation personnel using methyl methacrylate bone cement for fixation of metallic and plastic prostheses, and employees at pathology laboratories where methyl methacrylate may be used for embedding of histological preparations.

II. EXPOSURE LEVELS IN HEALTH CARE SETTINGS

The air levels of monomethyl methacrylate have been measured in several health care settings. The exposure periods are usually rather short, but levels may rise to several times the threshold limit value during the exposure. Diffusion through the skin after direct contact is in many cases the most important route of uptake and cause of symptoms.

When monitoring the breathing air concentrations of monomethyl methacrylate at different Swedish dental laboratories,[2] the highest concentrations (approximately 30% of the 8-h threshold limit value) were found in the mixing and handling of noncured methyl methacrylate dough. Much lower concentrations were measured in the finishing and polishing of cured denture material. Air concentrations of monomethyl methacrylate during work with cured denture material were also measured by Brune and Beltesbrekke.[3] They found that threshold limit values were not exceeded even in cases with no ventilation. The use of ventilated work areas resulted in considerably reduced concentrations of the monomer.

Measuring of methyl methacrylate during the mixing of orthopedic bone cement on an open trolley showed concentrations up to 1230 mg/m^3. The environmental concentrations of methyl methacrylate were nondetectable when mixing the same amount of bone cement in a fume cabinet.[4] Measurings of methyl methacrylate during an orthopedic operation showed concentrations from 1 to 5 mg/m^3 close to a mixing table equipped with a scavenger system.[5] Concentrations were high enough though to trigger a systemic reaction in an operation room nurse. Exposure levels in dental laboratories and orthopedic surgery are summarized in Table 1.

Surgical gloves are often used for protection when handling methyl methacrylate. Many times this protection is of limited value since methyl methacrylate rapidly penetrates most of the glove materials used.[6,7] The use of gloves may even increase the skin uptake by occlusion.

III. EFFECTS IN HEALTH CARE PERSONNEL

A. Dermatological Effects

Methyl methacrylate can cause both allergic and irritative contact dermatitis. The allergic contact dermatitis is caused by the monomethyl methacrylate that mainly occurs in noncured

Table 1
A SUMMARY OF EXPOSURE LEVELS OF METHYL METHACRYLATE IN DENTAL LABORATORIES AND OTHOPEDIC SURGERY

	Type of work	Exposure levels mg/m^3	Exposure levels ppm	Ref.
Dental laboratories				
	Handling of noncured methyl methacrylate	127 (45—347)	39 (14—106)	2
	Grinding and polishing polymerized details	7 (3—31)	2 (1—9)	2
	Finishing and polishing polymerized details	13 (6—20) 1.4[a]	4 (2—6) 0.4[a]	3
Orthopedic surgery				
	Mixing of orthopedic bone cement	410—1230 n.d.[b,d]	126—374 n.d.[b,d]	4
	Mixing of orthopedic bone cement	1.3—5[c]	0.4—1.5[c]	5

[a] Local ventilation.
[b] Fume cabinet.
[c] Scavenger system.
[d] n.d. Non-detectable.

materials. The methyl methacrylate products used contain several types of additives, some of which have been reported to be allergic sensitizers.[8] Methyl methacrylate was first reported as a cause of allergic contact dermatitis among dentists and dental technicians in 1954.[9] Allergic contact dermatitis has also been described in orthopedic personnel after the use of bone cement based on methyl methacrylate.[6,10]

In a Finnish study of 205 dental technicians and technical aides,[9] about 81% used methyl methacrylate denture materials daily. Only a few answered "yes" when questioned if they used protective gloves and 17% reported current hand dermatitis or previous local dermatological problems. About 40% of this group referred their problems to the handling of methyl methacrylate. Acrylate allergy had been diagnosed earlier in 2% of the respondents. No additional allergies were found in the clinical study.

B. Effects on the Respiratory Tract

Methyl methacrylate has been reported as the cause of occupational asthma. A case report describes a 40-year-old dental technician,[12] who after several years of work with methyl methacrylate, experienced chest tightness, dyspnea, and cough persisting for several hours after exposure. After having mixed mono- and polymethyl methacrylate for 20 minutes, he had a fall in peak expiratory flow (PEF) of 24%. The PEF values returned to normal 2 h after the exposure.

Occupational asthma has also been reported in an operation theater nurse[4] who handled bone cement based on methyl methacrylate daily for 7 years. Her symptoms were persistent cough, wheezing, and breathlessness. When challenged with methyl methacrylate at occupational exposure levels up to 1500 mg/m^3 2 to 3 min, she developed a late type of asthmatic reaction. The reaction started 6 h after the challenged, and a maximal fall in forced expiratory volume in 1 (FEV_1) was recorded 11 h after the exposure.

In a Polish study,[13] the incidence of chronic obstructive lung disease (COLD) was found to be almost twice as high in a group of 454 workers exposed to methyl methacrylate and styrene when compared to a group of 683 workers with no exposure to these compounds, but with similar working assignments. The diagnosis of lung obstruction was made if the ratio of observed to expected values, calculated from age and height, of FEV_1 was less than

80%. The frequency of chronic bronchitis and asthmatic symptoms was not found to be higher in the exposed group than in the nonexposed group. The environmental factors were concluded to be much more harmful to lung function than smoking.

C. Local Neurotoxic Effects

Methyl methacrylate has been implied to cause local neurotoxic effects of the hand. In a group of 205 Finnish dental technicians and technical aides, 25% was reported to have experienced at least one of the following symptoms: whitening of fingers and feelings of numbness, coldness, and pain.[11] Cases with similar symptoms have been reported in orthopedic surgeons.[10]

As a follow-up to the findings in the Finnish study, a comparison of different neurological functions was made between 20 dental technicians with either the above mentioned symptoms or hand dermatitis and 18 healthy volunteers.[14] The motor and sensory velocities in the forearms were similar and normal in both groups. The dental technicians had significantly slower distal sensory conduction velocities from digits I, II, and III on the right hand and also from the radial aspects of digits II and III on the left hand than did the controls. The findings were concluded to be consistent with mild axonal degeneration on the areas with the closest and most frequent contact with methyl methacrylate.

D. Effects on the Central Nervous System (CNS)

There are a few references in the literature to works of Russian origin[15,16] where CNS symptoms are reported.[17-20] Increased incidences of headache, irritability, fatigue, and loss of memory are described in industrial workers with long-time exposure to methyl methacrylate. These works unfortunately lack clear definitions of symptoms; no control groups have been used and connections between exposure levels and observed effects are missing. No reports of CNS effects in health care personnel have been found except for an indicative connection between exposure to methyl methacrylate and nausea and loss of appetite in dental students.[21]

E. Other Effects

Methyl methacrylate is described as having caused systemic reaction in a female operating room nurse.[5] The reaction appeared after only a few contacts with methyl methacrylate and was marked by hypertension, dyspnea, and generalized erythroderma. The nurse recovered uneventfully.

REFERENCES

1. American Conference of Governmental Industrial Hygienists, Documentation of the Threshold Limit Values and Biological Exposure Indices, 5th ed., ACGIH, Cincinatti, Ohio, 1986, 406.
2. **Ekenvall, L., Ancker, K., Gustavsson, P., and Göthe, C. J.,** Aspects of occupational medicine on work with room temperature cured methyl methacrylate (MMA) in dental care, *Tandteknikern,* 50, 444, 1981 (in Swedish).
3. **Brune, D. and Beltsebrekke, H.,** Levels of methylmethacrylate, formaldehyde and asbestos in dental workroom air, *Scand. J. Dent. Res.,* 89, 113, 1981.
4. **Pickering, C. A. C., Bainbridge, D., Birtwistle, I. H., and Griffiths, D. L.,** Occupational asthma due to methyl methacrylate in an orthopaedic theatre nurse, *Br. Med. J.,* 292, 1362, 1986.
5. **Scolnick, B. and Collins, J.,** Systemic reaction to methylmethacrylate in an operating room nurse, *J. Occup. Med.,* 28, 196, 1986.
6. **Pegum, J. S. and Medhurst, F. A.,** Contact dermatitis from penetration of rubber gloves by acrylic monomer, *Br. Med. J.,* 1, 141, 1971.
7. **Waegmaekers, T. H., Seutter, E., den Arend, J. A., and Malten, K. E.,** Permeability of surgeons' gloves to methyl methacrylate, *Acta Orthop. Scand.,* 54, 790, 1983.

8. **Fregert, S.,** *Manual of Contact Dermatitis,* 2nd ed., Munksgaard, Copenhagen, 1981, 42.
9. **Fisher, A. A.,** Allergic sensitization of the skin and oral mucosa to acrylic denture materials, *JAMA,* 156, 238, 1954.
10. **Fries, I. B., Fischer, A. A., and Salvati, E. A.,** Contact dermatitis in surgeons from methyl methacrylate bone cement, *J. Bone Joint Surg.,* 57, 547, 1975.
11. **Rajaniemi, R. and Tola, S.,** Subjective symptoms among dental technicians exposed to the monomer methyl methacrylate, *Scand. J. Work Environ. Health,* 11, 281, 1985.
12. **Lozewicz, S., Davison, A. G., Hopkirk, A., Burge, P. S., Boldy, D. A. R., Riordan, J. F., McGivern, D. V., Platts, B. W., Davies, D., and Newman Taylor, A. J.,** Occupational asthma due to methyl methacrylate and cyanoacrylates, *Thorax,* 40, 836, 1985.
13. **Jedrychowski, W.,** Styrene and methyl methacrylate in the industrial environment as a risk factor of chronic obstructive lung disease, *Int. Arch. Occup. Environ. Health,* 51, 151, 1982.
14. **Seppäläinen, A. M. and Rajaniemi, R.,** Local neurotoxicity of methyl methacrylate among dental technicians, *Am. J. Ind. Med.,* 5, 471, 1984.
15. **Cromer, J. and Kronoveter, K.,** A Study of Methyl Methacrylate Exposure and Employee Health, DHEW Publ No. 77-119, National Institute for Occupational Safety and Health, Cincinnati, OH, 1976.
16. **Innes, D. L. and Tansy, M. F.,** Central nervous effects of methyl methacrylate vapor, *Neurotoxicology,* 2, 515, 1981.
17. **Blagodatin, V. M., Golova, I. A., Bladokatkina, N. K., Rumyantseva, Y. P., Goryacheva, L. A., Aliyeva, N. K., and Gronsberg, Y. S.,** Establishing the maximum permissible concentration of the methyl ester of methacrylic acid in the air of a work area, *Gig. Tr. Prof. Zabol.,* 6, 5, 1976 (in Russian).
18. **Dobrinskij, S. I.,** Data concerning problems of industrial hygiene and occupational pathology in the manufacture of acrylic polymers, *Gig. Tr. Prof. Zabol.,* 14, 53, 1970 (in Russian).
19. **Karpow, B. D.,** The effect of small concentrations of methyl methacrylate on the inhibition and stimulation processes of the cortex of the brain, *Tr. Leningr. Sanit. Gig. Med. Inst.,* 14, 43, 1953 (in Russian).
20. **Raines, L. A.,** Toxicity of methacrylate in the conditions of a dental supplies factory, *Gig. Tr. Prof. Zabol.,* 1, 56, 1957 (in Russian).
21. **Tansy, M. F., Benhayem, S., Probst, S., and Jordan, J. S.,** The effects of methyl methacrylate vapor on gastric motor function, *J. Am. Dent. Assoc.,* 89, 372, 1974.

Chapter 9

HISTOPATHOLOGICAL LABORATORIES

A. O. Myking, E. Røynstrand, and O. D. Laerum

TABLE OF CONTENTS

I. INTRODUCTION

Occupational hazards connected to histopathological laboratories have been generally known for a long period.[1] Since the introduction of aniline dyes and further refinement of processing of biological materials, the procedures for histological work have been relatively standardized over the whole world for the last 100 years. However, in the last decades, with the introduction of new methods, new types of equipment, and laboratory instruments, as well as automation of the procedures, the picture has changed. On the one hand, automated procedures reduce the direct contact to chemicals during daily work. On the other hand, new chemicals and methods imply new types of exposures, partly with unknown hazards. Furthermore, there has been an increasing awareness of health problems related to long-term exposure to small doses of chemicals which have been in common use for years. In some cases, this may in the long run be more dangerous than hazards connected to more acute, high-dose exposures.

The spectrum of chemical agents used in modern histopathology is very broad. It is, therefore, beyond the scope of this chapter to cover all aspects of chemical hazards known to occur. For many chemicals like heavy metals, toxic side effects have been well known for years, and the reader is therefore referred to standard textbooks in toxicology, chemistry, and staining technology.[2-6] Other aspects, such as the danger of infections in the handling of fresh or frozen tissue and health hazards associated with plastic embedding for light or electron microscopy, are covered in Chapters 5 and 8 of this book. In this chapter, we shall review some agents which in recent years have attracted special attention, the hazards they imply, and how to avoid potential dangers to health.

II. CHEMICALS USED FOR FIXATION

In general, fixatives are chemical agents which react with tissues and thereby prevent their degradation. All fixatives should be considered to be toxic. Since several of the fixatives bear names after persons or short names for daily use, workers are often not aware of all their chemical components. An exact knowledge of this is a good background for avoiding health hazards. Some fixatives which are commonly used in histological work are listed in Table 1. In addition to being toxic, some of them can cause allergy or are potential carcinogens. Some are also inflammable organic solvents. Furthermore, fixatives usually occur in high concentrations, thus increasing their potential toxicity.

A. Formaldehyde

Aqueous solutions of formaldehyde (HCHO), either in water or in buffered solution, are the most commonly used fixative in histopathology. The structural formula is

$$\begin{array}{c} O \\ \| \\ H{-}C{-}H \end{array} \text{ (molecular weight of 30)}$$

Pure monomeric formaldehyde is a colorless gas which readily dissolves in water at concentrations up to 37% at room temperature. The aqueous solution of the substance is also called formalin. When solutions with more than 30% by weight are standing for some time, they usually become cloudy due to precipitation of polymers. HCHO is also available as polymer, of which the best known are paraformaldehyde and trioxane.[7,8]

At low concentrations, the main effect of HCHO is irritation of the eyes and mucous membranes. A summary of its effects is shown in Tables 2 and 3. Usually, persons are able to smell HCHO at about 1 ppm, although both eye irritation and identification of its smell

Table 1
CHEMICALS FOR FIXATION

Names	Composition		Main occupational hazards
Formaldehyde (H-CHO)	Formaline (40 g of H-CHO in 100 ml of	10 ml	Formalin vapor is highly irritating to mucous membranes; acts as an allergen to skin and airways; and is carcinogenic in test systems and animals; splash from aqueous solutions may cause serious injuries to the eyes; extensive tissue damage or death may occur if swallowed.
	H_2O)	90 ml	
	Water		
	Buffering agent added, ca. pH 7		
Glutaraldehyde $[(CH_2)_3CHO\cdot CHO]$	Glutaraldehyde (25 g/100 ml of H_2O)	16 ml	Mainly as for formalin.
	Phosphate buffer, pH 7.4	84 ml	
Zenker's fluid (Helly's)	Mercuric chloride ($HgCl_2$)	5g	See standard textbook on heavy metal toxicology.
	Potassium dichromate ($K_2Cr_2O_7$)	2.5 g	Dichromates may act as allergens to the skin and may have carcinogenic properties.
	Sodium sulfate (Na_2SO_4)	1 g	
	Distilled H_2O	100 ml	
	Conc CH_3COOH added before use	5 ml	(Zenker)
	40% formalin added before use	5 ml	(Helly)
Bouin's fluid	Picric acid (C_6H_2 $(NO_2)_3$-OH) saturated	75 ml	See standard textbooks of toxicology. Stock picric acid is highly explosive in crystalline state.
	aqueous solution	25 ml	
	Formalin 40%	5 ml	
	CH_3COOH conc		
Carnoy's fluid	Abs ethanol (C_2H_5OH)	60 ml	It is highly inflammable; narcotic effects (chloroform) which in high doses may lead to unconsciousness; chloroform may have carcinogenic properties.
	Chloroform ($CHCl_3$)	30 ml	
	CH_3COOH conc	10 ml	
Flemming's fluid	Chromic acid (CrO_3), 1% aqueous	15 ml	Chromic acid may have carcinogenic properties.
	Osmium tetroxide (OsO_4), 2% aqueous	4 ml	Vapors of osmium tetroxide are highly irritating and toxic to mucous membranes. High concentrations may cause extensive tissue necroses.
	CH_3COOH conc	1 ml	
Newcomer's fluid	Isopropanol ($CH_3CHOHCH_3$)	60 ml	It is highly inflammable. Dioxane is cauterizing and may have carcinogenic properties.
	Propionic acid (CH_3CH_2COOH)	30 ml	
	Petrolum ether (naphta, C_5-C_6-alcans)	10 ml	
	Acetone (CH_3COCH_3)	10 ml	
	Dioxane ($C_4H_8O_2$)	10 ml	

Table 2
REPORTED HEALTH EFFECTS OF FORMALDEHYDE AT VARIOUS CONCENTRATIONS[7]

Health effects reported	Approximate HCHO conc[a] (ppm)
None reported	0—0.05
Odor threshold	0.05—1.0
Eye irritation	0.01—2.0[b]
Upper airway irritation	0.10—25
Lower airway and pulmonary effects	5—30
Pulmonary edema, inflammation, pneumonia	50—100
Death	100 +

[a] Range of thresholds for effect listed.
[b] The low concentration (0.01) was observed in the presence of other pollutants that may have been acting synergistically.

Table 3
HEALTH EFFECTS AND REPRESENTATIVE EXPOSURE LEVELS[7]

Formaldehyde conc (ppm)	Health effects (exposure time)[a]
≤0.05	—
0.1	Human eye irritation begins in some people (minutes, hours)
0.5	Human mucociliary inhibition and squamous metaplasia; mid-point of range in one study (0.1—1.1 ppm) (years)
1.0	Human nose and throat irritation begins; most people have eye irritation (minutes, hours)
2.0	Rat squamous metaplasia and mucociliary system LOEL[b] (months)
3.0	Humans (most) experience nose and throat irritation (minutes)
	Monkey squamous metaplasia LOEL (weeks)
5.0	Rat, observed 1% cancer incidence (years)
	Human lower airway effects begin (minutes, hours)
15.0	Rat, observed 50% cancer incidence (years)
	Mouse, observed 1% cancer incidence (years)

[a] Duration of exposure causing the effect is indicated in parentheses.
[b] LOEL, lowest observed effect level.

has been reported at levels as low as 0.05 ppm. Above 1 ppm, nose, throat, and bronchial irritations are seen, and at a concentration of 5 ppm, nearly all persons will experience such irritation. Low-grade occupational exposure has also been reported to produce transitory neuropsychological symptoms.[9] Above 50 ppm, severe pulmonary reactions are seen, including pneumonia, bronchial inflammation, pulmonary edema, and sometimes death.

Accidental exposure of human eyes to aqueous solutions of HCHO can result in different injuries, ranging from discomfort and minor local damage to permanent opacity and loss of vision. The type of injury depends on the concentration and treatment. At levels above 1 ppm, lacrimation from the eye is usual. Direct exposure due to a splash of HCHO into the eyes should be treated with immediate flushing with water to prevent serious injury, including conjunctival and lid edema as well as more serious damage.[10-12] On longer exposure above the odor threshold of formaldehyde, so-called olfactory fatigue with increasing threshold for sensing smells to different agents, has been reported.

The same exposure may result in symptoms from the upper airways, including irritation with a feeling of dry or sore throat, mainly in the concentration range of 1 to 11 ppm. Tolerance to this type of irritation may occur after a few hours of exposure, but the symptoms may reappear after 1 to 2 h of interruption of exposure. This exposure can also cause more permanent alterations in the local defense mechanisms of the nasal mucosa with decreased mucosal ciliary clearance.[10]

In the lower airways, the symptoms are cough, chest tightness, and wheezing, usually at exposure doses around 5 to 30 ppm. The more permanent pulmonary damage from long-term exposure to formaldehyde has been debated, although some studies could indicate this. At very high doses, i.e., 50 to 100 ppm, pulmonary edema and pneumonitis can result. It is also anticipated that concentrations above 100 ppm can be fatal.[8,11,13]

HCHO can also cause asthma-like symptoms in humans by direct exposure. In addition, sensitization may lead to asthma. It also acts as a direct airway irritant in persons who already suffer from bronchial asthma of other causes.[14,15] It is well established that HCHO can sensitize the skin and cause allergic contact dermatitis. The dermatitis occurs by a delayed-type hypersensitivity mechanism[7,12] (for concentrations, see Table 3).

1. Carcinogenicity Studies

Recently, the Office of Pesticides and Toxic Substances at the U.S. Environmental Protection Agency (EPA) has published an assessment of health risks to garment workers and certain home residents from exposure to formaldehyde.[7] This was based on a large series of publications in the field, including both epidemiological and experimental studies.[15-19] The EPA has classified formaldehyde as a "probable human carcinogen" under its guidelines for carcinogenic risk assessment. The conclusion was that there is limited evidence to indicate that formaldehyde may be a carcinogen in humans. Nine studies reported statistically significant associations between respiratory neoplasms and exposure to formaldehyde or formaldehyde-containing products.

In animals there is a conclusion of "sufficient" evidence of carcinogenicity by the inhalation route.[20] Formaldehyde induces nasal squamous cell carcinomas both in rats and in mice. In addition, formaldehyde is mutagenic in numerous test systems using bacteria, fungi, and insects. It can also induce malignant transformation of cells in culture, cause DNA cross-linking, sister chromatid exchange, and chromosome aberrations. Furthermore, formaldehyde has been shown to form adducts with DNA and with proteins both *in vivo* and *in vitro*. It may also interfere with the capacity of DNA repair in mammalian cells.[21]

Based on structure-activity correlations, a potential carcinogenicity is supported.[23-25,27] Thus, formaldehyde is one of several aldehydes which have been shown to exhibit carcinogenic activity in experimental animals. For example, acetaldehyde, the closest structural analogue to formaldehyde, induces the same type of malignant tumours in rats as formaldehyde.

On the basis of rat studies, one has tried to estimate human cancer risks by anticipating a linear dose response relationship. However, some of the existing information indicates that a nonlinear risk as an assessment model is a better way to extrapolate the real cancer risk to humans.[17] Still, the predicted excess lifetime cancer risk estimates from a linear model are about equivalent to the excess cancer incidence observed in epidemiological studies.[7]

From 28 epidemiological studies examined by the EPA, it was concluded that there was a significant excess in total nasopharyngeal and lung cancer mortality. However, only a few of the studies have corrected for smoking and drinking habits. There has also been an indication of increasing incidence of leukemia and neoplasms of the brain and colon, although this has not been supported by other biological data.[7]

2. Conclusions

The body of data on hazardous effects of formaldehyde resulting from direct exposure of the respiratory tract and skin in humans is now so overwhelming that in combination with its extensive use it should be treated as a potentially dangerous substance in practical histopathological work. Therefore, both construction of the laboratory and the instructions for the workers should aim at keeping the exposure levels at a minimum.

B. Other Fixatives

Both glutaraldehyde, acrolein, and other aldehydes used for fixation are volatile and irritate skin and respiratory tracts at low concentration in the same way as formaldehyde. They should, therefore, be treated in the same way as formaldehyde, as indicated above. Of the other fixatives, such as ethanol, methanol, acetone, chloroform, acetic acid, picric acid, and different heavy metal salts, their toxicity is well documented and generally known.[2,22] Also, these should be treated with the same care as formaldehyde, although their potential dangers mainly apply to acute toxic effects.

Abuse of ethanol should also be considered as a possibility, as well as accidental ingestion of methanol instead of ethanol. It should be kept in mind that chrome can be the source of allergic reactions through contact sensitization to the skin. In electron microscopical laboratories, osmium tetroxide is the most commonly used agent for postfixation. In the toxicological literature, the threshold limit value is set at very low concentrations, i.e., 0.002 mg/m^3 of air. At this level, the odor of the substance is difficult to identify.[26]

The vapors of osmium tetroxide react with the respiratory tract, the cornea, and also the skin. Conjunctivitis and rhinitis are common; the nose and throat may become dry or sore, and the vocal chords may be affected. These reactions usually disappear in less than a day.[26] It should also be remembered that droplets of osmium tetroxide can cause necrosis by direct exposure to the skin or cornea. Osmium tetroxide should therefore be treated with even more care than formaldehyde and other fixatives.

III. SOLVENTS AND EMBEDDING MATERIAL

In histopathology, organic solvents are mainly used in combination with ethanol as clearing agents or intermedium before paraffin infiltration. It is also used as an admixture to the mounting medium for coverslipping.

The commonly used substances are aromatic compounds or derivatives of benzene.[22] While benzene itself because of its hematotoxic side effects should be avoided, its methylated (toluene) and dimethylated compounds (xylene) are widely used. Aliphatic carbohydrates have been suggested as substitutes because of a higher boiling temperature and possibly a lower general toxicity.[28,29] However, they have not gained general application and are used in just a few laboratories. Thus, in routine histopathology, organic solvents mean xylene or toluene. Both compounds give excellent results for paraffin embedding, but differ slightly in some physical and chemical properties.

A. Some Historical Remarks

During the 1950s and the first part of the 1960s, the acute and chronic toxicity of benzene, xylene, and toluene was considered to be more or less equal. Today, this is only considered to be true for the acute actions of these compounds. It has become clear that the similarity of their chronic side effects was caused by contamination of benzene. Benzene has a well-

Table 4
ORGANIC SOLVENTS

Name	Main occupational hazards
Toluene	The mixture of vapor and atmospheric air may be explosive. Airways are the main entrance to the body in occupational exposure. Absorption through the skin may occur. There is a prominent narcotic effect. High-dose exposure may cause unconsciousness. Injuries to the liver and kidneys as well as cardiac symptoms have been described. In chronic exposure,permanent symptoms from the central and peripheral nervous systems may develop. Teratogenic effects are suspected. The commercial product usually contains small amounts of benzene (below 2%). Special precautions are required to prevent inhalation of vapor (ventilated hoods, gas mask, etc.).
Xylene	Entrance to the body and acute side effects are mainly the same as those for toluene. In chronic exposure, disturbances from the central and peripheral nervous systems may develop. The commercial product usually contains 4 to 10% ethylbenzene. Precautions are the same as those for toluene.
Benzene	Entrance to the body and acute side effects are the same as those for toluene and xylene. In chronic exposure, severe hematologic complications may occur. There are carcinogenic effects in humans. Whenever possible, benzene should be substituted with other solvents. Precautions are in principle the same as those for toluene and xylene.
Ethanol	Inflammability and explosiveness of vapors seem to be the main occupational hazards. Social drinking may, however, influence the effects of occupational solvent exposure. In addition to other side effects, heavy drinking alone may be the cause of chronic neurological symptoms and brain atrophy.
Methanol	Methanol is highly toxic if swallowed. Acute intoxications may lead to blindness and permanent brain damage or death.
Trichlorethene (Trichlorethylene)	Entrance to the body and acute toxic side effects are mainly the same as those for toluene and xylene. It does not penetrate intact skin. Liver toxicity is especially important. It is carcinogenic in test animals, but this has not been not proven in humans. Precautions are the same as those for toluene and xylene.

documented hematotoxic effect, while xylene and toluene in their pure form are not considered to cause depression of blood formation.[30-32] The most important complication of chronic exposure to organic solvents in general is slowly development of permanent symptoms from the nervous system. This has mainly been substantiated during the late 1970s and the 1980s and has produced a high degree of attention to prevent or evacuate the vapor of solvents at workplaces.

B. Occupational Hazards of Organic Solvents in General

The occupational hazards of organic solvents in general have in recent years attracted considerable interest and concern. The many reports are mainly based on industrial use of solvents. However, in recent years, reports on toxic side effects among employees exposed to solvents in hospital environment have been published.[33-35] Some of these deal with possible teratogenic effects which are especially important at workplaces where younger women are in the majority.[36-39] Organic solvents are supposed to have toxic effects on a wide variety of organ systems of which the most important are the central and peripheral nervous systems, the cardiovascular system, the respiratory tract, and the skin.

The health effects of various solvents are shown in Table 4. In spite of many common aspects, important differences exist. Toxic side effects may be caused by intermediate metabolites and may, therefore, vary from compound to compound.[40-43] Unfortunately, there are gaps in our knowledge on such differences and likewise there are gaps between the degree of chronic exposure and health risks. Mutagenic effects of solvents have been demonstrated in biological test systems, but are for many reasons difficult to evaluate in man. These are important topics for further research.

C. Properties and Health Hazards of the Commonly Used Clearing Agents, Xylene and Toluene

The properties and health hazards of clearing agents are in most respects common (see Table 4). Reviews are given by the World Health Organization[30] (WHO) and in a Nordic expert group evaluation of health-based occupational exposure limits.[31] The short survey given below is based mainly on these two reports.

Xylene (dimethylbenzene) — Xylene exists in three isomeric forms. The commercial product, a colorless liquid with a characteristic odor known as xylol, is a mixture of all three. Xylene is produced from crude oil and may in its commercial form contain small amounts of impurities, mainly benzene derivatives. In occupational exposure, xylene is mainly absorbed through the lungs. Absorption of the vapor is similar for all three isomers and constitutes about 60 to 70% of the inhaled amount. Xylene may also be absorbed through the intact skin. The rate is estimated to be 2.0 to 2.5 $\mu g/cm^2/min$. In comparison, approximately 75 mg of xylene is absorbed per minute through the lungs at a resting state at 100 ppm. Xylene is rapidly metabolized and excreted in the urine, mainly in the form of methylhippuric acid. Only 3 to 6% is eliminated through the respiratory air.

Toluene (methylbenzene) — Toluene is also produced from crude oil. Similar to xylene, it is colorless fluid with a characteristic odor. Small amounts of impurities from benzene derivatives occur in the commercial product. Impurities from benzene itself are kept at low concentrations (below 2% in purified solutions). In occupational exposure, toluene vapor is in the same way as xylene absorbed through the lungs in quantities of 40 to 60%. About 20% of absorbed toluene may be eliminated through ventilation. Only small quantities are excreted unchanged in the urine; the remainder is eliminated by the kidneys in its conjugated form (hippuric acid). Skin absorption is probably less important than for xylene, but this point is somewhat controversial.

D. Evaluation of Health Risks

Many factors have to be considered, such as the amount of solvents used, the transportation and storing systems, and, most important, the air exchange capacity in the laboratory and the use of ventilated hoods. For considering sources of health risks, modes of transportation to and from the laboratory are important; likewise, routines for changing the clearing agents in the tissue processors are important. The annual consumption of toluene and xylene in a large laboratory is approximately 40 ml per tissue block or 3000 l annually in a laboratory taking 20,000 biopsies per year.

Our knowledge of hazards associated with clearing agents is mainly derived from industrial workplaces. The compounds are among other applications widely used as an admixture for engine fuels and as thinners, in the production of rubber compounds, and in rotogravure plants. Important sources of information are reports of accidents and chronic illness among sniffers.[44-48] However, these reports mainly express a high-dose risk, while exposure in modern histopathology laboratories should be negligible or preferably zero. Xylene and toluene will in the following be mentioned together as their negative side effects are supposed to be similar, but not definitely proven to be so in every respect.

E. Short-Term Occupational Exposure

With air concentrations of toluene from 190 to 1125 mg/m^3, symptoms vary from drowsiness and very mild headache to severe fatigue, muscular weakness, and incoordination. The subject is less extensively studied for xylene. EEG changes may be demonstrated on exposure to concentrations as low as 1 mg/m^3. Both solvents have narcotic effects and fatal cases have been reported in accidental exposure.[49] In severely affected individuals, permanent damage to the central nervous system (CNS) has been reported,[44,45] but it is difficult to document whether this is a specific effect or not.

F. Long-Term Occupational Exposure

Long-term occupational exposure causing mild discomfort and subclinical effects is more difficult to evaluate, but several well-documented reports point to permanent symptoms from the CNS as the most important complication.[50-52] Brain atrophy has been described, but is still a controversial issue.[53]

The symptoms may vary from insomnia, fatigue, and irritability to psychoorganic syndromes characteristic for presenile dementia. Progression of symptoms is supposed to stop when exposure ceases. Peripheral neuropathies have been described with preferential involvement of the long somatic nerves. In addition to symptoms from the central and peripheral nervous systems, there is reason to believe that clearing agents in common with other organic solvents may induce arrhythmia of the heart.[54] Ventricular fibrillation is suspected in some cases of sudden death in chronic exposure as well as in acute intoxication. Irritability of the respiratory system is well known and likewise there is a decreased sense of smell. Teratogenic effects are suspected and reports indicate a higher percentage of miscarriage and abortion among exposed women.[36-39] In spite of the close chemical relationship to benzene, hematological disorders have so far not been documented.

Disease mechanisms are poorly understood, especially in chronic exposure. Intermediate metabolites are suggested to play an important role, possibly by interfering with the cellular transportation system for oxygen.[40] In the peripheral nervous system, axonal degeneration has been suggested to play an important role.[40,41]

The relationship between the degree of exposure and the prevalence of disease is of great practical importance. The question is for obvious reasons difficult to answer. However, epidemiological studies may give some indications. In the years from 1973 to 1977, the Institute of Occupational Health in Helsinki has annually registered an average of five new cases of occupational disease caused by chronic exposure to toluene.[55] In the same period of time, there was an addition of 44 new cases per year due to exposure to solvent mixtures often containing toluene. In a review from 1984, approximately 250 new cases of solvent-induced disease are annually expected out of a total population of 4 million in Norway.[56] However, reservation must be made for diagnostic shortcomings. However, in spite of this, the problem is above any doubt serious enough to cause concern. Present knowledge of occupational hazards with organic solvents points to some main problems to be solved in the future: their possible role in carcinogenesis, the combined effects of several solvents, and diagnostic criteria which more precisely connects solvent exposure and disease.

G. Occupational Hazards of Ethanol and Paraffin

The only complication which rarely may be encountered in the handling of ethanol in the laboratory is the danger of explosion from mixtures of ethanol vapor and atmospheric air. Paraffin is a mixture of saturated high molecular aliphatic hydrogen carbon compounds with a melting point at 55 to 60°C. The paraffin often contains admixtures of other agents to facilitate penetration of tissue or to modify other physical properties. Admixtures of such compounds will often be the secret of the producer and the possible hazards to health will therefore be difficult to evaluate.

Dimethylsulfoxide (DMSO) (sulfinylbismethane) is one such compound added to facilitate tissue infiltration. It penetrates quickly through intact skin and for this reason is often added to creams and ointments. For review the reader is referred to the excellent paper by Brayton.[57] DMSO is considered to have low toxicity in man. The median lethal dose in laboratory animals is high. By intravenous administration in mice, rats, cats, dogs, and monkeys, it is between 2.5 and 8.9 g/kg of body weight. Combinations of DMSO with other toxic compounds account for its greatest toxic potential. The effects of carcinogenic substances are enhanced; likewise the toxic effects of different therapeutic agents applied to the skin are enhanced. Aromatic hydrocarbons and azo compounds in combination with DMSO penetrate rubber gloves. Whether the same is true for aliphatic hydrocarbons like paraffin remains to be shown.

Histotechnicians are mainly exposed to vapor from warm paraffin in the embedding process. The health effects of these compounds are not well understood. Care should be taken to do the work under ventilated hoods. This should be a requirement especially when working with DMSO.

IV. MISCELLANEOUS

Health risks may arise from contact with stains,[27] by handling fresh or frozen tissue, by accidents from thermic or mechanical injuries, or from cauterizing chemicals or ionizing irradiation. This shall be only briefly mentioned.

A list of commonly used stains is given in Table 5. Stains are usually dissolved in water and ethanol; only a few are dissolved in other organic solvents (mainly Sudan stains). Protection will therefore in most instances be achieved by the use of thin, waterproof gloves. A few stains and some other commonly used chemicals are classified as carcinogens (Table 6). These and the fat-soluble stains should be handled according to requirements for such compounds (see Section V).

The risk of infection in histopathology work outside the autopsy room is in general very low. Fixation renders the tissue germ free and reduces the risk of infection to the mere exposure of fresh tissue for which procedures for handling infectious material should be followed (see Chapter 5 of this book).

Thermic and mechanical accidents may occur from explosions of chemicals (crystalline picric acid or periodic acid) or mixtures of solvent vapors and air; likewise, they may occur from high- or low-pressure installations. Thermic injuries may also be induced by fire in highly inflammable material or by frost injuries from liquid nitrogen or carbon dioxide. The cauterizing effect of strong alkali or acids is well known. Various isotopes may be the sources of ionizing irradiation; more rarely the source may be technically neglected electron microscopes.[58] These topics are partly covered elsewhere in this book and in standard handbooks of laboratory techniques.

V. SOME PRACTICAL ASPECTS OF THE WORK IN THE LABORATORY

In practical laboratory work, the main hazardous exposures are contact with neurotoxic substances (solvents) and with allergens and carcinogens. These hazards have been partially treated in previous chapters. This chapter will deal with some exposed laboratory routines and with some aspects of the handling of carcinogenic chemicals.

By good air exchange capacity and by the use of ventilated hoods and gloves for protection of the skin, exposure to hazardous chemicals may be avoided. Two working positions which are illustrated in Figures 1 and 2 are in this respect difficult, namely, coverslipping of the slides and tissue embedding. Both procedures require very fine adjustments of movements and gloves are for this reason often not used.

In coverslipping, direct contact with solvents and embedding material can be avoided by using suitable instruments (Figure 1). Care should therefore be taken to give histotechnicians instructions and training on this particular point. In the near future, reliable equipment for automated coverslipping will be available and should be used in larger laboratories. In the embedding process, direct contact with splash from melted paraffin (55 to 60°C) cannot be avoided and gloves should therefore be used (Figure 2).

Storing of large quantities of fixed specimens often creates problems by the leakage of fixatives and pollution of the environment. Care should be taken to store the specimens in nondiffusable containers or heat-sealed plastic bags. A high air exchange capacity is likewise important.

Table 5
COMMONLY USED STAINS AND SOLVENT

Groups of stains	Main staining compounds (Color index number)	Stain usually dissolved in
Nitro stains	Naphtol yellow (10316)	Water
	Picric acid (10305)	Water
Azo stains	Monoazo	
	Azophloxin (11640)	Water
	Orange G (16230)	Water
	Ponceau 2R (de Xylidine) (16150)	Water
	Diazo	
	Congo red (22120)	Water, ethanol
	Scarlet red (26105) Sudan IV	Ethanol, acetone
	Sudan black (26150)	Ethanol, acetone
Triphenylmethane stains	Basic	Water, often with 5% phenol sol (Ziehl-Nielsen)
	Fuchsin (42510)	
	Crystal violet (42555)	Water, ethanol (aniline, phenol)
	Gentian violet (42535)	Water, ethanol (aniline, phenol)
	Malachite green (42000)	Water
	Acid	
	Aniline blue (42780/42755)	Water
	Light green (42095)	Water
	Methyl blue (42780)	Water
	Acid fuchsin (42685)	Water, often mixed with picric acid (Van Gieson)
Xanthene stains	Acridine red (45000)	Water, ethanol
	Eosin Y (45380)	Water, ethanol, acetic acid to lower pH
	Erythrosin B (45530)	Water, ethanol, acetic acid to lower pH
	Fluorescein (45350)	Water (1:10,000 in PBS-water)
	Phloxin B (45410)	Water
	Pyronin Y (45005)	Water (often mixed with methyl green which has been purified by the use of chloroform)
Acridine stains	Acridine orange (46005)	Water (1:10,000 in PBS-water) fluorescence
	Phosphin 3R (46045)	Water (1:10,000 in PBS-water) fluorescence
Anthrachinone stains	Alizarin red S (58005)	Ethanol
	Anthracene blue (58605)	Ethanol, water
	Nuclear fast red (none)	Ethanol, water
Azin stains	Azocarmine	Water
	Neutral red (50040)	
	Safranin (50240)	Water, ethanol
Oxacin stains	Celestine blue B (51050)	Water (Heavy metals are often added as mordants to these stains.)
	Gallocyanine (51030)	Water (Heavy metals are often added as mordants to these stains.)
	Cresyl fast violet (51010)	Water (Heavy metals are often added as mordants to these stains.)
Thiazin stains	Azur A (52005)	These stains are strongly basic and when mixed with eosin they produce salts giving the Giemsa effect. They are dissolved in ethanol/methanol.
	Thionin (5200)	
	Methylene blue (52015)	
Cu-phtalocyanin stains	Alcian blue (74240)	Water (to all mixtures of alcian stains, acids are added to control pH)
	Alcian green (74242)	
	Alcian yellow (none)	Water (to all mixtures of alcian stains, acids are added to control pH)
		Ethanol 96%
Stains from nature	Hematoxylin	Mainly water (mordants, acids, oxidants, glycerol, ethanol)
	Safran (Saffron)	Pure ethanol

Table 6
SOME COMMONLY USED COMPOUNDS CLASSIFIED AS CARCINOGENS

Chromium acid
Dichromates
Formaldehyde
Chloroform
Benzidine
Carbazole
Cacodylate
Congo red
Ponceau de Xylidine
Trypan blue

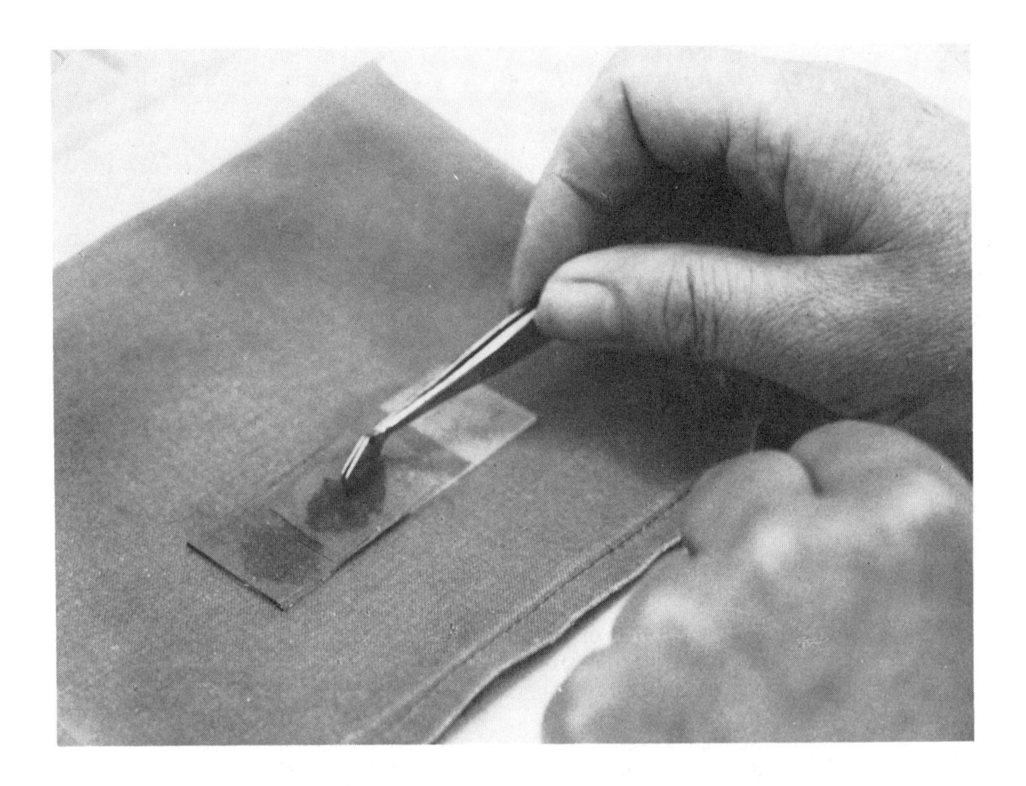

FIGURE 1. Photograph to show part of the coverslipping process. By use of simple instruments to avoid contact with solvents and resins, the work may be done without the use of protective gloves. Efficient ventilated hoods should always be used.

Concerning work with known carcinogenic compounds in the histopathological laboratories, this should follow general guidelines which have been proposed on an international basis.[23-25] During work with carcinogenic compounds, the same general safety rules apply as for toxic substances. Otherwise, the guidelines recommend medical surveillance of the personnel, education of the employees, and strict control of laboratory practices, where the daily work is defined by high-risk and low-risk situations. Provided that the laboratory facilities are satisfactory, the combination of protective clothing and good laboratory practice when handling the compounds should imply no significant risk during work with known carcinogens.

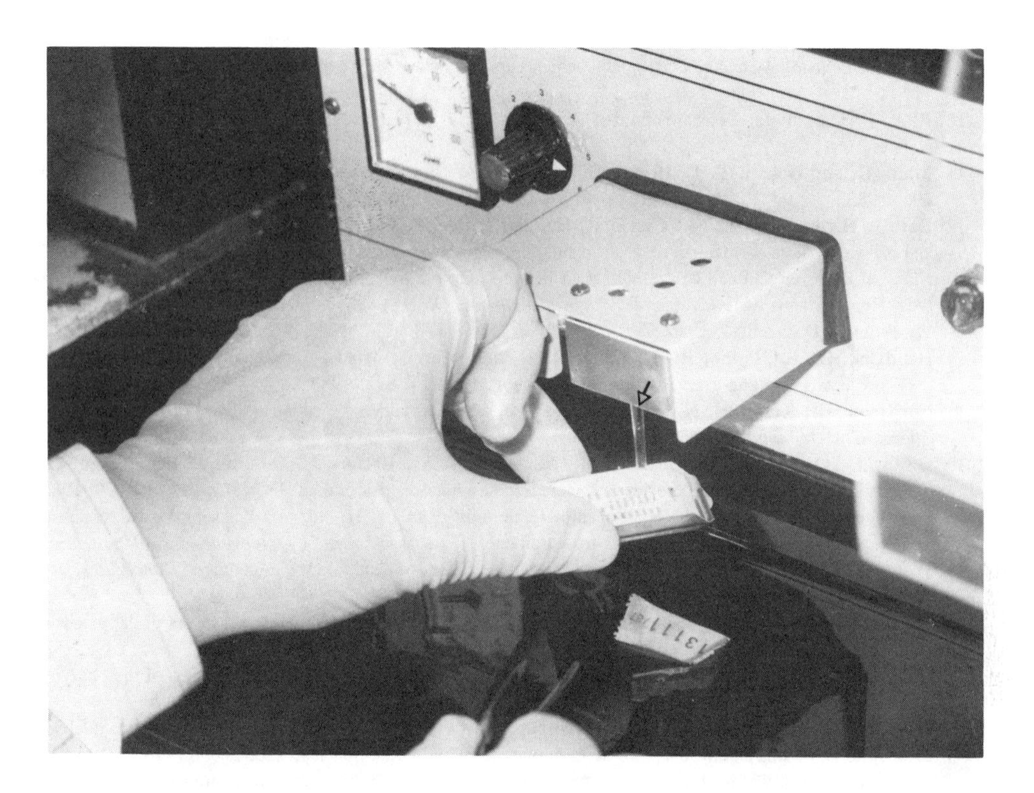

FIGURE 2. Photograph to show the tissue embedding process. The arrow points to the flow of melted paraffin from the heater. In this process it is not possible to avoid splash from the embedding medium and protective gloves should therefore be used. It is recommended that the work be done under ventilated hoods.

It should be emphasized that practical work with carcinogenic compounds is not dangerous as long as the risks are identified and safety measures can be taken.[26] On the other hand, neglecting such risks or unknown hazards, due to insufficient testing of various compounds, still implies a potential risk factor. The handling of waste products requires the same precautions as the original chemicals. Ecological consequences should also be borne in mind when evacuation and destruction of such compounds are planned.

REFERENCES

1. **Clark, G. and Kasten, F. H.,** *History of Staining,* Williams & Wilkins, Baltimore, 1983.
2. **Doull, J., Klaassen, C. D., and Amdur, M. O., Eds.** *Toxicology,* 2nd ed., Macmillan, New York, 1980.
3. ACGIH, Threshold Limit Values for Chemical Substances and Physical Agents in the Workroom Environment, American Conference of Governmental Industrial Hygienists, Cincinnati, 1980 (issued annually).
4. **Bretherisk, L.,** *Handbook of Reactive Chemical Hazards,* 2nd ed, Butterworths, London, 1979.
5. National Research Council, *Prudent Practices for Handling Hazardous Chemicals in Laboratories,* National Academy of Sciences, Washington, D.C., 1981.
6. **Clayton, G. D. and Clayton, F. E.,** *Patty's Industrial Hygiene and Toxicology, 3rd ed., Vol. 1, General Principles,* Wiley-Interscience, New York, 1979.
7. Office of Pesticides and Toxic Substances, Assessment of Health Risks to Garment Workers and Certain Home Residents from Exposure to Formaldehyde, U.S. Environmental Protection Agency, Washington, D.C., 1987.
8. **Clay, J. J., Gibson, J. E., and Waritz, R. S., Eds.,** *Formaldehyde: Toxicology — Epidemiology — Mechanisms,* Marcel Dekker, New York, 1983.

9. **Kilburn, K. H., Warshaw, R., and Thornton, J. C.,** Formaldehyde impairs memory, equilibrium, and dexterity in histology technicians: effects which persist for days after exposure, *Arch. Environ. Health,* 42, 117, 1987.
10. **EPA,** Formaldehyde, Determination of Significant Risk, Advance Notice of Proposed Rulemaking and Notice, 1984, 49.
11. **Main, D. and Hogan, T. J.,** Health effects of low-level exposure to formaldehyde, *J. Occup. Med.,* 25, 896, 1983.
12. **Day, J. H., Lees, R. E. M., Clark, R. H., and Pattee, P. L.,** Respiratory responses to formaldehyde and off-gas of urea formaldehyde foam insulation, *Can. Med. Assoc. J.,* 131, 1061, 1984.
13. National Research Council Committee on Aldehydes, Board on Toxicology and Environment Health Hazards, Health effects of formaldehyde, *Formaldehyde and Other Aldehydes,* National Academy of Sciences, Washington, D.C., 1981, chap. 7.
14. **Hendrick, D. J., Rando, R. J., Lane, D. J., and Morris, M. J.,** Formaldehyde asthma: challenge exposure-levels and fate after five years, *J. Occup. Med.,* 24, 893, 1982.
15. **Nordman, H., Keskinen, H., and Tuppurainen, M.,** Formaldehyde asthma — rare or overlooked?, *J. Allergy Clin. Immunol.,* 75, 91, 1985.
16. **Olsen, J. H., Plough Jensen, S., Hink, M., Faurbo, K., Breum, N. O., and Möller Jensen, O.,** Occupational formaldehyde exposure and increased nasal cancer risk in man, *Int. J. Cancer,* 34, 639, 1984.
17. **Swenberg, J. A., Barrow, C. S., Boreiko, C. J., and d' Heck, H. A.,** Non-linear biological responses to formaldehyde and their implications for carcinogenic risk assessment, *Carcinogenesis,* 4, 945, 1983.
18. **Walrath, J. and Fraumeni, J. F.,** Cancer and other causes of death among embalmers, *Cancer Res.,* 44, 4638, 1984.
19. **Vaughan, T. L., Strader, C., Davis, S., and Darling, J. R.,** Formaldehyde and cancers of the pharynx, sinus and nasal cavity. I. Occupational exposures, *Int. J. Cancer,* 38, 677, 1986.
20. **Vaughan, T. L., Strader, C., Davis, S., and Darling, J. R.,** Formaldehyde and cancers of the pharynx, sinus and nasal cavity. II. Residential exposures, *Int. J. Cancer,* 38, 685, 1986.
21. **Grafstrom, R. C., Fornace, A. J., Jr., Autrup, H., Lechner, J. F., and Harris, C. C.,** Formaldehyde damage to DNA and inhibition of DNA in human bronchial cells, *Science,* 220, 216, 1983.
22. IPCS Program on Chemical Safety, Selected Petroleum Products, Environmental Health Criteria 20, World Health Organization, Geneva, 1982.
23. IARC, Some industrial chemicals and dye stuffs, *IARC Monogr.,* 29, 345, 1982.
24. **Montesano, R., Bartsch, H., Boyland, E., Della Porta, G., Fischbein, L., Crismer, R. A., Swan, A. B., and Tomatis, L., Eds.,** *Handling Chemical Carcinogens in the Laboratory, Problems of Safety,* IARC Sci. Publ. No. 33, International Agency for Research on Cancer, Lyon, 1979.
25. **Douglas, B. and Waters, L. H., Eds.,** *Safe Handling of Chemical Carcinogens, Mutagens, Teratogens and Highly Toxic Substances,* Ann Arbor Science, Ann Arbor, MI, 1980.
26. IARC monographs on the evaluation of the carcinogenic risk of chemicals to humans, Chemicals and industrial processes associated with cancer in humans, IARC Monogr. Suppl. 1, International Agency for Research on Cancer, Lyon 1979.
27. **Afzelius, B. A.,** Occupational hazards, in *Electron Microscopy in Human Medicine,* Johannesen, J. V., Ed., McGraw-Hill, New York, 1978, 328.
28. **Visfeldt, J., Strange, L., and Larsen, J.,** Udskiftning af xylen med redestilleret petroleum i den histologiske teknik, *Ugeskr. Læg.,* 13, 943, 1982.
29. **Ladefoged, O.,** Udskiftning af xylen med redestilleret petroleum i den histologiske teknikk, *Ugeskr. Læg.,* 5, 2103, 1982.
30. WHO, Recommended Health-Based Limits in Occupational Exposure to Selected Organic Solvents, Tech. Rep. Ser. 664, World Health Organization, Geneva, 1981.
31. Nordisk ekspertgruppe for hygiejniske grænseværdier, Toluen, Arbete & Hälsa, 5, 1979.
32. Riskerna med lösningsmedel, The Swedish Work Environment Fund, 1987.
33. **Magnussen, Z. and Fossan, G. O.,** Nevrasteni og polynevropati, *Tidsskr. Nor. Lægeforen.,* 103, 2039, 1983.
34. **Bakke, J. V. and Brekke, P.,** Påvisning av løsemiddelskader i bedriftshelsetjenesten, *Tidsskr. Nor. Lægeforen,* 105, 680, 1985.
35. **Christenson, W. N., Brennan, C. F., and DeVito, J.,** Even "healthy" environments can harbor many hidden hazards, *Occup. Health Saf.,* 54, 40, 1985.
36. **Hansson, E., Jansa, S., Wande, H., Källèn, B., and Østlund, E.,** Pregnancy outcome for women working in laboratories in some of the pharmaceutical industries in Sweden, *Scand. J. Work Environ. Health,* 6, 131, 1980.
37. **Axelsson, G., Jeansson, S., Rylander, R., and Unander, M.,** Pregnancy abnormalities among personnel at a virological laboratory, *Am. J. Ind. Med.,* 1, 129, 1980.
38. **Hemminki, K., Mutanen, P., Saloniemi, I., Niemi, M.-L., and Vainio, H.,** Spontaneous abortions in hospital staff engaged in sterilising instruments with chemical agents, *Br. Med. J.,* 285, 1461, 1982.

39. **Axelsson, G. and Rylander, R.,** Exposure to anaesthetic gases and spontaneous abortion: response bias in a postal questionnaire study, *Int. J. Epidemiol.*, 11, 250, 1982.
40. **Spenser, P. S. and Schaumburg, H. H.,** Experimental neuropathy produced by 2,5-hexanedione — a major metabolite of the neurotoxic industrial solvent methyl *n*-butyl ketone, *J. Neurol. Neurosurg. Psychiatr.*, 38, 771, 1975.
41. **Spencer, P. S., Sabri, M. I., Schaumburg, H. H., and Moore, C. L.,** Does a defect of energy metabolism in the nerve fiber underlie axonal degeneration in polyneuropathies?, *Ann. Neurol.*, 5, 501, 1979.
42. **Riihimäki, V. and Savolainen, K.,** Human exposure to *m*-xylene, kinetics and acute effects on the central nervous system, *Ann. Occup. Hyg.*, 23, 411, 1980.
43. **Savolainen, H.,** Toxicological mechanisms in acute and chronic nervous system degeneration. Occupational neurology, *Acta Neurol. Scand. Suppl.*, 92, 66, p. 23, 1982.
44. **Grabski, D. A.,** Toluene sniffing producing cerebellar degeneration, case report, *Am. J. Psychiatr.*, 118, 461, 1961.
45. **Knox, J. W. and Nelson, J. R.,** Permanent encephalopathy from toluene inhalation, *N. Engl. J. Med.*, 275, 1494, 1966.
46. **Goto, I., Matsumura, M., Inoue, N., Murai, Y., Shida, K., Santa, T., and Kuroiwa, Y.,** Toxic polyneuropathy due to glue sniffing, *J. Neurol. Neurosurg. Psychiatr.*, 37, 848, 1974.
47. **Malm, G. and Lying-Tunell, U.,** Cerebellar dysfunction related to toluene sniffing, *Acta Neurol. Scand.*, 62, 188, 1980.
48. **King, M. D.,** Neurological sequelae of toluene abuse, *Hum. Toxicol.*, 1, 281, 1982.
49. **Benignus, V. A.,** Neurobehavioral effects of toluene: a review, *Toxicol. Teratol.*, 3, 407, 1981.
50. **Axelsson, O., Hane, M., and Hogstedt, C.,** A case-referent study on neuropsychiatric disorders among workers exposed to solvent, *Scand. J. Work Environ. Health*, 2, 14, 1976.
51. **Arlien-Søborg, P., Bruhn, P., Gyldensted, C., and Melgaard, B.,** Chronic painters' syndrome, *Acta Neurol. Scand.*, 60, 149, 1979.
52. **Juntunen, J., Antti-Poika, M., Tola, S., and Partanen, T.,** Clinical prognosis of patients with diagnosed chronic solvent intoxication, *Acta Neurol. Scand.*, 65, 488, 1982.
53. **Juntunen, J., Hernberg, S., Eistola, P., and Hupli, V.,** Exposure to industrial solvents and brain atrophy, *Eur. Neurol.*, 19, 366, 1980.
54. **Hayden, J. W., Peterson, R. G., and Bruckner, J. V.,** Toxicology of toluene (methylbenzene): review of current literature, *Clin. Toxicol.*, 11, 549, 1977.
55. **Kalliokoski, P. J.,** Toluene Exposure in Finnish Publication Rotogravure Plants, Ph.D. dissertation, University of Minnesota, Minneapolis, 1979.
56. **Leira, A. L.,** Hvordan møter vi bølgen av løsemiddelskader?, *Nor. Bedr. H.Tj.*, 5, 274, 1984.
57. **Brayton, C. F.,** Dimethyl sulfoxide: a review, *Cornell Vet.*, 76, 61, 1986.
58. **Devanney, J. A. and Daniels, C. J.,** Radiation leakage from electron microscopes, *Health Phys.*, 30, 234, 1976.

Chapter 10

RADIATION HAZARDS

Bertil R. R. Persson

TABLE OF CONTENTS

I. INTRODUCTION

During all of their existence, human beings and all other life forms have been exposed to radiation in different forms. First of all, one can consider the radiation from the sun as a natural component of life. We need the light in the visual part of the spectrum to be able to see and we need infrared (IR) radiation to be able to keep warm. There are, however, components in the radiation of which we can get ''too much'', i.e., ultraviolet (UV) radiation which can burn the skin and cause skin cancer and part of the cosmic radiation which reaches the earth as ionizing radiation in the form of muons, neutrons, and gamma rays. On the earth, ionizing radiation is also emitted in the form of alpha, beta, and gamma radiation from all the radioactive materials in the ground (uranium, thorium, and their daughters), in the air (radon and radon daughters), and in the body (potassium).

At the end of the 19th century, more precisely 1895, the German professor Wilhelm Conrad von Röntgen discovered X-rays and in 1896 Becquerel discovered natural radioactivity. These new radiation sources were introduced with great enthusiasm into medical practice. X-rays were used for both diagnostic and therapeutic purposes. After the discovery of radium in 1898 by Pierre and Marie Curie, radioactive sources for therapeutic use became available. Very soon, however, the adverse effects of high exposure levels of ionizing radiation to the body were discovered. Up to 1922, about 100 radiological hospital workers died from radiation effects after the medical use of ionizing radiation. In order to increase the safety of radiological personnel, the International Commission on Radiological Protection (ICRP) was formed in 1928. The ICRP makes recommendations for radiation protection which are the basis for the laws and rules which control the handling and use of ionizing radiation and radioactive materials.

Today we have had much experience with radiation safety and the present recommendations provide safe working conditions for radiological workers. Not only are acute effects considered, but hypothetical risks for long-term cancer induction and genetic effects are also taken into account. In this respect, the radiation protection recommendations are much more strict than, for example, the regulations for work with hazardous chemicals.

The approaches and concepts used in the development of protection guidelines for nonionizing radiation have evolved over the past quarter of a century. For electromagnetic radiation, the incident power density was first used as the limiting parameter. Recent exposure limits, however, are based on a consideration of the relationship between bioeffects and the magnitude of the whole-body average specific absorption rate and current densities induced in the body.

This chapter will discuss the safety aspects of working with various kinds of radiation and radioactive compounds in a hospital. There are several medical specialities such as X-ray diagnosis, radiation therapy, and nuclear medicine which are entirely based on the use of ionizing radiation. There are also several new technologies in medicine based on nonionizing radiation such as nuclear magnetic resonance (NMR) examinations, hyperthermia cancer treatment, diagnosis and physical therapy with ultrasound, light amplification by stimulated emission of radiation (LASER) surgery, ultrasound crushing of kidney stones, psoriasis (PUVA) treatment, etc. It is therefore important for hospital workers to be well informed in the safety of all kinds of radiation, both ionizing and nonionizing.

II. PHYSICAL ASPECTS OF RADIATION

A. Ionizing and Nonionizing Radiation

Radiation is nothing else but transport of energy which means that energy is transferred from the place of its origin, which is called the radiation source, to another place, called the target. In this text, we are dealing with health effects which means we are interested in that amount of energy which is absorbed by the human body when it is hit by radiation in any form. The absorbed energy then affects the tissue and cells of the body and can result in various kinds of biological effects.

According to the amount of energy each quantum or package of radiation carries, we divide radiation into the following two different categories: (1) ionizing radiation, with energy quanta larger than 100 eV, and (2) nonionizing radiation, with energy quanta less than 100 eV. Ionizing radiation means such radiation which has enough energy to split the material of tissue into ions, which means electrons and electrically charged molecular fragments, i.e., ions.

In order to ionize a molecule in the human body an average of 100 eV (1.6×10^{-17} J) of absorbed energy is required. We have this amount of quantum energy in X-rays and gamma rays which both are electromagnetic radiation with energy quanta or photons with energies in the order of thousands to millions of electronvolts.

Particle radiation, such as electrons or beta rays from radioactive decay, is also ionizing radiation, as well as alpha, proton, and neutron radiation and many other kinds of subatomic particles in cosmic radiation. Radiation in which the quanta do not have enough energy to ionize atoms or molecules is called nonionizing radiation. UV light, visual light, IR light, microwaves, and radio frequency (Rf) radiation are electromagnetic radiation belonging to this category of radiation. Acoustic radiation (i.e., ultrasound and infrasound), which propagates through vibrations of the molecules in a medium, also belong to this category.

B. Electromagnetic Radiation

The most important type of radiation is electromagnetic (EM) radiation which is very common and which covers both ionizing and nonionizing radiation. From a physical point of view, EM radiation is a transport of energy by means of oscillating electrical and magnetic fields. It is the frequency of these oscillations which determines the energy and properties of the EM radiation.

EM radiation is transferred in energy quanta which are named photons. Each of these photons has a specific energy which is given by the following relation:

$$E_{ph} = h \cdot \nu \tag{1}$$

where ν is the frequency (in hertz) and h is Plank's constant ($h = 6.626 \cdot 10^{-34}$ J $= 4.136 \cdot 10^{-15}$ eV). EM radiation with the highest energy photons is gamma radiation, which originates from transitions between different energy levels in the atomic nucleus. Gamma radiation usually has energies around or above $1 \cdot 10^{6}$ eV which corresponds to frequencies of $3 \cdot 10^{19}$ Hz and wavelengths of 10^{-11} m.

X-rays which originate from transitions between different energy levels in the inner atomic electron shells have frequencies in the range of $3 \cdot 10^{16}$ to $3 \cdot 10^{19}$ Hz which correspond to energies between 100 to 100,000 eV. The lower limit to the X-ray zone corresponds to the border between ionizing and nonionizing radiation. EM radiation below 100 eV is UV radiation and visible light which originates from transitions between intermediate and outer atomic electron levels. At still lower frequencies, the EM radiation is called IR radiation, microwaves and radiowaves, or Rf radiation. EM radiation is physically defined by the electric field vector $\overline{E}$, and the magnetic field vector $\overline{H}$.

The electric field in a point in space is defined as the force, $\overline{F}$, which would act on a unit charge in that point.

$$\overline{E} = \overline{F}/e \tag{2}$$

A charge $q = n \cdot e$ (s) would at the same point be influenced by the electrostatic force:

$$\overline{F}_e = q \cdot \overline{E} \tag{3}$$

If the charge is moving in a magnetic field, it would also be influenced by a magnetic force which is proportional to the velocity. The direction of the force is in the z-direction if the velocity is in the x-direction and the magnetic field is in the y-direction. Thus, the magnetic force, F_m, can be given by a vectorial product:

$$\overline{F}_m = q \cdot v \times / \overline{B} \tag{4}$$

where $\overline{v}$ is the velocity of the charged particles in meters per second (m/s) and $\overline{B}$ is the magnetic induction in tesla (T) (1 T = 1 weber per square meter (Wb/m^2) = 10,000 gauss (G). The magnetic field $\mathcal{H}$ is related to the magnetic induction $\overline{B}$ by the relation:

$$\overline{B} = \mu \cdot \overline{\mathcal{H}} \tag{5}$$

where μ is the magnetic permeability constant which is $4\pi \cdot 10^{-7}$ Vs/Am in a vacuum.

The perpendicularly oscillating electric and magnetic fields of the EM radiation interact with the electrical charges and dipoles in biological tissue and give rise to forces acting on molecules and biological structures. The velocity of an EM wave through a certain medium depends on the electrical and magnetic properties of the medium. The electrical properties are described by the electric conductivity (σ) and the relative permittivity or dielectric constant (ϵ') which is the ratio between the capacity of a capacitor filled with the medium in question and in vacuum. The absolute dielectric constant is given by:

$$\epsilon_o \cong (1/36\pi) \cdot 10^{-9} \quad (As/Vm) \cong 8.85 \cdot 10^{-12} \quad (F/m) \tag{6}$$

The magnetic properties are described by the magnetic conductivity given by the relative permeability, μ':

$$\mu' = \mu/\mu_o \tag{7}$$

Iron is a material with high conductivity for magnetic fields and thus has a high value of μ'. In a medium not bounded by conductors, the product of the electric field, $\overline{E}$, and the magnetic field, $\mathcal{H}$, is equal to the power density, $\overline{S}$, the so-called "Poyntings vector".

$$\overline{S} = \overline{E} \times \overline{\mathcal{H}} \qquad (V/m \times A/m = VA/m^2 = W/m^2) \tag{8}$$

The propagation of EM waves in a medium depends on the dielectric constant, ϵ, and the permeability constant, μ. The energy density of EM radiation in a medium will be discussed in more detail in the following sections.

C. Particle Radiation

High-velocity charged particles which collide with atoms in a media will interact with the atomic electrons and release second electrons. Thus, a pair of ions is formed and this type of radiation is ionizing radiation. Charged particles of this kind are alpha particles, protons, and beta particles which are called direct ionizing radiation. Particles without charge like neutrons must first release protons and electrons and are thus called indirectly ionizing radiation.

D. Acoustic Radiation

Ultrasound is another type of energy transfer which takes place through vibrations in the molecules of the material. It is thus a mechanical type of energy transfer which requires a medium for its propagation.

III. HAZARDS OF IONIZING RADIATION

A. Radiation Protection Standards

1. Principles of Radiation Protection

All ionizing radiation from radiation therapy machines, X-ray tubes, or radioactive material is potentially hazardous. Radiation sources outside the body give rise to external radiation exposure of personnel, and radioactive material which enters the body irradiates its organs and tissue by internal exposure. Radiological personnel in hospitals working with ionizing radiation and radioactive materials must therefore be capable of preventing and controlling these hazards. There is a golden rule of radiation protection which is always applicable: **as many exposures may involve some degree of risk, any unnecessary exposure must be avoided, and all doses must kept as low as is readily achievable, economics and social consideration being taken into account.**

There are three principles which can be applied to prevent or control the exposure of personnel to radiation hazards: remove the hazard, **guard the hazard, and guard the worker.** These principles imply that working places are properly designed and that appropriate equipment and shielding are provided to ensure the maximum amount of protection. The last principle refers to the requirements of periodic measurements of the radiation level in the working environment and continuous personal monitoring.

2. Radiation Units and Standards

a. Absorbed Dose (D)

When ionizing radiation passes through matter, it interacts with the atoms and molecules in the medium it traverses, producing ionizations and excitations. Depending on the medium, the absorbed energy may give rise to observable effects, e.g., blackening of photographic films, release of electrical charges in gases, biological effects, and heating. The energy imparted may be expressed in joules per kilogram, which led to the concept of radiation absorbed dose.

The absorbed dose, D, is the quotient of $\delta\epsilon$ by δm, where $\delta\epsilon$ is the mean energy imparted by ionizing radiation to the matter in a volume element of the mass, δm:

$$D = \frac{\delta\epsilon}{\delta m} \tag{9}$$

The special unit of absorbed dose is the gray (abbreviated Gy):

$$1 \text{ Gy} = 1 \text{ J/kg} \ (= 100 \text{ rad})$$

The earlier special unit of absorbed dose was the rad.

b. Dose Equivalent (H)

From the biological point of view, evidence has accumulated that the effects of the various

Table 1
THE RELATIONSHIP BETWEEN THE COLLISION STOPPING POWER, L_∞, AND THE QUALITY FACTOR[2]

L_∞ in water (keV/μm)	Quality factor, Q
<3.5	1
7.0	2
23	5
53	10
>175	20

types of ionizing radiation are not the same. One can assume that radiation can bring about a change in an organism only by virtue of the energy that is actually absorbed. A biological effect, however, may also depend on the spatial distribution of the energy released along the track of the ionizing particle. It will therefore depend on the type and quality of the radiation, and equal energy imparted by different types of radiation may not produce the same biological effects. For radiation protection purposes, a separate quantity, the dose equivalent, is therefore used for comparison of risks of biological effects of radiation from different types of sources.

The dose equivalent, H, is defined as the product of the absorbed dose, D, and a quality factor, Q, which is different for various types of radiation:

$$H = D \cdot Q \tag{10}$$

The special unit for the dose equivalent is the sievert, when the absorbed dose is given in grays.[2]

$$1\ \text{J/kg} = 1\ \text{sievert} = 1\ \text{Sv}\ (= 100\ \text{rem})$$

The earlier unit was rems, when the absorbed dose was given in rads.

The linear energy transfer or collision-stopping power, L_∞, can be used to specify the radiation quality. The relationship between L_∞ and the quality factor, Q, which is recommended to be used for radiation protection purposes, is given in Table 1. For beta, gamma, and X-rays, the quality factor, Q, is equal to 1. For alpha particles, multicharged particles, the quality factor is 20. For neutrons the quality factor is recommended to be 20.[1] For protons and single-charged particles of rest mass greater than 1 amu of unknown energy, the quality factor is recommended to be equal to 10.

c. *Effective Dose Equivalent* (H_E)

The International Commission on Radiation Protection (ICRP) has recommended a quantity for allowing for the different mortality risks associated with irradiation of different organs, together with a proportion of the hereditary effects.[2] This quantity is called the effective dose equivalent, H_E, and is the sum of the mean dose equivalent, H_T, in tissue, "T", weighted with a factor, ω_T, which represents the proportion of the stochastic risk resulting from irradiation of tissue "T" to the total risk when the whole body is irradiated uniformly.

$$H_E = \sum_T \omega_T \cdot H_T \tag{11}$$

Table 2
WEIGHTING FACTORS RECOMMENDED BY ICRP FOR CALCULATION OF EFFECTIVE DOSE EQUIVALENT[2]

Organ or tissue	Weighting factor, w_T
Gonads	0.25
Breast	0.15
Red bone marrow	0.12
Lung	0.12
Thyroid	0.03
Bone surfaces	0.03
Remainder	0.30

To assess the effective dose equivalent, the dose equivalent in each tissue from all sources is assessed and multiplied by the appropriate weighting factor and the resulting products are then summed. If all the tissues in the body were uniformly irradiated, the result is assumed to be numerically equivalent to the whole-body dose equivalent. The values of ω_T recommended by the ICRP are shown in Table 2 and are appropriate for protection for individuals of all ages and both sexes, i.e., for workers and members of the public. The value for the gonads includes serious hereditary effects, as expressed in the first two generations.[2]

d. Committed Dose Equivalent

In order to take account of the time distribution of absorbed dose to internal organs exposed to the radiation from incorporated radionuclides, the ICRP has defined the "committed dose equivalent". This is the time integral of the dose equivalent rate in a particular organ following an intake of radioactive material into the body. The integration time is 50 years after intake, taken to correspond to the period of active work during a lifetime.

The formal definition of committed dose equivalent for a single intake of radioactive material at time, t_o, which causes a dose equivalent rate in an organ of $\dot{H}(t)$ at time t is given by the following equation:

$$H_{50} = \int_{t_o}^{t_o + 50\text{years}} \dot{H}(t) \cdot dt \qquad (12)$$

e. Collective Dose Equivalent

With the assumption that the effect is directly proportional to dose equivalent, it is sometimes useful to define a quantity to measure the total radiation exposure of a group of individuals. This quantity is called the collective dose equivalent and is given by the expression:

$$S = \int_o^{\infty} H \cdot N(H) \cdot dH = \sum_i \overline{H}_i \, N(\overline{H})_i \qquad (13)$$

where N(H)dH is the number of individuals receiving a dose equivalent between H and H + dH and $N(\overline{H})_i$ is the number of individuals in a population subgroup receiving an average dose equivalent of $\overline{H}_i$.

3. Biological Effects of Ionizing Radiation

a. Introduction

The biological effects of ionizing radiation in humans depend on the fact that energy is

Table 3
THE ACUTE EFFECTS OF WHOLE-BODY EXPOSURE TO IONIZING RADIATION[1]

<0.25 Gy

No detectable clinical effects; probably no delayed effects on individuals

0.50 Gy

Slight transient blood changes; no other clinically detectable effects; delayed effects possible, but serious effects on the average individual very improbable

0.5—1.0 Gy

Nausea and fatigue with vomiting possible above 1.25 Gy; marked changes in blood pictures with delayed recovery; shortening of life expectancy

2.0—4.0 Gy

Nausea and vomiting within 24 h; following latent period of about 1 week, epilation, loss of appetite, general weakness, and other symptoms, such as sore throat and diarrhea; possible death in 2 to 6 weeks for a small number of the individuals exposed; recovery likely, unless complicated by poor previous health, superimposed injuries, or infections

4.0—6.0 Gy

Nausea and vomiting after 1 to 2 h; after a latent period of 1 week, beginning of epilation, loss of appetite, and general weakness accompanied by fever; severe inflamation of mouth and throat in the third week; symptoms such as pallor, diarrhea, nose bleeds, and rapid emaciation in about the fourth week; some deaths in 2 to 6 weeks; possibly death can occur in about 50% of the exposed individuals

>6 Gy

Nausea and vomiting within 1 to 2 h; short latent period following initial nausea; diarrhea, vomiting, inflammation of mouth and throat toward end of the first week; fever, rapid emaciation, and death as early as the second week for all exposed individuals

imparted to the tissue. This results in chemical transformations of biologically important molecules like DNA. The physiochemical transformations take place within fractions of a second, while the biological effects appear after hours, days, and even years after exposure. If the transformation of the DNA of a cell is extensive, the cell will either die or be transformed to an outlaw cell which might develop to a cancer cell. Transformations in the genetic material of the testes or ovaries will not affect the exposed individuals, but rather will affect their progeny.

b. Hazards at High Radiation Doses

Exposure to high doses of ionizing radiation during a short period of time will cause acute effects which appear at certain periods after the exposure. In Table 3 are listed the acute effects of whole-body exposure with various levels of instantaneous exposure to a given absorbed dose.[1] The individual variation is large and the effects are reduced when only a part of the body is exposed. In radiation therapy against cancer, the total absorbed dose to part of the body is delivered in daily fractions of about 2 Gy during a period of several weeks until a total absorbed dose of 20 to 70 Gy is delivered.

c. Health Hazards of Low Radiation Doses

An interesting question in radiation protection is the risk for biological effects at very low absorbed dose levels. Our knowledge of the biological effects at low levels of radiation

Table 4
INCIDENCE RISKS AS NUMBER OF CASES PER 10,000 INDIVIDUALS EXPOSED TO IONIZING RADIATION CORRESPONDING TO AN AVERAGE DOSE EQUIVALENT OF 1 Sv IN EACH ORGAN[3]

Organ	Genetic effects	Fatal somatic effects	Morbid effects
Gonads	40	—	—
Breast	—	25	50
Red marrow	—	20	—
Lung	—	20	—
Thyroid	—	5	100
Bone	—	5	—
Other organ	—	50	—
Skin	—	1	100
Total	40	126	250

exposure is still very diffuse and limited. This is mainly due to the fact that, at low absorbed dose levels, the effects are so rare that it is extremely difficult to get significant experimental data. One thus has to extrapolate from experience at absorbed dose levels above 1 Gy down to milligray levels. We don't know for certain that the dose/effect relationship is linear down to zero, but this is the assumption used by the ICRP for establishing guidelines for permissible levels.[2] The linear extrapolated risk factors for different types of tissues or organs are given in Table 4.

B. System of Dose Limitation

1. Justification of the Practice

According to the recommendation of the ICRP, "No practice involving ionizing radiation shall be adopted unless its introduction process a positive net benefit." The expression "positive net benefit" invokes the idea of cost-benefit analysis. The choice of "positive net benefit" will, however, depend on many factors, only some of which will be associated with irradiation protection. Therefore, more general decision-making methodologies would need to be applied to decisions on the justification of practices. Often there is a need to ensure that the total detriment from a practice is very small in relation to the expected benefit of the practice. This is especially true in the use of ionizing radiation in hospitals.

2. Optimization of Radiation Protection

The ICRP has also recommended that "all exposures of ionizing radiation shall be kept as low as reasonably achievable, and social factors (be) taken into account." Since any exposure to ionizing radiation is assumed to involve some degree of hazard (i.e., risk of inducing genetic effects and cancer), all exposures should be kept as low as is reasonably achievable.

The optimization applies for doses below the recommended safety dose limits and therefore nonstochastic effects are precluded. For the stochastic effects, the mathematical expectation of the amount of harm in an exposed group of people is proportional to the collective effective dose equivalent.

The optimization procedure necessitates expression of cost of protection and detriment in the same unit. Since the health detriment is proportional to the collective effective dose equivalent, it is assumed that the cost of the health detriment is also proportional to that quantity. The cost of health detriment in monetary units is the cost that society is willing to pay to avoid this detriment. This cost is in the range of $1000 per man sievert (man Sv) to $100,000 per man sievert.

3. Recommended Safety Dose Limits for Individuals

In order to prevent acute effects (nonstochastic effects) of radiation exposure and to limit the occurrence of late effects (stochastic effects) to an acceptable level, the International Commission on Radiological Protection (ICRP) makes the following recommendations:[2]

- The dose-equivalent limit to all single tissues except the eye lens is 0.5 Sv (50 rem) in a year.
- The dose-equivalent limit to the lens of the eye is 0.15 Sv (15 rem) in a year.
- The limit for the annual effective dose equivalent for radiological workers, i.e., uniform irradiation of the whole body, is 0.05 Sv (50 mSv = 5 rem).
- In case of nonuniform irradiation, the dose limitation is based on an assumption of equal risks, i.e., that the whole body is irradiated uniformly.

The last condition will be met if the sum of all individual dose equivalents to each tissue, H_T, weighted by the relative sensitivity of the tissue in question is less than or equal to the dose equivalent for the body irradiated uniformly. This sum is called the effective dose equivalent. The recommended annual dose-equivalent limit for uniform irradiation of the whole body is 50 mSv (5 rem).

The hands and forearms, the feet and ankles, the skin, and the lens of the eye should be excluded from the computation of effective dose equivalent. For women of reproductive capacity, any necessary exposure should be as uniformly distributed in time as is practicable. The purpose of this is the protection of the embryo before a pregnancy is known. When a woman is known to be pregnant, she should work only in such working conditions where it is most unlikely that the annual exposure will exceed three tenths of the dose equivalent limits (i.e., 15 mSv). In order to be able to compare dose equivalent limits of instantaneous exposure from external sources with prolonged exposure received from intake of radioactive materials, the ICRP has introduced the concept of "annual limits of intake" (ALI). The ALI corresponds to the committed effective dose equivalent from an intake of a given radionuclide equal to the dose equivalent limit for radiological workers. Keeping the intakes in each year less than the ALI ensures that the maximum annual dose equivalent from that radionuclide will always be less than the dose equivalent limit even if the intake continues every year for 50 years.

When both external and internal exposures are received in a year, the annual dose limit will not be exceeded if the following condition is met:

$$\frac{H_E}{50\ (\mathrm{mSv})} + \sum \frac{I_J}{ALI_{j,L}} < 1 \tag{14}$$

where H_E is the actual annual effective dose equivalent, I_j is the annual intake of radionuclide j, and $ALI_{j,L}$ is the annual limit on intake for radionuclide j.

In case of dose to the skin, the following condition is met:

$$\frac{H_{skin}}{500\ (\mathrm{mSV})} < 1 \tag{15}$$

where H_{skin} is the shallow dose equivalent index. In case of inhaled radioactivity, "derived air concentration" (DAC) is of practical use. This is obtained by dividing the ALI value by a standard volume of air inhaled in a working year of 2000 h at a breathing rate of 1.2 m^3h^{-1}.

4. Radiation Hazards Correlated with the Safety Limit

The objective of radiation protection is to maintain nonstochastic effects at an acceptable

Table 5
AVERAGE INDIVIDUAL DOSES AND COLLECTIVE DOSES TO WORKERS IN VARIOUS TYPES OF OCCUPATIONS[3]

Type of occupation	Average individual dose (mSv/year)	Collective dose
Nuclear industry	1—20	30[a]
Medical	0.3—3	1[b]
Industrial radiography	1—3	—
Research	0.3—5	0.5[b]

[a] Man sievert/GW_e a.
[b] Man sievert/million population.

level. That means that even if the dose is very small, there is still a small probability of a detrimental health effect. This effect can be genetic and somatic. The average risk factor for hereditary effects, as expressed in the first two generations, is for radiation protection purposes given by the ICRP to be about 4 cases per 1000 man Sv. The additional damage to later generations is of the same magnitude (Table 4).

The somatic risks are distributed in time after each exposure. After the time of exposure, the probability of a detrimental effect increases from approximately zero for a few years (latent period) up to a maximum value and then declines to zero again. The cancers can be divided into leukemia and all other cancers because of their different latent periods. For leukemia the average annual risk is about eight cases during 25 years following an exposure corresponding to 10,000 man Sv. The lifetime risk of mortality from leukemia is two cases per 1000 man Sv. For other cancers, the shape of the risk curve is less certain, but if the average annual risk is taken to be two cases per 1000 man Sv during 40 years, this corresponds to a lifetime risk of eight cases per 1000 man Sv. The total lifetime risk of leukemia and other cancers is therefore about one case per 100 man Sv.

The safety limit for the effective dose equivalent for radiological workers is 50 mSv/year during 50 years, which will accordingly result in an annual risk of 0.05%/year and a lifetime accumulated risk of 1 to 3%. The U.N. Scientific Committee on Effects of Atomic Radiation (UNSCEAR) has recently summarized information on occupational exposure for various occupations. Table 5 shows a rough summary of average individual doses and collective doses to workers in various types of occupations.[3]

5. Management of Overexposed Workers

The recommended dose limits of the ICRP do not apply to accidents and emergencies, but apply only to those conditions where the source of exposure is under control. Measures to be taken in the case of overexposure of workers can be summarized in the following common elements:

a. Notification

Provisions are made for the notification of the worker, the immediate responsible person, the employer, the medical adviser, and the responsible authority under specified conditions, when a suspected overexposure occurs.

b. Assessment of Dose and Immediate Medical Attention

Immediate action is required to be taken to assess the dose incurred and to communicate this information to the medical advisor.

c. Investigation

A detailed investigation is required to identify and record the circumstances causing overexposure and recommendations are made for corrective action if indicated, to ensure that recurrence of the circumstances causing the accident is unlikely. An important element in corrective action is to ensure that information developed in the investigation is widely disseminated so that other activities can profit from the experience.

d. Recording of Exposure

Requirements are included for recording of the dose by the employer, the medical advisor, and, in some cases, a central state authority.

e. Medical Follow-Up and Future Employment Restrictions

Medical follow-up and future employment restrictions are handled on a case-by-case basis through consultations between the employer, worker, medical advisor, and competent authority. In one case, there would be generic requirements in regulations that, following an overexposure, the individual must not be assigned to tasks which are likely to result in the dose equivalent exceeding 1% of the annual limit during the remainder of the calendar year in which the exposure occurred.

IV. RADIOTHERAPY SAFETY

A. Introduction

Almost anyone working in a hospital may on some occasion have to deal with radiation equipment or a radioactive source. For some workers, radiation comprises part of their job and can be planned. This category of workers usually receives considerable training in the safety aspects of radiation and the safe use of radioactive materials. Below are given different categories of hospital workers expected to be involved with radiation therapy to various extents.[4]

Hospital Workers Trained in Radiation Safety (Radiological Personnel)

Hospital physicists
Radiotherapy technologists
Radiotherapists
Radiotherapy nurses
Nuclear pharmacists

This category is usually well trained in radiation safety and hospital physicists are usually the most educated and experienced in this field. The radiological workers are working in an environment where they might be exposed to ionizing radiation and safety precautions are very high. This category of Hospital workers must carry a personal dosimeter.

Hospital Workers Infrequently Involved with Radiation Usually Working Together with Personnel Trained in Radiation Safety

Nurses caring for brachytherapy patients
Operating room nurses
Recovery room nurses
Regular pharmacists
Patient transporters
Engineers
Housekeepers

Table 6
ANNUAL DOSE EQUIVALENT TO RADIATION THERAPY PERSONNEL

Profession	Average dose quivalent	
	mSv/year	Percent (%) of MPD[a] (50,000 mSv)
Therapy technologists	2000	4
Engineers	1800	3.6
Simulator technicians	600	1.2
Physicists	500	1.0
Radiotherapist	500	1.0
Nurses, clinical area	200	0.4

[a] MPD: maximum permissible dose.

This category of workers has little or no training in radiation safety. During their work together with radiological personnel, they might receive some training through casual conversation. The risk of being highly exposed to radiation is usually small, but if they are poorly prepared for it, they experience a high level of anxiety.

Hospital Workers not Expected to be Involved with Radiation

Housekeepers
Maintenance workers
Administrators
Kitchen staff
Security guards

This category of workers is not expected to be exposed to radiation in their daily work. In an emergency situation, however, they might be involved with radiation. It is thus important that all hospital employees are trained in radiation safety for emergency situations.

B. Radiation Safety in External Beam Therapy

1. Introduction

The radiation treatment room is shielded according to the requirements for the accelerator or remote-controlled radionuclide source in question. All treatment rooms must be fitted with electrical interlock to guard against entry when a radiation hazard exists. In the control room, where the personnel are located during the exposure of the patient, the radiation level should be well below and not exceed 25 μSv/h.

All personnel working with radiotherapy must be issued a radiation monitor or personal dosimeter (usually film or thermoluminescence dosimetry) to assess general body dose. The maximum permissible annual dose equivalent to radiotherapy personnel is 50,000 μSv. Under normal conditions, most employers will have exposures less than 1250 μSv per quarter, but in average personnel receive less than 1000 μSv/year. The average annual exposure of different radiation therapy personnel is in Table 6. These values clearly show that radiation safety in radiation therapy facilities is well designed.

2. Safety of Equipment for External Irradiation

a. Radiation Shield

For equipment producing X-rays, the shield usually consists of the lead lining of the tube. Leakage can, however, occur at joints, screw holes, power cable entry points, coolant hoses, and in the region of the X-ray port or aperture. For radionuclide gamma ray sources (^{60}Co, ^{137}Cs), a potential hazard exists even in the "beam-off" state. Therefore the radiation source must be embedded in a shield of a considerable thickness of lead or heavy alloy equipped with a source transport/shutter and interlock system.

For high-energy accelerators, the elements of high voltage and Rf-generating circuitry (e.g., magnetrons, klystrons, and thyratrons) may emit soft X-rays and should therefore be included in radiation surveys and posted with appropriate warning signs. Significant radiation leakage may be found originating from around the radiation head and from the region between the exit of the accelerating cavity and radiation head.

b. Induced Radioactivity and Neutron Production

Accelerators producing X-ray beams above 10 MV produced a significant number of neutrons by photonuclear reactions such as (γ,n) reactions. The neutrons thus produced do not contribute significantly to the dose to the patient, but they present an occupational radiation safety problem which must be considered in the design of the room shielding.

When running high-energy accelerators for extended periods of time (e.g., during commissions), components near the target become radioactive due to induced radioactivity and should be monitored before being handled by maintenance personnel. For example, in a lead-antimony wedge receiving an absorbed dose of 1 kGy at 6 MV, a surface dose rate of 2 mGy/h may occur. The half-life of these produced radionuclides are short and the problem is overcome by waiting for some time to "cool off".[5]

3. Radiation Safety Measures in Design of Hospital Buildings

Radiotherapy departments should be located in an area which is set apart for the purpose of providing appropriate shielding for all outside persons. The amount of shielding in the walls of the treatment rooms will vary from 0.2 to 2 m of concrete depending on type of equipment. It is important at an early stage in the planning of new installations to have a discussion between the architect, the radiation protection specialist (e.g., the hospital physicist), and the manufacturer of the chosen equipment. In case of an emergency, the local fire service must be aware of which radiotherapy machine contains radioactive sources and the local water authority will need to be consulted as to how to drain water in case of an accidental release of radioactivity.

C. Radiation Safety in Brachytherapy

1. Manual Techniques

The branch of radiotherapy in which closed radioactive sources are placed within the body or close to the surface to be treated is called brachytherapy. The use of this technique presents the greatest radiation safety problem in all medical uses of ionizing radiation.

In order to minimize the hazards of brachytherapy, it is necessary to establish a well-equipped laboratory with appropriate shielding and handling equipment. The closed radioactive sources should be stored and prepared for use in this laboratory to which access is restricted. The main storage facility should be locked except when in use, so that the issue and movement of sources can be rigorously controlled. Record keeping of the location of each source at any time is most important.

The most serious hazard in hospital radiation safety is when a brachytherapy source is lost. It is therefore most important to guard against loss or inadvertent disposal of brachytherapy sources. Nothing associated with the patient should be disposed of or otherwise dispersed, until it has been inspected for radioactive material with a suitable detector.

2. Afterloading Techniques

Afterloading techniques allow the insertion of applicators and preparation of the patient in the absence of radioactive material. Then the radioactive sources are inserted with special shielding conditions which reduce staff exposure to an acceptable level. This technique is therefore strongly recommended whenever it is clinically acceptable.

D. Emergency Arrangements and Procedures

The loss of radioactive sources is one of the most serious radiation emergencies which can occur in a hospital. All possible measures must therefore be taken to minimize the chance of such a loss. If the loss of a radioactive source does occur anyhow, it is first of all important to report it according to a previously prescribed action. The responsible radiation physicist should give immediate instructions to stop the movement of relevant articles and involved personnel for monitoring. A general search for the source should take place, including shielding materials and waste deposits. The inventory should be checked for the possibility that the source is placed in the wrong place. The possibility that the patient left the hospital with the source should also be investigated.

In the worst case, the lost radioactive source becomes seriously damaged and spread of radioactive material takes place. The general procedure of how to handle such a situation should aim at minimizing the spread of contamination. The practical means for handling radioactive contamination are envisaged in Section VI.

V. RADIATION SAFETY IN DIAGNOSTIC USE OF X-RAYS

A. Introduction

Radiation safety in the diagnostic use of X-rays follows the recommendations stated by the ICRP.[6-9] The ICRP recommends dose-equivalent limits for occupational exposure and assesses associated levels of risk which give maximum permissible limits of radiation exposure in the occupational environment. The planning of a diagnostic X-ray department is based on extreme or relatively pessimistic values of the parameters which determine the radiation protection devices or procedures. Therefore the real occupational radiation exposure is usually well below maximum permissible absorbed doses.

B. Fundamental Principles of X-ray Diagnostic Exposure

1. Sources of Radiation Exposure

a. Primary Beam

The primary beam which is used for the diagnostic exposure of the patient should always be avoided by the operator. When the field size and position of the primary beam are known, the working procedure must be such that the personnel always avoid the primary beam.

b. Leakage Radiation

Except for the primary beam emitted through the window of the X-ray tube, radiation may penetrate the X-ray tube housing. Diagnostic X-ray tubes must thus be shielded so that this leakage radiation at maximum output is below 1 mSv/h (about 100 mR/h in older units) at a distance of 1 m from focus. In daily work, maximum output is very seldom used, which makes the leakage of radiation a minor radiation safety problem.

c. Scattered Radiation from the Patient

Minimizing of the primary field size is the most important principle to reduce the radiation exposure from scattered radiation to both personnel and patient. It is therefore important that in all cases one uses the smallest possible field size and that the position is correctly known. Another way to decrease the absorbed dose to the patient as well as scattered radiation to the personnel is to use compression of the patient. In principle no personnel should be in the X-ray laboratory when the exposure takes place. Control room and control closets are effectively shielded and should be used as much as possible.

If it is necessary to be in the X-ray laboratory during the exposure (i.e., fluoroscopy), one should use protective clothing (lead rubber clothing, gloves, mobile shields, etc.). In critical situations when, for example, the patient must be held in position, it is also important that personnel are well shielded and that they do not put their hands in the primary beam.

It is important for personnel to realize that radiation safety is a part of the working procedures in the X-ray laboratory and it must be accepted that it takes some time. It is desirable to have radiation safety instruction for those who are working in the laboratory for which everyone working in the laboratory feels responsible.

C. Safety of X-ray Equipment

1. Aperture and Collimator

First of all, the X-ray tube must be enclosed in a radiation shield which allows the X-ray beam to pass only in a small aperture. The primary X-ray beam would in any situation be restricted to the area of the patient which is of interest to be examined. The equipment must therefore be equipped with an aperture or a collimator which defines the primary field limits.

2. Filter

The lowest energy photons generated in the X-ray tube do not penetrate the patient to contribute to the X-ray image and will therefore be considered as unnecessary. In order to limit the exposure to the patient as well as personnel, these low-energy photons must be filtered off the primary beam. The recommendation for the minimum filtration is as follows:[9] X-rays generated below 100 kV, at least 2 mm of Al; X-rays generated above 100 kV, at least 3 mm of Al. It is of course acceptable to use more filtering to harden the beam and improve the image quality in certain circumstances and also to lower the absorbed dose to the patient.

In equipment for fluoroscopy, there must be an X-ray absorber opposite the X-ray tube to absorb the primary beam. This means shielding of the image intensifier. The X-ray tube and the X-ray absorber must be connected in such a way that the primary beam is absorbed in any normal use of the X-ray equipment. The ICRP recommends the following thicknesses for the absorber:[9] lead equivalent of >1.5 mm for equipment with a maximum of 75 kV HV; lead equivalent >2 mm for equipment with a maximum of 100 kV high voltage (HV). Above 100 kV, the lead equivalent is increased with 0.01 mm/kV HV.

3. Minimum Focus-Skin Distance

The absorbed dose to the patient varies with the inverse square of the distance to the focus of the X-ray tube. Therefore if the X-ray tube comes too close to the patient, the absorbed dose will be unacceptably high. To avoid this, it would not be possible to move the X-ray tube and collimator closer to the patient than a certain distance. The minimum distance for both stationary and mobile X-ray equipment is recommended to be 30 cm.[8]

4. Exposure Control

X-ray equipment should be equipped with an arrangement (device) to control the exposure in both X-ray photography and fluoroscopy. This must be constructed so that the possibility of wrong or undesired exposure is minimum. This requirement means that the voltage (kilovolts), current (milliamperes), and exposure time(s) are the parameters which all must be adjusted, and at the same time the high voltage contact for exposure or fluoroscopy must be reliable.

5. Secondary Barriers, Distance Control

The X-ray equipment and gantry will be equipped with shielding barriers for secondary radiation. These barriers are usually of metal or lead rubber and can be modified to fit the

Table 7
PRINCIPLES OF PROTECTION FOR HAZARDS OF RADIOACTIVE MATERIALS

Minimize the Hazard

1. Keep the amount of radioactive material required to a minimum.
2. Choose radioactive material presenting the least possible hazard (avoid alpha emitters if possible).
3. Choose the safest and most practicable procedure.
4. Dispose of radioactive waste safely.
5. Restrict the movement of radioactive material to a minimum.

Guard the Hazard (Containment)

Prevent hazardous release of radioactive material to the environment by using sealed containers during transport and work with open sources in fumehoods or glove boxes.

Shielding

Use shielded transport container. Reduce radiation levels by working behind lead shields. Use protective clothing for routine radioactive operations.

Guard the Worker

Choose adequate materials, instruments, and facilities. Plan the working procedures and instruct assisting personnel. Give emergency instruction to minimize the consequences if emergency occurs.

Monitoring

Monitor working area and personnel for radiation exposure and radioactive contamination.

technique in question. It will also be possible to control the equipment at a panel well shielded in a cabin.

D. Personal Exposure

Diagnostic use of external beams of X-rays is the most widespread and common use of radiation in medicine. Average annual doses to hospital radiological workers using X-rays for diagnostics vary between 0.4 to 2.5 mGy in most reports. For dental workers, the annual doses are lower, between 0.05 to 0.5 mGy.[3] At the University Hospital in Lund, the average annual dose equivalent (1986) was 0.6 mSv, of which 60% was below 0.4 mSv and 99% was below 2.5 mSv. The annual dose equivalent to dental X-ray personnel was between 0 to 0.1 mSv.[10] Personnel exposure in coronary angiography has been studied at the University Hospital in Lund, and the average absorbed dose measured at different anatomic sites of the body was as follows:

Chest of the radiologist:	0.01	mGy
Head of the radiology assistant:	0.09 ± 0.03	mGy
Neck	0.13 ± 0.04	mGy
Left hand	0.9 ± 0.6	mGy
Right hand	0.4 ± 0.2	mGy

VI. POTENTIAL HAZARDS IN NUCLEAR MEDICINE

A. Introduction

Safe handling of radioactive materials in the nuclear medicine laboratory must be the compulsory concern of all nuclear medicine personnel. Major procedures involved in safe handling are shown in Table 7.

B. Control of the External Irradiation

1. General Principles

Reduction of radiation exposure to personnel and to patients in the nuclear medicine laboratory is accomplished by the judicious use of the principles of time, distance, and shielding. The goal of exposure control is to reduce exposure to levels as low as is reasonably achievable (ALARA) while maintaining a working environment in which tasks may be performed quickly and efficiently.

The easiest way to reduce exposure is to minimize the time spent in a radiation area. For example, when a new procedure is introduced, where high-activity sources must be handled, low-activity runs of the procedure will help ensure that it is performed quickly, yet correctly. Another way to minimize the time any one person is exposed to high levels of radiation is to rotate the duties of nuclear medicine personnel.[11]

Radiation exposure is inversely proportional to the square of the distance to the radiation source (''inverse square law''). For example, by increasing by a factor of two, the exposure is decreased by a factor of four. Although this reduction in exposure is exactly true only for point sources, the use of distance, especially in handling radioactive sources, will greatly reduce exposure. When transferring an unshielded radioactive vial, for example, the use of tongs or similar remote handling devices will reduce exposure significantly, particularly to the hands. If time and distance are not enough to reduce the radiation exposure, one has to shield the radiation source. Shielding involves positioning of an absorbing material between the source and the area where the exposure is measured (i.e., where the personnel will be located).

2. Exposure from Sources

The alpha emitters should always be handled in glove boxes. The glove and the window of the box give enough shielding for external alpha irradiation. The external beta irradiation is also rather easy to control due to its short range. In water, tissue, and plastic materials, the range is about 0.5 cm for beta radiation with a maximum energy of about 1 MeV, but one must always take the production of bremsstrahlung into consideration when materials of high atomic number such as lead are used.

Gamma radiation is more penetrating than both alpha and beta radiations, and therefore all of the above mentioned factors must be considered in order to obtain an optimal and economic shielding against gamma radiation. When planning an operation involving gamma-emitting radionuclides, it is necessary to estimate the exposure of personnel. This can often be done with sufficient accuracy by using Equation 16.

$$H_\gamma = \Gamma \frac{A \cdot t}{d^2} \cdot T_d \ (\mu Sv) \qquad (16)$$

where $H\gamma$ is the dose equivalent due to gamma ray exposure; Γ is the specific gamma constant ($\mu Sv\text{-}m^2\text{-}MBq^{-1}\text{-}h^{-1}$); A is the activity of the source (MBq); t is the time estimated to perform the preparation (hours); d is the distance between the source and the personnel (meters); T_d is the transmission through the shielding barrier of thickness d (meters), calculated from the following equation:

$$T_d = B_H \cdot 2^{-z/HVL} \qquad (17)$$

where B is the dose equivalent build-up factor; z is the shield thickness; and HVL is the half-value layer.

The thickness of the shielding material necessary to reduce the exposure by half is called the half-value layer (HVL). The HVL is a function of the energy of the radiation and the

Table 8
SPECIFIC GAMMA CONSTANTS FOR VARIOUS RADIONUCLIDES

Radionuclide	Specific gamma constant (at 15 cm)	
	mR/h-mCi	μSv/h-MBq
^{99m}Tc	3.0	80
^{131}I	9.3	250
^{133}X	0.6	16
^{99}Mo	6.1	165
^{203}Hg	6.0	162
^{75}Se	8.4	227
^{169}Yb	4.0	108

material used for shielding. For example, the HVL of lead for ^{99m}Tc (140-keV photons) is 0.31 mm, while for ^{131}I (364-keV photons) the HVL is 2.3 mm. The HVL of concrete for ^{99m}Tc is 19.3 mm. The total thickness of the shielding material necessary to reduce the dose equivalent also depends on the distance from the source and the output from the gamma ray sources. The latter is given by the specific gamma ray constant. In Table 8 the specific gamma constants for some frquently used radionuclides in nuclear medicine are given.

In addition to shielding the working area, all vials and syringes should be shielded to reduce exposure to the hands. Vial holders that allow only the rubber portion of the vial to be exposed while a dosage is withdrawn are preferred, since these holders minimize exposure. These vials may be made from lead, lead glass, or lead glass and steel. The vial shields should be clearly labeled to prevent accidental use of the wrong vial. If a large number of radiopharmaceuticals are used, a vial shield color coding system is preferred. Shielded vials may be stored at the rear of the working area and may be additionally shielded by placing them behind lead bricks.

The most critical radiation exposure to personnel in a nuclear medicine department is exposure to the fingers when handling syringes with a high amount of activity for injection. Filled syringes to be used for patient injections should be shielded during transport and, if practical, during injection. Holders may be used for individual syringes. If a large number of syringes are to be transported at one time, a shielded carrying tray may be used. A tray may provide the safest method of transportation if the hot lab is located a distance from the area where patients will be injected.

3. Exposure from Patients

Patients who have been injected with radioactive materials also represent radiation sources for potential exposure of personnel. For example, the exposure rate from a patient administered 20 mCi (740 MBq) of ^{99m}Tc-DTPA for a brain scan was about 1 to 2 mR/h (1000 to 2000 μSv/h) at a 1-m distance. Use of a lead apron (0.5-mm lead) to drape the trunk of the patient reduces this exposure to less than 0.5 mR/h (500 μSv). This procedure does not interfere with the study or inconvenience the patient. Although use of a lead apron is not appropriate in many situations, it may be used effectively with brain scans to provide additional reduction in the exposure of personnel to radiation. The technologist should also wear a lead apron when working with high activities and should make every effort to remain as far from the patient as is reasonable during the study.

4. Shielding of Radioactive Waste

All radioactive vials, used syringes, and contaminated materials (e.g., gloves, cotton swabs, and absorbent paper) must be stored immediately after use and prior to disposal.

The storage area should be shielded and located in a locked area away from the flow of traffic in the department; often a corner of the hot lab may be used. A bin of lead bricks lined with plastic garbage bags is adequate. The bin may contain receptacles for separation of vials, syringes, and needles and such items as contaminated gloves, swabs, and papers.

C. Contamination Control

1. Normal Conditions

With the transfer of large amounts of radioactive materials in the nuclear medicine laboratory, there exists the potential for radioactive contamination. This is a serious problem from the aspect of potential external and internal exposure of personnel and contamination of equipment. A number of general procedures should be followed whenever working with unsealed sources of radioactive materials. Knee-length labcoats should always be worn and checked regularly, especially if contamination is suspected. When handling radioactive materials' containers, gloves should be worn, preferably latex surgical gloves, since they are less permeable to contamination. Gloves should be changed after each operation and disposed of in a waste receptable to prevent cross-contamination. Also, gloves must be removed before touching objects such as a pencil, a telephone receiver, or when using the radionuclide calibrator. All working surfaces should be covered with plastic-lined absorbent paper. In the event of a spill, the paper will prevent the spread of radioactive liquid and may be easily removed as radioactive waste.

A survey meter should be used for detecting radioactive contamination that emits gamma rays and/or high-energy beta particles. Potential causes of contamination within the department should be identified and appropriate preventive measures should be taken. All cotton swabs used in maintenance techniques must be handled as possibly contaminated, including those used during dosage preparation and patient injection. Small receptacles near working areas may be used to hold such contaminated waste until transfer is made to the departmental waste storage area. Patients who have received ^{99m}Tc and ^{131}I orally present a critical source of contamination, since saliva and mucous secretions contain radioactivity which may contaminate other objects in the room. Care should be taken to cover surfaces that the patient may touch (e.g., the gamma camera collimator) with disposable material such as a plastic wrap. There is also a risk of contamination from urine and feces. A particularly critical contamination problem is the care of ^{131}I therapy patients. Special procedures for care of these patients should be available in every department where ^{131}I therapy is performed.[12]

2. General Decontamination Procedures

Radioactive accidents with contamination of working areas and personnel occasionally occur in the nuclear medicine laboratory. Identification of potential problems and the subsequent corrective measures will help prevent such accidents. In the event of a radioactive accident, decontamination techniques are needed that will minimize exposure to personnel and reduce contamination of areas of the lab, particularly to prevent contamination of radiation detection equipment.[13]

Before a radiation accident occurs, decontamination procedures such as those given below must be developed. All personnel should be familiar with and trained in dealing with the procedures. A package containing these supplies for decontamination procedures should be easily accessible in each lab using radioactive materials:

1. Detergent and scouring powder for cleansing
2. Generous supply of paper towels for absorbing waste liquid waste and covering clean areas
3. Radioactive warning tape and signs for isolating the area
4. Plastic bags, both large and small, for containing waste and serving as shoe covers

5. Generous supply of disposable latex gloves and small plastic bags to cover hands and survey instruments and disposable smocks, jump suits, and caps for personnel involved in decontamination procedures
6. Scrub brushes for scrubbing the contaminated lab areas and skin
7. Scissors
8. Marking pen to label waste containers
9. Masking tape for sealing bags
10. Filter paper for wipe tests during and following decontamination

Survey instruments either should be available and/or the nearest location of survey instruments should be posted.

3. Procedure for a Major Radioactive Accident

a. Radioactive Liquid Spill

A major accident in the nuclear medicine laboratory will probably be a liquid radioactive material spill. A satisfactory procedure to follow is[13]

1. Contain the spill with absorbent paper, shut off ventilation fans, evacuate the area, and prevent reentrance.
2. Immediately notify the lab supervisor and radiation safety officer. If personally contaminated, either have someone else perform the notification while you proceed with personnel decontamination procedures or remove contaminated clothing, cover hands with gloves if contamination is suspected, and proceed with notification. Every effort must be made to prevent spread of contamination.
3. Survey all persons who may be contaminated. Proceed with the personnel decontamination procedure. Do not allow anyone to leave the area until it has been determined that they are not contaminated.
4. Estimate the exposure level. With high exposure levels, more than one occupational worker may need to be enlisted to aid with the initial decontamination procedure, thus limiting the amount of exposure to any one individual. This should be done under the supervision of the lab superviser or radiation safety officer.
5. Using protective clothing, footwear, and gloves, start decontamination procedures. Work from the outside of the spill toward the center to prevent spreading the spill over a larger area. Working quickly, yet carefully, absorb all liquid and place it in a plastic bag. Tape the bag shut and place it in another bag to prevent further contamination. Tape the second bag and label and store it in a shielded area. This procedure will remove much of the radioactivity, allowing the decontamination to proceed at a slower, more deliberate pace. One person should remain uncontaminated to transfer waste as needed and to operate survey instruments.

b. Area Decontamination — Low-Level Exposure

The following applies after contaminated personnel and the major contamination have been confined and removed as outlined in the five points above:[13]

1. With a survey meter, or using wipes of areas with suspected contamination, define the border of the contamination area with tape. Cover the surrounding clean area to help prevent accidental tracking of contamination. This covering should be changed often in order to prevent the spread of the contamination to larger areas. Start at the outside of the contaminated area and clean toward the center, using detergent and small amounts of liquid. Absorb contaminated liquid with paper towels and place them in a marked waste bag. Abrasive powder and a scrub brush may be used to remove

persistent contamination. Care should be taken not to splatter the material by use of the brush. Contaminated wax on tiles may be removed with a strong wax remover. Check frequently with a survey meter to determine when an area is below twice the background. If another person is not available to help with surveying, remember to change gloves before operating the survey meter. A disposable glove placed over the detector probe will help prevent accidental contamination.

2. When the entire area is thought to be decontaminated, survey thoroughly by obtaining wipes and checking these with the survey meter or in a well counter. If no contamination above twice the background is present, remove the masking tape and place in waste.
3. Survey protective clothing and place it in a storage or waste area if contaminated. Check for personal contamination and proceed with personal decontamination procedures if necessary.
4. Remove all contaminated waste to the waste storage area.
5. Develop ways by which a repeat of the incident may be prevented.

Some contamination may not be reduced to twice background by cleaning procedures, and other methods may be required. Tile may be removed and replaced. Since most radioactive accidents in the nuclear medicine lab involve radionuclides with a short half-life, areas of persistent low-level contamination may be covered with paper or plastic and a lead sheet if necessary and the radioactive contamination may be allowed to decay. These areas must be labeled and checked with a survey meter after an appropriate time has elapsed.

4. Personnel Decontamination

All persons who may have been contaminated during a spill or area decontamination procedure must be checked for contamination before leaving the area. A survey meter may be used for this purpose.[13]

1. Check clothing and belongings for contamination. Remove contaminated articles and determine the extent of skin contamination.
2. Local areas of skin contamination may be decontaminated by spot cleaning. Adjacent areas of uncontaminated skin may be covered with plastic and taped to prevent spread of contamination. Wash the skin thoroughly with soap or mild detergent and water, scrubbing gently with a soft brush and rinsing frequently.
3. If skin contamination is extensive, a thorough shower is necessary. Clean hair, hands, and fingernails carefully. Cut contaminated hair and fingernails if necessary.

VII. SAFETY OF ULTRAVIOLET RADIATION

A. Ultraviolet (UV) Radiation

UV radiation is that part of the EM spectrum lying between the softest ionizing radiation (X-rays) on one side and the visible radiation on the other. This range of the spectrum is classified into different zones according to Table 9.

Most sources of UV radiation in the hospital can be grouped together in the categories shown below:

1. Gas discharges
   Mercury lamps (low, medium, and high pressure)
   Mercury lamps with metal halides
   Noble gas lamps
   Flash tubes
   Hydrogen and deuterium lamps
   Welding arcs

Table 9
CLASSIFICATION OF UV RADIATION INTO DIFFERENT REGIONS

Region	Characteristic	Wavelength (nm)	Photonenergy (eV)
UV-A	Black light region	315—400	3.94—3.10
UV-B	Skin erythemal region	280—315	4.43—3.94
UV-C	Germicidal region	100—280	12.40—4.43
	Vacuum UV	<190	

2. Incandescent sources
   Tungsten halogen lamps
3. Flourescent lamps
   Flourescent tubes
   Flourescent sun UV emitters
   Black light UV emitters
4. Mixed sources
   Carbon arc

To permit the transmission of UV when the discharge does not take place in free air, gas discharge arcs and other UV sources must be contained within an envelope of quartz or UV-transmitting glass. On the other hand, sources that are designed primarily to emit visible radiation, but which also produce significant but unwanted amounts of UV, should be provided with an external screen of glass that does not permit the passage of UV-B and UV-C radiation.

The sun is the most important UV source. The broad spectrum and the intensity of the UV from the sun are due to the high temperature at its surface and its size. The intensity is such that the UV radiation reaching the earth atmosphere would probably be lethal to most living organisms on the surface. Fortunately, they are shielded by the atmosphere. The ozone layer in the upper atmosphere is particularly important in this connection.

1. Biological Effects and Safety Standards

a. Transmission and Absorption in Biological Tissue

Penetration of the human body by shorter wavelength UV is restricted to the epidermis. Penetration is somewhat deeper at longer wavelengths and in nonpigmented subjects, where there is some penetration into the dermis, especially at wavelengths greater than 300 nm. Most of the UV which hits the eye will be absorbed by the cornea. The lens and the tissues in the anterior part of the eye may, however, be exposed to UV at wavelengths above 295 nm. The final absorption takes place in the lens, and the retina can be exposed only under special circumstances.

b. Absorption and Photochemical Effects

The absorption of UV depends on the wavelength. Nucleic acids which are the biomolecules of greatest importance have their main absorption peak close to 265 nm, due to the pyrimidine structure. The absorbing sites in protein are the aromatic amino acids with absorption peaks at 275 nm for tyrosine and 280 nm for tryptophane. *In vivo* absorption of UV produces DNA interstrand cross-links and DNA strand breaks. In the UV-B region, the production of covalent dimers of thymidine is of greatest importance.

c. Pathological Effects in Man

Because of its limited penetration, the effects of UV in man will essentially be restricted to the skin and eyes; only under special circumstances may such effects also extend to the oral cavity.

Skin — Four immediate changes which occur in the skin are darkening of preexisting melanin pigment; production of erythema (sunburn); upward migration and production of melanin granule (suntanning); and changes in epidermal cell growth.

Mouth — As a consequence of recent dental practices, in which plastic materials are hardened (or cured) by UV treatment, the mucous membrane of the mouth may be exposed to unwanted UV irradiation when defective equipment is used. To date, only severe erythema of the skin around the mouth has been observed under such conditions.

Eye — The main clinical effects of UV on the eye are photokeratitis and conjunctivitis, which appear 2 to 24 h after irradiation. The symptoms are acute hyperemia, photophobia, and blepharospasm, which last from 1 to 5 d. In general, there is no residual lesion. Photokeratitis is caused preferentially by UV-B, but is also caused by UV-A, though with decreasing sensitivity. The peak sensitivity of the cornea appears at wavelengths of 270 or 288 nm. The effect depends on the total energy absorbed. The threshold has a minimum at 270 nm of 50 J/m^2, rising to 550 J/m^2 at 310 nm, and followed by a steep increase to 22,500 J/m^2 at 315 nm.

d. Late Effects

After prolonged exposure to sunlight over a period of years, the dermis will begin to degenerate, with a decrease in elasticity due to degeneration of the collagen fibers combined with other histological changes. The visible symptoms will be deep furrows in the skin, giving an appearance of premature aging.

The epidermis may also be involved, with the development of actinic keratosis. The importance of this lesion is difficult to evaluate, but the occurrence of an increased cellular proliferation rate and a certain amount of cellular atypia suggests that it may represent a precancerous stage in development of squamous cell carcinoma.

Cancer of the skin is a well-recognized effect of UV irradiation, both in experimental animals and in man. The three types of cancer concerned are, namely, basal cell carcinoma, squamous cell carcinoma, and malignant melanoma. The evidence for the production of cancer by UV is good for all types of malignancies. In man, the evidence for UV induction of skin cancer is both clinical and epidemiological. Basal cell and squamous cell carcinomas appear preferentially on the uncovered skin of lightly pigmented individuals. The latency period is in the order of 10 to 15 years for malignant melanomas.

e. Safety Standards

The standards established by the American Conference of Governmental Industrial Hygienists (ACGIH), which specify a threshold limit value, have also been used in preparing guidelines for other countries. This standard is based on the action spectrum for the MED (the minimal eythema dose) and the minimal photokeratitis dose in normal white-skinned individuals. This means that acute effects alone have been taken into consideration. For the UV-B region and at lower wavelengths, it states that the radiant exposure in an 8-h period must not exceed a given value (1000 J/m^2 for 200 nm, 30 J/m^2 for 270 nm, and 10,000 J/m^2 for 315 nm). For the wavelength range of 320 to 400 nm, the total irradiance on unprotected skin or eye must not exceed 10 W/m^2 for periods exceeding 1000 s (about 17 min). For radiant exposures of shorter durations, it should not exceed 10,000 J/m^2.

2. Hospital Use of Ultraviolet (UV) Radiation

Medical irradiation in the form of phototherapy is at present expanding. In the phototherapy of infants with neonatal jaundice (hyperbilirubinemia), incorrectly selected flourescent tubes have given rise to erythema. The use of certain photodynamic dyes and of light for the treatment of herpes, or of psoralens and UV-A for other skin diseases, such as psoriasis (PUVA), will expose the patient to a number of risks, but there is so far only experimental evidence as to their nature. Up to the present, the action spectrum of erythema is the only effect that has been well studied quantitatively. Neither the action spectrum of the effect on psoriasis nor the potential carcinogenic spectrum is known. If adequate safety precautions are not taken, the treatment personnel may also be at risk. The radiation facilities used in dermatology for the treatment of psoriasis (PUVA) provide an example where containment is necessary for the protection of the hospital employee. This appears to be difficult, but, as the UV emitted is mainly UV-A, it may be of minor importance.

VIII. SAFETY OF LASER (LIGHT AMPLIFICATION BY STIMULATED EMISSION OF RADIATION)

A. Introduction

The acronym "LASER" (light amplification by stimulated emission of radiation) is commonly applied to devices that emit an intense, coherent, directional beam of "light". This very special type of optical radiation is a result of a process whereby an electron or molecule undergoes a quantum jump from a higher to a lower energy state, causing a spatially and temporarily coherent beam of light to be emitted. The extremely collimated character and generally high degree of monochromaticity of the LASER beam make this device of potential value in the treatment of the eye and skin, in various diagnostic techniques, in surgery of the skin and internal organs and in dentistry, enamel scaling, bridgework, etc.

B. Physical Principles for Biological Interaction

The effect of LASER irradiation in biological tissue depends on the time distribution of the LASER pulse. There are three main types of effects characterized by the length of exposure time, t_{exp}, as follows:

$t_{exp} < 10\ \mu s$	(mechanical shock wave effect)
$10\ \mu s < t_{exp} < 10\ s$	(thermal effect)
$t_{exp} > 10\ s$	(photochemical effect)

When an object is exposed to very short LASER pulses ($<10\ \mu s$), the energy is absorbed within a very small volume. This small volume becomes extremely hot and gives rise to a microexplosion with a shockwave which can give rise to mechanical damage.

At longer exposure times ($10\ \mu s < t_{exp} < 10\ s$), the absorbed energy becomes distributed within a larger volume and the effect becomes milder and is in principle a conventional local heat effect. At still longer exposure times ($t_{exp} > 10\ s$), the primary effect will depend on the wavelength of the LASER. For a long-wave LASER, the thermal effect is still dominating, but for a short-wave LASER in the UW range, the primary effect can be of photochemical nature.

C. Biological Effects

1. Effects on the Eye

When LASER radiation hits the eye, it first penetrates the cornea and then the lens. These parts are transparent to light in the visual region, 380 to 750 nm. Only a small part, about 5%, of the radiation that passes through the eye media is used for vision; the greater part is absorbed in the pigment granules in the pigment epithelium and the choroid which underlie the photoreceptors (the rods and cones). The absorbed energy is converted to heat which can cause tissue damage or retinal burn.

The eye is normally adjusted to parallel incoming light which is focused to the "yellow spot". If in such a case the eye is hit by LASER light which is highly parallel, the intensity in focus can be extremely high. This can result in a retinal burn followed by considerable loss of vision. The eye does not record any blinding effect or sensation of pain, which is

Table 10
MINIMAL SKIN REACTION LEVELS FOR DIFFERENT LASER TYPES

Type of LASER	Radiant exposure (kJ/m^2)	Exposure duration (t)
Ruby LASER		
White skin	110—200	2.5 ms
Pigmented	22—69	2.5 ms
Q-switched ruby	2.5—3.4	75 ns
Argon LASER	40—82	1 s
Carbon dioxide	28	1 s
Neodynium glass		
Q-switch	25—57	75 ns
Neodynium-YAG	460—780	1 s

very treacherous. If the eye is hit directly by the light from a high-power LASER, serious bleeding and tissue explosion can occur which give rise to considerable damage.

LASER light in the far end of the visual region and in the infrared (IR) region can give rise to both lenticular and corneal damage. For the longer wavelengths, the IR damage mechanism appears to be thermal. The CO_2 LASER at 10.6 μm in its action on all materials containing water exemplifies the thermal nature for damage.

2. LASER Effects on the Skin

The biological effects of LASER on the skin are less serious for the eye because damage in the skin can easily be repaired and healed, but very intensive exposure can cause the pigments to disappear as well as cause the appearance of burns. Certain pharmaceuticals and cosmetics might increase the sensitivity of the skin. The exposure levels required to produce minimal reactions in human skin for six common LASER types emitting LASER light in the visible and IR regions are in Table 10.

3. Medical Use of LASER

For surgical use, the Co_2 LASER with a wavelength of 10.6 μm is the most used. With maximum power, i, the order of 60 W and focal spot of 0.1 mm in diameter, the CO_2 laser can be used for cutting tissue. The tissue can be coagulated locally by heating and in some cases it is heated so heavily that it is burned to black coal (carbonized). By defocusing the beam, it can be used for stopping the blood flow in blood vessels as large as 1 mm in adults and 1.5 mm in children.

The neodym-yttrium aluminium granat (YAG)-LASER with a wavelength of 1060 nm (1.06 μm) has lower absorption in water and tissue and penetrates therefore to larger depths. Light of this wavelength can also be led through optical fibers of quartz and used on various types of endoscopy. It is therefore particularly used in stopping bleedings in the esophagus and the ventricle. It is also very useful for tumor treatment in the bladder. By adjusting the power and exposure time, it is possible to coagulate the tumor without damage to the surrounding tissues and organs. The great benefit for patients who have undergone LASER surgery is that they are without pain after the surgery and are therefore easy to mobilize. Another benefit is that there is much less bleeding and LASER surgery is therefore the method of choice for hemophilic patients.

D. Hazard Evaluation and Safety Standards

1. General Procedures

The application of exposure limits in evaluating high-intensity optical sources requires information regarding the use and the frequency of exposure. Three aspects of the use of

Table 11
THE HAZARD CLASSIFICATIONS DEFINED BY THE OUTPUT PARAMETERS AND ACCESSIBLE LEVELS OF RADIATION

Class No.	Specification
1: nonrisk	LASER systems that are not hazardous
2: low risk	LASER systems (visible only) that are normally nonhazardous by virtue of normal aversion responses
3: moderate risk	LASER systems where intrabeam viewing of the direct beam and specular reflections may be hazardous
4: high risk	LASER systems where even diffuse reflections may be hazardous or where the beams produce fire hazard or serious skin hazard

the LASER influence the total hazard evaluation and thereby determine the application of control measures:

1. The intrinsic capability of the LASER to injure personnel
2. The environment in which it will be used
3. The personnel who operate the LASER and the personnel who may be exposed

The most practical general means for both evaluation and control of LASER radiation hazards are to classify the LASER systems according to their relative hazards and to draw up specifications for the appropriate controls for each class.

2. Classification of LASER Device Hazards

The hazard classifications given in Table 11 are defined by the output parameters and accessible levels of radiation. The basis of the hazard classification scheme is the ability of the primary LASER beam or reflected beam to cause biological damage to the eye or skin. A class 1 LASER is considered to be incapable of producing damaging radiation levels and is therefore exempt from any control measures or other forms of surveillance. A class 2 LASER or low-power LASER system may be viewed directly under carefully controlled conditions, but must have a cautionary label affixed to the external surface of the device. The moderate-risk class 3 category (or medium-power system) requires control measures to prevent viewing of the direct beam. Class 4 high-risk (or high-power) systems require the use of controls which prevent exposure of the eye and skin to the direct and diffusely reflected beam. In addition to the possibility of eye damage, exposure to optical radiation from such devices could constitute a serious skin hazard.

3. Practical Protection Measures

a. Measures to Prevent Damage by LASER Exposure

1. Never look into a LASER beam. An invisible LASER beam from a Nd-YAG-laser (1.06 μm) can be as dangerous as a visible red beam of a rubin laser (0.69 μm).
2. Protection measures must be made which to as large extent as possible protect human exposure to LASER radiation. In laboratory work, the LASER and the object which will be exposed must be shielded in such a way that the LASER cannot be switched on if the shielding is removed.

3. Areas where the LASER is placed must not be viewed by binoculars.
4. Special protection glasses with side protection must be used when working with the LASER.
5. No unprotected areas of the body are allowed to be exposed to direct LASER irradiation.
6. All persons in the vicinity of the LASER beam must be informed about protection measures.

b. Control of LASER Operating Personnel

Personnel who are operating LASER instruments should undergo an eye examination at least once a year. Observed changes would be photographed and registered. At any time when an accidental exposure is suspected, the person should also be examined.

IX. MICROWAVE AND RADIOFREQUENCY (Rf) RADIATION

A. Introduction

Microwaves and radio frequencies (Rf) are EM radiant energy in the frequency ranges of 300 kHz to 300 MHz for Rf and 300 MHz to 300 GHz for microwaves. In this frequency range, the EM interaction with the environment is described by the well-known Maxwell equations. In this formalism the EM radiation is described in terms of interrelated electric (E) and magnetic ($\mathcal{H}$) fields propagated through space in a wave-like pattern. The radiation intensity, which for MW/Rf is called "power density", is given by the vector product of the E and $\mathcal{H}$ field strengths.

The microwave or Rf radiation absorbed in biological tissue causes thermal effects by increasing the body or tissue temperature. Some nonthermal biological effects might be related to direct interaction of the EM fields and tissue. Most studies of the biological effects or health hazards from EM fields are conducted at frequencies in the range of 13 to 20,000 MHz.

B. Fundamental Properties of Radiofrequency (Rf) Fields and Their Interaction with Tissue

1. Propagation of Electromagnetic (EM) Fields in Free Space

The fundamental physical representation of EM fields is given by the well-known Maxwell equations. These equations also describe the interaction of EM fields with matter. An EM wave is described by its electric field vector ($\overline{E}$) and its corresponding perpendicular magnetic field vector ($\overline{\mathcal{H}}$). In free space, EM waves propagate uniformly in all directions from a theoretical point source. From a Rf or microwave source, the EM waves can propagate along a coaxial cable or inside a hollow metal pipe (wave guide). The EM waves can also be propagated outward into the space that surrounds a transmitting antenna, as in communications broadcasting. The exposure rate of Rf radiation in free space is given by the absolute value of S in watts per square meter (see Equation 8).

2. Propagation of Electromagnetic (EM) Waves in Tissue

The propagation of EM waves in a medium depends on the permittivity constant, ϵ (dielectric constant), and the permeability constant μ, which describe the electric and magnetic properties of the medium, respectively. These constants are often given relative to vacuum.

- vacuum permittivity constant

$$\epsilon_o = 10^7/4\pi c^2 = 8.85418 \cdot 10^{-12}\ C^2/Nm^2\ (F/m)$$

- relative permittivity or dielectic constant

$$\epsilon' = \epsilon/\epsilon_o$$

- vacuum permeability constant

$$\mu_o = 4\pi \cdot 10^{-7}\ \mathrm{T \cdot m/A\ (H/m)}$$

- relative permeability constant

$$\mu' = \mu/\mu_o$$

The total energy density in a medium is given by Equation 9. The power density in tissue in the direction of the wave propagation is given by:

$$S = \overline{E} \times \overline{\mathcal{H}} = \frac{E^2}{Z} = \frac{1}{377} \cdot \frac{E^2}{\sqrt{\mu'/\epsilon'}}\ (\mathrm{W/m^2}) \tag{18}$$

When an EM wave propagated in air impinges normally on any material having dielectric properties different from those of air, several complex events take place such as reflection, diffraction, absorption, interference, and polarization. These events cause changes in the intensity, wavelength, and direction of propagation of the wave.

3. Electromagnetic (EM) Fields Close to Antennas

Close to the antenna, there are very complicated relations between power density, electrical field strength, and magnetic field strength. In most cases, it is not possible to calculate the power density from analytical expressions of $\overline{E}$ and $\overline{\mathcal{H}}$. For a small antenna, these conditions are assumed to be within a distance of one wavelength from the antenna. At a greater distance, free space conditions are assumed. If the antenna is large in relation to the wavelength, free space conditions are assumed farther away than at a distance, R, which is approximately given by the following equation:

$$R = 2 \cdot a^2/\lambda \tag{19}$$

where a is the maximum geometrical extension and λ is the wavelength (meters).

4. Specific Energy Absorption Rate (SAR)

The specific absorption rate (SAR) corresponds to the rate of specific absorbed EM power in tissue. The absorbed power in the irradiated body depends on the electric conductivity (σ) and the density (ρ) of the medium and the square of the electric field strength (E). For a sinusoidal wave, the effective field strength is $E_{eff} = E_o/\sqrt{2}$ where E_o is the peak value. Thus, introducing this, the rate of specific absorbed power is given by:

$$P \equiv \frac{\sigma}{2 \cdot \rho} |E_o|^2\ [\mathrm{W/kg}] \tag{20}$$

The quantities ϵ and σ vary with the frequency of the EM wave and with the type of tissue as shown in Tables 12 and 13.[14-16]

In radiation protection literature, the specific power density is often given as the specific power density rate related to the incident power density, S (watts per square meter):

$$SAR_{rel} = \frac{P}{S_i} \left[\frac{\mathrm{W/kg}}{\mathrm{W/m^2}}\right] \tag{21}$$

where S_i can be either 1 or 10 $\mathrm{mW/cm^2}$ (10 or 100 $\mathrm{W/m^2}$). Thus, the value of SAR_{rel} can differ by a factor of 10 depending on which reference level is used. Therefore one has to be observant in reading the literature on this subject.

It is in principle possible to calculate the internal EM fields and the rate at which energy

Table 12
DIELECTRIC PERMITTIVITIES FOR VARIOUS TISSUES[14,15]

Type of tissue	Dielectric value	Frequency (mHz) 0.1	1	10	50	100	10^3	10^4
Muscle	ϵ'	$3 \cdot 10^4$	$2 \cdot 10^3$	220	90	75	54	40
	ϵ''	10^5	10^4	$1.2 \cdot 10^3$	400	80	27	21
Brain	ϵ'			240	110	80	40	35
	ϵ''		$2 \cdot 10^3$	500	180	95	17	15
Liver	ϵ'	10^4	$1.5 \cdot 10^3$	240	90	78	46	36
	ϵ''	$5 \cdot 10^4$	$5 \cdot 10^3$	800	195	110	18	11
Bone	ϵ'	8.0			7.2		5.8	4.9
	ϵ''	0.5			10.8		1.3	0.7
Fat	ϵ'				12	10	6.0	4.0
	ϵ''				17	12	2.0	0.7

Table 13
ELECTRIC CONDUCTIVITY IN MUSCLE/SKIN TISSUE AT VARIOUS FREQUENCIES[16]

Frequency, ν (MHz)	Conductivity, s (Sm^{-1})
1	0.400
2	0.455
5	0.556
10	0.625
20	0.625
30	0.625
100	0.885

is absorbed throughout the interior of the irradiated tissue. For such calculations, one has to know in detail the electrical and geometrical characteristics of the irradiated body and the external exposure conditions. Both such theoretical calculations and experimental measurements of the SAR show that the total amount and distribution of energy absorbed by an exposed animal or human will vary greatly depending on object size, field frequency, and field orientation.

The most significant result of such work is the demonstration that there is a characteristic resonant frequency for each subject at which plane wave exposure will result in a significantly higher level of total absorbed energy than that which occurs at other frequencies. This level is highest when the $\overline{E}$ field vector of the incident field is parallel to the long axis of the body.

The absorption of EM waves is very unevenly distributed among various parts of the body. Regions with small cross-sections get the highest SAR value, i.e., the neck and the lower parts of the extremities. At whole-body resonance, which occurs at about 47 MHz, the SAR localized in the legs is about five times that of the whole-body average, indicating its intense relative absorption in the legs.

Partial-body resonances also exist for the arms and for the head at frequencies of 150 and 375 MHz, respectively. Localized SAR values for the neck are seen to be relatively high throughout the frequency range of 100 to 400 MHz. The significance of the head resonance effect has been emphasized by Hagman et al.,[17] who modeled the head and neck with much improved resolution. In this study, the data show strong localized absorption at the front and back of the neck and in the center-base region of the skull; localized SAR values are two to four times the whole-body average.

C. Thermal Effects of Electromagnetic (EM) Field Exposure

1. Biological Response to Thermal Effects

Absorption of the Rf magnetic fields in biological tissue results in increased oscillation of molecules and generation of heat. If this occurs in the human body, a compensatory dilation of blood vessels causes an increase of blood flow and the removal of excess heat, which is dissipated largely through the skin. The lens of the eye is a notable exception, as it cannot lose heat as quickly as other tissues.

A measurable rise in temperature will occur if the rate of heat production or absorption exceeds the rate of heat loss. Changes of up to 1°C can occur in the body as a result of changing environmental or physiological conditions. If, however, a temperature rise of several degrees occurs and is maintained, then adverse biological changes result. The majority of biological consequences from high exposure levels of Rf radiation can be assessed in terms of elevated body temperature. This is the least controversial of the Rf radiation effects and can be predicted by applying classical biophysical and EM theory.

The maximum expected temperature increase is obtained without considering any heat loss to the surroundings and can be calculated from the following equation:

$$\Delta T = \frac{(SAR) \cdot t}{c_p} \tag{22}$$

where c_p is the specific heat of tissue (~3.47 $kJ \cdot kg^{-1} \cdot {}^\circ C$). Thus, with a SAR value of 2 W/kg applied for 10 min, the maximum temperature rise would be 0.35°C. The basal metabolic heat production increases with temperature according to the following relationship:

$$BMR(T_{CORE}) = 1.21 \cdot 1.07^{2(T_{CORE}-37)} \cdot (W/kg) \tag{23}$$

Below a 0.5°C rise in core temperature, no effects have been observed on reproduction, fetal weight, growth, development, hematological and immunological end points, hormone levels, and clinical blood chemistry.[19-28] An increase of 0.5°C is within normal oscillations of human temperature between the sleeping state and maximal level of daily physical activity.[29]

The physiology of cardiovascular response to externally applied heat in relation to thermoregulation has been reviewed by Rowell.[30] Cardiovascular changes as well as increased cutaneous blood flow with accompanying increase in cutaneous blood volume may occur. This is associated with an increased heart rate up to about 120 to 180 beats per minute, as well as changes in blood pressure and peripheral vascular resistance. The occurrence of tachyarrhythmias, ventricular arrhythmias, and premature ventricular contractions (PVC) has been reported as a result of clinical whole-body hyperthermia.[31,32]

Short-wave and microwave diathermy have been in therapeutic use for over half a century. The main effect of this type of therapy is to produce heat in deep tissues. The necessary SAR value is of the order of 50 to 100 W/kg, leading to local tissue temperatures of 40 to 42°C. One of the most common indications for treatment is pain relief.[33] This is, however, not an effect specifically induced by the EM field, but merely a result of local heating.

Teratogenic effects have been reported in animals according to O'Connor.[34] Experiments have shown that whole-body SARs in excess of 20 W/kg are required to cause anatomical deformities in the offspring of pregnant animals. These levels of absorbed power caused an increase of several degrees in the body temperature of the mothers and would have been fatal if prolonged. It is thus of special importance to protect pregnant women from such high Rf exposure levels which can cause temperature elevation in the fetus.

D. Biological Nonthermal Effects of Electromagnetic (EM) Field Exposure

1. Introduction

Neurophysiological and other effects resulting from exposure to EM fields at levels con-

Table 14
NONTHERMAL BIOLOGICAL EFFECTS OF ELECTROMAGNETIC (EM) FIELDS AT VARIOUS FREQUENCIES SEEN AT VARIOUS BIOLOGICAL LEVELS OF ORGANIZATION

Level of organization	Biological effects	Frequency	Ref.
Man	Spark discharge perception	50 Hz	35
	Nerve excitation by induction		36
	Microwave hearing	GHz	37
	Neurastenic syndrome	MHz—GHz	38
Animal	Peripheral nerve function and synaptic transmission in rat	50 Hz—kHz	39
	EEG changes in rabbit	MHz	40
	Cloudy swelling, vacuolization brain	GHz	41
Organ	Ca^{2+} efflux in chicken brain	50 Hz	42
	Ca^{2+} efflux in chicken and cat brains	MHz	42
	Bradycardia rat heart	GHz	43
Cellular	On-off signals in frog retinal cells	50 Hz	44
	Nerve cell fatigue	MHz	45
	Nerve-muscle osmotically available water	GHz	46

sidered to be nonthermal are seen at various levels of organization of the living organism. The most pertinent effects of EM fields are summarized in Table 14.[35-46] As can be seen from this table, nonthermal effects have been reported in the whole frequency range from 50 Hz up to several gigahertz. The question of whether the interaction of low-intensity EM fields with human beings on a nonthermal basis is to be considered in safety regulations is, however, still a matter of debate.

2. *Mutagenic Effects*

The mutagenic effects of microwaves and Rf radiation have been studies in various organisms with different results which have been evaluated by Leonard et al.[47] Yao observed chromosomal aberrations that were induced in the corneal epithelium of Chinese hamsters exposed to 2450-MHz microwaves at a power of 100 mW/cm^2.[48] Alam et al.[49] observed a significant increase in the frequency of chromosomal aberrations in Chinese hamster cells exposed *in vitro* (2450 MHz, 0 to 200 mW/cm^2). McRee et al.,[50] however, did not find any increase in the frequency of sister-chromatid exchange in bone marrow cells of a mouse after chronic exposure (2450 MHz; 20 mW/cm^2; 8 h each day for 28 consecutive days). Several investigators found no genetic effects from exposure of microorganisms to microwaves.[51-53]

No mutagenic effects of microwaves are also reported from a single exposure of *Drosophila melongaster*.[54,55] A recent study by Marec et al.[56] shows that repeated 2375-MHz microwave exposures have no effect on the frequency of sex-linked recessive lethal mutations in *D. melongaster*. The highest frequency of lethal mutations, 3 in 1118 tested chromosomes (0.27%), was noted at the power density of 20 W/cm^2 repeated 5 times in 10 min.

3. *Effects at the Cellular Level*

Lövsund et al.[44] used frog retina for studies of ganglion cell activity and its dependence on the magnetic field to which the preparation was exposed. The intensity of the cellular discharge was found to be dependent on the frequency and intensity of the magnetic field. The frequency of the field giving the maximal poststimulus discharge (20 Hz) agreed with that giving maximal sensitivity for magnetophosphenes in human volunteers.

The effect of continuous wave (CW) exposure to microwave radiation (2450 MHz) on isolated frog sciatic nerves has been studied by McRee and Wachtel.[45] They used SARs ranging from 0 to 100 mW/g and measured the effect on the viability of the nerves in terms of the ability of the nerve trunks to sustain a high firing rate over prolonged periods without suffering appreciable changes in excitability. The gradual loss of viability thus observed was suggested to depend on interference with the maintenance of the ionic concentration gradient across the membranes. Nerve-muscle preparations from frog skeletal muscle were exposed *in vitro* for 2 h at 10 mW/cm^2 by Portela et al.[46] They detected transient changes in a number of passive and dynamic electrical parameters of the cell membrane, in the propagation velocity of the action potential, and in the water permeability.

4. Effects at the Organ Level

Adey and collaborators have, in an extensive series of papers, described changes in the efflux of calcium in brain tissue following the application of an EM field.[42,57,58] Electrical gradients in the order of 50 mV/cm pulses increased the calcium efflux in the brain of the cat. Similar experiments were carried out with extremely low-frequency fields with electric field strengths between 5 and 100 V/m and frequencies between 6 and 32 Hz. The results were a decreased efflux of calcium for 6 and 16 Hz and nearly unchanged conditions for the rest of the frequencies. Frequency as well as amplitude ''windows'' were observed.[42,57,58] Recently, Blackman et al.[59] found that 50 MHz modulated with 19 Hz at certain power levels produced changes in the Ca^{2+} efflux from chicken brains exposed *in vitro.*

It has been found that microwave irradiation at an absorbed power level of about 2 W/kg can influence the heart rate of isolated perfused hearts.[43] The interaction mechanism causing the effect is not known, but the authors hypothesized that the microwave energy interacts with the remaining portion of the autonomic nervous system within the heart to produce the chronotropic effect.

5. Effects at the Organism Level

The effect of an amplitude-modulated Rf field on mammalian EEGs has been investigated by Takashima et al.[60] Acute irradiation of rabbits with 1 to 30-MHz Rf fields at levels of 0.5 to 1 kV/m modulated at 60 Hz had no effect. Abnormal EEG patterns did, however, develop in rabbits chronically irradiated for 2 h/day for 6 weeks at 1 to 10 MHz, 15-Hz modulation. The exposure took place between two large aluminium plates separated by a distance of 20 cm. The electrical field strength of the Rf radiation ranged from 0.5 to 1 kV/m. In EEG recording, silver electrodes were placed on the skull. After about 2 or 3 weeks of exposure, an abnormal EEG pattern developed. A low-frequency component around 4 Hz was enhanced and the high-frequency components decreased. These results were later corroborated by experiments performed by Takashima and Schwan.[60] In particular, after a long irradiation at 100 to 200 V/m, the EEG was characterized by the presence of large-amplitude spindle-like signals with a frequency of 14 to 16 Hz, which appeared at a mean interval of 20 to 30 s.

Histological studies of brain tissues after microwave exposure have shown pathological changes. Tolgskaya and Gordon reported on morphological changes in rabbits, rats, and mice.[41] One common feature in almost all cases was a swelling of the cytoplasm of nerve cells with a marked vacuolation in the hypothalamic region. Since the dosimetry of this investigation is not clear, it cannot be excluded that the effects may have been thermally generated.

E. Effects on Humans

1. Microwave Hearing

It is now well established that when humans are exposed to pulse-modulated microwaves above certain intensity levels, an audible sound occurs which appears to originate from within the head. The threshold for various pulse widths and frequencies have recently been given by Lin.[37] The estimated peak SAR in the head at the threshold was of the order of 0.5 to 1 W/kg. The time averaged incident power density was about 0.1 mW/cm^2. The interaction mechanism behind the phenomenon is thought to be purely thermal. The pulsed microwave energy intitiates a thermoelastic wave of pressure in the brain tissue which activates the inner ear receptors via bone conduction. It is not likely that the sound results from a direct interaction between the microwaves and the cochlear nearve or the neurons along the auditory pathway.[37] At the moment, the potential health hazard of this effect is under discussion.

2. The Neurasthenic Syndrome

Several investigators have performed epidemiological studies on workers exposed to Rf and microwave radiation. The complex situations of various occupational environments make it very difficult to determine the level of exposure. In most cases only a rough estimate is possible because of the wide variation in field intensity over the dimension of the body. It is furthermore difficult to specify exposure time since most processes have intermittent duty cycles. Many of the epidemiological studies have been criticized on the grounds that other environmental factors, which might influence the outcome, have been neglected. Specifically, such factors as noise, vibration, ventilation, lighting, and shift work should be given attention. Information about age distribution and occupation in the control groups is also inadequately provided in most studies.[38] The symptoms and signs commonly found more frequently in the exposed groups than in controls include headache, irritability, dizziness, loss of appetite, sleeplessness, sweating, difficulties in concentration or memory, depression, emotional instability, dermographism, thyroid gland enlargement, and tremor of extended fingers. These disturbances are often called ''the neurasthenic syndrome''.[38]

3. Epidemiological Studies

Epidemiological studies of the effects of Rf radiation exposure carried out in the Soviet Union and other eastern European countries report effects attributable to changes in the central nervous and cardiovascular systems.[61] The behavioral changes described are subjective in nature. They include headache, disturbance of sleep, fatigue, general weakness, lowering of sexual potency, depression, and irritability. Associated with these symptoms are endocrine and metabolic changes. These studies are, however, based primarily on clinical studies, rather than on examinations of specific diseases with subsequent correlation with Rf radiation exposures. Their utility is therefore limited and similar shortcomings are seen in the U.S. and western European studies.

Most of the studies that provide sufficient technical details have dealt with human populations exposed to low levels of Rf radiation without detecting any effect. Lilienfeld et al.[62] found no effect on life span or cause of death in persons exposed to Rf radiation over a period of years. Robinette et al.[63] also found no effect on life span, morbidity, or cause of death in adult males occupationally exposed to radar for periods up to 4 years. Cleary et al.[64] found no association between cataract formation and occupational exposure, and Cohen et al.[65] report no correlation between paternal exposure to Rf radiation for several years and Down's syndrome (mongolism) in the offspring. A comprehensive medical surveillance program conducted by an aircraft manufacturer did not detect any significant changes in the health status of 335 empolyees exposed to 400 to 8000-MHz radar of average power densities of >4 mW/cm^2 and a SAR of >0.12 W/kg.[66]

In a review of several epidemiological studies of workers and military and naval personnel in the U.S., Silverman concluded that there are no pathological changes related specifically to exposure to Rf or microwave radiation.[67] An association between microwave exposure and cancer has been examined in two prospective epidemiological studies, but neither has revealed an excess of any form of cancer that can be interpreted as microwave related.[67]

Although human epidemiological studies are few in number and have a principal weakness of inadequate information on the degree of Rf radiation exposure, there is no indication of any obvious relationship between low-level exposure and increased mortality or mobidity. No reasonable estimate of an upper exposure limit can, however, be derived from present human studies.

4. Oncogenic Effects

A mechanism for a relationship between exposure to magnetic fields and the induction of cancer has been proposed by Esterly.[68] According to this hypothesis, alterations in the mitotic processes caused by exposure to magnetic fields can provide a proliferative stimulus to latent tumor cells, thereby leading to the expression of malignant neoplasia.

Experimental animal studies have shown a proliferative reaction of microglia in the brain (gliosis) following exposure to EM fields in various ranges of frequencies.[69] Furthermore, recent experimental studies by Szmigielski et al.[70] indicate that EM fields (2450 - MHz microwaves) may promote the development of spontaneous and benzopyrene-induced skin cancer in mice. There are, however, also epidemiological studies which have not revealed an excess of any form of cancer that can be interpreted to be microwave related.[67]

a. Leukemia

Several reports have shown an increased risk of leukemia among electricians and other workers regularly exposed to EM fields.[71-74] It is, however, not yet fully understood if the EM fields or any other occupational exposure factor is the cause of this increased risk.

b. Brain Tumors

An epidemiological study was recently conducted to explore the association between occupational exposure to EM fields and the occurrence of brain tumors.[75] The study used data from the death certificates of 951 men who died of brain tumors. Death certificates were also obtained for a control group of men who died of causes other than malignancy and whose age and date of death matched with the case subjects. When the case group was compared with the control group, men employed in an electricity-related occupation, such as electricians, electric or electronic engineers, and utility company servicemen, were found to experience a significantly higher proportion of primary brain tumors, especially of glioma/astrocytoma. Furthermore, the mean age at death was found to significantly younger among cases in the presumed high EM exposure group. These findings suggest that EM exposure may be associated with the pathogenesis of brain tumors, particularly in the promoting state. In another study by Milham, a significant increase of mortality from benign brain tumors (5 observed vs. 0.7 expected) was found among aluminium reduction plant workers, particularly among those engaged in mechanics and electrical maintenance and metal production.[76]

In a study of brain tumor mortality among asbestos insulation workers, Seidman et al.[77] found that the excess of brain tumors among insulators was concentrated among the younger employees (those under 50) and during the early period after onset of work (15 to 24 years), in contrast with the age distribution and latency observed for other asbestos-associated neoplasms.[77] In recent decades there has been a rapid increase in the use of Rf EM fields in industrial, military, and civilian settings. During the same time (1940 to 1970), the mortality rate for brain tumors in the U.S. population has risen more than 50%.[78]

Despite the accumulation of epidemiologic evidence, it remains uncertain whether the increased risk of leukemia and brain tumors observed among electrical workers is due to the magnetic or electric fields themselves. Further epidemiologic and experimental studies are needed to explore the relationship between EM field exposure and tumorigenesis.

Table 15
SUMMARY OF ESTIMATES OF BIOLOGICAL EFFECTS OF ELECTROMAGNETIC Rf RADIATION AT DIFFERENT SPECIFIC ABSORPTION RATES (SAR)

SAR (W/kg)	Type of biological effect expected or reported
100	Increased embryonic and fetal resorptions, birth defects, postnatal weight decrements, and reduced survival on reirradiation of the offspring of mice exposed during pregnancy
50	No mutation in bacteria
(45)	No enzyme effects
	No change in lymphocyte transformation
	Increased K^+ efflux and Na^+ influx
	Intact mammalian systems affected
	Increased response of the immune system
	No effects on postnatal survival, adult body weight, or longevity
10	
<6	Gray area: "effects" with inconsistent results (90% of "effects" involves behavioral, hematological, or immunological elements, CNS structure and/or function, or hormone levels)
1.4	Core temperature <0.5°C
0.8—0.3	Onset of thermoregulatory response at 25% of resting metabolic rate (RMR)
0.4	SAR not likely to be associated with "effects"
<0.4	No significant change in health status

F. Summary of Biological Effects at Different Specific Energy Absorption Rate (SAR) Levels

For lower Rf exposure levels which do not increase the body temperature significantly, it is not possible to use the core temperature as an indicator of absorbed external energy. The whole-body-averaged absorbed Rf power (SAR) is then a concept typically used. The SAR is currently the parameter used most frequently to correlate nonthermal biological responses of Rf radiation energy absorbed by a biological system.

The relation between measured or estimated SARs and reported effects is summarized in Table 15. Experiments on rabbits have shown that the energy absorption in the eye sufficient to cause cataract is about 100 W/kg, resulting in a temperature rise of several degrees.[79] Values near 100 W/kg are associated with effects on mammalian systems, such as increased embryonic and fetal resorptions, birth defects, postnatal weight decrements, and reduced survival on reirradiation of the offspring of mice exposed during pregnancy.[34,80,81]

No effects are usually reported at SARs between 50 and 100 W/kg in biological systems such as *Drosophila* investigated by Hamnerius et al.,[55] bacterial cells studied by Blackman et al., or biopolymers examined by Hamrick and Allis.[83,84] At SARs in the range of 10 to 45 W/kg, *in vivo* studies have shown few effects on the irradiated systems if temperature is properly controlled. Some investigators reported no mutation induction in bacteria;[52,85-87] others found no effects on enzyme activities[88-90] or no effects on physical characteristics or on the structure of biomolecules such as nucleic acids and proteins[84,87,91] and no change in lymphocyte transformation.[92] Postnatal survival, adult body weight, and longevity appear to be unaffected at SARs between 10 to 45 W/kg.[23,93] No incidence of chromosome aberrations is reported in this SAR level by McRee et al.[94] and Huang et al.,[95] but Varma and Traboulay report no changes in the physical properties of DNA extracted from the testes of mice exposed to 15 to 65 W/kg for 80 min.[96]

Other studies report that the human red blood cells exhibit increased electrophoretic mobility, K^+ efflux, Na^+ influx, and hydrogen exchange at SARs between 10 and 45 W/kg.[97,98] Seaman and Wachtel observed a threshold at 7 W/kg for the increased firing rate of *Aplysia* pacemaker neurons.[100]

Table 16
POWER DISTRIBUTION FOR HUMAN INFANTS, CHILDREN, AND ADULTS CORRESPONDING TO A SAR VALUE OF 0.4 W/kg AT RESONANT FREQUENCY[115]

			Resonant frequency	
Age (years)	Average mass (kg)	Power density (mW/cm²)	Free space (MHz)	Ground plane (MHz)
1	10.0	1.2	190	95
5	19.5	1.1	140	70
10	32.2	1.2	95	50
Adult female	61.0	1.7	80	40
Adult male	70.0	1.7	80	40

Intact mammalian systems have generally been found to be affected by SARs in the 10 to 45-W/kg range, even at exposures as short as minutes or hours. Lethal effects have been reported by Chernovetz et al.,[101] along with fetotoxic effects by Berman et al.[19] Some components of the immune system appear to react with an increased response, whereas others demonstrate diminished responses. Those components reported to increase are levels of neutrophils or polymorphonuclear leucocytes and levels of splenic lymphocytes.[102-109] Furthermore, lymphocyte response to mitogen stimulation was observed by Huang and Mold[110] and Wiktor-Jedrejczak et al.[106-108] and an increase in lymphocyte transformation to the lymphoblast stage is reported by Huang et al.[95]

Biological effects at SAR levels below 10 W/kg can be arranged in two groups: "observed effects" and "no effects", respectively. The current literature deals with a limited range of biological variables which are summarized in Table 15. Almost all of the "no effects" data found in the literature are associated with SARs ≤6 W/kg regardless of species, exposure duration, or mode of exposure. This might indicate a break point at about 6 W/kg above which biological effects are more likely.

Below 6 W/kg, there seems to exist a "gray" area in which both positive and negative results are reported by investigators examining the same or similar end points. Positive effects reported with SARs below 6 W/kg involve behavioral, hematological, or immunological elements, central nervous system (CNS) structure and/or function, or hormone levels.

One of the most sensitive biological effects of Rf exposure is the reduction of an animal's ability to perform tasks for which it has been trained. This effect has been observed in rats and monkeys when the SARs exceeds 4 W/kg, resulting in a rise in body temperature of 1°C.[111,112] The threshold for such reversible effects seems to occur between 4 and 8 W/kg, in spite of considerable differences in frequency, mode of irradiation, and animal species.[113]

There are reports in the literature that quote biological effects at levels of Rf absorbed power below 4 W/kg. One example is the increased efflux of radiolabeled calcium ions from chick brain tissue exposed *in vitro* for 20 min to sinusoidally modulated 147-MHz radiation at a SAR as low as 0.0023 W/kg.[114] It is, however, uncertain how these findings relate to changes in human health.

The increased variability of the results of observed effects in the SAR range of 0.4 to 6 W/kg may suggest that this zone is at the lower limit of the effective SAR for some biological end points. In order to ensure an adequate margin of safety, it is proposed to limit whole-body Rf exposure to give a SAR of 0.4 W/kg. The power densities for human infants, children, and adults corresponding to a SAR of 0.4 W/kg is given in Table 16.[115] The guidelines for Rf exposure in medical nuclear magnetic resonance (NMR) examinations will be discussed in Section XI.

Table 17
STANDARDS FOR OCCUPATIONAL EXPOSURE (8 h/d) AND EXPOSURE OF THE GENERAL PUBLIC (24 h/d) TO RF RADIATION IN VARIOUS COUNTRIES[117,118]

Frequency range (MHz)	Exposure limit: Electric field strength (RMS) (V/m)	Exposure limit: Equivalent plane-wave power density (W/m²)	Exposure duration	Ref.
Czechoslovakia—1970				
10—30	50	—	8 h/d	
30—300	10	—	8 h/d	
30—300	1	—	24 h/d	
Poland—1972, 1977				
10—300	20	—	8 h/d	
30—300	—	2	10 h/d	
10—300	7	—	24 h/d	
30—300	—	0.1	24 h/d	
U.S.S.R.—1976				
10—30	20	—	8 h/d	
30—50	10 (0.3 A/m)	—	8 h/d	
50—300	5	—	8 h/d	
Canada—1966				
>10—300		100 (10 Wh/m²)	24 h/d (6 min pulsed)	
Sweden—1976				
10—300		10	8 h/d	
U.K.—1971				
30—300		100 (10 Wh/m²)	24 h/d (6 min pulsed)	
U.S.—1974				
10—300	200 (0.5 A/m)	100 (10 Wh/m²)	24 h/d (6 min pulsed)	
IRPA—1984				118
0.1—1	194	100	8 h/d	
1—10	$194/f^{1/2}$	100/f	8 h/d	
10—400	61	10	8 h/d	
0.1—1	87	20	24 h/d	
1—10	87/f*	20/f	24 h/d	
10—400	27.5	2	24 h/d	

G. Guidelines

The first standards for controlling exposure to Rf radiation were introduced in the 1950s in the U.S. and the U.S.S.R. The maximum permissible exposure levels proposed then have remained substantially unchanged; for continuous exposure these are, respectively, 10 mW/cm² and 10 μW/m². Most other countries developed national standards based on either the U.S. or the U.S.S.R. values.[113,116] Subsequently, however, some countries have proposed standards that fall between these extremes.[117] Standards for occupational exposure (8 h/d) and exposure of the general public (24 h/d) to Rf radiation in various countries[117,118] are summarized in Table 17.

These standards, however, do not apply to practitioners of the healing arts. For near field exposure, which is mostly used in medical applications, the only applicable Rf protection guides are the mean squared electric and magnetic field strengths which are given in Tables 18, 19, and 20. For convenience, these guides may be expressed in equivalent plane wave power density which is also given in the tables.[113,118-120]

Table 18
WEST GERMAN NATIONAL STANDARDS GUIDELINES FOR RF RADIATION[119]

Frequency, f (MHz)	Electric field, E^2 (V^2/m^2)	Magnetic field, H^2 (A^2/m^2)	Power density, $S = E \times H$ (W/m^2)
Exposures Greater than 6 min			
Federal Republic of Germany			
0.01—0.03	$(2000—1500)^2$	$(500—350)^2$	
0.03—2	$(1500)^2$	$(7.5/f)^2$	
2—30	$(3000/f)^2$	$(7.5/f)^2$	
30—3000	$(100)^2$	$(0.25)^2$	25
Exposures of Less than 6 min			
0.03—2	$\leq 1.35 \cdot 10^7$	$<337.5/f^2$	
2—30	$\leq 5.4 \cdot 10^7/f^2$	$<337.5/f^2$	
30—3000	$\leq 6 \cdot 10^4$	<0.375	≤ 150

Table 19
RECENT GUIDELINES FOR OCCUPATIONAL EXPOSURE TO RF RADIATION[113,120]

Frequency, f (MHz)	Electric field, E^2 (V^2/m^2)	Magnetic field, H^2 (A^2/m^2)	Power density, $S = E \times H$ (W/m^2)
Canada			
10—300	3,600 (1 h)	0.026 (1 h)	10 (1 h)
	90,000 (1 min)	0.64 (1 min)	250 (1 min)
U.S.S.R.			
0.06—1.5		25	
1.5—3	2,500		
3—30	400		
30—50	100	0.09	
50—300	25		0.01 (entire workday)
U.S.			
0.3—3	400,000	2.5	10
3—30	4,000	0.25	$90/f^2$
30—300	4,000	0.025	0.1

X. SAFETY OF EXTREMELY LOW-FREQUENCY (ELF) ELECTRIC AND MAGNETIC FIELDS

A. Introduction

Extremely low-frequency (''ELF'') electric and magnetic fields are generated by 50- and 60-Hz power lines as well as by electronic equipment of various kinds, in particular video terminals. The concern of biological effects of ELF fields started when future power transmission networks were planned to exceed 800 and reach up to 1200 kV.

Electric and magnetic fields are also generated whenever electric energy is used, whether in transportation (subway), industrial, hospital, or domestic activities. In the interest of health protection, the question has been raised as to the effects, if any, that these fields may have on man. At present the question of electrical and magnetic fields from video terminals is a matter of great concern in occupational health. There are indications that pregnant women working at video terminals have an increased risk of spontaneous abortion.

Table 20
INTERNATIONAL STANDARDS — GUIDELINES FOR RF RADIATION[118]

Frequency, f (MHz)	Electric field, E^2 (V^2/m^2)	Magnetic field, H^2 (A^2/m^2)	Power density (W/m^2)
General public exposure			
0.1—1	7569	0.0529	20
1—10	7569/f	0.0529/f	20/f
10—400	756	0.0052	2
400—2000	1.89 · f	0.0000137 · f	f/200
Occupational exposure			
0.1—1	37,636	0.26	100
1—10	37,636/f	0.26/f	100/f
10—400	3,721	0.026	10
400—2000	9 · f	0.000064 · f	f/40

Note: For occupational exposure, the limits in frequency ranges above 10 MHz may be exceeded for specific applications provided the SAR remains below 0.4 W/kg, when averaged over the body, and the spatial average SAR remains below 4 W/kg, when averaged over any 1 g of tissue.

B. Origin of Extremely Low-Frequency (ELF) Electric and Magnetic Fields

1. Power Transmission Lines

Electric energy is distributed from power stations to consumer centers in a high-voltage power transmission line network. In order to decrease the transmission power losses and make the use of the power more efficient and safe for the consumer, one tries to increase the voltage as high as possible. The electrical field strength beneath a high-voltage power line decreases rapidly toward the ground and away from the powerline. The magnetic field under a high-voltage alternating current overhead line has, at ground level, the properties of an elliptical rotational field. The maximum flux density of this field is to be found either along the trace axis of the line or along the outer conductors. The maximum flux density is approximately 10 μT/kA of loading current independent of the voltage.

2. Natural and Domestic Fields

Naturally occurring electric fields of low frequencies are usually of very low intensity. Intense electric fields, however, are created by various weather conditions with field strengths according to the following examples:

Fair weather fields	0.1 kV/m
Fields with atmospheric disturbances	1.5 kV/m
Thunderstorm fields	3—20 kV/m

By introducing electrical power in our daily world, we create an artificial electric environment. Table 21 shows examples of some levels of electric and magnetic fields in our domestic environment.

3. Videoterminals

a. Electrostatic fields

The high voltage which accelerates the electrons in a cathode-ray tube can give rise to a strong electrostatic field between the screen and the operator. A simular but much weaker field can originate from the electrostatic charge of the operator. Such electrostatic fields have been correlated to irritations and skin damage on the face of videoterminal operators. Detailed mechanisms about how this damage arises are not yet fully understood. One theory

Table 21
ELF ELECTRIC AND MAGNETIC FIELDS (60 Hz AT DISTANCES OF 30 AND 3 cm, RESPECTIVELY)

Equipment	Electric field strength at 30 cm (V/m)	Magnetic flux density at 3 cm (mT)
Grill	130	—
Stereo	90	—
Refrigerator	60	—
Flatiron	60	—
Toaster	40	—
Hairdryer	40	0.001—0.1
Color television	30	1.0—2.5
Vacuum cleaner	16	0.1—0.5
Electric bulb	2	0.01—0.1

is that air dust particles are accelerated in the electrostatic field and bombard the face of the operator who becomes irritated.

The electrostatic fields from 147 different types of video terminals have been studied by Paulsson.[121] He found a distribution between 0 (the limit of detectability) and up to 50 kV/m. Because of the potential risk of harmful effects due to the electrostatic field, it should be as low as possible. To protect the operator, it is rather easy to reduce the electrostatic field the videoscreen gives rise to by introducing a protective screen between the videoterminal and the operator. This might, however, decrease the image quality and make it more difficult to read the text on the screen. Guidelines for the electrostatic field from videoterminals recommend values below −1 and +1 kV/m. It is recommended that the keyboard should also be grounded in order to discharge the operator.

b. Magnetic Fields

The magnetic fields which control the electron beam in the cathode-ray tube also reach outside the screen. Since the magnetic field varies with time, the physical parameter which is of interest is the magnetic induction (i.e., the time derivative of the flux density) dB/dt (tesla per second). The values of the magnetic induction were also measured by Paulson, and he found values from 0 up to 350 mT/s.[121] Videoterminals working with higher sweep frequency (above 30 kHz) usually have high values of the magnetic field, but from a technical point of view, this is not necessary. Based on the experiences from these measurements, the guidelines recommend the following maximum levels:

B-value (peak to peak)	200 (nT)
dB/dt (peak to peak)	50 mT/s

These values refer to measurements in the horizontal plane at a distance of 30 cm or more away from the screen.

It is assumed that rapid development can permit still lower levels in the future. From radiation safety, this is a desirable development because the magnetic fields represent unnecessary radiation. For the same reason, the terminal would be equipped with an automatic switch which would switch off the magnetic field when the terminal is not in use.

C. Effects of Extremely Low-Frequency (ELF) Fields

1. Physical Effects

a. Electrical Discharges

Electric fields produce a capacitive charging of unground metallic objects. Discharges can occur if a grounded person touches such objects, the effect of such discharges depending on the amount of stored energy at the moment of contact. The limit of sensitivity is about 1 mJ, and to be dangerous to humans, the accumulated energy must be approximately 1 J.

b. Ion Concentration in Air

In enclosed rooms, e.g., laboratories, containing an internal electric field, the natural ion concentration in the air decreases. In contrast, changes in concentration of air ions around high-voltage lines cannot be readily observed, because of free air circulation.

c. Corona

Corona discharges from high-voltage electrodes occur if the voltage exceeds the corona inception voltage; this depends on the field gradient (electrode geometry) and irregularities at the electrode surface (particles, rain drops). Corona discharges may cause audible noise, the production of ozone (O_3), and possibly UV radiation. If the design of high-voltage equipment is such that the field gradient is sufficiently low, audible noise can be kept within tolerable limits and UV radiation can be avoided completely. Ozone is very unstable, rapidly reduces to oxygen, and is therefore difficult to detect.

2. Biological Effects

ELF magnetic fields generate internal electric currents according to the induction law. The induced current density "J" in a body with radius "r" and electrical conductivity "σ" can be written as:

$$\bar{J} = \frac{d\vec{B}}{dt} \cdot r \frac{\sigma}{2} = \text{const} \frac{d\vec{B}}{dt} \tag{24}$$

where σ is the electrical conductivity of the tissue. If we assume the radius of the head to be 0.1 m, the constant in equation becomes:

$$\text{const} = 10 \text{ mAs/m}^2\text{T}$$

For the trunk of the whole body, we assume r = 0.15 m which gives the value of:

$$\text{const} = 15 \text{ mAs/m}^2\text{T}$$

The energy imparted per time unit with a current density of $\bar{J}$ is then given by the expression:

$$P = \vec{E} \cdot \vec{J} = |\vec{E}|^2 \cdot \sigma \text{ (W)} \tag{25}$$

Where $\vec{E}$ is the electrical field and σ is the electrical conductivity of the tissue. The induced average electric field is $\bar{E} = |\vec{E}|\sqrt{2}$. Thus, the average absorbed power per unit mass or the specific absorbed power "SAR" is

$$SAR = \left(\frac{|\vec{E}|}{\sqrt{2}}\right)^2 \frac{\sigma}{\rho} = \frac{\sigma}{2 \cdot \rho} \cdot |\vec{E}|^2 \text{ (W/kg)} \tag{26}$$

Ventricular fibrillation is the most serious adverse response that may be anticipated during exposure to rapidly changing magnetic fields. The safe threshold for heart fibrillation has been found to be a complex function of waveform, amplitude, time duration, and repetition rate. The threshold is reported to be 300 Am^{-2}, with an electrode area of 1.8 cm^2.[122-124] According to Roy, the change in field with time for the induction of fibrillation is probably over 500 T/s, with the assumption of approximately 10 mAm^{-2} for 1 T/s scaling.[123]

In tissues ELF magnetic fields induce electric currents which can be sufficiently large to interfere with the normal function of neurons.[44,125] From experimental data, it is inferred that the threshold value will be lowest when the pulse width exceeds about 10 ms.[126-128] The best documented biological effect of ELF magnetic fields is the induction of visual light flashes known as magnetophosphenes. This phenomenon was first described by D'Arsonval in 1893 and has later been studied by other investigators.[44,129-132] In a recent experimental study, the induction threshold for magnetophosphenes was found to be about 1.3 T/s (rise time of 2 ms) for younger males and about 1.9 T/s for older males.[133] The induction of magnetophosphenes also requires a minimum reached B_{max}, which for frequencies of 20 to 30 Hz has been reported to be approximately 10 mT.

Developmental effects have been reported in fertilized chicken eggs incubated for 48 h exposed to weak (0.12 to 12 μT), 10 to 10,000-Hz magnetic fields.[134] These experiments have given rise to suspicions that magnetic fields from videoterminals might cause spontaneous abortions and malformation even in humans. There is, however, no definite evidence from conducted epidemiological studies that there is any correlation between the ELF fields recorded in front of videoterminals and pregnancy complications in humans.

The biological effects of ELF magnetic fields of various intensity have been studied by various investigators.[134-136] They have found indications of a linear relationship between the rate of change of the field and the induced current density. The effects seem to also be related to the length of exposure.

D. Recommendations

There are no specific recommendations for occupational exposure to ELF electric and magnetic fields other than the guidelines for work at videoterminals:

Electrostatic fields:	In the interval −1 to +1 kV/m; electrostatic discharge of the working place
Magnetic fields:	dB/dt less than 50 mT/s; B less than 200 nT; Switch off the field when not in use

XI. SAFETY ASPECTS OF CLINICAL NUCLEAR MAGNETIC RESONANCE (NMR) EXAMINATIONS

A. Introduction

In "standard" whole-body NMR-scanning equipment, the static magnetic field is generated by an electric current driven through large solenoid coils. Dynamic magnetic gradient fields are generated by electric current pulses in coils located at various orientations, thus

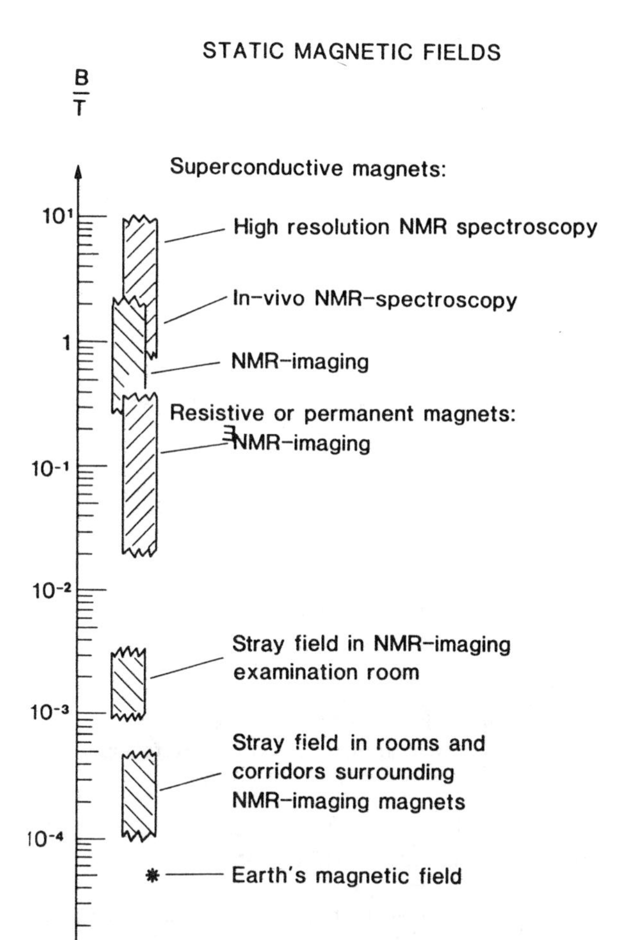

FIGURE 1. Levels of static magnetic fields used in various NMR examination situations compared with levels in surroundings and earth magnetic field.

producing magnetic gradients in x-, y-, and z-directions. The Rf radiation is transmitted through a specially shaped coil which also serves as an antenna receiving the NMR signals.

The maximum static field strength which can be reached with resistive whole-body magnets based on coils without iron is about 0.15 T. By using an iron core or shields, it is possible to reach about 0.3 T for resistive magnets which is also the upper limit for permanent magnets. Superconductive magnets, however, are being used with field strengths of up to about 2 T for whole-body NMR imaging and spectroscopy, respectively. Smaller magnets for laboratory purposes can reach field strengths up to about 10 T.

The levels of magnetic fields used in whole-body NMR magnets are shown in Figure 1. The magnitude of magnetic fields in the vicinity of NMR installations is in the range of 1000 to 10,000 μT. The stray fields in the corridors surrounding NMR imaging magnets is about 500 μT (5 G) which is one magnitude higher than the earth magnetic field, which is about 50 μT (0.5 G).

In NMR imaging the time of exposure to time-varying magnetic fields is very short, but of high magnitude. During rapidly rising and falling gradients, the time derivate (dB/dt) can be about 1 T/s inside the magnet. Levels of time-varying magnetic fields used in NMR imaging compared to those occurring in other applications are shown in Figure 2.

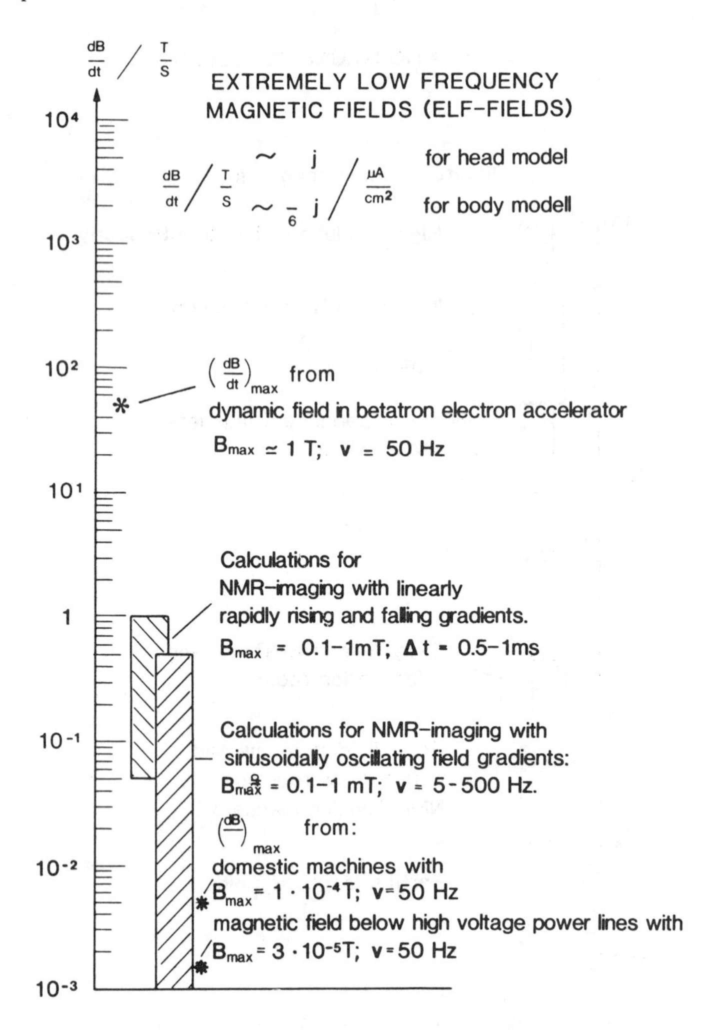

FIGURE 2. Levels of time-varying magnetic fields in NMR examination situations compared with various extremely low-frequency (ELF) magnetic fields in the environment.

In clinical NMR examinations, the proton resonance is at present the most commonly used, but in the future *in vivo* NMR spectroscopy of both sodium and phosphorous will be used. A comparison between the proton resonance frequencies obtained at various magnetic fields and the corresponding frequency bands of EM radiation that are used for broadcasting and for various industrial and medical purposes is shown in Figure 3. It is evident that the most commonly used NMR proton imaging frequencies are also used for several other applications. Therefore the Rf noise level is quite high and it is important to shield the NMR equipment from external Rf noise in a Faraday's cave.

The biological effects of Rf fields are discussed in Section IX and the biological effects of gradient switching which is equivalent to ELF magnetic fields are discussed in Section X. The most specific exposure of NMR examinations is the static magnetic field and this will therefore be considered in this section.

B. Possible Health Effects of the Exposure to Static Magnetic Fields in Nuclear Magnetic Resonance (NMR) Examinations

1. Static Magnetic Fields

The distortions caused by magnetic fields on cells or cellular components surrounded by

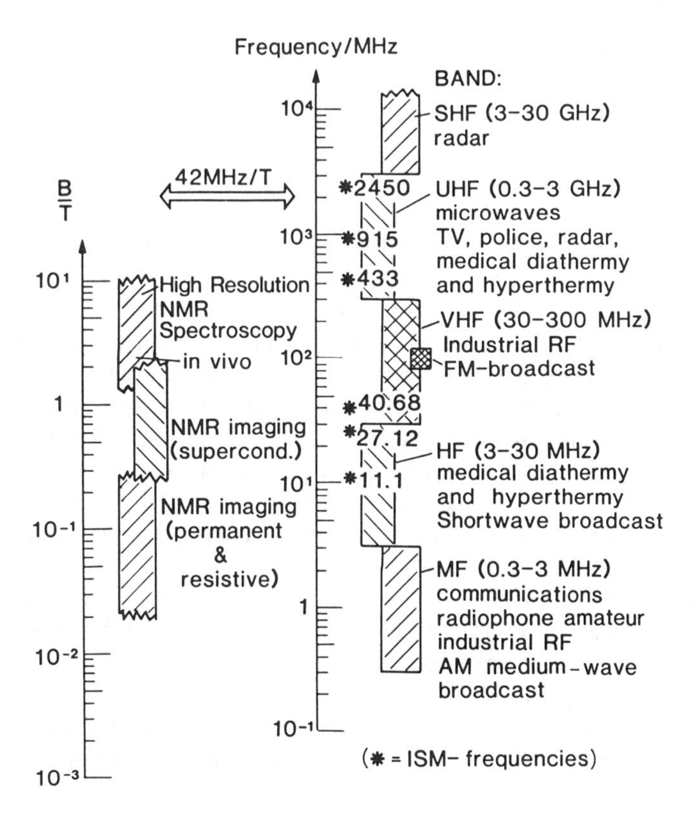

FIGURE 3. Levels of static magnetic fields used in various NMR examination situations with corresponding proton resonance Larmor frequency compared with other uses of these frequency bands.

a liquid of different permeability give rise to elastic forces (stress) associated with binding forces of the matter comprising the liquid and the body. In addition to these tensile strengths, molecules or subcellular structures with anisotropy in magnetic susceptibility will be influenced by a twisting force dependent on the orientation to the applied magnetic field and proportional to the field squared. It is thus theoretically possible that the forces exerted on molecules by applied magnetic fields could induce conformal changes in enzyme structure, and thereby altering biochemical reaction rates.

Among the effects of static magnetic fields on living matter at the cellular level growth retardation, changes in cell metabolism, increase in cell permeability, and increased excretion of Na^+ and K^+ in mice have been reported.[137,138] On the organ level, effects have been reported on heart function in squirrel monkeys, on abnormalities in the adrenal cortex in mice, and on decrease of potassium and calcium content in mice liver.[139-142]

In nuclear physics accelerator laboratories and in the vicinity of controlled thermonuclear reactors, people can be occupationally exposed to quite high magnetic fields. Various, but not harmful, effects have been reported, i.e., when the hand is placed in a static field of 10 T, one experiences a feeling of bitter coldness and aching bones accompanied by a sensation of ants moving on the hand.[143] People with metallic fillings in their teeth who are exposed to magnetic fields of up to 1.5 T experienced a strange taste sensation in the mouth. Similar reports of pain in filled teeth has been reported in an individual exposed to 2 T for about 15 min.[143] These effects, however, disappear when the exposure ceases.

On the organism level, steady magnetic fields induce a flow potential. Thus, in a magnetic field applied to man, the flow of blood perpendicular to the field creates a potential "U" across the diameter "d" of the blood vessels proportional to the flux density "$\bar{B}$" and to the rate of blood flow "$\bar{v}$":

$$U = \bar{B} \cdot d \cdot \bar{v} \qquad (27)$$

The amplitude and polarity of the flow potential depends on the orientation of the blood flow and hence the body, with respect to the magnetic field.

In man, the induced flow potential will be most significant in the major blood vessels around the heart and in the heart itself. With a peak aortic blood velocity of 0.63/s and an aorta diameter of 0.025 m, the theoretical peak flow potential generated in the aorta wall is estimated to be 16 mV/T. This potential will be applied across the diameter of the vessel and will therefore be very much lower across individual cells for which the depolarization threshold is 40 mV. Only a small fraction of this potential will be expressed across individual cells. The exposure conditions used in NMR imaging has not resulted in any change in the ECG of humans during or after exposure, but the margin of safety is as yet unknown.[144]

A Russian survey of occupationally exposed personnel (0.5 to 5 mT) working in the permanent magnet industry and in the machine-building industry reported general symptoms, including headaches, chest pains, rapid development of fatigue, blurred vision, dizziness, loss of appetite, insomnia, itching and burning sensations in the wrists, and sweating.[145] Two U.S. surveys on exposure of man to low magnetic fields (20 mT), however, found no significant differences between exposed groups and control groups.[133,146,147]

The most significant reported effects of static magnetic field exposure to biological systems are summarized in Figure 4.[148-160] There seems to be no pronounced level below which no effects occur, and trends in the data suggestive of a dose response are hard to find.

2. Experiments in Nuclear Magnetic Resonance (NMR) Imaging Exposure Conditions

a. Body Heating from Radiofrequency (Rf) Exposure

In 1981, Bottomley and Edelstein presented a general expression for Rf power deposition in a body during NMR imaging:[161]

$$P_{max} = K \cdot \frac{s \cdot \nu^2 \cdot R^2 \cdot D}{\rho \cdot \tau^2} \cdot (2\theta/\pi)^2 \text{ (W/kg)} \qquad (28)$$

The equation above, however, does not consider attenuation of the Rf radiation. At higher frequencies, the surface SAR is less than what would be expected from a low-frequency extrapolation because reduction in magnetic field penetration is reduced, thus giving rise to a decreased flux linkage.[162] This attenuation factor is negligible below a frequency of some 10 MHz, but increases from 0 to 90% in the frequency range of 10 to 100 MHz. The surface SAR has been calculated for different imaging situations, using Equation 28 and a correction factor for Rf attenuation derived from results presented by Bottomley and Andrew.[162] The SAR values highly depend on the way the dissipated energy is averaged over the entire length of the sequence, as can be seen from Figure 5.

Experimental results indicate, however, that the actual power deposited is about twofold lower than that estimated from Equation 28. The disagreement can partially be explained by differences in geometric configurations, in the heterogeneity and compartmentation of the human sample, and by the use of the worse case resistivity (muscle) for computing in the curves. The equation above assumes a uniform Rf field across the sample and the observed differences might also depend on the above mentioned Rf attenuation effects.

A measurable rise in body temperature will occur if the rate of heat production or absorption

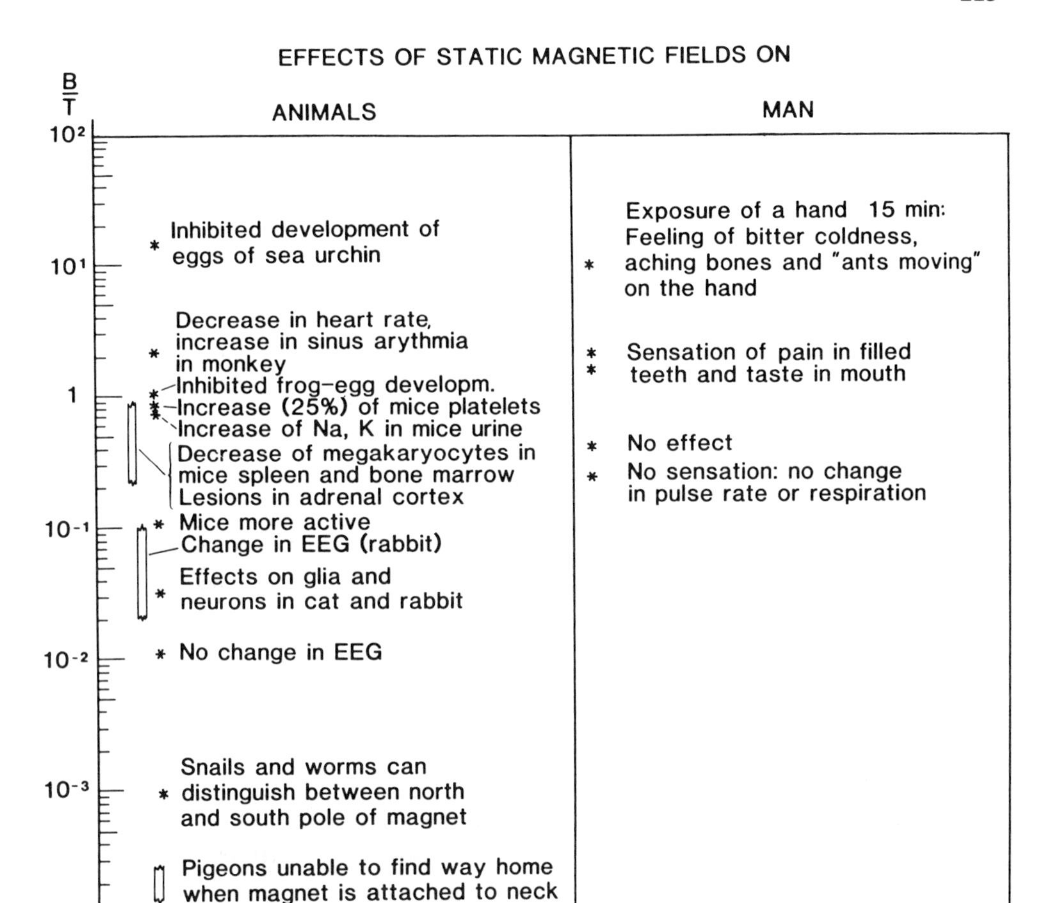

FIGURE 4. A short summary of the reported effects of static magnetic fields on animals and man. More details are given in the text.

exceeds the rate of heat loss. Changes of up to 1°C can occur in the body as a result of changing environmental or physiological conditions. If, however, a temperature rise of several degrees occurs and is maintained, then adverse biological changes occur. Below a 0.5°C rise in core temperature, no detectable effects on reproduction, fetal weight, growth, development, hematological and immunological end points, hormone levels, or clinical blood chemistry have been observed so far.[20-28,92]

b. Other Biological Effects of Nuclear Magnetic Resonance (NMR) Exposure

The effect of NMR exposure, i.e., the combination of static magnetic fields, time-varying gradient magnetic fields, and Rf radiation, has been investigated by several authors and thus far no effects have been found on *Escherichia coli,* on human lymphocytes, on the DNA of Chinese hamster cells, or on mutation and cytotoxicity of mammalian cells.[164-167]

C. Physical Hazards

1. Metallic Implants in Patients

A potential hazard for patients examined with NMR arises from the longitudinal forces and torques exerted on ferromagnetic implants such as aneurysm clips. Stainless steel aneurysm clips and other clips are regarded by many as being nonferromagnetic, but many such clips have revealed considerable ferromagnetism within the magnetic fields considered in

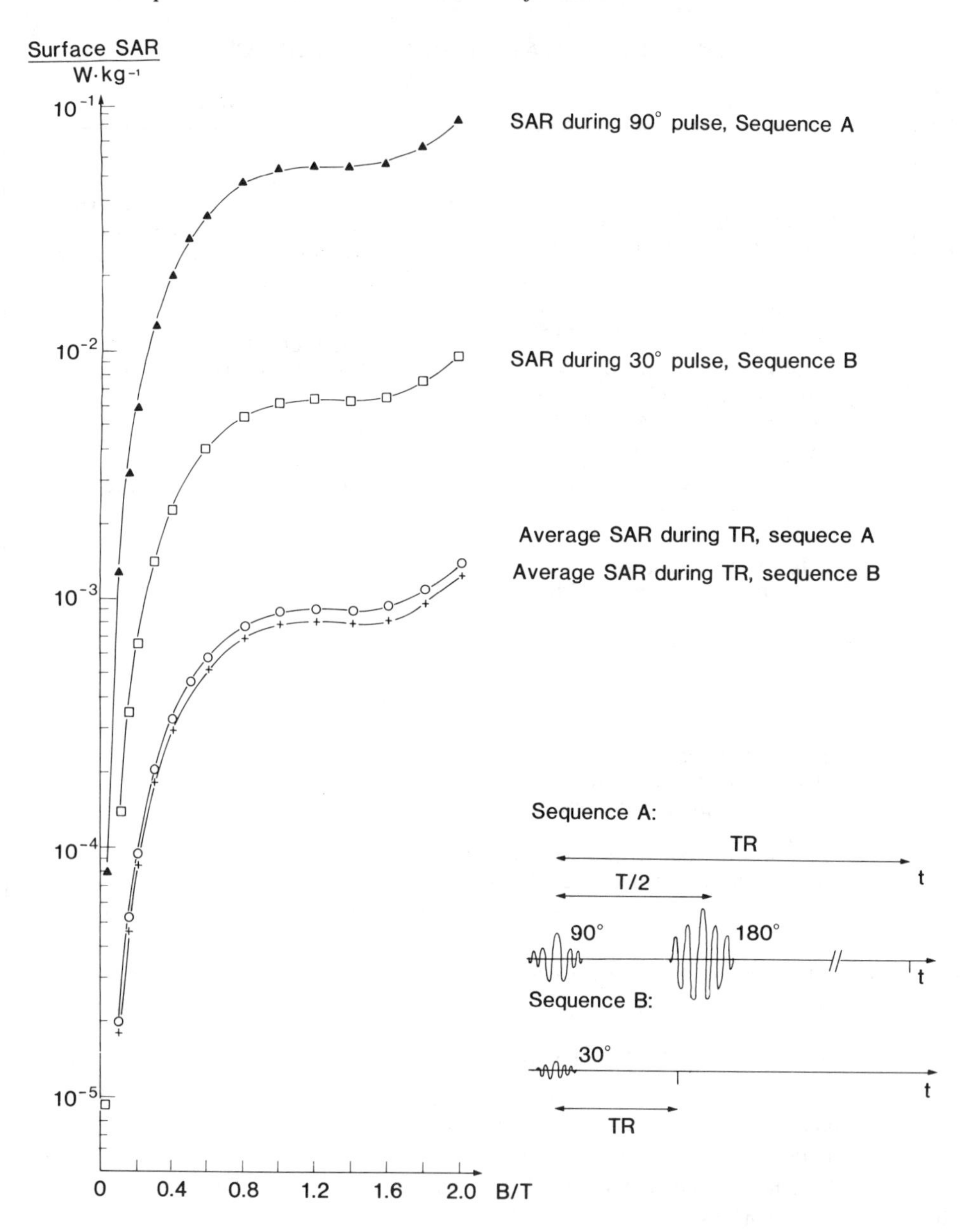

FIGURE 5. Calculated surface specific energy absorption rate (SAR) values for 90°-τ-180°-T pulse sequence with spin-echo time, TE = 50 ms, and recovery time, TR = 2.0 s, for three different averaging methods including radio frequency (Rf) attenuation effects. The length of the 90 and 180° pulses were assumed to be 250 and 500 s, respectively.

NMR imaging. It can, therefore, reasonably be expected that surgical aneurysm clips will rotate intracranically, with a corresponding risk to the NMR patient.

Surgical clips are often manufactured from martensitic stainless steel that is magnetically prone to stress corrosion failure and is quite sensitive to magnetic flux. Highly martensitic clips thus present a risk to the patient during an NMR examination and should therefore be avoided in patients expected to be examined with NMR.[168,169] Clips made of austenitic stainless steel with high nickel content (16 to 35% Ni) are, however, found to have no ferromagnetism and are thus recommended for use in NMR patients, as well as in clips made of tantalum or titanium.[170,171]

Table 22
RECOMMENDED EXPOSURE LIMITS TO STATIC MAGNETIC FIELD AT STANFORD LINEAR ACCELERATOR[158,173]

Period	Body area	Magnetic field strength	
		Tesla (T)	Gauss (G)
Extended periods (h)	Whole body or head	0.02	200
	Arms and hands	0.2	2,000
Short periods (min)	Whole body or head	0.2	2,000
	Arms and hands	2	20,000

In order to determine whether or not hip prostheses during NMR imaging procedure pose a significant hazard, Davis et al.[172] exposed them to changing magnetic fields and Rf radiation greater than those associated with an NMR imaging procedure. It was found that relatively large temperature rises were recorded when both hip protheses, in saline, were exposed to Rf radiation. This effect was caused by the large conducting paths created when the two prostheses were in contact and surrounded by a conducting solution. The Rf exposure caused the solution around the prostheses to rise 2 to 6°C, depending on the orientation angle. The results suggest that individuals with large implants should only be imaged with NMR when highly necessary until further investigations on this subject are completed.

2. Effects on Patients with Pacemakers

As a person with an implanted pacemaker approaches the NMR imaging system, the static magnetic field will possibly become strong enough to close the reed relay. Measurements have shown that this will occur when the field strength at the position of the pacemaker is greater than approximately 1 mT (10 G).

Aside from the reed relay, which is designed to interact with a magnetic field, pacemakers have numerous ferromagnetic components which will experience an undesirable force or torque in a magnetic field. In particular, the lithium-cell battery, which occupies most of the volume within the pacemaker, experiences a substantial torque, tending to align it with the magnetic field.

In addition, an unipolar pacemaker lead acts as an antenna in which the time-varying gradient and Rf magnetic field of a NMR imaging system can induce currents. In spite of the fact that the pacemaker of a patient lying within the scanner normally would be switched to an asynchronous mode in which it should not respond to external pulses, the induced currents might be strong enough to cause the nonlinear operation of critical electronic components and to thereby disrupt the normal output of the pacemaker.

The gradients are pulsed at a rate comparable to that at which the heart beats, and the induced current might be sufficient to cause fibrillation. Taking the above effects under consideration, it is not advised that patients with implanted pacemakers be examined with NMR.

D. Guidance Limits of Exposure During Nuclear Magnetic Resonance (NMR) Clinical Imaging

1. Static Magnetic Fields

Safety limits of occupational exposure to static magnetic fields in the U.S. and the U.S.S.R. are given in Tables 22, 23, and 24.[145,158,173,174] For NMR imaging conditions, the National Radiological Protection Board in the U.K. has recommended limiting the clinical exposure of patients to static magnetic fields below 2.5 T.[175,176] The Bureau of Radiological Health (U.S.) recommends limiting the whole or partial body exposure to static magnetic

Table 23
RECOMMENDED EXPOSURE LIMITS TO STATIC MAGNETIC FIELD AT THE BROOKHAVEN NATIONAL LABORATORY[174]

Period	Body area	Magnetic field strength	
		Tesla (T)	Gauss (G)
Less than 15 min	Whole body	>1	>10,000
Less than 1 h	Whole body	0.5—1	5,000—10,000
Unlimited	Whole body	0.01—0.5	100—5,000

Table 24
RECOMMENDED EXPOSURE LIMITS TO UNIFORM AND GRADIENT STATIC MAGNETIC FIELD FOR CONTINUOUS OCCUPATIONAL EXPOSURE IN THE U.S.S.R.[145]

Period (8 h/d)	Body area	Magnetic field strength	
		Tesla (T)	Gauss (G)
Continuous	Whole body: uniform	0.03	300
	Whole body: gradient	0.05—0.2 (T/m)	5—20 (G/cm)
Continuous	Hands: uniform	0.07	700
	Hands: gradient	0.1—0.2 (T/m)	10—20 (G/cm)

fields to below 2 T and the corresponding guideline given by the West German Health authority, ''Bundesgesundheitsamt'', is 2 T for whole or partial body exposure (BGA). The various guidelines for exposure to static magnetic fields in different countries are summarized in Figure 6 and Table 25.[177-179]

2. Dynamic Magnetic Gradient Fields

The guideline levels of exposure to time-varying gradient fields in NMR imaging have been determined after considering existing standards and recommendations for occupational situations where repeated or chronic exposure is expected. The guidelines for NMR exposure to time-varying magnetic gradient fields for the U.K., the U.S., Canada, and the Federal Republic of Germany are summarized in Figure 7 and Table 25.

The National Radiological Protection Board of the U.K. recommends as a guideline for NMR imaging and *in vivo* spectroscopy that exposure to time-varying magnetic fields be below a 20-T/s variation.[175,176] The Bureau of Radiological Health in the U.S., however, is more conservative.[178] They recommend a maximum value of 3 T/s which is in agreement with the suggestion by Budinger.[180] He supports his suggestion with a rather extensive review of the literature on bioeffects and an in-depth discussion of the mechanisms of interaction. The German authority gives guidelines in terms of induced currents, and their recommendations are still more conservative. They are a factor of about 10 lower than the British guidelines in terms of time variation of magnetic flux density.

3. Radiofrequency (Rf) Radiation

The guidelines for Rf radiation in NMR imaging are preferentially given in terms of the SAR instead of in terms of field levels for the following reasons:

1. The fields in NMR exposure are primarily magnetic.
2. The fields are pulsed individually for each equipment.
3. The field level is not an independent variable of the system (a specific relationship must be obtained for field level, pulse characteristics, and frequency).

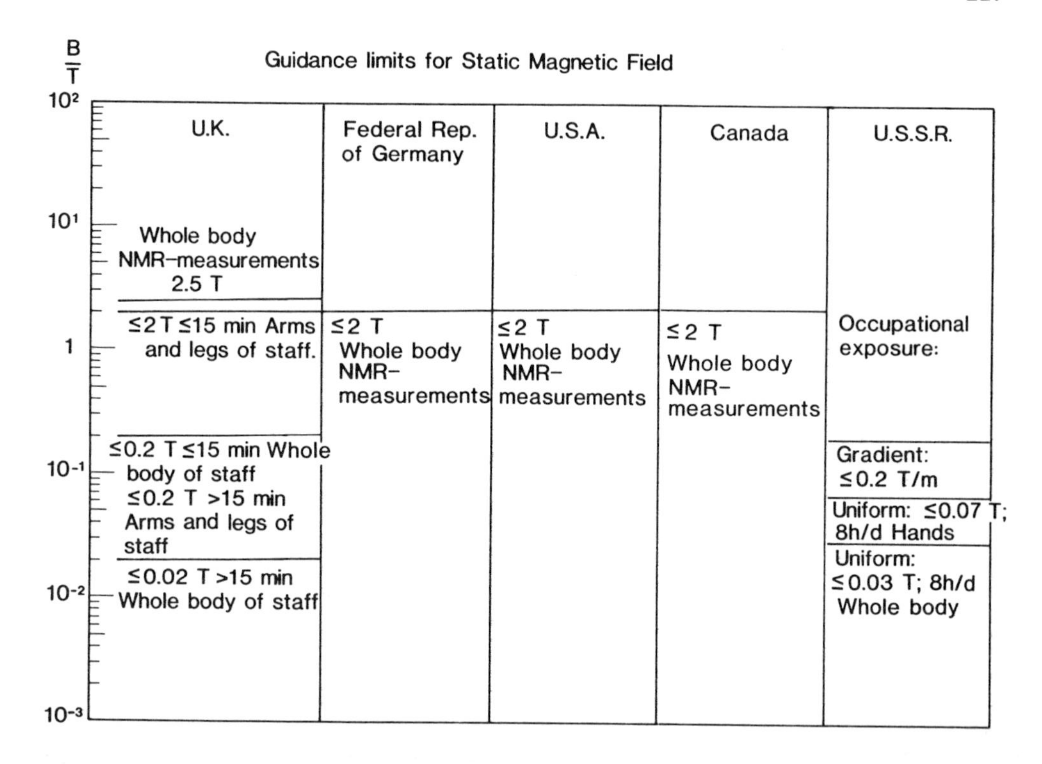

FIGURE 6. Guideline limits for static magnetic fields given by authorities in different countries for the various types of field exposure involved in clinical NMR examinations.

Table 25
SUMMARY OF GUIDELINES FOR STATIC MAGNETIC FIELDS AND DYNAMIC MAGNETIC FIELDS — RF-RADIATION EXPOSURE IN THE CLINICAL USE OF NMR IMAGING AND *IN VIVO* SPECTROSCOPY[175-179]

Static magnetic fields (T)	Dynamic magnetic fields (T/s)	RF radiation	Ref.
U.K.			175, 179
2.2	20	Rise in temperature of 1°C SAR of <0.4 W/kg in whole body SAR of <4 W/kg in 1 g	
U.S.			177
2.0	3	SAR of <0.4 W/kg in whole body SAR of <2 W/kg in 1 g	
FRG			178
2.0	For periods of >10 ms Current density in whole body of <30 mA/m² Field strength of <30 V/m For periods <10 ms 300/τ mA/m² or 300/τ V/m (τ in ms)	SAR of <1 W/kg in whole body SAR of <5 W/kg in 1 kg	
Canada			179
2.0	3	SAR of <1 W/kg, >15 min SAR of <2 W/kg, <15 min, in 25% of body weight	

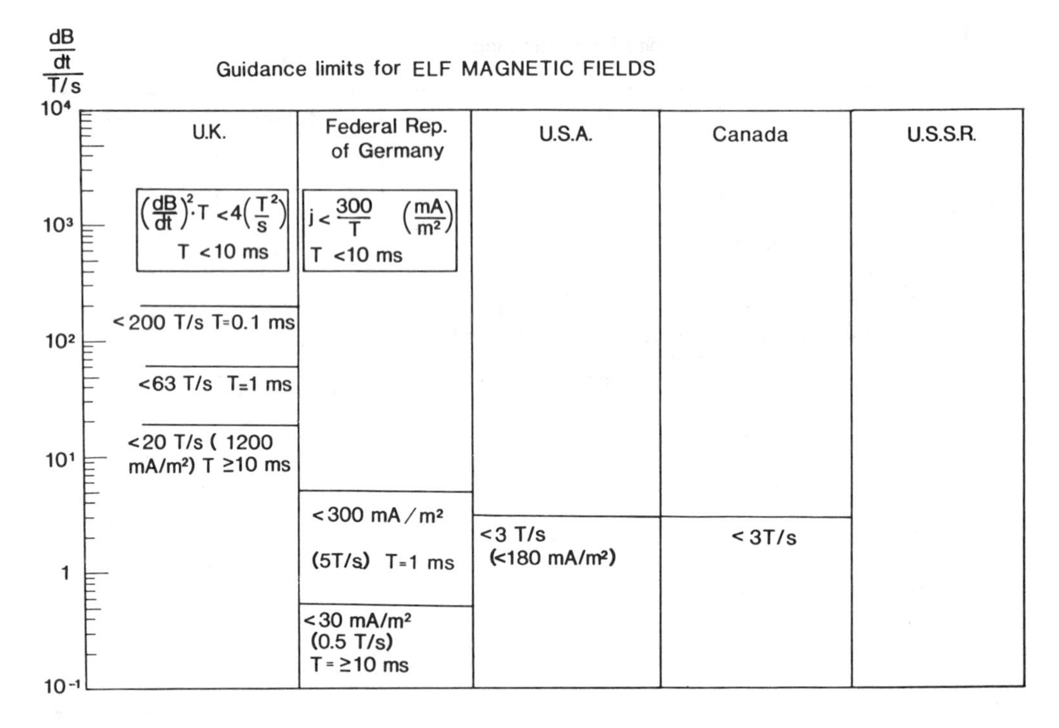

FIGURE 7. Guideline limits for time-varying magnetic gradient fields given by authorities in different countries for the various types of field exposure involved in clinical NMR examinations.

The guidelines proposed for the U.S. allow an average power deposition which is about half the basal metabolic rate for adults and which should not result in significant evaluations of local core or body temperature.[177] In order to ensure an adequate margin of safety, it is proposed that the whole body exposure be limited to 0.4 W/kg. A maximum SAR limit is also included in the guidelines because heating of the body by NMR systems is uneven, being greatest near the surface and approaching zero at the center of the body if the coil is concentric with the body.

A summary of the guidelines for Rf radiation in NMR imaging is given in Figure 8 and Table 25. The maximum average SAR is 0.4 W/kg for both the U.S. and the U.K., but it is 1.0 W/kg in the Federal Republic of Germany. The guidelines for local SAR also vary, being highest in Federal Republic of Germany at 5 W/kg and lowest in the U.S. and Canada at 2 W/kg.

4. *Supervision of Exposed Persons*

British and Canadian guidelines also give recommendations about supervision of volunteers and patients undergoing NMR examinations.[176,181] They recommend that volunteers participating in experimental trials of NMR imaging techniques should be medically assessed and pronounced to be suitable candidates before exposure.

Patients should be exposed only with the approval of a registered medical practitioner who should be satisfied that the exposure is likely to contribute to the treatment of the patient or is part of a research project approved by a local research ethical committee. It is also recommended that pregnant women be excluded during the first trimester. The exposure of individuals with large metallic implants such as hip protheses should be performed only if highly needed. The Rf and gradient switching should be stopped immediately if discomfort is experienced around the implant. Intracranial metallic clips, particularly those used to treat aneurysms, may present a hazard if the clips are made of magnetic materials. Patients fitted

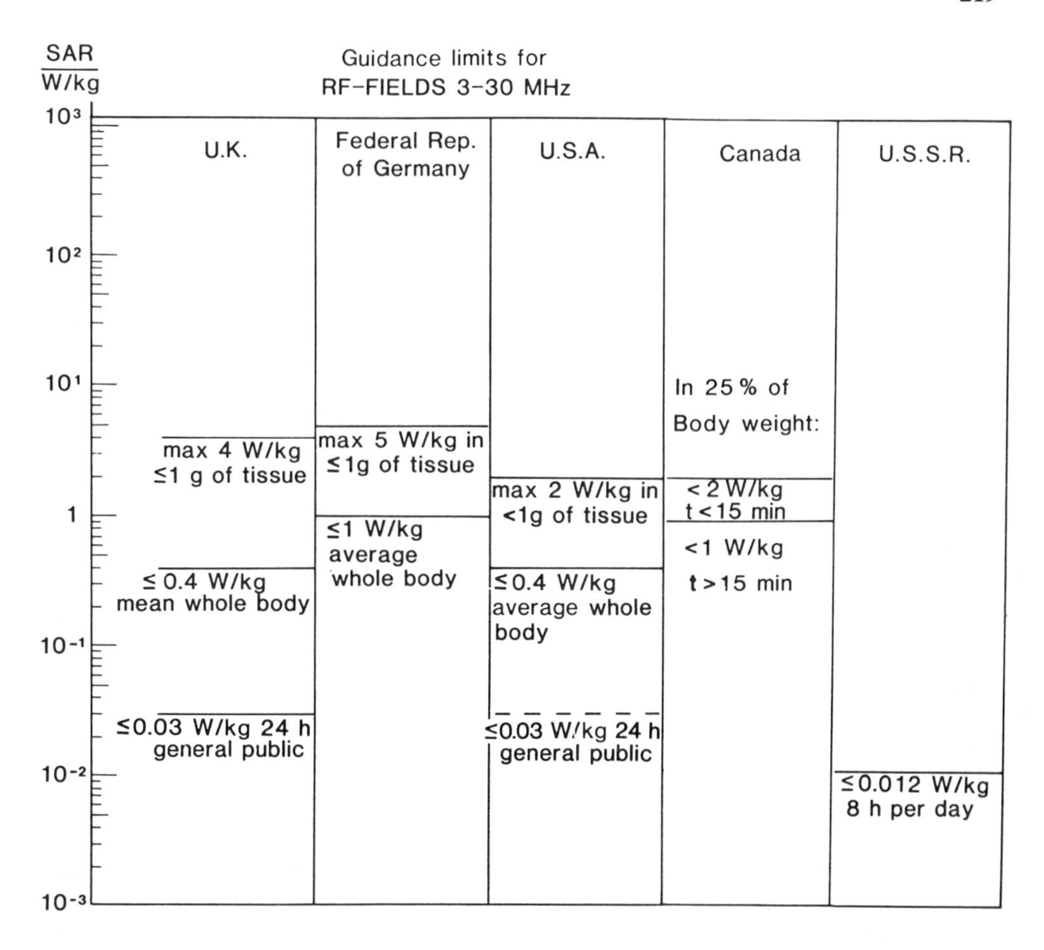

FIGURE 8. Guideline limits for radio frequency (Rf) field in terms of "specific energy absorption rate" (SAR, watts per kilogram) given by authorities in different countries for the various types of field exposure involved in clinical NMR examinations.

with such clips should be excluded. Appropriate rescue equipment should be available during imaging and a registered medical practitioner trained in the techniques of resuscitation should be available at short notice, although there is no need for him to be present during the exposure.

E. Safety Considerations on Siting and Shielding of Clinical Nuclear Magnetic Resonance (NMR) Installations

When a NMR imaging unit is to be installed in an hospital area, two major problems arise: (1) the environment must be protected from the machine and (2) the machine must be protected from the environment.[182]

1. Protecting the Environment

The area surrounding the NMR magnet must be supervised and restricted. A proposal for restriction limits is given below:[183]

>1.5 mT: "Caution — High Magnetic Field"

Inside this area, pacemakers might be disturbed, large ferrous objects may move toward the magnet, and small ferrous objects might become projectiles.

0.5—1.5 mT: "Caution — Magnetic Field"

Area in which supervision of the entrance into the high magnetic field area should be maintained.

<0.5 mT: ''No supervision''

Outside the 0.5—mT area there is little or no risk for health and safety problems and this area thus does not need to be supervised.

In order to maintain these limits, many NMR installations require magnetic shielding. Without active shielding, the $\vec{B}$ field is inversely proportional to the cube of the distance and thus a sufficient shield effect can be accomplished by means of using large rooms for the NMR unit. If large areas are not available, shielding with steel plates reduces the $\vec{B}$ field in the environment to an extent depending on shield thickness. Computer simulations are necessary to calculate the shielding efficiency, and the shield must be symmetrically placed around the magnet in order to disturb the main field homogeniety as little as possible.[183]

In addition, precautions must be taken in the event that a quench occurs in a superconductive unit. The coolant must be directed away to the outdoor atmosphere or the room must be of sufficient volume to accommodate the coolant and still leave sufficient air volume at ground level to support life. In addition, the room must have adequate ventilation to prevent the pressure rising to values at which windows might blow out, with a consequent risk of injuries from flying glass.[184]

Another aspect on protecting the environment comes from the fact that a large amount of medical equipment is sensitive to magnetic fields. Thus, image intensifiers, cathode-ray tubes, computers, disk drives, computerized axial tomography (CT) scanners, etc. must be either shielded or placed at a proper distance from the NMR unit.

2. Protecting the Machine

The NMR imaging unit needs two kinds of protection:

1. The main magnetic field must be protected from homogeniety disruptions, which might be caused by ferrous objects placed near the magnet. If the magnet is shielded, the steel must be symmetrically placed around the magnet. These efforts having been undertaken, shim coils can compensate for smaller variations in the magnetic field. The homogeniety requirements are below 50 ppm for imaging and below 1 ppm for spectroscopy.
2. Most users of NMR imaging systems agree that Rf shielding is necessary. The recommended attenuation for Rf fields is between 80 (99.99% attenuation) and 100 dB (99.999% attenuation).[183]

Shields of copper mesh or solid copper are commonly used, and the shielding effectiveness (the sum of all reflection and absorption of the Rf wave) depends on the frequency of the Rf wave, the conductivity and the thickness of the shield, and on leakage effects. The effects of leakage can be severe if the holes are small (i.e., in the same size order as the Rf wavelength) and such holes must therefore be carefully soldered, i.e., with copper tape. Rf energy transport in larger holes, necessary for ventilation, etc., can be hindered with wave guides and electrical lines can be shielded by means of proper Rf filtering.[183]

XII. POTENTIAL HAZARDS OF ACOUSTIC RADIATION

A. Ultrasound

1. Physical Properties and Generation of Ultrasound

The phenomenon which we experience as sound is a periodic change in the air pressure which influences the eardrum. Thus, sound or acoustic radiation can be physically described as a pressure wave which propagates through mechanical vibrations of particles in the propagating media. Wavelength is the distance between two identically vibrating particles and frequency is the number of waves which are passing a certain point in space per unit of time.

Table 26
ULTRASOUND PROPERTIES OF VARIOUS MEDIA

Medium	Sound velocity, c (m/s)	Acoustic impedance, $Z = \rho \cdot c$ (kg/s · m²)	Exponential attenuation coefficient at 1 MHz, μ (mm⁻¹)	Acoustical attenuation coefficient at 1 MHz, α (dB/mm)
Air	330	$4.5 \cdot 10^{-10}$	0.03	0.13
Water	1500	$1.50 \cdot 10^{-6}$	0.000023	0.0001
Liver	1540	$1.54 \cdot 10^{-6}$	0.016	0.07
Fat	1440	$1.40 \cdot 10^{-6}$	0.01—0.05	0.05—0.2
Lens of the eye	—	$1.84 \cdot 10^{-6}$	—	—
Blood	1520	$1.61 \cdot 10^{-6}$	0.0023	0.01
Skeletal muscle	1590	$1.70 \cdot 10^{-6}$	—	—
Skull bone	3360	$6.00 \cdot 10^{-6}$	0.13	0.55
Perspex (Lucite®)	2680	$3.16 \cdot 10^{-6}$	0.021	0.09
Aluminium	2700	$17.0 \cdot 10^{-6}$		
Brass	8400	$40.0 \cdot 10^{-6}$		

of 20 kHz is about 0.1 dB/m in dry air and about 1 dB/m in air with 37% relative humidity. At frequencies around 200 kHz, the attenuation is about 10 dB/m. For still higher frequencies, airborne ultrasound is absorbed so heavily in air that the scattering in air at these frequencies normally is not an occupational problem.

3. Biological Effects of Ultrasound

a. Biophysical Mechanisms

When ultrasound energy is absorbed in tissue, some of it becomes heat which gives rise to a local increase of temperature. If the temperature becomes too high, above 45°C, it can coagulate proteins and give rise to local burns. The absorption of ultrasound energy also gives rise to mechanic vibrations in the cell membranes and other structures which can cause molecular and biological structures to disrupt.

Under certain circumstances, the absorption of ultrasound in liquids can give to formation of gas bubbles or cavities which collapse when the ultrasound is shut off. This effect is called "cavitation" and has a great effect on biological structures because it can break membranes and cause microcirculation inside the cells. In the frequency range above 1 MHz, the threshold for cavitation in pure water is about 10,000 W/m² (1000 mW/cm²) for a continuous beam. In biological and viscous media, no cavitation appears around 1 to 20 MHz. Also, ultrasound in short pulses of 1 to 0.1 μs does not give rise to cavitation.

b. Effects on Biomolecules and Cells

Effects of ultrasound on the molecular level are partly chemical effects due to radical formation in case of cavitation and partly mechanical degradation of large molecules which have a fragile structure. The possibility of mechanical degradation of DNA molecules might be of importance in causing chromosome breakage. This has, however, not yet been proven at moderate frequencies and power.

High-power ultrasound exposure of cell cultures can lead to acute death of the cells due to mechanical disruption of the cell membrane, but this does not occur at the moderate levels which actually occur in occupational exposure. Less severe structural changes of membranes are usually reversible and not lethal for the cell.

c. Effects on Animals

Ultrasound is used to a large extent in the examination of pregnant women. It is therefore of special interest to study the effect of ultrasound on animals which are exposed *in utero*.

Table 27
RESULTS OF SOME EXPERIMENTS WHICH HAVE BEEN CARRIED OUT TO FIND EVIDENCE OF CHANGES OF A GENETIC OR TERATOGENIC NATURE AND FOR CONSEQUENT EFFECTS IN THE FETUS[185-191]

Biological system exposed	Exposure conditions[a]	Biological end points assayed	Findings	Ref.
Pregnant mice (297 exposed animals)	1—3 Mhz, 5 min I (tp, sp) = 0.2—4.9 MW/m^2 I (ta, sp) = 7.5—270 kW/m^2 Pulse duration of 10 μs—10 ms	Litter size resorption rate, abnormalities	No significant effect	185
Mouse gonads (M + F)	1.5 MHz, 15 min I (tp, sa) to 450 kW/m^2 I (ta, sa) to 16 kW/m^2 Pulse duration of 30 μs—1 ms	Dominant-lethal mutations, testis weight, sperm count, chromosome abnormalities, sterility	No significant effect, except in female sterility	186
Pregnant mice, two strains	2.25 MHz, 5 h, continuous I (ta, s?) = 0.4 kW/m^2	Fetal weight, fetal death, malformations	Significant effects on late fetal death in one strain	187
Pregnant rats	1.9 Mhz, 10 min I (tp, sa) to 1 MW/m^2 I (ta, sa) to 10 kW/m^2	Fetal viability, resorptions	No significant effect	188
Pregnant mice	1 MHz, 3 min, continuous I (sa) to 5 kW/m^2	21-d survival of offspring	Significant decrease in survival at 1.25 kW/m^2	189
Pregnant mice	1 MHz, continuous (?) for I (sa) 5—7 kW/m^2, 300 s 20—30 kW/m^2, 20 s 30—55 kW/m^2, 10 s	Fetal weight	Significant decrease in average result, all intensities	190
Pregnant mice	2.3 MHz, 5 or 60 min continuous, 1.6—10 kW/m^2	Structural changes, congenital malformations, chromosome aberrations	No significant effect	191

[a] I(ta) = temporal average intensity; I(tp) = time peak intensity; I(sp) = spatial peak intensity; I(sa) = spatial average intensity.

The results of some experiments which have been carried out to find evidence of changes of a genetic or teratogenic nature and of the consequent effects in the fetus are given in Table 27.[185-191] It may be difficult to draw straightforward conclusions from these results, but no evidence was found of any genetic effect, even though the exposure levels were in some cases sufficient to cause a degree of heat-induced sterilization. The search for chromosome aberrations resulting from other *in vivo* irradiation experiments has in all cases proved negative.[192-194]

Some of the strongest evidence in favor of the existence of a teratogenic effect is provided by the results obtained by Shoji et al.[187] These results, however, have been the subject of some controversy and may be explained, as suggested by Lele, as resulting from prolonged induction of moderate temperature rise.[195] The findings of Curto on neonatal survival and O'Brien on fetal weight reduction evidently call for careful consideration.[189,190]

A summary of different kinds of biological effects obtained at ultrasound exposure of various power levels is given in Figure 10.[189,190,195,201] Most of these data were deliberately selected as indicating the lowest known exposure level at which bioeffects have been reported. Below the energy density of 50 J/cm^2, there seems to be no effects of ultrasound exposure. This is equivalent to power densities below 50/t W/cm^2, where t is the time of exposure in seconds. At very low-power density, the upper limit for biological effects is 100 mW/cm^2.

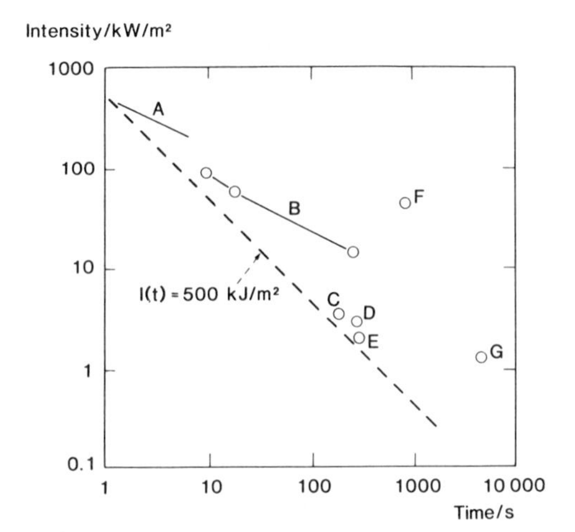

FIGURE 10. A summary of different kinds of biological effects obtained at ultrasound exposure of various power levels.[206] (A) Paralysis at 37°C;[196] (B) fetal weight reduction;[190] (C) postpartum mortality;[189] (D) wound healing;[197,198] (E) altered mitotic rate (variable results);[199] (F) genetic damage (negative result);[200] (G) fetal abnormalities ''postulated'' by Lele;[195] negative results reported by Miller et al.[202]

d. Effects on Humans

The most common complaints made by persons exposed to high levels of high-audible-frequency noise, such as that generated by processes involving cavitation, are of an unpleasant ''fullness'' or pressure in the ears; headaches, which do not persist after the exposure has ceased; tinnitus; and perhaps also nausea and mild vertigo.[202,203] These effects may also be caused by subharmonic distortion products of ultrasound frequencies at higher levels. The effects do not appear to be closely dependent on the duration of exposure.

Neither temporary nor permanent losses of hearing have been found in industrial workers exposed to levels of up to about 110 to 120 dB at low ultrasound frequencies, but temporary shifts in the threshold of hearing were noted at 17 to 37 kHz at levels of 148 to 154 dB.[204] No epidemiological or comparable studies appear to have been made in relation to occupational exposure to ultrasound, but several investigations have been performed on women who have been examined with ultrasound during their pregnancy. No evidence has been found so far that the exposure to ultrasound at diagnostic levels increases the risk of abnormal births. A large epidemiological investigation was made of 1114 apparently normal pregnant women examined by ultrasound in 3 different centers and at various stages of pregnancy.[205] A 2.7% incidence of congenital abnormalities was found by physical examination of newborns in the group, as compared with a figure of 4.8% reported in a separate and unmatched survey of women who had not been examined with ultrasound. Neither the time of gestation at which the first ultrasound examination was made nor the number of examinations seemed to increase the risk of fetal abnormality.

4. Criteria for Safe Exposure Levels

Ultrasound is used in a wide range of power levels in applications from mechanical industry to medical examinations. High-power applications use power densities above 100 W/cm² and some examples are given below. Ultrasound cleaning uses levels of 100 to 1000 W/cm² with frequencies around 20 to 80 kHz. The objects to be cleaned are submerged in a bath of cleaning liquid in contact with a ultrasound generator. The cleaning process takes place by cavitation and mechanical scrubbing of the objects. The scattering of ultrasound into the air is insignificant in this process, but in direct physical contact with the ultrasound cleaner, efficient transfer of ultrasound energy to the hands is obtained. Therefore thick

Table 28
SAFETY LIMITS FOR EXPOSURE OF AN UNPROTECTED EAR FOR ULTRASOUND DURING 8 h/d AT DIFFERENT FREQUENCIES ABOVE 20 kHz

Middle frequency, 1/octave band (kHz)	Sound pressure level relative to 20 μPa (dB)		
	8 h/d	≤4 h/d	≤1 h/d
20	105	108	114
25	110	113	119
≥31	115	118	124

protective gloves are recommended. Ultrasound plastic welding uses levels of about 10 to 100 W/cm^2 at 20 kHz. Other uses of ultrasound are production of emulsions and aerosols, process control, burgler alarms, material testing, remote controls for televisions, and other domestic electronics.

In medical ultrasound therapy, equipment is operated at frequencies from about 0.8 to 3 MHz, with a spatial average output not exceeding 3 W/cm^2 (30,000 W/m^2). This power level is in the range where undesirable bioeffects have been reported at an exposure time above 10 s. It is therefore important that the equipment used for therapy be calibrated to provide the operator with the capacity of delivering a prescribed energy.

5. Guidelines for Safe Use of Ultrasound

In order to protect the worker from unnecessary exposure to ultrasound, one can apply the following measures:

1. Use the lowest power level and the shortest time possible for the operation.
2. Avoid unnecessary contact between ultrasound equipment and the user. Use protective gloves in ultrasound cleaning.
3. Remember that possible damage due to airborne ultrasound can give auditory effects. Use hearing protection.

Different kinds of biological effects, obtained at ultrasound exposure at various power levels,[206] are summarized in Figure 10. Continuous ultrasound exposure produces no effects below 100 mW/cm^2 and exposure to pulsed ultrasound produces no effects below the power level of 50,000/t mW/cm^2, where t is the exposure time in seconds.

The following guidelines can thus apply to the safe occupational use of ultrasound: 10 mW/cm^2 for continuous ultrasound and 5,000/t mW/cm^2 for pulsed ultrasound, where t is the exposure time in seconds. Guidance limits for exposure to airborne ultrasound are given relative to a sound pressure of 20 μPa: sound pressure level $= \log_{10} \frac{P\ (\mu Pa)}{20}$. The safety limits for exposure of an unprotected ear for ultrasound during 8 h/d at different frequencies above 20 kHz are given in Table 28.

B. Infrasound

1. Physical Properties and Origin of Infrasound

Acoustic radiation of sound with frequencies below 20 Hz is called infrasound. The wavelength in air is 17 m at 20 Hz and 170 m at 2 Hz. The attenuation of infrasound in air is negligible, so it can propagate over large areas. The extension does not stop by barriers or walls because of the large wavelength. The intensity of infrasound is given by the sound pressure level in decibels which is defined in the same way as heavy sound, i.e., with the reference pressure of 20 μPa equal to $20 \cdot 10^{-6}$ N/m^2.

Measurements of infrasound can be made with conventional sound level meters equipped with a special low pass filter which attenuates signals with frequencies above 20 Hz. The infrasound level (IL) from such a measurement is given in the decibels (IL). In the environment, infrasound originates from thunderstorms, volcanic eruptions, large waterfalls, and air turbulence. In the working environment, infrasound originates from machines, engines, and ventilation installations. By resonance phenomena, the infrasound can be amplified in rooms and ventilation channels. Ventilation fans can produce infrasound through turbulence around the rotating wheels and fan wings. The sound from exhausting steam and gases from valves, turbines, jet engines, etc. has a strong component of infrasound. Also piston engines and compressors can give rise to infrasound which can be amplified in various spaces.

2. Biological Effects of Infrasound

Humans can hear sound within a frequency range of 20 to 20,000 Hz. It is, however, possible to have a hearing sensation below 20 Hz, although without tonal characteristics. The hearing threshold at the frequency of 20 Hz is about 74 dB, while at 4 Hz the threshold is as high as 130 dB. One starts to experience tickling and unpleasant pressure in the ears at 140 dB.

In infrasound exposure, not only is the ear cochlea stimulated, but the whole inner ear with the balance mechanism is also stimulated. This can eventually explain symptoms like a feeling of indisposition and uneasiness which occur at exposure to infrasound. At frequencies of 10 Hz, there is a wheeze in the voice which causes a cough and hawking effects.

3. Safety Aspects on Infrasound

In the working environment, people are generally exposed to a spectrum of frequencies of hearing sounds as well as infrasound. This is why it is so difficult to verify the inconvenience and damage caused by infrasound. In order to prevent damage as well as subjective feelings of uneasiness which can arise indirectly in connection with tiredness or irritation, there are defined maximum limits of exposure to infrasound. In the frequency range of 2 to 20 Hz, the maximum sound pressure allowed during 8 h/d is

$$100 \text{ dB relative to } 20\ \mu\text{Pa } (20 \cdot 10^{-6}\ \text{N/m}^2)$$

For a shorter time of exposure below 1 h, 130 dB can be permitted. This safety limit concerns the total contribution from all frequencies in the range from 2 to 20 Hz.

REFERENCES

1. International Atomic Energy Association, Radiation Protection Procedures, IAEA Safety Ser. No. 38, International Atomic Energy Agency, Vienna, 1973.
2. International Commission on Radiation Protection, *Annals of the ICRP*, Vol. 1, Publ. 26, International Commission on Radiation Protection, Pergamon Press, New York, 1977.
3. U.N. Scientific Committee on the Effects of Atomic Radiation, Ionizing Radiation: Sources and Biological Effects, Report to the General Assemby, U.N. Sales Publ. No. E.82.IX.8, United Nations, New York, 1982.
4. American Association of Physicists in Medicine, Radiotherapy Safety, Thomadsen, B., Ed., American Assoc. of Physicists in Medicine Symp. Proc. No. 4, American Institute of Physicists, New York, 1984.
5. Institute of Physical Sciences in Medicine, *Commissioning and Quality Assurance of Megavoltage X-ray Equipment*, The Institute of Physical Sciences in Medicine, London, 1986.

6. International Commission on Radiologic Protection, *Protection Against X-rays up to Energies of 3 MeV and Beta- and Gamma-Rays from Sealed Sources,* ICRP Publ. 3, International Commission on Radiological Protection and Measurements, Pergamon Press, Oxford, 1960.
7. International Commission on Radiological Protection, *Protection Against Ionizing Radiation from External Sources,* ICRP Publ. 15, International Commission on Radiological Protection and Measurements, Pergamon Press, Oxford, 1970.
8. International Commission on Radiological Protection, *Protection of the Patient in X-ray Diagnosis,* ICRP Publ. 16, International Commission on Radiological Protection and Measurements, Pergamon Press, Oxford, 1970.
9. International Commission on Radiological Protection, *Protection Against Ionising Radiation from External Sources Used in Medicine,* ICRP Publ. 33, International Commission on Radiological Protection and Measurements, Pergamon Press, Oxford, 1968.
10. **Holje, G.,** *Annual Report of Personnel Exposure,* Department of Hospital Physics, University Hospital, Lund, Sweden, 1987.
11. Nuclear Regulatory Commission, Information Relevant to Ensuring that Occupational Radiation Exposures at Medical Institutions will be as low as Possibly Achievable, NRC Regulary Guide 8.18, Nuclear Regulatory Commission, 1978.
12. NCRP, Precautions in the Management of Patients who have Received Therapeutic Amounts of Radionuclides, NCRP Rep. No. 37, National Council on Radiation Protection and Measurements, Washington, D.C., 1978.
13. Bureau of Radiological Health, Workshop Manual for Radionuclide Handling and Radiopharmaceutical Quality Assurance, HHS Publ. FDA 82-8191, Bureau of Radiological Health, National Technical Information Service, U.S. Department of Commerce, Springfield, VA, 1982.
14. **Schwan, H. P. and Foster, K. R.,** RF-field interactions with biological systems: electrical properties and biophysical mechanisms, *Proc. IEEE,* 68, 104, 1980.
15. **Stuchly, M. A. and Stuchly, S. S.,** Dielectric properties of biological substances — tabulated, *J. Microwave Power,* 15(1), 19, 1980.
16. **Guy, A. W., Lehman, J. F., and Stonebridge, J. B.,** Therapeutic applications of electromagnetic power, *Proc. IEEE,* 62, 55, 1974.
17. **Hagman, M. J., Gandhi, O. P., D'Andrea, J. A., and Chatterjee, I.,** Head resonance: numerical solutions and experimental results, *IEEE Trans. Microwave Theory Tech.,* 27(9), 809, 1979.
18. **Law, H. T. and Pettigrew, R. T.,** Heat transfer in whole body hyperthermia, *Ann. N.Y. Acad. Sci.,* 335, 298, 1980.
19. **Berman, E., Kinn, J. B., and Carter, H. B.,** Observations of mouse fetuses after irradiation with 2.45 GHz microwaves, *Health Phys.,* 35, 791, 1978.
20. **Berman, E., Carter, H. B., and House, D.,** Tests of mutagenesis and reproduction in male rats exposed to 2450-MHz (CW) microwaves, *Bioelectromagnetics,* 1, 65, 1980.
21. **Johnson, R. B., Mizumori, S., and Lovely, R. H.,** Adult behavioral deficit in rats exposed prenatally to 918-MHz microwaves, in Developmental Toxicology of Energy-Related Pollutants, Mahlum, D. D., Sikov, M. R., Hacket, P. L., and Andrew, F. D., Eds., DOE Symp. Ser. 47, Department of Energy, Washington, D.C., 1978, 281.
22. **Michaelson, S. M., Guillet, R., and Heggeness, F. W.,** Influence of microwave exposure on functional maturation of the rat, in Developmental Toxicology of Energy-Related Pollutants, Mahlum, D. D., Sikov, M. R., Hacket, P. L., and Andrew, F. D., Eds., DOE Symp. Ser. 47, Department of Energy, Washington, D.C., 1978, 300.
23. **Guillet, R. and Michaelson, S. M.,** The effects of repeated microwave exposure on neonatal rats, *Radio Sci.,* 12(659), 125, 1977.
24. **Djordjevic, Z., Lazarevic, N., and Djorkovic, V.,** Studies on the hematologic effects of long-term, low-dose microwave exposure, *Aviat. Space Environ. Med.,* 48, 516, 1977.
25. **Smialowicz, R. J., Kinn, J. B., and Elder, J. A.,** Perinatal exposure of rats to 2450-MHz CW microwave radiation: effects on lymphocytes, *Radio Sci.,* 14(6S), 147, 1979.
26. **Milroy, W. C. and Michelson, S. M.,** Thyroid pathophysiology of microwave radiation, *Aerosp. Med.,* 43, 1126, 1972.
27. **Lovely, R. H., Myers, D. E., and Guy, A. W.,** Irradiation of rats by 918-MHz microwaves at 2.5 mW/cm^2: delineation the dose-response relationship, *Radio Sci.,* 12(6S), 139, 1977.
28. **Wangemann, R. T. and Cleary, S. F.,** The *in vivo* effects of 2.45 GHz microwave radiation on rabbit serum components and sleeping times, *Radiat. Environ. Biophys.,* 13, 89, 1976.
29. **Cahn, H. A., Folk, G. E., Jr., and Huston, P. E.,** Age comparison of human day-night physiological differences, *Aerosp. Med.,* 39, 608, 1968.
30. **Rowell, L. B.,** Cardiovascular aspects of human thermoregulation, *Circ. Res.,* 52, 367, 1983.
31. **Houdas, Y. and Ring, E. F. J.,** *Human Body Temperature: Its Measurement and Regulation,* Plenum Press, New York, 1982.

32. **Hugander, A., Gillis, W. K., and Robbins, I. H.,** Whole body hyperthermia in the treatment of neoplastic disease, in *Hyperthermia Monograph,* Bhaba Atomic Research Center, Bhaba, India, 1985.
33. **Stillwell, G. K.,** General principles of thermotherapy, in *Therapeutic Heat and Cold,* 2nd ed., Licht, S., Ed., Waverly Press, Baltimore, MD, 1965, 232.
34. **O'Connor, M. E.,** Mammalian teratogenesis and radiofrequency fields, *Proc. IEEE,* 68, 56, 1980.
35. **Deno, D. W.,** *Electrostatic and Electromagnetic Effects of Ultrahigh-Voltage Transmission Lines,* EL-802, Research Project 566-1, Electric Power Research Institute, Palo Alto, CA, 1978.
36. **Lövsund, P. and Hansson-Mild, K.,** Low Frequency Electromagnetic Fields Near Some Induction Heaters, Invest. Rep. 1978:38, National Board of Occupational Safety and Health, Sweden, 1978.
37. **Lin, J. C.,** The microwave auditory phenomena, *Proc. IEEE,* 68, 67, 1980.
38. **Silverman, C.,** Nervous and behavioral effects of microwave radiation in humans, *Am. J. Epidemiol.,* 97, 219, 1973.
39. **Jaffe, R. A., Laszewski, B. L., Carr, D. B., and Phillips, R. D.,** Chronic exposure to a 60 Hz electric field: effects on synaptic transmission and peripheral nerve function in the rat, *Bioelectromagnetics,* 1, 131, 1980.
40. **Takashima, S., Onaral, B., and Schwan, H. P.,** Effects of modulated RF energy on the EEG of mammalian brain. Effects of acute and chronic irradiations, *Radiat. Environ. Biophys.,* 16, 15, 1979.
41. **Tolgskaya, M. S. and Gordon, Z. V.,** *Pathological Effects of radio waves,* Consultants Bureau, New York, 1973.
42. **Adey, W. R.,** Frequency and power windowing in tissue interaction with weak electromagnetic fields, *Proc. IEEE,* 68, 119, 1980.
43. **Olsen, R. G., Lords, J. L., and Durney, C. H.,** Microwave-induced chronotropic effects in the isolated rat heart, *Ann. Biomed. Eng.,* 5, 395, 1977.
44. **Lövsund, P., Nilsson, S. E. G., Reuther, T., and Öberg, P. Å.,** Magnetophosphenes. A quantitative analysis of thresholds, *Med. Biol. Engl. Comput.,* 18, 326, 1980.
45. **McRee, D. I. and Wachtel, H.,** The effects of microwave radiation on the vitality of isolated rat heart, *Ann. Biomed. Eng.,* 5, 395, 1980.
46. **Portela, A., Llobera, O., Michaelson, S. M., Stewart, P. A., Perez, J. C., Guerrero, A. H., Rodriguez, C. A., and Perez, R. J.,** Transient effects of low-level microwave irradiation on bioelectric muscle cell properties and on water permeability and its distribution, in *Fundamental and Applied Aspects of Nonionizing Radiation,* Michaelson, S. M., Miller, M., Magin, M. W., and Carstensen, E. L., Eds., Plenum Press, New York, 1970, 93.
47. **Leonard, A., Bertrand, A. J., and Bruyere, A.,** An evaluation of the mutagenic, carcinogenic and teratogenic potential of microwaves, *Mutat. Res.,* 123, 31, 1983.
48. **Yao, K. T. S.,** Microwave radiation-induced chromosomal aberrations in corneal epithelum of Chinese hamsters, *J. Hered.,* 69, 409, 1978.
49. **Alam, M. T., Barthakur, N., Lambert, N. G., and Kasatiya, S. S.,** Cytological effects of microwave radiation in Chinese hamster cells *in vitro, Can. J. Genet. Cytol.,* 20, 23, 1978.
50. **McRee, D. I., MacNichols, G., and Livingston, G. K.,** Incidence of sister chromatid exchange in bone marrow cells of the mouse following microwave exposure, *Radiat. Res.,* 85, 340, 1981.
51. **Fielitz, J., Bogl, W., Stockhausen, K., and Kossel, F.,** Der Einfluss von UV- und Mikrowellenstrahlung auf biologisches Material, *STH Ber.,* 17, 63, 1979 (in German).
52. **Dutta, S. K., Nelson, W. H., Blackman, C. F., and Brusick, D. J.,** Lack of microbial genetic response to 2.45-GHz CW and 8.5 to 9.6-GHz pulsed microwaves, *J. Microwave Power,* 14, 275, 1979.
53. **Dardalhon, M., Averbeck, D., and Berterud, A. J.,** Studies on possible genetic effects of microwaves in procaryotic and eucaryotic cells, *Radiat. Environ. Biophys.,* 20, 37, 1981.
54. **Pay, T. L., Beyer, E. C., and Reichelderfer, C. F.,** Microwave effects on reproductive capacity and genetic transmission in *Drosophila melangaster, J. Microwave Power,* 7, 75, 1972.
55. **Hamnerius, Y., Olofsson, H., Rasmuson, A., and Rasmuson, B. A.,** Negative test for mutagenic action of microwave radiation in *Drosophila melangaster, Mutat. Res.,* 68, 217, 1979.
56. **Marec, F., Ondrácek, J., and Brunnhofer, V.,** The effect of repeated microwave irradiation on the frequency of sex-linked recessive lethal mutations in *Drosophila melongaster, Mutat. Res.,* 157, 163, 1985.
57. **Adey, W. R.,** Long-range electromagnetic field interactions at brain cell surfaces, in *Magnetic Field Effect on Biological Systems,* Tenforde, T. S., Ed., Plenum Press, New York, 1979, 57.
58. **Adey, W. R.,** Tissue interactions with nonionizing electromagnetic fields, *Physiol. Rev.,* 61, 435, 1981.
59. **Blackman, C. F., Benane, S. G., Joines, W. T., Hollis, M. A., and House, D. E.,** Calcium-ion efflux from brain tissue: power density versus internal field-intensity dependencies at 50-MHz RF-radiation, *Bioelectromagnetic,* 1, 277, 1980.
60. **Takashima, S. and Schwan, H. P.,** Effects of radio frequency fields on the EEG of rabbit brains, in *Program and Abstracts from the 1979 Spring Meeting,* International Union of Radio Science, Seattle, 1979.

61. **Sadchikova, M. N.,** Clinical manifestations of reactions to microwave irradiation in various occupational groups, in *Biologic Effects and Health Hazards of Microwave Radiation,* Proc. Int. Symp., Polish Medical Publishing, Warsaw, 1974, 261.
62. **Lilienfeld, A. M., Tonascia, J., Tonascia, S., Libauer, C. A., and Cauthen, G. M.,** Foreign Service Health Status Study — Evaluation of Health Status of Foreign Service and Other Employees from Selected Eastern European Posts, Final Rep. Contract No. 6025-619073 (NTIS PB-288163), U.S. Department of State, Washington, D.C., 1978, 436.
63. **Robinette, C. D., Silverman, C., and Jablon, S.,** Effects upon health of occupational exposure to microwave radiation (radar), *Am. J. Epidemiol.,* 112, 39, 1980.
64. **Cleary, S. F., Pasternack, B. S., and Bebe, G. W.,** Cataract incidence in radar workers, *Arch. Environ. Health,* 11, 179, 1965.
65. **Cohen, B. H., Lilienfeld, A. M., Kramer, S., and Hyman, L. C.,** Parental factors in Down's syndrome: results of the second Baltimore case-control study, in *Population Cytogenetics — Studies in Humans,* Hook, E. B. and Porter, I. H., Eds., Academic Press, New York, 1977, 301.
66. **Barron, C. I. and Baraff, A. A.,** Medical considerations of exposure to microwaves (radar), *JAMA,* 168, 1194, 1958.
67. **Silverman, C.,** Epidemiologic studies of microwave effects, *Proc. IEEE,* 68, 78, 1980.
68. **Esterly, C. E.,** Cancer link to magnetic field exposure: a hypothesis, *Am. J. Epidemiol.,* 114, 169, 1981.
69. **Kholodov, Y. A.,** The effects of electromagnetism and magnetic field on the central nervous system, *NASA Tech. Transl.,* F-465, 1967.
70. **Szmigielski, S. A., Szydzinski, A., Pietraszek, A., Bielec, M., Janlak, M., and Wrembel, J. K.,** Accelerated development of spontaneous and Benzopyrene-induced skin cancer in mice exposed to 2450-MHz microwave radiation, *Bioelectromagnetics,* 3(2), 179, 1982.
71. **Milan, S.,** Mortality from leukemia in workers exposed to electrical and magnetic fields, *N. Engl. J. Med.,* 307, 249, 1982.
72. **Wright, W. E., Peter, J. M., and Mack, T. M.,** Leukemia in workers exposed to electrical and magnetic fields, *Lancet,* 2, 1160, 1982.
73. **McDowall, M. E.,** Leukemia incidence in electrical workers in England and Wales, *Lancet,* 1, 982, 1983.
74. **Coleman, M., Bell, J., and Skeet, D.,** Leukemia incidence in electrical workers, *Lancet,* 1, 246, 1983.
75. **Lin, R. S., Dischinger, P. C., Conde, J., and Farrell, K. P.,** Occupational exposure to electromagnetic fields and the occurrence of brain tumors, *J. Occup. Med.,* 27, 413, 1985.
76. **Milham, S.,** Mortality in aluminium reduction plant workers, *J. Occup. Med.,* 21, 475, 1979.
77. **Seidman, H., Selikoff, I. J., and Hammond, E. C.,** Mortality of brain tumours among asbestos insulation workers, in the United States and Canada, *Ann. N.Y. Acad. Sci.,* 381, 73, 1982.
78. **Garfinkel, L. and Sarokhan, B.,** Trends in brain cancer tumor mortality and morbidity in the United States, *Ann. N.Y. Acad. Sci.,* 381, 1, 1982.
79. **Kramar, P. O., Harris, C., Gay, A. W., and Emery, A. F.,** Mechanism of microwave cataractogenesis in rabbits, in *Biological Effects of Electromagnetic Waves,* Vol. 1, Johnson, C. C. and Shore, M. L., Eds., HEW Publ. (FDA) 77-8010, U.S. Department of Health, Education and Welfare, Washington, D.C., 1976, 49.
80. **Rugh, R.,** Are mouse fetuses which survive microwave radiation permanently affected thereby?, *Health Phys.,* 31, 33, 1976.
81. **Rugh, R., Ginns, E. I., Ho, M. S., and Leach, W. M.,** Responses of the mouse to microwave radiation during gestrous cycle and pregnancy, *Radiat. Res.,* 62, 225, 1975.
82. **Blackman, C. F., Benane, S. G., Weil, C. M., and Ali, J. S.,** Effects of nonionizing electromagnetic radiation on single-cell biologic systems, *Ann. N.Y. Acad. Sci.,* 247, 352, 1975.
83. **Hamrick, P. E. and Butler, B. T.,** Exposure of bacteria to 2450 MHz microwave radiation, *J. Microwave Power,* 8, 227, 1973.
84. **Allis, J. W.,** Irradiation of bovine serum albumin with a crossed-beam exposure-detection system, *Ann. N.Y. Acad. Sci.,* 247, 312, 1975.
85. **Blackman, C. F., Surles, M. C., and Benane, S. G.,** The effect of microwave exposure on bacteria: mutation induction, in *Biological Effects of Electromagnetic Waves,* Vol. 1, Johnson, C. C. and Shore, M. L., Eds., HEW Publ. (FDA) 77-8010, Department of Health, Education and Welfare, Rockville, MD, 1976, 406.
86. **Dutta, S. K., Nelson, W. H., Blackman, C. F., and Brusick, D. J.,** Cellular effects in microbial tester strains caused by exposure to microwaves or elevated temperatures, *J. Environ. Pathol. Toxicol.,* 3, 195, 1980.
87. **Corelli, J. C., Gutmann, R. J., Kohazi, S., and Levy, J.,** Effects of 2.6-4.0 GHz microwave radiation on *E. coli B., J. Microwave Power,* 12, 141, 1977.
88. **Ward, T. R., Allis, J. W., and Elder, J. A.,** Measure of enzymatic activity coincident with 2450 MHz microwave exposure, *J. Microwave Power,* 10, 315, 1975.

89. **Bini, M., Checcucci, A., Ignesti, A., Millanta, L., Rubino, N., Camici, C., Manao, G., and Ramponi, G.,** Analysis of the effects of microwave energy on enzymatic activity of lactate dehydrogenase (LDH), *J. Microwave Power,* 13, 95, 1978.
90. **Allis, J. W. and Fromme, M. L.,** Activity of membrane bound enzymes exposed to sinusoidally modulated 3450-MHz microwave radiation, *Radio Sci.,* 14(65), 85, 1979.
91. **Allis, J. W., Fromme, M. L., and Janes, D. E.,** Pseudosubstrate binding of ribonuclease during exposure to microwave radiation at 1.70 and 2.45 GHz, in *Biological Effects of Electromagnetic Waves,* Vol. 1., Johnson, C. C. and Shore, M. L., Eds., HEW Publ. (FDA) 77-8010, U.S. Department of Health, Education and Welfare, Rockville, MD, 1976, 366.
92. **Smialowicz, R. J.,** The effect of microwaves (2450 MHz) on lymphocyte blast transformation *in vitro, Biological Effects of Electromagnetic Waves,* Vol. 1, Johnson, C. C. and Shore, M. L., HEW Publ. (FDA) 77-8010, U.S. Department of Health, Education and Welfare, Rockville, MD, 1976, 472.
93. **Spalding, J. F., Freyman, R. W., and Holland, L. M.,** Effects of 800-MHz electromagnetic radiation on body weight, activity, hemapoiesis and life span in mice, *Health Phys.,* 20, 421, 1971.
94. **McRee, D. I., Livingston, G. K., and MacNichols, G.,** Incidence of sister chromatid exchange in bone marrow cells of the mouse following microwave exposure, in *Symp. Electromagnetic Fields in Biological Systems,* The Institute of Electrical and Electronic Engineers/International Medical Physics Institute, Ottawa, 1978, 15 (abstract).
95. **Huang, A. T., Engle, M. E., Elder, J. A., Kinn, J. B., and Ward, T. R.,** The effect of microwave radiation (2450 MHz) on the morphology and chromosomes of lymphocytes, *Radio Sci.,* 12(65), 173, 1977.
96. **Varma, M. M. and Traboulay, E. A., Jr.,** Evaluation of dominant lethal test and DNA studies in measuring mutagenicity caused b non-ionizing radiation, in *Biological Effects of Electromagnetic Waves,* Vol. 1, Johnson, C. C. and Shore, M. L., Eds., HEW Publ. (FDA) 77-8010, U.S. Department of Health, Education and Welfare, Rockville, MD, 1976, 386.
97. **Ismaliov, E. Sh.,** Mechanism of effects of microwaves on erythrocyte permeability for potassium and sodium ions, *Biol. Nauki. (Moscow),* 21, 58, 1971 (English translation, NTIS Rep. No. JPRS 72606, Springfield, VA, 1971, 38.)
98. **Ismaliov, E. Sh.,** Infrared spectra of erythrocyte ghosts in the region of the Amide I and Amide II bands on microwave irradiation, *Biofizika,* 21, 940, 1976 (English translation, *Biophysics,* 21, 961, 1977).
99. **Ismaliov, E. Sh.,** Effect of ultrahigh frequency electromagnetic radiation on the electrophoretic mobility of erythrocytes, *Biophysics,* 22, 510, 1978 (*Biofizika,* 21, 940, 1977).
100. **Seaman, R. L. and Wachtel, H.,** Slow and rapid responses to CW and pulsed microwave radiation by individual *Aplysia* pacemakers, *J. Microwave Power,* 13, 77, 1978.
101. **Chernovetz, M. E., Justesen, D. R., and Oke, A. F.,** A teratological study of the rat: microwave and infrared radiation compared, *Radio Sci.,* 12(65), 191, 1977.
102. **Kitsovskaya, I. A.,** The effect of centimeter waves of different intensities on the blood and hemopoetic organs of white rats, *Gig. Tr. Prof. Zabol.,* 8, 14, 1964.
103. **Michaelson, S. M., Thompson, R. A. E., El Tamami, M. Y., Seth, H. S., and Howland, W.,** The hematological effects of microwave exposure, *Aerosp. Med.,* 35, 824, 1964.
104. **Lappenbusch, W. L., Gillespie, L. J., Leach, W. M., and Anderson, G. E.,** Effects of 2450-MHz microwaves on radiation response of X-irradiated chinese hamsters, *Radiat. Res.,* 54, 294, 1973.
105. **Lidbury, R. P.,** Effects of radio frequency radiation on inflammation, *Radio Sci.,* 12(6S), 179, 1977.
106. **Wiktor-Jedrzejczak, W., Ahmed, A., Czerski, P., Leach, W. M., and Sell, K. W.,** Immune response of mice at 2450-MHz microwave radiation: overview of immunology and empirical studies of lymphoid splenic cells, *Radio Sci.,* 12(6S), 209, 1977.
107. **Wiktor-Jedrzejczak, W., Ahmed, A., Czerski, P., Leach, W. M., and Sell, K. W.,** Increase in the frequency of Fc-receptor (FcR) bearing cells in the mouse spleen following a single exposure of mice to 2450 MHz microwaves, *Biomedicine,* 27, 250, 1977.
108. **Wiktor-Jedrzejczak, W., Ahmed, A., Sell, K. W., Czerski, P., and Leach, W. M.,** Microwaves induce an increase in the frequency of complement receptor-bearing lymphoid spleen cells in mice, *J. Immunol.,* 118, 1499, 1977.
109. **Sulek, K., Schlagel, C. J., Wiktor-Jedrzecjzak, W., Ho, H. S., Leach, M., Ahmed, A., and Woody, J. N.,** Biological effects of microwave exposure. I. Threshold conditions for the induction of the increase in complement receptor positive (CR) mouse spleen cells following exposure to 2450-MHz microwaves, *Radiat. Res.,* 83, 127, 1980.
110. **Huang, A. T. and Mold, N. G.,** Immunologic and hematopoietic alterations by 2450-MHz electromagnetic radiation, *Bioelectromagnetics,* 1, 77, 1980.
111. **DeLorge, J.,** Operant behaviour and rectal temperature of squirrel monkeys during 2.45 GHz microwave irradiation, *Radio Sci.,* 14(65), 217, 1979.
112. **Michaelson, S. M.,** Microwave biological effects. An overview, *Proc. IEEE,* 68, 40, 1980.

113. ANSI, *American National Standards Safety Level with Respect to Human Exposure to Radiofrequency Electromagnetic Fields,* 300 kHz-100 GHz (ANSI C95.1-1982), American National Standards Institute, New York, 1982.
114. **Heller, J. H.,** Cellular effects of microwave radiation, in Biological Effects and Health Implications of Microwave radiation, Cleary, S. F., Ed., HEW Publ. BRH/DBE 70-2, Bureau of Radiological Health, U.S. Department of Health, Education and Welfare, Rockville, MD, 1970, 116.
115. **Durney, C. H., Johnson, C. C., Barber, P. W., Massoudi, H., Iskander, M. F., Lords, J. L., Ryser, D. K., Allen, S. J., and Mitchell, J. C.,** Radiofrequency Radiation Dosimetry Handbook, 2nd ed., Rep. SAM-TR-78-22, U.S. Air Force School of Aerospace Medicine, Brooks AFB, Texas, 1978, 141.
116. U.S.S.R., Temporary Sanitary Rules for Working with Centimeter Waves, Ministery of Health Protection of the U.S.S.R., 1958.
117. **Michaelson, S. M.,** Microwave and radiofrequency radiation, in *Nonionizing Radiation Protection,* Suess, M. J., Ed., WHO Regional Publ. Eur. Ser. No. 10, World Health Organization, Copenhagen, 1982, 97.
118. IRPA, Interim guidelines on limit of exposure to radiofrequency electromagnetic fields in the frequency range from 100 kHz to 300 GHz, *Health Phys.,* 46(4), 975, 1984.
119. **Rozzel, T. C.,** West German EMF exposure standard, *Bioelectromagn. Soc. Newsl.,* 55, 1, 1985.
120. **Michaelson, S. M.,** Microwave/Radiofrequency protection guide and standards, in *Biological Effects and Dosimetry of Nonionizing Radiation,* Grandolfo, M., Michaelson, S. M., and Rindi, A., Eds., Plenum Press, New York, 1983, 645.
121. **Paulsson, L. E.,** Videoterminals, Internal Report, National Institute of Radiation Protection, Stockholm, Sweden, 1984.
122. **Roy, O. Z., Park, G. C., and Stott, J. R.,** Intracardiac catheter fibrillation thresholds as a function of duration of 60 Hz current and electrode area, *IEEE Trans. Biomed. Eng.,* 24, 430, 1977.
123. **Roy, O. Z.,** Technical note: summary of cardiac fibrillation thresholds for 60 Hz currents and voltage applied directly to the heart, *Med. Biol. Eng. Comput.,* 18, 657, 1980.
124. **Watson, A. B., Wright, J. S., and Loughman, J.,** Electrical thresholds for ventricular fibrillation in man, *Med. J. Aust.,* 1, 1179, 1973.
125. **Polson, M. J. R., Barker, A. T., and Freeston, I. L.,** Stimulation of nerve trunks with time-varying magnetic fields, *Med. Biol. Eng. Comput.,* 20, 243, 1982.
126. **Noble, D. and Stein, R. R.,** The threshold conditions for initiation of action potentials by excitable cells, *J. Physiol.,* 187, 129, 1966.
127. **Ranck, J. B., Jr.,** Which elements are excited in electrical stimulation of mammalian central nervous system: a review, *Brain Res.,* 98, 417, 1975.
128. **Prosser, C. L., Ed.,** *Comparative Animal Physiology,* W. B. Saunders, Philadelphia, 1973.
129. **D'Arsonval, M. A.,** Action physiologique des courants alternatifs a grande frequence, *Arch. Physiol.,* 5, 401, 1893.
130. **Barlow, H. B., Kohn, H. I., and Walsh, E. G.,** The effect of dark adaption and of light upon the electric threshold of the human eye, *J. Physiol. (London),* 148, 376, 1947.
131. **Tucker, R. D. and Schmitt, O. H.,** Tests for Human Perception of 60 Hz Moderate Strength Low Frequency Magnetic Fields, Ph.D. thesis, University of Minnesota, Minneapolis, 1976.
132. **Budinger, T. F.,** Nuclear magnetic resonance (NMR) in vivo studies: known thresholds for health effects, *J. Comput. Assisted Tomogr.,* 5, 800, 1981.
133. **Budinger, T. F., Cullander, C., and Bordow, R.,** Switched magnetic field thresholds for the induction of magnetophophenenes, in *Proc. 3rd Ann. Meet. Soc. Magnetic Resonance in Medicine,* Society of Magnetic Resonance in Medicine, Berkeley, CA, 1984, 118.
134. **Delgado, J. M. R., Leal, J., Montegudo, J. L., and Gracia, M. G.,** Embryological changes induced by weak, extremely low frequency electromagnetic fields, *J. Anat.,* 134(3), 533, 1982.
135. **Brighton, C. T., Ed.,** *Electric and Magnetic Control of Musculo-Skeletal Growth and Repair,* Grune & Stratton, New York, 1979.
136. **Fink, M.,** *Convulsive Therapy: Theory and Practice,* Raven Press, New York, 1979.
137. **Kim, Y. S.,** Some possible effects of static magnetic fields on cancer, Tower International Technomedical Institute, Inc., *J. Life Sci.,* 6, 11, 1976.
138. **Hanneman, G. D.,** Changes in sodium and potassium content of urine from mice subjected to intense magnetic fields, in *Biological Effects of Magnetic Fields,* Vol. 2, Barnothy, F., Plenum Press, New York, 1969.
139. **Beischer, D. E. and Knepton, J. C., Jr.,** Influence of strong magnetic fields on the electrocardiogram of squirrel monkeys, *Aerosp. Med.,* 35, 939, 1964.
140. **Beischer, D. E.,** Vectorcardiogram and aortic blood flow of squirrel monkeys (*Saimire sciureus*) in a strong super conductive magnet, in *Biological Effects of Magnetic Fields,* Vol. 2, Barnothy, M. F., Ed., Plenum Press, New York, 1969, 241.

141. **Toroptsev, I. V., Garganeyev, G. P., Gorshenina, T. I., and Teplyakova, N. L.,** Pathologoanatomic characteristics of changes in experimental animals under the influence of magnetic fields, in Influence of Magnetic Fields on Biological Objects, Kholodov, Yu, A., Ed., NTIS Rep. No. JPRS 63038, National Technical Information Service, Springfield, VA, 1974, 95.
142. **Markuze, I. I., Ambartsumyan, R. G., Chibrikin, V. M., and Piruzyan, L. A.,** Investigation of the PMP action on the alteration of electrolyte concentration in the blood and organs of animals, *Izv. Akad. Nauk. SSSR Ser. Biol.,* 2, 281, 1973.
143. **Ketchen, E. E., Porter, W. E., and Bolton, N. E.,** The biological effects of magnetic fields on man, *Am. Ind. Hyg. Assoc.,* 39, 1, 1978.
144. **Saunders, R. D.,** Biologic effects of NMR clinical imaging, *Appl. Radiol.,* Sept/Oct, 43, 1982.
145. **Vyalov, A. M.,** Clinico-hygienic and experimental data on the effects of magnetic fields under industrial conditions, in Influence of Magnetic Fields on Biological Objects, Rep. JPRS 5303S, Kyaoloov, Y., Ed., National Technical Informational Service, Springfield, VA, 1971.
146. **Marsh, J. L., Armstrong, T. J., Jacobson, A. P., and Smith, R. G.,** Health effect of occupational exposure to steady magnetic fields, *Am. Ind. Hyg. Assoc. J.,* 43, 387, 1982.
147. **Rockette, H. E. and Arena, V. C.,** Mortality studies of aluminium reduction plant workers: potroom and carbon department, *J. Occup. Med.,* 25(7), 549, 1983.
148. **Beischer, D. E.,** Survival of animals in magnetic fields of 140,000 Oe, in *Biological Effects of Magnetic Fields,* Vol. 1, Barnothy, M. F., Ed., Plenum Press, New York, 1964.
149. **Neurath, P. W.,** The effect of high gradient, high strength fields on the early embryonic development of frogs, in *Biological Effects of Magnetic Fields,* Vol. 2, Barnothy, M. F., Ed., Plenum Press, New York, 1969, 177.
150. **Barnothy, M. F. and Sumegi, I.,** Effects of the magnetic fields on internal organs and endocrine system of mice, in *Biological Effects of Magnetic Fields,* Vol. 1, Barnothy, M. F., Ed., Plenum Press, New York, 1969, 263.
151. **Barnothy, M. F. and Barnothy, J. M.,** Magnetic field and the number of blood platelets, *Nature (London),* 225, 1146, 1970.
152. **Hanneman, G. D.,** Changes produced in urinary sodium, potassium, and calcium excretion in mice exposed to homogeneous electromagnetic field, *Aerosp. Med.,* 38, 275, 1967.
153. **Russel, D. R. and Hedrick, H. G.,** Preferences of mice to consume food and water in an environment of high magnetic field, in *Biological Effects of Magnetic Fields,* Vol. 1, Barnothy, M. F., Ed., Plenum Press, New York, 1969, 233.
154. **Kholodov, Ju. A., Alexandrovskaya, M. M., Lukjanova, S. N., and Udarova, N. S.,** Investigation of the reactions of mammalian brain to static magnetic fields, in *Biological Effects of Magnetic Fields,* Vol. 2, Barnothy, M. F., Ed., Plenum Press, New York, 1969, 215.
155. **Aleksandrovskaya, M. M. and Kholodov, Yu. A.,** The reactions of brain neuroglia to exposure to a constant magnetic field, in *Questions of Hematology, Radiobiology and the Biological Actions of Magnetic Fields,* U.S.S.R. Academy of Science, Tomsk, U.S.S.R., 1965.
156. **Brown, F. A., Jr.,** Effects and after-effects on planarians of reversals of the horizontal magnetic vector, *Nature (London),* 209, 533, 1966.
157. **Keeton, W. T.,** The mystery of pigeon homing, *Sci. Am.,* 231, 96, 1974.
158. **Beischer, D. E. and Reno, V. R.,** Magnetic fields and man; where do we stand today, in *Special Biophysical Problems in Aerospace Medicine,* Conf. Proc. No. 95, Part III, Pfister, A. M., Ed., AGARD Medical Panel Specialist Meeting, Luchon, France, 1971.
159. **Beischer, D. E.,** Human tolerance to magnetic fields, *Astronautics,* 7, 24, 1962.
160. **Peterson, F. and Kennelly, A. E.,** Some physiological experiments with magnets at the Edison Laboratory, *N.Y. Med. J.,* 56, 729, 1892.
161. **Bottomley, P. A. and Edelstein, W. A.,** Power deposition in whole-body NMR imaging, *Med. Phys.,* 8, 510, 1981.
162. **Bottomley, P. A. and Andrew, E. R.,** RF magnetic field penetration, phase shift and power dissipation in biological tissue: implications for NMR imaging, *Phys. Med. Biol.,* 23, 630, 1978.
163. **Bottomley, P. A., Rowland, R. W., Edelstein, W. A., and Schenck, J. F.,** Estimating radiofrequency power deposition in body NMR-imaging, *Magn. Reson. Med.,* 2, 336, 1985.
164. **Thomas, A. and Morris, P. G.,** The effects of NMR exposure on living organisms. I. A microbial assay, *Br. J. Radiol.,* 54, 615, 1981.
165. **Cooke, P. and Morris, P. G.,** The effects of NMR exposure on living organisms. II. A genetic study of human lymphocytes, *Br. J. Radiol.,* 54, 622, 1981.
166. **Wolff, S., Crooks, L. E., Brown, P., Howard, R., and Painter, R. B.,** Test for DNA and chromosomal damage induced by nuclear magnetic resonance imaging, *Radiology,* 136, 707, 1980.
167. **Schwartz, J. L. and Crooks, L. E.,** NMR imaging produces no observable mutations or cytotoxicity in mammalian cells, *Am. J. Roentgenol.,* 139, 583, 1982.

168. **Dujovny, M., Kossovsky, N., Kossovsky, R., Diaz, F. G., and Ausman, J. I.,** Magnetic aneurysm clips. Correlation with martensite content and implications for nuclear magnetic resonance examination, *Am. Coll Surg. Forum,* XXXIV, 525, 1983.
169. **Dujovny, M., Kossovsky, N., Kossowsky, R., Perlin, A., Gatti, E. F., Segal, R., and Diaz, F. G.,** Mechanical and metallurgical properties of vascular clips designed for temporary use, *Microsurgery,* 4, 124, 1983.
170. **New, P. F. J., Rosen, B. R., Brady, T. J., Buonanno, F. S., Kistler, J. P., Burt, C. T., Hinshaw, W. S., Newhouse, J. H., Pohost, G. M., and Taveras, J. M.,** Potential hazards and artifacts of ferromagnetic and nonferromagnetic surgical and dental materials and devices in nuclear magnetic resonance imaging, *Radiology,* 147, 139, 1983.
171. **Zimmermann, B. H. and Faul, D. D.,** Artifacts and hazards in NMR imaging due to metal implants and cardiac pacemakers, *Diagn. Imag. Clin. Med.,* 53, 53, 1984.
172. **Davis, P. L., Crooks, L., Arakawa, M., McRee, R., Kaufman, L., and Margulis, A. R.,** Potential hazards in NMR imaging: heating effects of changing magnetic fields and RF fields on small metallic implants, *Am. J. Roentgenol.,* 137, 857, 1981.
173. SLAC, *Limits on Human Exposure in Static Magnetic Fields,* Stanford Linear Accelerator Center, Stanford, CA, 1970.
174. National Accelerator Laboratory, Interim Standards for Occupational Exposure to Magnetic Fields, National Accelerator Laboratory, Berkeley, CA, 1979.
175. National Radiological Protection Board, (Great Britain), Exposure to nuclear magnetic resonance clinical imaging, *Radiography,* 47, 258, 1981.
176. National Radiological Protection Board (Great Britain), Revised guidance on acceptable limits of exposure during nuclear magnetic resonance clinical imaging, *Br. J. Radiol.,* 56, 974, 1983.
177. Bureau of Radiological Health, Guidelines for Evaluating Electromagnetic Risk for Trials of Clinical NMR Systems, FDA HFX-460, Bureau of Radiological Health, U.S. Food and Drug Administration, Rockville, MD, 1982.
178. BGA, (Bundesgesundheitsamt), Empfehlungen zur Vermeidung gesundheitlicher Risiken verursacht durch magnetisch und hoch frequente electromagnetische Felder bei der NMR-tomographie und in-vivo NMR spectroskopie, *Bundesgesundheitsblatt,* 27(3), 92, 1984 (in German).
179. Environmental Health Directorate, Guidelines on Exposure to Electromagnetic Fields from Magnetic Resonance Clinical Systems, EHD Publ. No. 86-EHD-127, Environmental Health Directorate, Ottawa, 1986.
180. **Budinger, T. F.,** Thresholds for physiological effects due to RF and magnetic fields in NMR imaging, *IEEE Trans. Nucl. Sci.,* 26, 2821, 1979.
181. Environmental Health Directorate, Recommendations to Ensure Protection of Patients and Operational Personnel from Potential Hazard in Proton and NMR Imaging, Publ. No. 85-EHD-124, Environmental Health Directorate, Ottawa, 1985.
182. **Einstein, S. G., Maudsley, A. A., Mun, S. K., Simon, H. E., Hilial, S. K., Sano, R. M., and Roeschman, P.,** Installation of high field NMR systems into existing clinical facilities: special considerations, in *Technology of Nuclear Magnetic Resonance,* Esser, P. D. and Johnston, R. E., Eds., Society of Nuclear Medicine, New York, 1984, 217.
183. **Pavlicek, M., Mackintyre, W., Go, R., O'Donnel, J., and Feiglin, D.,** Special architectural consideration in designing a NMR facility, in *Technology of Nuclear Magnetic Resonance,* Esser, P. D. and Johnston, R. E., Eds., Society of Nuclear Medicine, New York, 1984, 252.
184. **Bore, P. J. and Timms, W. E.,** The installation of high field NMR equipment in a hospital environment, *Magn. Reson. Imaging,* 1, 387, 1984.
185. **Warwick, R., Pond, J. B., Woodward, B., and Connolly, C. C.,** Hazards of diagnostic ultrasonography — study with mice, *IEEE Trans. Sonic Ultrasonics,* 17, 158, 1970.
186. **Lyon, M. F. and Simpson, G. M.,** An investigation into the possible genetic hazards of ultrasound, *Br. J. Radiol.,* 47, 712, 1974.
187. **Shoji, R., Momma, E., Shimizu, T., and Matsuda, S.,** Influence of low intensity ultrasonic irradiation on prenatal development of two inbred mouse strains, *Teratology,* 12, 227, 1975.
188. **Garrison, B. M., Bo, W. J., Krueger, W., Kremkau, F., and McKinney, W.,** The influence of ovarian sonication on fetal development in the rat, *J. Clin. Ultrasound,* 1, 316, 1973.
189. **Curto, K. A.,** Early postpartum mortality following ultrasound radiation, in *Ultrasound in Medicine,* Vol. 2, White, D. N. and Barnes, R., Eds., Plenum Press, New York, 1976, 535.
190. **O'Brien, W. D.,** Ultrasonically induced fetal weight reduction in mice, in *Ultrasound in Medicine,* Vol. 2, White, D. D. and Barnes, R., Eds., Plenum Press, New York, 1976, 531.
191. **Mannor, S. M., et al.,** The safety of ultrasound in fetal monitoring, *Am. J. Obstet. Gynecol.,* 113, 653, 1972.
192. **Abdulla, U., et al.,** Effect of diagnostic ultrasound on material and foetal chromosomes, *Lancet,* 2, 829, 1971.

193. **Ikeuchi, T., et al.,** Ultrasound and embryonic chromosomes, *Br. Med. J.,* 1, 112, 1973.
194. **Levi, S., et al.,** In vivo effect of ultrasound at human therapeutic doses on marrow cell chromosomes of golden hamster, *Humangenetik,* 25, 133, 1974.
195. **Lele, P. P.,** Ultrasonic teratology in mouse and man, in *Ultrasound in Medicine,* Kanzner, E., et al., Eds., Excerpta Medica, Amsterdam, 1975, 22.
196. **Fry, F. J., et al.,** Threshold ultrasonic dosages for histological changes in mammalian brain, *J. Acoust. Soc. Am.,* 48, 1413, 1970.
197. **Dyson, M., et al.,** Stimulation of tissue regeneration by pulsed plane wave ultrasound, *IEEE Trans. Sonic Ultrasonic,* 17, 133, 1970.
198. **Dyson, M., et al.,** Stimulation and healing of varicose ulcers by ultrasound, *Ultrasonic,* 14, 232, 1976.
199. **Kremkau, F. W. Witcofski, R. L.,** Mitotic reduction in rat liver exposed to ultrasound, *J. Clin. Ultrasound,* 2, 123, 1974.
200. **Lynnworth, L. C.,** Industrial applications of ultrasound — a review. II. Measurements, tests and process control using low intensity ultrasound, *IEEE Trans. Sonic Ultrasonic,* 22, 71, 1973.
201. **Miller, M. W., et al.,** Absence of mitotic reduction in regenerating rat livers exposed to ultrasound, *J. Clin. Ultrasound,* 4, 169, 1976.
202. **Acton, W. I.,** A criterion for the prediction of auditory and subjective effects due to ultrasonic sources, *Ann. Occup. Hyg.,* 11, 227, 1968.
203. **Skillern, C. P.,** Human response to measured sound pressure levels from ultrasonic devices, *Am. Hyg. Assoc. J.,* 26, 132, 1965.
204. **Parrack, M. O.,** Effects of airborne ultrasound on humans, *Int. Audiol.,* 5, 294, 1966.
205. **Hellman, L. M., et al.,** Safety of diagnostic ultrasound in obstetrics, *Lancet,* 1, 1133, 1970.
206. **Hill, C. R. and terHaar, G.,** Ultrasound, in *Nonionizing Radiation Protection,* Suess, M. J., Ed., WHO Regional Publ. Eur. Ser. No. 10, Regional Office for Europe, World Health Organization, Copenhagen, 1982, 217.

Chapter 11

BIOMECHANICAL ASPECTS ON BODY LOADING DURING WORK AND SPORT*

Roland Örtengren

TABLE OF CONTENTS

* This chapter is a translated and expanded version of a paper presented at the Swedish Sports Research Council Conference on Injuries in Sport in Stockholm, January 22 to 24, 1987.

I. INTRODUCTION

During evolution, the body of man has been adapted to the need for mobility and force that is specific for the human way of life. Perhaps "was specific" should be said rather than "is" since for humans of today with modern customs and habits, the purposefulness of the evolutionary development is not so obvious. As a result of progress in science and technology, the conditions of life have changed, and humans of today have acquired many artificial means for ambulation and force production. Therefore man is not so dependent on his own physical resources for survival as during earlier times. Also, in many jobs the need for mobility and ambulation has become restricted. Still, the body is designed for physical activity and needs exercise to maintain health and fitness. That is what many people feel and they try to compensate for their sendentary jobs by devoting themselves to jogging and other sporting activities during their spare time.

It is easy to be highly impressed by the ability and capacity of the system for locomotion when one sees the achievements that able male and female athletes can accomplish, and no doubt the loading on the body must be extremely high in many athletic activities at top-level performance. The strength and suppleness of the different tissues in the musculoskeletal system and the interplay between them are of crucial importance for accomplishments to be achieved. Normally the strength characteristics of the different tissues are adapted to each other and to the need of force that the weight of the body demands so that movements and ambulation can take place without risk of damages. However, sporting activities at an elite level demand top performance and strain to the very limit. It occurs not so seldom that the loading becomes larger than what the tissue can withstand with more or less serious damages as a result.

Injuries during sporting activities can occur in many ways. Tissue rupture can be caused by a sudden high loading due to high force demand or to a slipping incident, and wear can be the result of repetitions of loadings at a level that the body cannot or does not have sufficient time to adapt to. Except for a few rare cases, the injuries are unintentional — even in fighting sports — and part of the training is normally devoted to instructions on how to avoid injuries. How this is accomplished can be specific for a particular type of game in which certain loadings occur, but there are also general recommendations. Thus it is necessary to train oneself and to increase ones' strength gradually; one should warm up before particularly strenuous activities, and one should vary loadings and movements. Just as in working life, it must be remembered that monotonously repeated movements as well as single high loadings are likely to cause overload disorders. It is the resistance of the tissues against wear and their strength that determine whether an injury will occur or not.

The loadings that the body is subject to in working life are normally much lower than in sport. Still, the injury rate is high. About half the number of occupational injuries and diseases reported in Sweden every year are classified as so-called overload disorders. It has been calculated that overload disorders cause losses in Sweden in the order of $2.8 billion (1983), about 2.3% of the gross national product. To that must be added the personal suffering of those affected which cannot be valued in money. It is obvious that much can be gained if the overload injury rate could be reduced. It is also obvious that preventive measures are more effective than treatment. Work training, teaching of work practices, and practical ergonomics are means of preventive measures for individuals. Ergonomic design and continuous ergonomic maintenance of work places and work procedures are measures to be required from management.

The aim of this chapter is to give a biomechanical background to why and how the body is loaded during work and other physical activities and to how injurious effects can be caused. Biomechanical principles are illustrated and a few examples of biomechanical calculations are given. How physical overload can be avoided in many practical situations is discussed in Chapters 12 and 17.

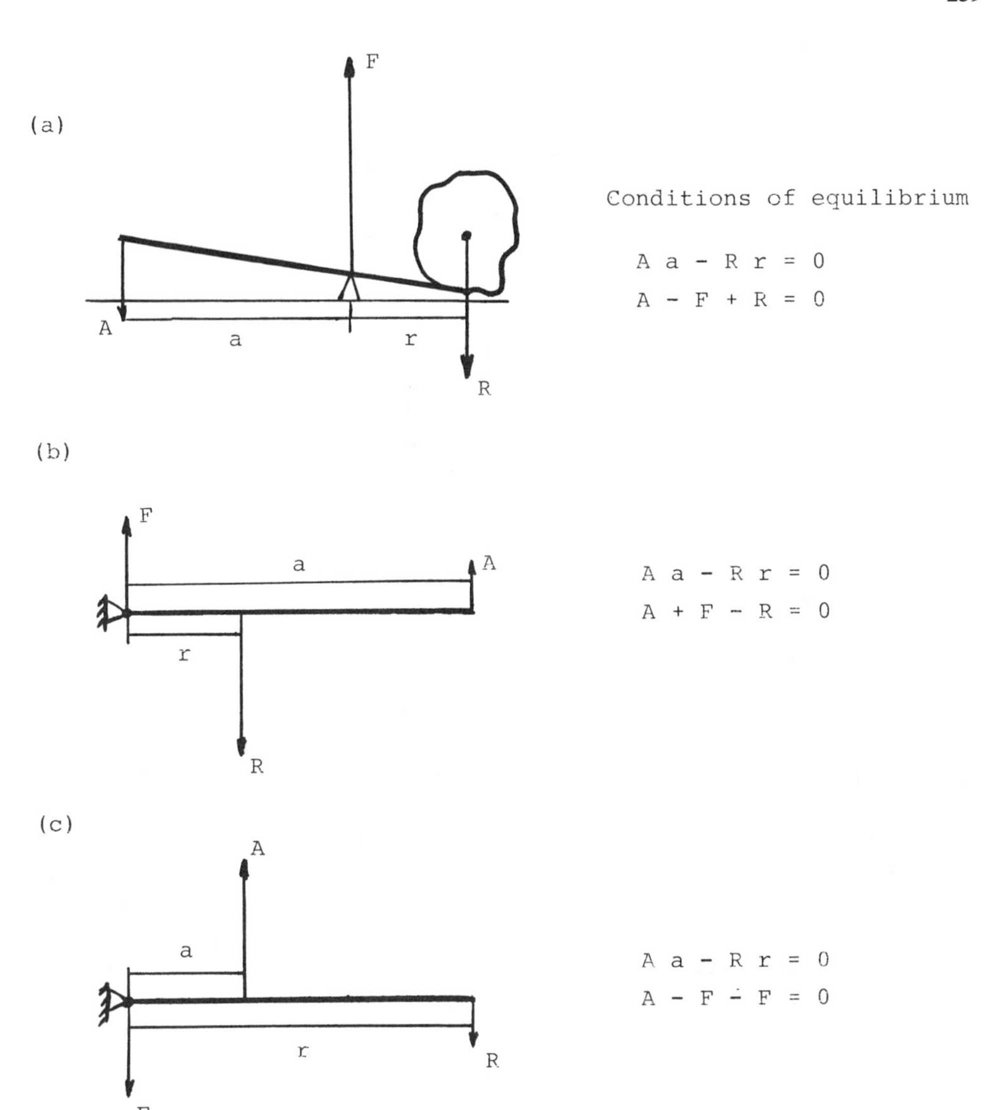

FIGURE 1. Three cases of lever action. In Type I (a), the action and the resistance attack on each side of the fulcrum; in Type II (b) and Type III (c), they attack on the same side. Note the differences in reaction force at the fulcrum between the three cases. Type III is representative of the conditions in the musculoskeletal system.

II. LEVER ACTION

Lever action can be illustrated by a rod or pole hinged around a pivot point, a fulcrum. A force applied causes the lever to rotate. The turning action is determined by the turning moment, or torque, which is equal to the length of the lever arm multiplied by the force component perpendicular to the lever, or equivalently, the force multiplied by the perpendicular distance to the pivot point (Figure 1). Because of the multiplying action of the lever, either a gain in force or in distance can be accomplished. One can say that ''what is gained in force is lost in distance'' and vice versa. For a given force, a longer lever gives a larger turning moment than a shorter one. The lever is one of the simplest machines known.

In kinesiology it is convenient to distinguish between three types of levers to describe the action. The type depends on where the acting force or effort and the resistance are located in relation to each other and the fulcrum. The three types are illustrated in Figure 1. The

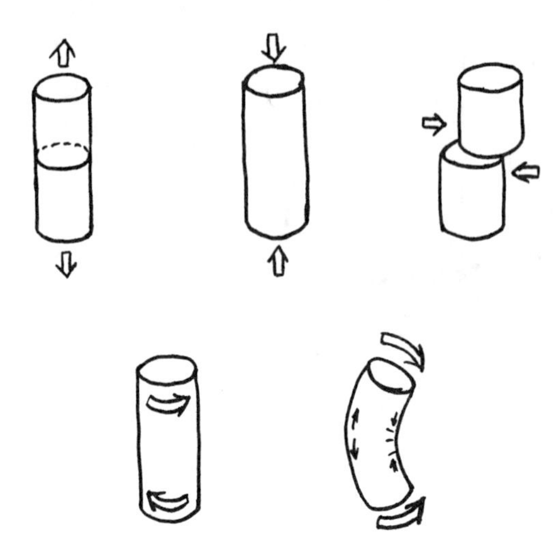

FIGURE 2. Types of loading to which a material can be exposed. The specific loading, the stress, is calculated by dividing the acting force by the cross-sectional area. The stress is denoted by σ for tension and compression and by τ for shear. Torsion results in a shear stress. Bending causes tension on one side of the material and compression on the other side.

muscles in the body have lever arms that are shorter than the lever arms of the resistances. One consequence of this is that the muscle forces must be larger than the forces of the resistances to cause movements, often four to five times larger, which in turn causes the joint forces to become large. Another geometrical consequence of the ratio between the lever arms is that a relatively small shortening of a muscle leads to a large change of position at the end of the lever arm. One can conclude from this that mobility and speed have been favored during evolutional development, at least for hunting animals. Such animals also have a light build and need only a little of the available muscle forces to support their own bodies against gravity. Instead, the available muscle power is used for increased mobility and fast acceleration. It is a different matter for heavy animals such as the hippopotamus and the rhinoceros, which have rather short and sturdy legs to support their weight and accelerate slowly.

III. LOADING AND MATERIAL STRENGTH

Biological material — generally living tissue — exposed to forces and pressures is studied in one branch of biomechanics. The purpose is to determine stress, distribution of forces, and concomitant movements. When a force acts on a biological tissue, two things can happen: the tissue can be put into motion or it can become deformed, often both. The deformation is determined by the material properties of the tissue. If they are known, it is in principle possible to predict the deformation that a given force will cause. However, in contrast to, for example, engineering materials, the properties of biological tissue are much more difficult to determine and describe since the behavior can be quite different depending on such factors as temperature, speed of deformation, and direction of force.

Depending on the direction of the force in relation to the extension and the dimensions of the material, the loading can be characterized as tension, compression, and shear or as bending and torsion (Figure 2). The local loading the material is exposed to due to an external force depends on the dimensions (and also the structure) of the material. To be able to relate the loading to the specific properties of the material, i.e., the properties independent of the dimensions, a specific loading — stress — is calculated by dividing the external force by the cross-sectional area of the material as Figure 2 shows in some cases. In torsion the

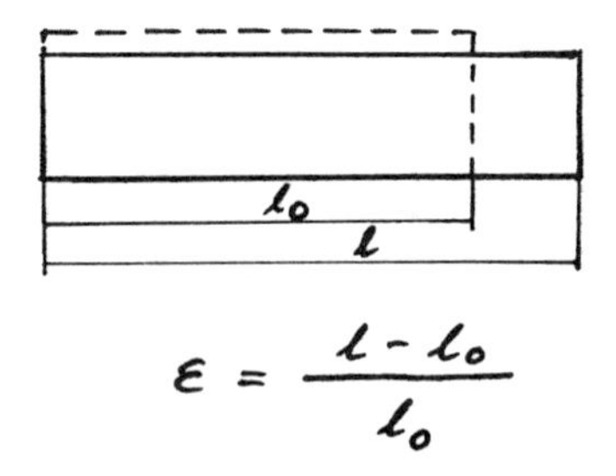

FIGURE 3. Calculation of tensile deformation, i.e., strain ϵ, for a material exposed to a tensile force. The extension causes the cross-sectional area to decrease a little, but the change is negligible in most cases.

material is locally exposed to shear stress, while in bending there is a tensile stress on one side of the material and a compressive stress on the other (compare with Table 2).

To simplify the calculations in mechanics, it is often necessary to assume that a material does not deform when it is exposed to forces. Bodies consisting of such material are called rigid bodies. The results are sufficiently accurate when only forces and gross movements are of concern. In reality, conditions are different. All materials that are exposed to forces deform more or less. Tension causes the material to elongate. The specific measure of elongation is called strain and is calculated as shown in Figure 3.

The relationship between stress and strain for a material shows how compliant the material is and is important for judging the properties of the material. For many engineering materials, such as steel, the strain is proportional to the stress within a large range of stress levels, i.e., the material obeys Hook's law. This means that the material is elastic and resumes its original form when the stress is removed. The constant of proportionality is called modulus of elasticity or Young's modulus.

With the exception of bone tissue, biological materials do not obey Hook's law; instead, the relation between stress and strain is nonlinear. Examples of stress-strain diagrams for some tissues are shown in Figure 4.

In addition to being nonlinear, most biological materials have another characteristic that makes them deviate from an ideal material. When a biological material is subject to a constant strain, it continues to lengthen — creep, and when the strain is removed, a remaining lengthening has occurred (see Figure 4b). A material that deforms permanently under strain is said to be viscous since its behavior resembles that of a liquid. Most biological materials that are both elastic and show this liquid-like behavior are said to be viscoelastic. The deformation of a living viscoelastic tissue is normally not permanent, but the original shape is restored after some time. For example, the height of the intervertebral disks of the spine decreases continuously during waking hours, causing the body height to shorten by 10 to 15 mm. During sleep, when the weight of the trunk and the back muscle tension are relieved, the height loss of the spine is recovered.

The most important tissues in the locomotor system are bone, cartilage, ligament, tendon, and muscle. As shown in Table 1, the tissues have differing capacities to withstand loading. Bone and cartilage have high strength for compressive loading, while ligament, tendon, and muscle are adapted to tensile loading. As the other tissues, bone has different strength in different directions, but the strength is high enough to make loading possible in all directions. Strength data for cortical bone are given in Table 2.

Muscle tissue holds an exceptional position among the tissues compared in Table 1 because muscle tissue alone has the ability to develop force by active shortening. Since force can only be developed under contraction, the muscles in the body must be organized in pairs or mostly in groups that work together alternatingly as agonists and antagonists in relation to

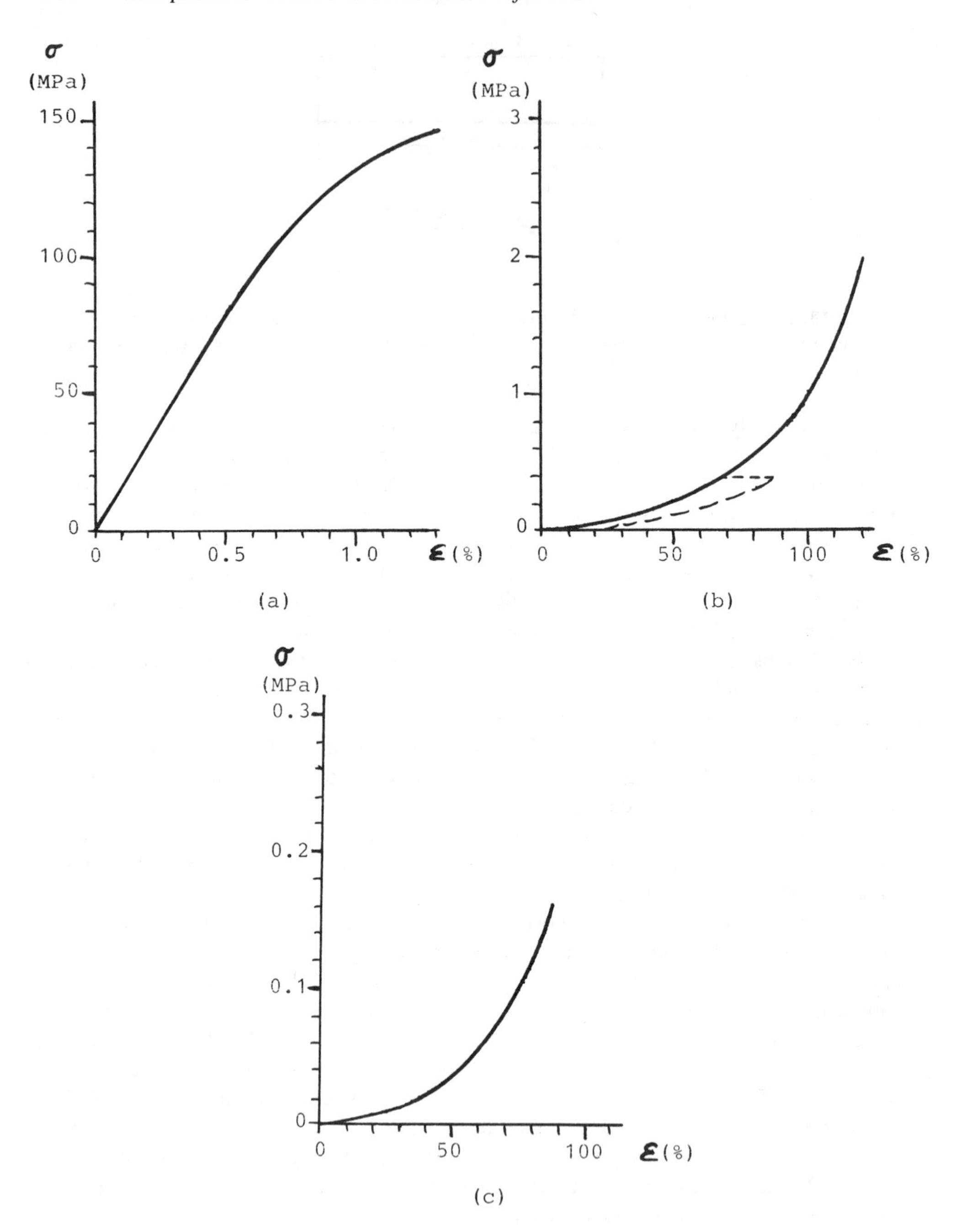

FIGURE 4. Stress-strain curves in tension for human compact bone (a), ligamentum nuchae in cattle (b), and human muscle tissue (c). In (a) the stress increases linearly with the strain in the first part of the curve which means that the material is linear in that interval. The dashed curve in (b) shows how the strain increases due to creep when the stress is kept constant for a period of time. After unloading, a residual lengthening lasts for some time. Note the differences in scale between the diagrams. (Average data are from Yamada, H., *Strength of Biological Materials,* Williams & Wilkins, Baltimore, 1970.)

each other. For example, the biceps muscle of the upper arm causes elbow flexion when it contracts, while the triceps causes extension, although other muscles in the region also contribute to the movements.

Data in Table 2 show that bone tissue has larger strength in compression than in extension.

Table 1
RELATIVE STRENGTH OF THE TISSUES OF THE LOCOMOTOR SYSTEM WHEN SUBJECTED TO DIFFERENT LOADING

	Type of loading			
Tissue	**Compression**	**Tension**	**Torsion**	**Bending**
Bone	+	+	+	+
Cartilage	+	—	—	—
Ligament	—	+	—	—
Tendon	—	+	—	—
Muscle	—	+[a]	—	—

[a] Muscle tissue is the only tissue that can develop tension actively.

Table 2
FAILURE STRENGTH DATA FOR ADULT HUMAN BONE SPECIMENS (IN MPa)

Loading	Cortical bone	Cancellous bone
Compression	200	10
Tension	132	10
Shear	70	—

Note: 1 MPa = 10^6 N/m^2 = 1 N/(mm)^2.

Average data are from Reilly, D. and Burstein, A., *J. Biomech.*, 8, 393, 1975.

By contracting particular muscles, it is possible to utilize this property to reduce the effect of an unfavorable loading. Figure 5 illustrates how it can be possible for a down-hill skier to reduce the risk for three-point bending fracture of the tibia by contracting the calf muscles. This action results in a prestress that reduces the bending and increases the compressive loading of the tibia for which bone tissue has a higher strength.[3] It is doubtful, however, whether this measure really works in practice in down-hill skiing since the course of events is so fast during a falling accident that there will probably not be enough time for a voluntary muscle contraction to reach sufficient strength.

IV. FRICTION AND WEAR

Friction is a very important concept in mechanics and refers to the resistance against movement that takes place when two surfaces slide against each other. Generally, friction represents a loss, but friction is nevertheless an important prerequisite for many activities. Friction permits us to get a hold for our feet when we walk and run; it prevents the wheels of a car from spinning and skidding during acceleration and braking. Friction acts as a loss in the resistance against the movement that is present in bearings and also in flowing liquids. Frictional losses consume energy which is dissipated as heat.

Effects of friction are illustrated in Figure 6 where also the coefficient of friction is defined. The force of resistance due to friction can be calculated using the coefficient of friction and the normal force, i.e., the contact force perpendicular to the contact surface. The coefficient of friction, which can assume values between 0 and 1, tells how smooth and even the surfaces are that slide against each other. Examples of coefficients of friction for some common materials are listed in Table 3.

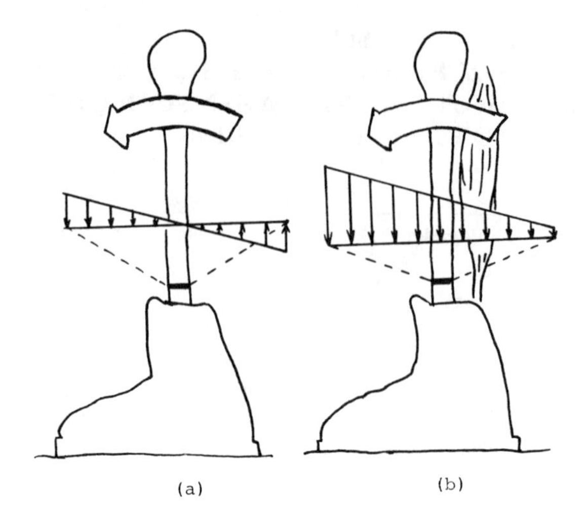

FIGURE 5. In three-point bending of the tibia, there is a compressive stress frontally and a tensile stress posteriorly (a). By contracting the triceps surae, it is possible to preload the tibia so that the tensile stress is neutralized (b), thus making use of the large compressive strength of the bone. (Adapted from Frankel, V. and Nordin, M., *Basic Biomechanics of the Skeletal System,* Lea & Febiger, Philadelphia, 1980. With permission.)

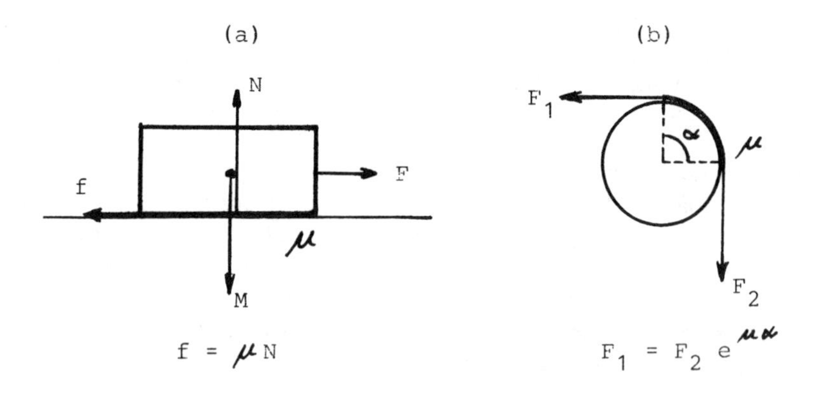

FIGURE 6. Friction between plane surfaces (a) and for a line sliding on a cylinder (b).

Table 3
COEFFICIENTS OF FRICTION FOR COMMON MATERIALS

Surfaces	Coefficient of friction (μ)
Wood on wood (dry)	0.23—0.50
Metal on metal (dry)	0.15—0.20
Metal on metal (greased)	0.03—0.05
Rubber on concrete (dry)	0.60—0.70
Rubber crutch tip on rough wood	0.70—0.75
Rubber crutch tip on tile	0.30—0.40
Cartilage on cartilage	0.005—0.02

Adapted from Williams, M., Lissner, H., and Le Veau, B., *Biomechanics of Human Motion,* W. B. Saunders, Philadelphia, 1977.

From the data it is seen that the lowest friction occurs between the biological material that covers the joint surfaces. Obviously, nature surpasses engineering technology in its power to accomplish low friction when two surfaces slide against each other. The low friction between cartilage surfaces is due to two factors: (1) the surfaces are very sleek and (2) the surfaces are lubricated by the synovial fluid, which not only is an excellent lubricant, but also is stored in the cartilage so that more liquid is set free when the surface pressure is increased. Thus, in spite of the fact that the compressive forces in a joint can become very large, the frictional forces stay small. This is, of course, necessary; otherwise a large and damaging wear would result between the surfaces. Still the frictional forces in a joint can be too large, and this is believed to be one possible cause of the joint disease arthrosis, "worn-out joint". In arthrosis the joint surfaces have been broken down and become rough and rugged, and the friction can cause ache and pain when the joint is loaded during motion.

It is not only joint surfaces that slide against each other during locomotion. The muscles are often not located where the force exertion is needed. The force is then transmitted by means of tendons which can run across more than one joint. The bones of the joints then act as break points which the tendons slide over during the movement. The amount of friction when a tendon slides over a bone edge depends on the angle of circumflection as illustrated in Figure 6. The example of the figure shows that about 3% of the pulling force is lost when the coefficient of friction is 0.02 and the angle of circumflection is 90°. If the coefficient of friction were instead 0.15, as for metal on metal without lubrication, the loss of force would be as large as 27%. Such a large loss could never take place in the locomotor system without causing severe damage and pain.

In the same way as in engines and other machinery, a wear-out takes place in the body due to the frictional sliding. However, in contrast to machines, the body is composed of living tissue and the wear-out is counteracted by a continuously ongoing adaption and growth, and when damage has occurred, a healing process starts immediately. Depending on the rate of these competing processes, which differs between individuals, the result can become a progressing wear-out, an equilibrium at some level, or a healing. Since it is not possible to say anything certain about the rate of these processes, ergonomic recommendations for work design are to avoid repeated loadings at high levels as much as possible. Tissues such as tendons, ligaments, and cartilage, which are poorly vascularized, can take a very long time to heal after a damage, years rather than weeks.

V. FATIGUE FAILURE

Fatigue failure takes place when a material breaks after having been exposed to a certain number of loadings, each at a level clearly below normal failure strength. For example, a nail does not break if it is bent back and forth a couple of times, but it breaks if the bending is repeated 10 to 15 times. The level of the loading as well as the number of loading cycles determine when fatigue failure will occur. The relationship is illustrated in Figure 7. Bone tissue in particular can break due to fatigue failure, which is then called stress fracture. It is generally believed that the fracture is caused by a summation of microscopic ruptures and microtrauma that have occurred at each loading cycle. Thus, fatigue failure can be caused by long-distance running or intensely repeated strength training. Also, in bone tissue a healing takes place, but the process is rather slow.

VI. BIOMECHANICAL MODELS

Statements about loadings on the body should be based on experimental evidence, i.e., measurements. Unfortunately, this is not always possible in biomechanics. For a number of measurements, the proper transducer is simply not available. The attachment of a transducer within the body may require surgery, and that cannot be done for ethical reasons other than

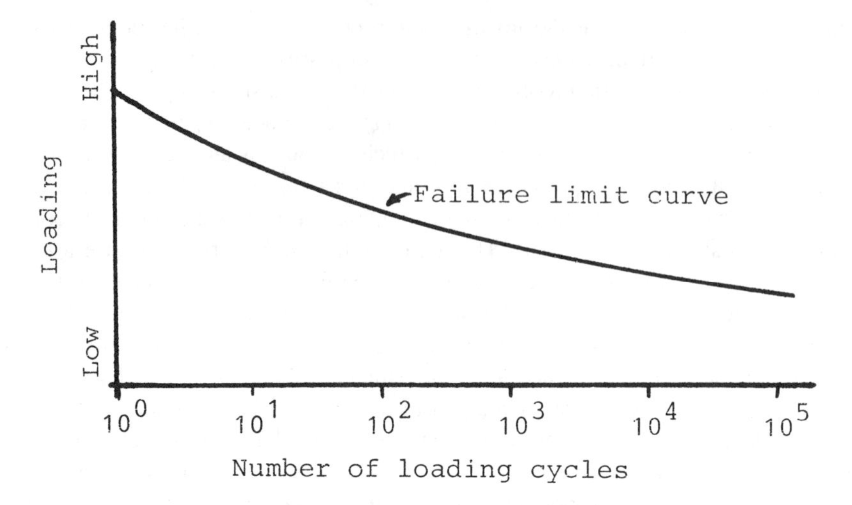

FIGURE 7. Failure strength. The diagram shows the relationship between load level and a number of loading cycles until failure occurs. The failure strength after one cycle is equal to the ordinary tensile failure strength of the material. Since for repeated loadings failure occurs at a lower value, every loading must cause some a small damage, the effect of which is accumulated.

under certain circumstances. Therefore one has to rely on indirect measurements of, for example, electrical muscle activity instead of direct muscle force.

Another possibility is to design a biomechanical model in which anthropometric data and known properties of tissues have been integrated together with assumptions on how the forces and pressures of interest are generated. The model is then used to calculate quantities that can be measured, and by comparing the model predictions with the measured results, it is possible to make conclusions about the validity of the assumptions and thus of the whole model. Models are also used to integrate results from many different experiments. In this way, the validity of a single experiment can be checked against current knowledge. Several biomechanical models are now available for load calculations, particularly for the back, but also for the shoulder and other joints.[5,6]

A model often refers to a physical object. In biomechanics it is more often a set of mathematical equations such as those used in the examples below. By writing the equations in symbolic form, i.e., using variables and parameters, they express general relationships. By selecting proper values for the parameters, the equations can be adapted to a specific situation and solved for that case. This and other techniques to increase the generality make models enormously useful.

The equations can also be used in computer programs, and several of the models mentioned above have been adapted to computerized calculations. One example is the lifting strength prediction model designed by Martin and Chaffin.[7] For sagittally symmetrical, static lifts, the required turning moments in all relevant joints and the muscle forces necessary to generate the moments are calculated. These forces are compared with population strength data, and for each joint the percentage of the population capable of the exertion is determined. For many lifts carried out according to ergonomic recommendations, it can be seen that hip-extension strength is limiting the capacity.

Investigations of the relationship between loadings and disorders can benefit from model calculations. For example, it has been shown that there is an increased risk for back disorders in lifting tasks for which the biomechanically calculated load on the low back amounts to 2500 N* or more and there is a considerably increased risk when the calculated load exceeds

* The newton (N) is the force unit in the International System of Units (SI); 1 N = 0.102 kilopound = 0.224 pound force.

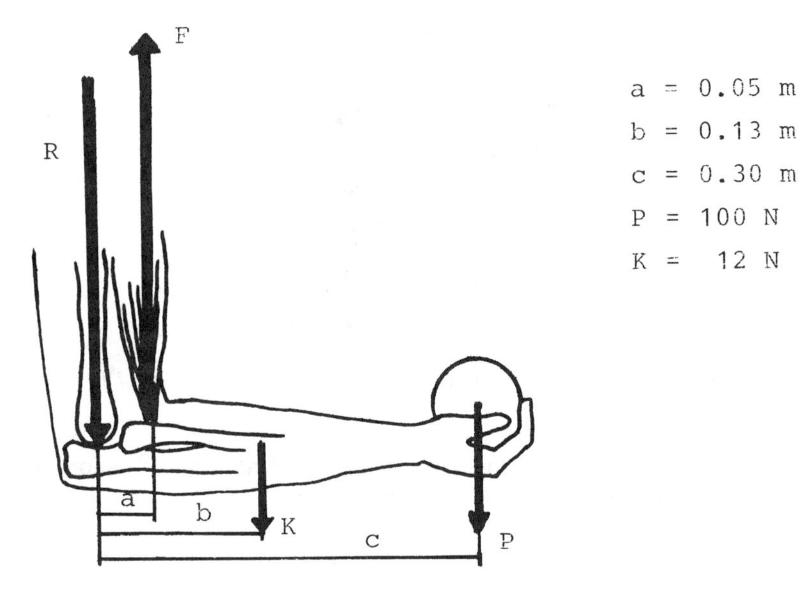

FIGURE 8. Simplified anatomical representation of elbow and lower arm when the arm is kept horizontal and a load is held in the palm.

4500 N. Such data can be used in recommendations of limits for loads on the back in lifting. The most extensive guide in that respect has been established by the National Institute for Occupational Safety and Health (NIOSH).[8,9] The guide gives formulas for calculation of one weight limit for administrative action (AL) at the work place and one maximum permissible limit (MPL). Such factors as horizontal lifting distance, starting height and vertical travel height, and average frequency of lifting are considered in the calculations. Under ideal conditions, the action limit weight is 392 N.

VII. EXAMPLES OF BIOMECHANICAL CALCULATIONS

Static muscle loading — A simplified picture of the geometry when the lower arm is loaded in a horizontal position is shown in Figure 8. A 100-N load is held in the supinated palm. Assuming that the triceps muscle is not active, the required force of contraction in the biceps muscle can be calculated for static equilibrium according to Newton's first law. This law says that the sum of all forces and the sum of all turning moments acting on a body in static equilibrium must equal zero. With notions according to the figure, the condition of moment equilibrium yields

$$Fa - Kb - Pc = 0 \tag{1}$$

Solving for F gives

$$F = Kb/a + Pc/a \tag{2}$$

resulting in, with the numerical data,

$$F = 631 \text{ (N)}$$

Thus, the contraction force in the muscle is more than six times larger than the weight of the object being held, a relationship reflecting the ratio between the lengths of the lever

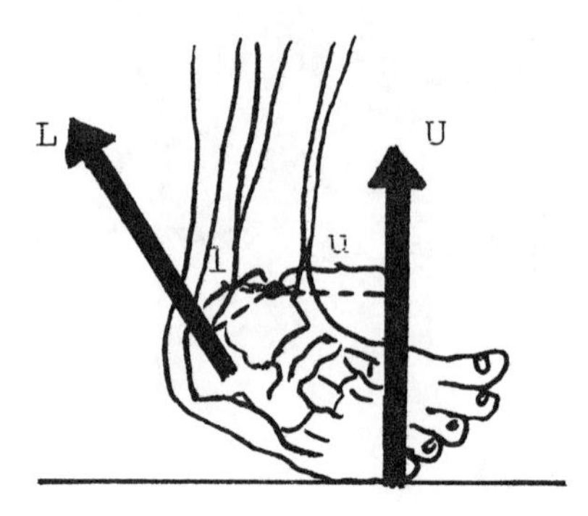

FIGURE 9. Simplified anatomical representation of foot and ankle for assessment of the loading due to sprain.

arms. The large contraction force in the biceps must be balanced by a reaction force in the joint. The condition of vertical force equilibrium yields

$$F - R - K - P = 0 \tag{3}$$

Using the numerical data, R can be calculated

$$R = 519 \text{ (N)}$$

The force acts downward on the lower arm. Thus, for force equilibrium, the better part of the contraction force in the biceps must be compensated for by a compression force in the joint. This force determines the contact pressure.

Sprain — A common form of sprain occurs when a step is missed so that ground contact is made with the outside of the foot. Then the ligament between the calcaneus and the fibula is stretched. The sprain can occur when walking down stairs and when a step is missed. With geometrical conditions according to Figure 9 and given that the person's weight is 75 kg and the height of a step is 20 cm, it is possible to calculate the force to which the ligament is subjected. The condition for turning moment equilibrium applied to the foot yields

$$Uu - Ll = 0 \tag{4}$$

where U is the force from the ground, L is the force in the ligament, u is the moment arm of the ground force, and l is that of the ligament force. Hence,

$$L = Uu/l \tag{5}$$

When the step is missed, the body accelerates under free fall for 20 cm. The final velocity, v, can be calculated according to

$$v = \sqrt{(2gs)} \tag{6}$$

where g is the acceleration of gravity and s is the height of the step, yielding

$$v = \sqrt{(2 \cdot 9.81 \cdot 0.2)} = 1.98 \text{ (m/s)}$$

This velocity must be braked when contact with the ground is made. The contact force, U, can be calculated using the law of momentum if it is assumed that the force, U, is constant and acts during the time, t, equal to, say, 0.3 s. The law of momentum yields

$$U = m \cdot v/t \tag{7}$$

where m is the person's weight, resulting in

$$U = 75 \cdot 1.98/0.3 = 495 \text{ (N)}$$

Taking into account the static weight of the person ($75 \cdot 9.81 = 736$ N), the ligament force, L, is obtained using Equation 5

$$L = (736 + 495) \cdot 0.15/0.03 = 6154 \text{ (N)}$$

This force is large and it is likely that the ligament would rupture if the sprain occurred according to the calculations, i.e., if the foot were hitting the ground without any bending of the knee. The equations show that one way of reducing the load on the ligament is to increase the contact time, t, during which the braking occurs, and this is what happens in practice when the stumbling is counteracted by bending the knee joint.

REFERENCES

1. **Yamada, H.,** *Strength of Biological Materials,* Williams & Wilkins, Baltimore, 1970.
2. **Reilly, D. and Burstein, A.,** The elastic and ultimate properties of compact bone tissue, *J. Biomech.,* 8, 393, 1975.
3. **Frankel, V. and Nordin, M.,** *Basic Biomechanics of the Skeletal System,* Lea & Febiger, Philadelphia, 1980.
4. **Williams, M., Lissner, H., and Le Veau, B.,** *Biomechanics of Human Motion,* W. B. Saunders, Philadelphia, 1977.
5. **Schultz, A. B., Andersson, B. J. G., Örtengren, R., Haderspeck, K., and Nachemson, A.,** Loads on the lumbar spine, *J. Bone Jt. Surg.,* 64-A, 713, 1982.
6. **Tsotsis, G.,** Entvicklung eines biomechanischen Models des Hand-Arm-Systems, IPA-IAO Forschung und Praxis, Band 108, Springer-Verlag, Berlin, 1987.
7. **Martin, J. B. and Chaffin, D. B.,** Biomechanical computerized simulation of human strength in sagittal plane activities, *AIIE Trans.,* 4(1), 18, 1972.
8. National Institute for Occupational Safety and Health, A Work Practices Guide for Manual Lifting, Tech. Rep. No. 81-122, U.S. Department of Health and Human Services, Cincinnatti, 1981.
9. **Chaffin, D. B. and Andersson, G. B. J.,** *Occupational Biomechanics,* John Wiley & Sons, New York, 1984.

Chapter 12

ERGONOMICS FOR HEALTH PROFESSIONALS IN HOSPITALS AND THE COMMUNITY*

J. D. G. Troup

TABLE OF CONTENTS

I. ERGONOMICS: A DEFINITION

Ergonomics is the study of the relationship between human operators, the work they have to do, and the environment in which the work is done. It can be applied in two ways: initially, to the analysis of an existing working process, but in due course, to its redesign. Ergonomic improvements to current workplaces can sometimes be introduced, but a satisfactory solution may only be possible with a totally fresh approach to the design.

If a job is ergonomically satisfactory, then the work place will be safe, the work satisfying to the operators, and efficient by any standards of good management. An ergonomic intervention at the workplace should be cost-beneficial not only in terms of increased productivity, but also with respect to minimizing work-related morbidity, job turnover, the need for recruiting and training new workers, and any work stoppages, whether caused by humans or machines.[1-3]

An ergonomic study depends first on a clear understanding of the overall aims of the work. The questions "why is this job being done?" and "what overall purpose is it serving?" must always be asked and clearly answered. Then the tasks can be analyzed and specific questions can be posed about whether the demands of the work match the capabilities of the operators. In the majority of jobs, the demands of work can be defined in reasonably precise terms with regard to the physical and mental workload. There are others in which there is a degree of uncertainty and this applies to many of the interactions at work, certainly between human operators, but also between the human operator and other beings: for farmers and veterinary surgeons, e.g., farm animals, and for health professionals, e.g., patients.

II. ERGONOMICS, MEDICINE, AND HEALTH CARE: AN EXTENDED DEFINITION

In many apparently routine and mundane working systems, there are elements of uncertainty: even in a simple stock control system which, though dependent on reliable deliveries, must allow for some uncertainty in delivery dates. In the practice of medicine and health care, uncertainties abound. Though it may be possible to undertake a task analysis of, for example, giving an injection, every health professional has to be prepared to perform this task, or indeed any other, in a variety of different ways, depending on the condition of the patient.

There are three main variables that contribute to the uncertainties of the work for health professionals: the physical dependence of the patient, the medical constraints, and the patient's behavior. None of them are static and behavior, in particular, is often unpredictable. For this reason, every working system in a hospital or in community health care must be adaptable. Not only may the individual patient present problems of uncertainty, so may the medical or health care needs of a community.

The definition must therefore be extended. Ergonomics applied to hospital or community health services is the study of the relationships between the health professionals and their working capacity, the physical and mental nature of their work, the patients in hospitals and in the community at whom it is directed, and the environment in which the work is done.

III. IDENTIFICATION OF THE NEED FOR ERGONOMICS

The creation of a new hospital, a new health center, or even a new department is the signal for the introduction of ergonomics. At the earliest stages of the design process, the need for ergonomics is paramount. Ergonomics can also be applied to existing working systems, though clearly less rigorously and less satisfactorily, to achieve some improvements on an *ad hoc* basis. New working procedures, new communications, new equipment, and simple adaptations in the existing environment can be introduced without major structural changes and often without great expenditure. The problem in an untutored society is that the need may not be recognized.

Advice on ergonomics has recently been listed by the International Labor Conference as one of the functions of an occupational health service. Part of the onus of recognizing the need for ergonomic intervention therefore falls on occupational health professionals. The indications can be listed as follows:

- Work-related morbidity
- Accidents
- Work stoppages
- Job turnover
- Maintenance problems

Wherever the prevalence and incidence rates are high or show signs of increasing, ergonomic investigation may be a necessity. This is abundantly true for back pain which is one of the most common sources of work-related morbidity in health professionals.[4] Recognition therefore depends on a comprehensive and reliable information system that includes

1. Occupational history of employees with data on qualifications, training, and experience
2. Preemployment medical screening data, including previous medical history
3. Job descriptions and demands on operators
4. Epidemiological data on the prevalence of work-related symptoms, the incidence of injuries, and the effect of existing morbidity on working capacity, including sickness and absence data
5. Job turnover
6. Accident data using an accident model based on identification of the first unforeseen event and the environmental and individual factors relevant to it[5]
7. Attacks on health professionals in psychiatric and psychogeriatric wards and homes
8. Environmental data, including occupational hygiene data
9. Productivity data, including information about work stoppages (both human and mechanical)

IV. ASSESSMENT OF WORKLOAD

There are a number of methods for assessing workload, using job profiles in which factors such as "environmental" or "the physical load" are defined. The "Renault" method takes account of 8 such factors subdivided into 27 criteria including 4 criteria concerning job design, each criterion being rated subjectively from "very arduous" to "very satisfactory".[6] A more comprehensive system, AET,[7] has been developed and, in addition to industrial use, it has been applied to the needs of the disabled.[8] A more limited approach is to categorize working postures: the OWAS system uses a method of classification with four possible back postures, three for upper limbs and five for lower limbs,[9] while the ARBAN system[10] combines a more extended range of postures with ratings of perceived exertion.

None of these methods gives a satisfactory assessment of either the dynamic stresses arising from manual force exertions, e.g., during patient handling or lifting, or of postural stress as such. Although various attempts have been made to monitor trunk inclination in the sagittal plane, no information is provided by such systems about what the hands are doing, i.e., whether or not they are being used to lift, pull, push, or support, and therefore off-load, the spine. A system was therefore developed whereby every change of posture or activity could be recorded, using a microcomputer and a simple two-digit coding system to create a real-time record for subsequent analysis of the frequency and duration of each posture and activity.[11] This method proved to be both repeatable and easy to learn. It was used initially to observe 12 nurses throughout 2 full working shifts. Although the method is expensive in that only one worker can be observed at a time, it may prove fruitful, as a purely descriptive method, in determining the overall frequency and duration of handling and postural stresses.

V. TASK ANALYSIS: THE DEMANDS OF WORK VS. THE CAPABILITIES OF WORKERS

"The purpose of 'task analysis', and its related technique of 'function analysis', is to make a step-by-step comparison of the demands an operation makes on the operator with the capabilities of the operator."[12] The analysis is begun by describing every step that is taken in completing the task. The description is followed by the task analysis itself in order to derive the demands made on the operator by the task. The "functions" are the basic units of behavior needed to accomplish the task: some of them can be allocated to the human operator, while others are allocated to a machine.

A task or process can be described sequentially, assuming that it follows a set pattern with a minimum of choices or alternatives in its course. The start-up procedure for a piece of equipment would be an example of this (and likewise turning the patient in bed). Each step in the sequence can be described, and with each step, the ergonomic problems can be identified.

Often a task cannot be described by means of a simple sequence because of the alternative strategies that may have to be adopted. Then the description demands a flow chart with a choice of sequences to be followed, depending for instance on a patient's response to treatment. In some of the more complex tasks for health professionals, as in an intensive care unit, the description may have to be based on that for a process control operation in which the operators' tasks are analogous to a hierarchical computer program for which each successive step depends on the information available. Yet, however complex the descriptive model has to be, there will be a temporal sequence to be followed and the ergonomic problems of each can be defined. These problems may be visual, auditory, kinesthetic, or concerned with the exertion of forces. There may be problems of reach or spatial constraint, and the human operators may be exposed to a variety of environmental stresses or hazards.

Hitherto, most systems of task analysis proposed by ergonomists have been applied by independent, professionally trained observers. A complementary and potentially fruitful approach is to consult the workers themselves.

VI. ERGONOMIC SELF-ASSESSMENT BY THE OPERATOR

In order to provide a method to allow individual workers to evaluate their own workplaces, Koskela and colleagues have designed a structured form.[13] Fourteen questions are asked about ergonomic factors and the form allows a sketch plan of the layout of the workplace to be made for each worker. The aim of the method is to teach people how to modify their working environment to make it ergonomically more satisfactory and thus more comfortable for the worker as well as more efficient in operation. It has been applied to groups of sedentary workers such as typists and cashiers and could well be applied to the work of many health professionals.

The freedom of individual workers to bring about ergonomic improvements to their workplace is subject to a number of obvious constraints. The attitude of the employers may create major obstacles and financial restrictions are inevitable. However, in the case of typists, for example, there are a variety of modifications that individuals can make. They

FIGURE 1. Modification of working postures to improve comfort and efficiency: putting shoes on. (Reproduced from Lloyd, P., Tarling, C., Wright, B., and Troup, J. D., *The Handling of Patients: a Guide for Nurses,* The Back Pain Association in collaboration with the Royal College of Nursing, London, 1987. With permission.)

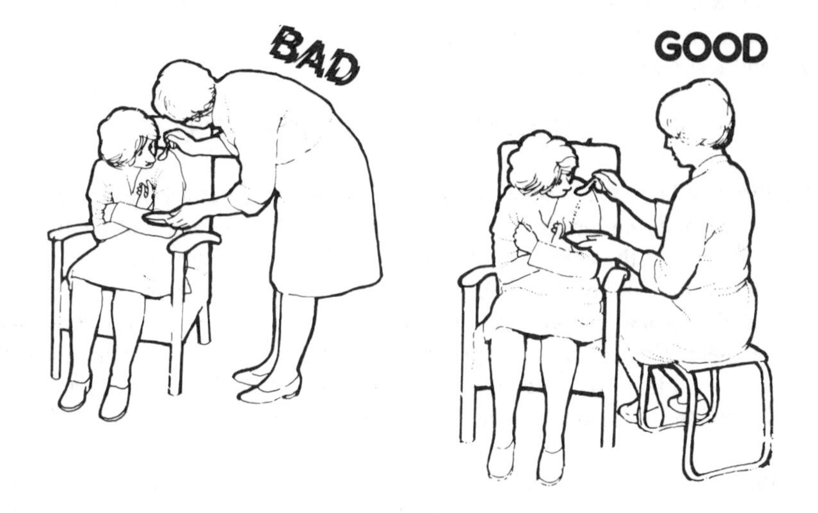

FIGURE 2. Modification of working postures to improve comfort and efficiency: feeding the patient. (Reproduced from Lloyd, P., Tarling, C., Wright, B., and Troup, J. D., *The Handling of Patients: a Guide for Nurses,* The Back Pain Association in collaboration with the Royal College of Nursing, London, 1987. With permission.)

can adjust and vary the relative heights of work surfaces as well as their chairs and they can explore a number of options for the layout of their work at no great cost. What matters is that individual workers are given an insight into basic ergonomics which allows them to make their workplace more comfortable, and, incidentally, improve their working efficiency (Figures 1 and 2). In the process, they may become aware of the need for a wider range of ergonomic improvement in the design of the workplace. However, given the premise that ergonomic improvement should be cost-beneficial, constructive attitudes can be encouraged at every hierarchical level. To give some insight into the range of ergonomic problems that can readily be appreciated, the following principles were offered during a brief instructional course at the University of Nottingham on ergonomics for nurses:

1. The worker should be able to maintain an upright and forward-facing posture during work.
2. Where vision is a task requirement, the necessary work points must be readily visible, with the head and trunk upright or with the head just inclined slightly forward.
3. All work activities should permit the worker to adopt several different but equally safe and comfortable postures without limitation of working capacity.
4. Work should be arranged so that it may be done, at will, in either a seated or standing position.
5. The weight of the body, when standing, should be carried equally on both feet, and any foot pedals should be designed accordingly.
6. Work activities should be performed with the joints at about the midpoint of their range of movement. This applies particularly to the head, trunk, and upper limbs.
7. Where muscular force has to be exerted, the largest appropriate muscle groups should be used, in a direction colinear with the natural movement of the limbs concerned.
8. Work should not be performed consistently at or above the level of the heart. Even the occasional performance in which forces are exerted above heart level should be avoided. Whenever light hand work has to be performed above heart level, rests should be provided for the arms.
9. When forces have to be repeatedly exerted, it should be feasible to exert them with either of the arms, or either of the legs, without adjustment of the equipment.
10. Rest pauses should be allowed after all prolonged spells of work, whether the stresses be mental, physical, environmental, or informational, and the optimal durations for work and rest should be determined.

VII. ARCHITECTURAL CONSIDERATIONS

The structure that houses a hospital or health center tends to outlast the specific functions and aims for which it was originally designed. Changes in nursing and medical practice emerge at intervals, but the existing walls remain and somehow the new working practices are accommodated though not, invariably, with much regard for ergonomic considerations. In one 50-year-old hospital in northwest England in which the original, single-story buildings are still in use, there is no way in which patients in the orthopedic wards can use the bathroom unless they can walk there unaided. In another, new hospital, the introduction of new fire regulations led to the installation of double doors that are too heavy for the weaker patients to open, while both doors have to be held open to give passage to a wheelchair. By any standards, this is inexcusable. The likelihood of such blunders being repeated and left uncorrected is, mercifully, receding as the volume of ergonomic data available to architects and planners increases.[14-16]

Architects have not, in the past, been taught ergonomics in the course of their training. Unless they can demonstrate their knowledge of, or at least their sensitivity to, the subject, it must be presumed they are ignorant of it. Much the same is true of those who commission the architect in the first place. It may well be up to the health professionals themselves to ask the question of planners and architects: "do these people have any idea of what we have to do or of the problems we have to tackle in this fine new building which they are planning?"

VIII. ERGONOMICS AND THE PRINCIPLES OF DESIGN

In ergonomics, the process of design is a discipline which begins with a clear definition of the aims and purposes to be served. The process encompasses the design of the building itself, the equipment to be used, and all the furniture and fittings as well as taking account of the environmental factors. It must include the design of every working procedure together with the selection and training of the operators.

In hospitals and in the community where patients are cared for, their needs should dominate the aims and purposes to be served. An approach as comprehensive as this has seldom been attempted, let alone achieved. Nonetheless, models for such a design process do exist, albeit on a modest scale.

The starting point for the design models is set by the requirements of the user, i.e., the patient in a hospital or the disabled person in the community.[17] It would be wrong, however, to focus solely on user requirements. For example, in an otherwise valuable study of the design requirements for seating elderly and disabled people, there was little written about associated ergonomic problems for nurses or care assistants: although mention was made of access to the seated person, there was no analysis of the effect of chair design on the techniques of transferring dependent patients to and from the chairs.[18]

A classical example of the design process was initiated in 1963 when the King Edward's Hospital Fund, in consultation with the Ministry of Health in the U.K., set up a working party to study the design of hospital bedsteads.[19] They began by finding out what had already been done, to avoid duplication. Previous work being deemed to have been inadequate, the next step was to determine the characteristics required of a general hospital bedstead. This was tackled by the Royal College of Art, but the research was supplemented by approaching about 20,000 members of hospital staff via the medium of national television. In the second stage of the design process, a specification was drawn up, and this was followed by the design and construction of prototypes. These were installed in a women's general surgical ward in 1965 and were assessed by means of constant observation between 600 and 2200 h for 5 months. At the same time, sociological studies were conducted to obtain the opinions of the patients and staff. The results were analyzed and, following revision of the design, the King's Fund bed was put into production in 1967. It was a great success and is still in use.

A second example concerns the means to permit disabled persons to carry out the activities of washing, bathing, and toiletting independently.[20] This project began by surveying groups of young, middle-aged, elderly, pregnant, and disabled people to determine their patterns of hygienic activity, including care of the hair and nails. From these data, performance specifications were drawn up for the evaluation of conventional hygienic facilities in terms of access, postural support, visual needs, the supply of water to all parts of the body, and drainage. Thus, the mismatch between users' requirements and traditional accommodation and equipment was revealed. The next stage was the design and construction of a prototype for a "low handicap" alternative to standard hygienic equipment (Figure 3). Since this was reported in 1981, further trials have proved successful and the unit is now in production.

Such examples are comparatively rare and the intrinsic needs of the disabled appear to have attracted more ergonomic attention than those of medical and paramedical personnel, e.g., wheelchair users.[21,22] A couple of studies of the ergonomics of interactions between the surgeon and the operating theater nurse and of the design and arrangement of the instrument trolley and tables were made, not by an ergonomist, but by a professor of surgery.[23,24]

Recently, however, there has been a series of studies carried out in Sweden by the National Board for Technical Development, with particular reference to the design and equipment of hospitals and homes for the elderly and disabled. These studies have begun with extensive inquiries into the needs of the patients or residents for whom the care is offered and into the problems of the nursing staff and care assistants:[25,26] they are preliminary to the development of a transfer system based on horizontal rails along with the walls behind the beds, thus allowing hinged shafts to be lowered beside the patient who can then be transferred from bed to chair, etc. This development is of major interest as it exemplifies how a department of a governmental office can function to the benefit of a health service, sponsoring an ergonomically satisfactory standard of design in manufacturing industry for the good of the health service.

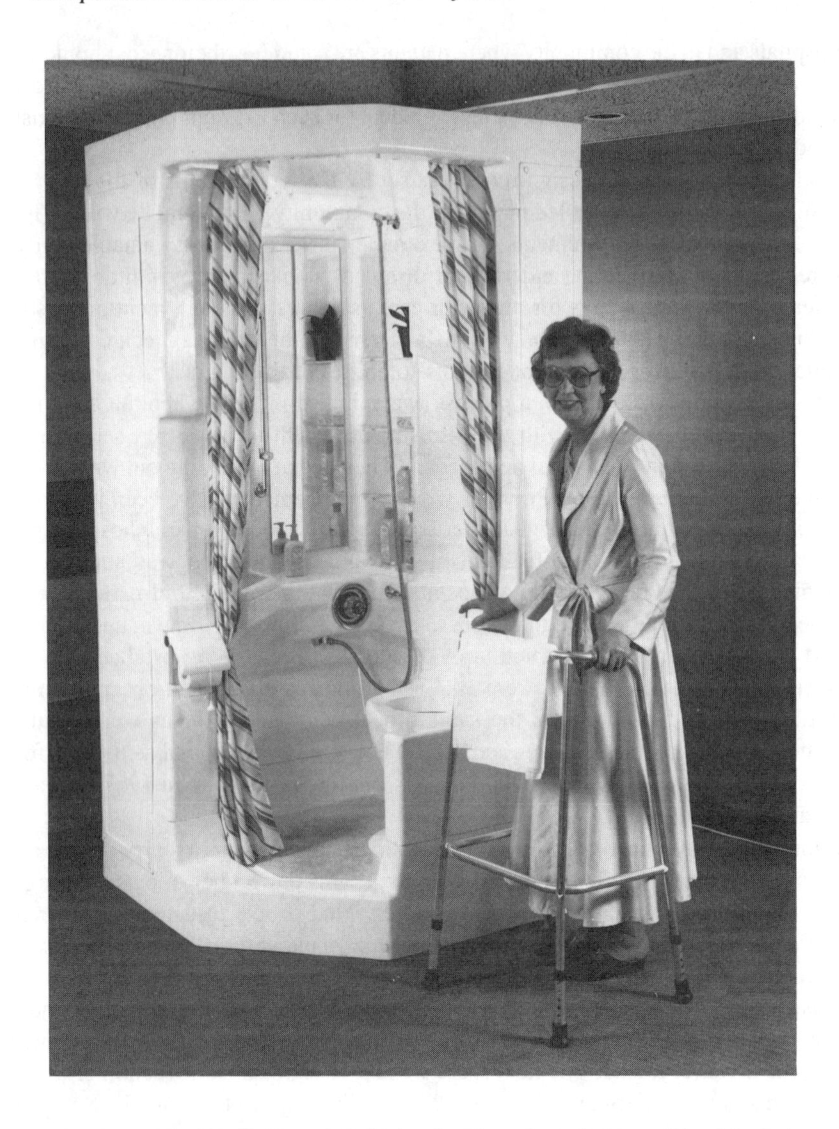

FIGURE 3. The Triad® (Stannah Ltd.) low-handicap alternative to traditional hygienic equipment.

IX. BACK PAIN IN HEALTH PROFESSIONALS

Although nursing, in particular, is thought by the general public to be exceptionally hard work, both physically and mentally, on analysis it equates with "light industrial".[27] Yet, this is probably misleading for the prevalence of back pain — generally regarded as an indicator of the physical workload — is unquestionably high in nurses and it appears to be related to patient handling.[28-34] The same is true for ambulance personnel.[35] Though the actual time engaged in patient handling and lifting may not be exceptional,[11] the magnitude of individual loads may occasionally be inordinate. The other factor is the postural stress to which nurses are exposed: because of the layout of their working environment, the organization of work, and the equipment they are obliged to use.[36]

Ergonomics is undoubtedly the foremost preventive approach to back pain; nevertheless the need to train nurses to increase their patient-handling skills cannot be ignored. However, the choice of preventive approach to back problems in individual groups of health professionals depends on the magnitude of the stresses induced in the spine by lifting, pushing, pulling, carrying, and by the posture at work; on the effects of those stresses on their health; on the freedom they have to adapt their working environment to make the task safer and easier; and on the strength and physical skill they bring to lifting and handling.

X. GUIDELINES ON LIFTING AND HANDLING

The capacity to push, pull, or press vertically down depends mainly on body weight, but, equally, on the stability of the posture. In pushing and pulling postures, much depends on the frictional resistance between feet and floor. Lifting, on the other hand, depends on the strength of the extensor muscles of the back and lower limbs. There are several problems. How much can they safely lift or handle? How much force is safe for an individual to exert? In the case of a group of workers, how much force can the weakest member of the group exert?

Lifting is essentially a dynamic process, but there is an inherent static force element that varies with the posture in which the lifting forces are exerted and with the duration of the task. It is theoretically possible to measure the maximal dynamic lifting forces that an individual can exert, but this exposes him to obvious risks of overload. The alternative is to measure the maximal strength that can be exerted either isokinetically or isometrically, though neither method of testing is truly physiological. Isometric lifting strength has, however, been used to predict back injuries by comparing individual strength with the actual lifting forces required at work,[37] those with the least margin of strength in reserve being the ones who were most at risk of musculoskeletal injury.[38] This is potentially valuable where lifting work is standardized and isometric force simulation is valid. The disadvantage for some populations is that isometric strength testing itself carries a small risk.[39]

A different approach is to rely on psychophysical tests of lifting strength — in other words, on tests in which the individuals determine for themselves the weights they feel they can lift comfortably and without undue exertion or strain. This can be done relatively quickly and reliably[40] and the methodology can be applied to a specific workplace.[41] The evidence up to now suggests that psychophysical tests of lifting strength relate more closely to the experience of back pain than maximal isometric strength tests.[40] Psychophysically acceptable weights of lift, lower, push, and pull for males and females have been published by Snook.[42] These tabulated data provide valuable ergonomic guidelines for symmetrical applications of force.

Another criterion for the weight of lift that can be deemed to be safe is based on biomechanical considerations. In the ''Work Practices Guide for Manual Lifting'', two limits were proposed by NIOSH:[43] the ''action limit'' (AL) below which no regulations or guidelines were necessary and the ''maximum permissible limit'' (MPL) above which no lifting should be allowed. They were based on epidemiological, biomechanical, physiological, and psychophysical criteria. Figure 4 shows the AL and MPL for infrequent vertical lifts from floor to knuckle height at different horizontal distances from the ankles. This offers a simple and useful model, though it must be recognized that it does not take into account the increasing instability when attempting to lift at greater distances from the body.

The biomechanical basis for the two limits was that the compressive load on the lumbar spine for the AL should not exceed 350 kg and for the MPL, it should not exceed 650 kg. If this criterion is to be applied in practice, it requires that an attempt be made to calculate lumbosacral compression for a given lift[44] and it is not easy, as Figures 5, 6, and 7 will show. The problem becomes more or less insoluble when considering what lumbosacral compression is induced when, for example, a nurse is helping to move a patient. It may be reasonable to make a rough calculation based on the initial lifting posture, the weight to be lifted, and on the body segment weights of the nurse. However, this ignores the dynamic

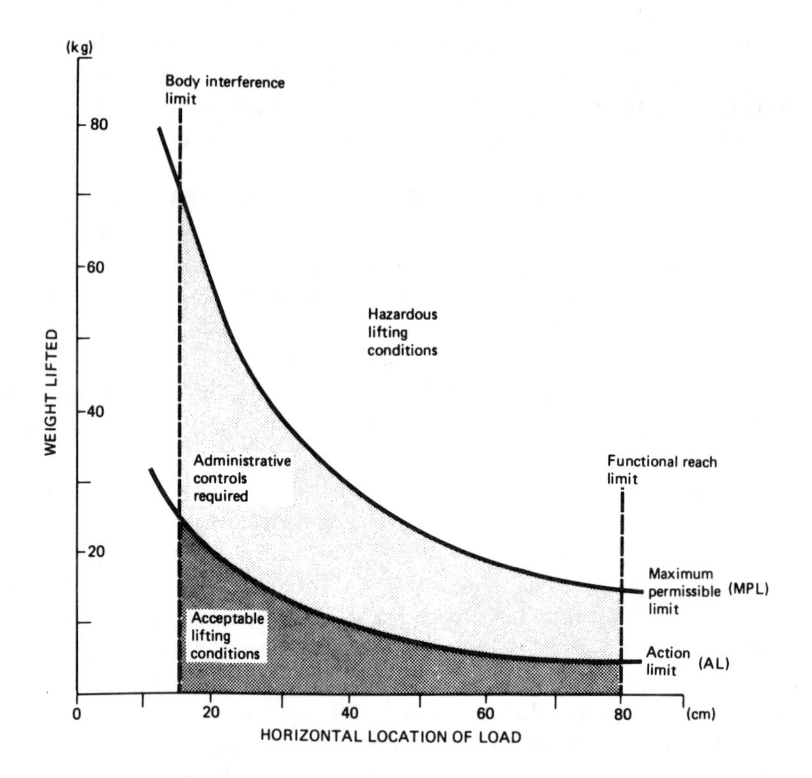

FIGURE 4. "Action limit" (AL) and "maximum permissible limit" (MPL) for infrequent lifts from floor to knuckle height. (Reproduced from Troup and Edwards[44] by permission of HMSO: Crown Copyright — based on figure 8.1, p. 125 of NIOSH Guide[43]).

factor and the estimate may be too low. On the other hand, it may overestimate the spinal stress if it is assumed that a vertical lift takes place when in fact the nurse is applying kinetic energy as in the "rocking lifts" (Figures 8 and 9) for which the actual lifting force required is minimal.

Regrettably, none of the mathematical models used in the published guidelines on safe limits are without some disadvantages. If these are ignored or overlooked, people may be exposed to overload. Often, the duration of application of lifting force or the duration for which a given level of lumbosacral compression has to be sustained is excluded from the guidelines. This is potentially dangerous, for whenever the compressive load exceeds the osmotic pressure in the tissues of the lumbar spine, fluid may be expelled, the fibers of the annulus fibrosus may be reoriented, and the motion segment becomes stiffer and more susceptible to injury, changes which result from the so-called "creep effects".[45]

For these reasons, no numerical guidance is offered in this chapter as to what is or is not safe to lift. If, for the sake of argument, it was declared that, under satisfactory lifting conditions, all employees could lift up to 25 kg with safety, this would immediately be invalidated for a proportion of employees whenever the load was asymmetrical, had no handles, required a twisted posture, was placed at arm's length, etc. Any such limit becomes unsafe if it is applied uncritically. Safe lifting depends on strength and mobility, but, more importantly, depends on stability and skill, and these, at the time of writing, are insusceptible to objective measurement in the field.

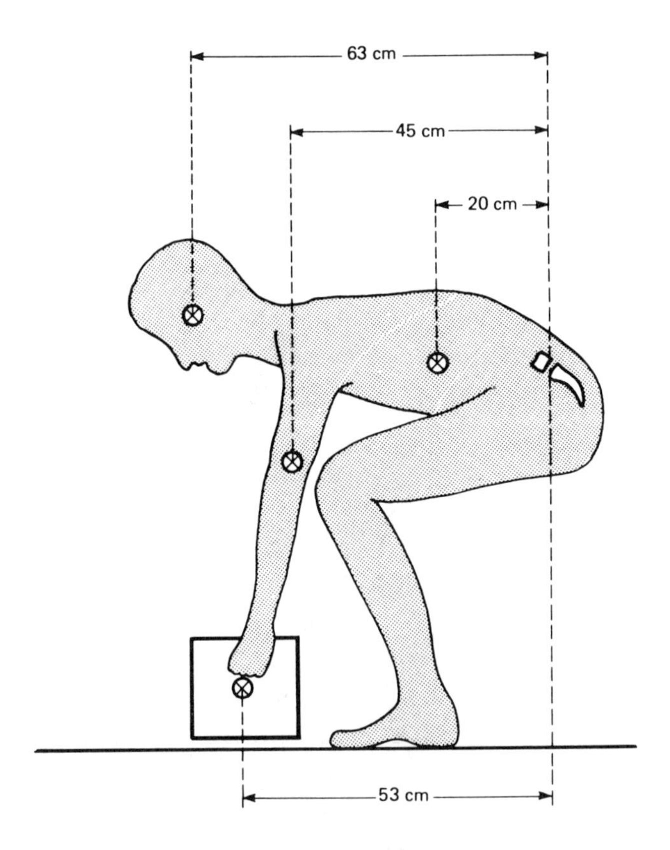

FIGURE 5. Assessment of lumbosacral compression: vertical lift from stooping position with box in front of knees. The circled crosses indicate the centers of mass for the head, upper limbs, trunk, and load: the horizontal distances from the lumbosacral disk are shown. (Reproduced from Troup and Edwards[44] with permission from HMSO: Crown Copyright).

XI. LIFTING, HANDLING, AND TRANSFERRING PATIENTS: THE USE OF HOISTS, MECHANICAL AIDS, AND THE DEVELOPMENT OF MANUAL SKILLS

Some financially well-endowed hospitals, fully equipped with hoists and mechanical aids to lifting, are beginning to issue edicts that nurses must never lift a patient vertically. This is a very rational and laudable ideal. Certainly it is applicable to the majority of the chronically disabled. However, it does raise one or two questions.

If a ban on lifting leads to a generation of nurses with no lifting experience or skill, they might readily be hurt if confronted with a lifting emergency. The other problem is that patients who are partially independent or who are getting better will continue to need some assistance and support in their movements and transfers, particularly in situations in which the continued use of hoists and mechanical lifting aids may interfere with rehabilitation.

Although it is essential that nurses are warned of the dangers of lifting vertically and that the use of such techniques as the ''orthodox lift'' (in which patients are lifted vertically at arm's length by nurses on either side of the patient with their hands linked under him) is banned, it has to be recognized that a vertical component remains as a feature in most patient transfers, however brief the duration. In 1981, the Royal College of Nursing, in collaboration with the Back Pain Association, published *The Handling of Patients: A Guide for Nurse Managers*,[46,47] a revised and updated edition was published in 1987. The need arose for two reasons: (1) from the prevalence of low back pain and the incidence of back injuries, which

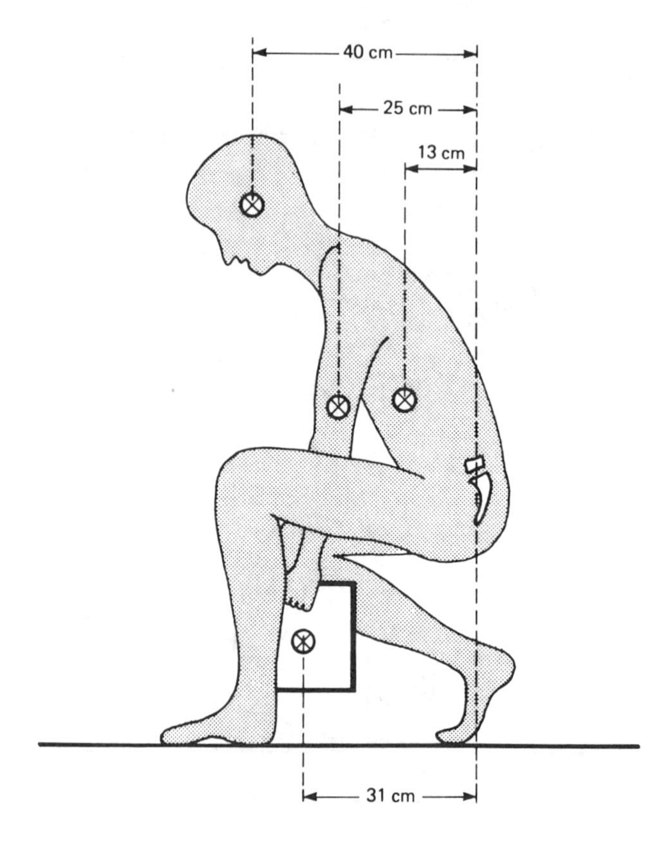

FIGURE 6. Assessment of lumbosacral compression: vertical lift from crouching position with box between knees. Details are as for Figure 5. (Reproduced from Troup and Edwards[44] with permission from HMSO: Crown Copyright).

particularly affected nurses in training, and from the number of accidents arising from patient handling; and (2) from the poor standards of instruction in safe lifting and handling methods which then prevailed. In the summary of the principles of handling, the first rule was "Always consider the aim and overall objective to be served by every handling task. Never lift the patient unless you have to."

This publication provided a review of the epidemiology of back pain in nurses, the causes of back pain, and the principles of biomechanics, followed by the principles of handling, the techniques of handling, and the use of hoists and lifting aids. In conclusion, there was a set of recommendations directed at hospital employers: recommendations which, later, served to improve the common law rights of injured nurses when pleading negligence on the part of their employers. The publication is widely used as a training manual by schools of nursing in the U.K. It has been translated into Finnish by the Institute of Occupational Health, Helsinki, where it also has been used as the basis for training in a study of patient-handling skills.[48]

It is often argued that if hospitals and the apparatus and furniture with which they are equipped were properly designed, there would be no need for specific training in patient handling and that no consideration would need to be given to the question of preemployment screening methods when recruiting people to a career in nursing. Logically, the argument may be tenable. Practicably, it is unrealistic because of the general lack of ergonomically satisfactory conditions for the movement and transfer of patients and also because of the unlikelihood that the ergonomic problems will be promptly solved as soon as they are

The total torque attributable to the upper part of the body is divided by the distance (5 cm) between L5/S1 and the line of action of the back muscles to give the LCF attributable to the upper part of the body. This figure is used to derive the loads that would, added to this, produce LCFs at the AL and MPL.

	Mass (kg)	*Figure 11* Distance from centre of mass to L5/S1 (cm)	*Figure 11* Torque: mass × distance (kg cm)	*Figure 12* Distance from centre of mass to L5/S1 (cm)	*Figure 12* Torque: mass × distance (kg cm)
head	5	63	315	40	200
upper limbs	9	45	405	25	225
trunk	32	20	640	13	416
Total torque attributable to upper part of body			1360		841
LCF attributable to upper part of body (kg)		272 (1360/5)		168 (841/5)	
LCF allowance for load (kg) to bring total LCF to 350 kg (AL)		78 (350–272)		182 (350–168)	
LCF allowance for load (kg) to bring total LCF to 650 kg (MPL)		378 (650–272)		482 (650–168)	
Horizontal distance between load and lumbosacral disc		53		31	
Leverage ratio: Mechanical disadvantage (load — disc distance) / (back muscle — disc distance)		10.6 (53/5)		6.2 (31/5)	
Estimated load (kg) to bring total LCF to 350 kg (AL)		7.4 (78/10.6)		29.4 (182/6.2)	
Estimated load (kg) to bring total LCF to 650 kg (MPL)		35.7 (378/10.6)		77.7 (482/6.2)	

Assumptions
(a) body weight = 70 kg
(b) distance between centre of lumbosacral disc and line of action of back muscles = 5 cm
(c) values of LCF at AL (350 kg) and MPL (650 kg) suggested by NIOSH (1981)

FIGURE 7. Calculation of lumbosacral compression force (LCF) for the lifting postures in Figures 5 and 6. (Reproduced from Troup and Edwards[44] with permission from HMSO: Crown Copyright).

recognized. There is, moreover, the argument that there has been, until recently, very little solid evidence that training does any good. Yet in countries in which there is a statutory requirement for employers to offer training or where employers are at risk of personal injury litigation, some consideration must be given to the quality of training in patient handling. The skilled and experienced instructors have relied mainly on their ability to teach nurses techniques of patient handling that feel substantially easier and require less effort to perform.

Recently, a new system of training has been developed in Helsinki in which student nurses receive a total of 20 h of instruction in the first semester. It includes instruction in the basic biomechanics and ergonomics of nursing practice. Students are required to practice teaching the handling techniques they have learned and use is made of videotape recordings for self-evaluation. Independent assessment of handling skills, comparing the skills of students trained in the traditional way with students who had received the new system of training, revealed significantly better handling by the latter group.[49] Preliminary results from a group of 54 student nurses showed that, in the year following graduation when their patient-handling skills were assessed, of those who reported back accidents all had been rated as either "bad"

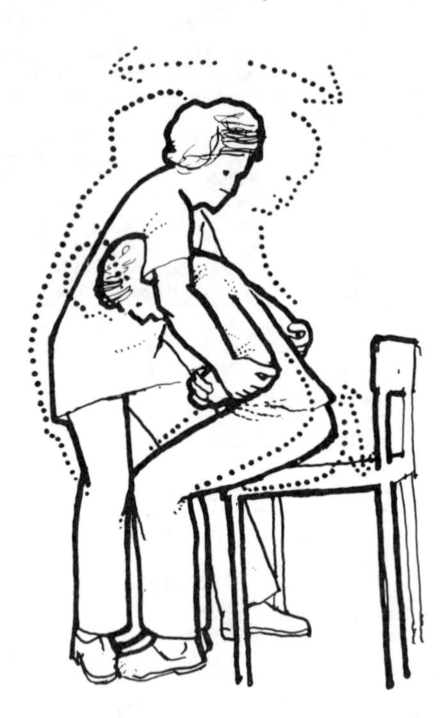

FIGURE 8. "Rocking lift" for patients who can cooperate and control their head and arm positions, the rocking giving the patient kinetic energy to relieve the nurse of the lifting effort: using "elbow lift grip" for horizontal transfers. (Reproduced from Lloyd, P., Tarling, C., Wright, B., and Troup, J. D. G., *The Handling of Patients: a Guide for Nurses,* The Back Pain Association in collaboration with the Royal College of Nursing, London, 1987. With permission.)

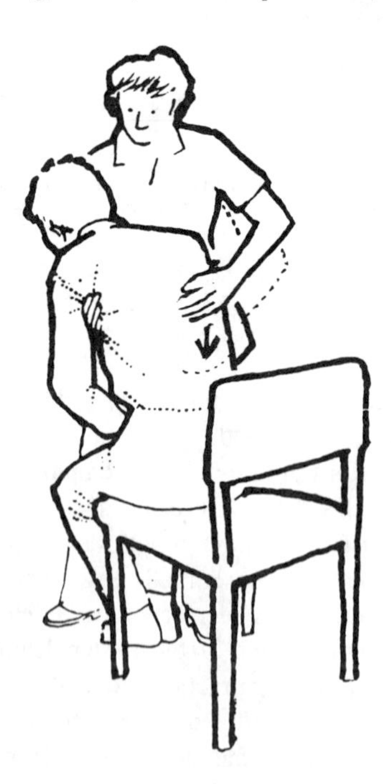

FIGURE 9. "Rocking lift" for patients who can cooperate and control their head and arm positions, the rocking giving the patient kinetic energy to relieve the nurse of the lifting effort: using "axillary hold" for horizontal transfers or for standing the patient up. (Reproduced from Lloyd, P., Tarling, C., Wright, B., and Troup, J. D. G., *The Handling of Patients: a Guide for Nurses,* The Back Pain Association in collaboration with the Royal College of Nursing, London, 1987. With permission.)

or "poor" (chi-square = 5.76, $p < 0.05$),[33] results which have since been confirmed and amplified in a larger sample of nurses. There is little doubt, therefore, that training can have a preventive effect on back injuries to nurses.

Pending the ergonomic revolution for nurses, training in safer and easier methods of patient handling is therefore worthwhile. However, attention still has to be paid to the development of better methods of transfer, not only hoists or adjustable height trolleys, but some of the simpler aids. For example, in the course of developing criteria for training,[50] it became apparent that even with the most suitable techniques for patient transfer, there was often a mismatch between the size of the patient and the length of the nurse's arms. In effect, they needed longer arms. A solution was duly developed and the resulting patient-handling sling has proved adaptable to a variety of techniques (Figure 10). A number of other techniques have been developed, based on simple sliding boards, sliding quilts, etc.

XII. ERGONOMICS, THE CARE PLANS FOR INDIVIDUAL PATIENTS AND THE ORGANIZATION OF WORK

Ideally, a "care plan" is drawn up for every patient, taking into account every medical and nursing consideration for the initial program of therapeutic management and rehabilitation. The question of patient handling is central to this care plan. As soon as the medical condition permits, plans are made for rehabilitation and recovery of physical independence. As the patient makes progress, the need for patient handling by the nursing staff recedes and their physical burdens are reduced. In effect, the "care plan" can serve, ergonomically, as a broadly based "task analysis" and the ergonomic problems that are likely to arise can be identified in advance. However, such a plan has to be agreed to by not only the medical and nursing staffs and physical and occupational therapists, but by the administrators. Though in theory there is no reason for failing to draw up a "care plan", in practice there are innumerable circumstances that may arise and hinder its implementation.

At the simplest level, some patients are resistant to the prospect of having to fend for themselves, and the nursing staff must be flexible in its approach. It is of little use to spend time encouraging people to walk, to the toilet for example, if they are struggling against incontinence; it is better to encourage them to walk on the way back.

The more serious ergonomic problems stem from inadequacies of the management system. When a patient is admitted as an "emergency", no preliminary planning is possible and sometimes the hospital staff is exposed to great difficulties. For example, a patient was brought to a London teaching hospital after having suffered a cerebrovascular accident (CVA). She was confused and helpless and she weighed 220 kg. The only hoist available in the hospital had a safe working load limit of 160 kg. No lifting aids were apparently available. Bed sores being a major problem, the decision was made that ten people should be assembled each time the patient was moved, but as the patient was short in height and wider than the bed, there was no way in which her weight could be evenly shared between ten people. More than one nurse was injured. It has to be accepted that, while no preliminary planning was possible, the want of foresight for such an eventuality raises a number of questions concerning administrative competence.

Even when preliminary planning is feasible, there can be unexpected, though avoidable, problems. One example which is not uncommon is lack of consultation between the medical staff, the admissions department, and the nursing staff. Those in charge of a ward have no knowledge of his condition until a patient appears; they have only the diagnostic label with no information about physical dependency or about the needs for postoperative management. Thus, no adaptations concerning ward equipment or levels of staffing can be foreseen. However, even when foresight and planning are feasible, reasonable solutions are not always possible. Once in an orthopedic ward, the question was asked, "Why are you not nursing

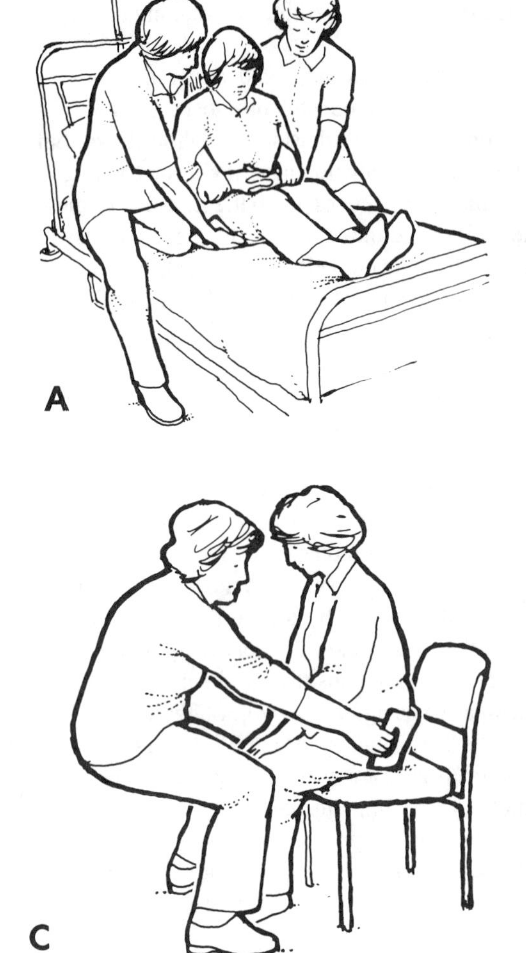

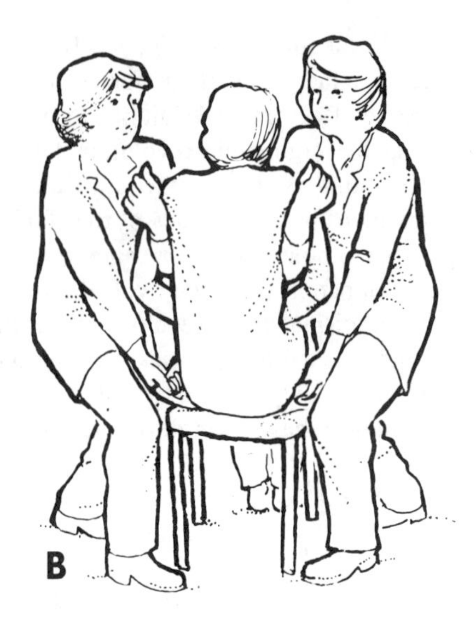

FIGURE 10. Using the patient-handling sling to enable nurses to move patients, keeping the load close to their bodies if any lifting action is involved and, if necessary, kneeling on the bed. (A) ''Through-arm grip'' with sling to move patient up the bed; (B) ''cross-arm grip'' for carrying patient or for transferring from chair to bed, etc.; (C) ''pelvic or sling hold'' to stand a patient up, the nurse gripping the patient's knees with her own and simply using her body weight as a counterbalance to bring the patient to the erect position. (Reproduced from Lloyd, P., Tarling, C., Wright, B., and Troup, J. D. G., *The Handling of Patients: a Guide for Nurses,* The Back Pain Association in collaboration with the Royal College of Nursing, London, 1987. With permission.)

this patient on a turning frame?'' The answer was that the ward would lose the adjustable-height bed which the turning frame was replacing and that eventually when the turning frame was returned, it would be replaced by an old, nonadjustable bed. The nursing officer in charge decided that the temporary disadvantages of coping without a turning frame were outweighed by those that would follow the loss of an adjustable-height bed for an indefinite period. Such occurrences are commonplace in many poorly administered, underfinanced hospitals. However, in this instance at least, the nurse in charge was free to make a rational decision.

The ''care plan'' provides a framework in which ergonomic problems can be foreseen and action can be taken. It is not applicable only to the program of therapeutic management in the hospital. It should offer a bridge between management in the hospital and management in the community when the patient is discharged home. Long before a chronically disabled patient leaves hospital, preparations can be made so that such rehabilitation as is possible may proceed uninterrupted. ''Care plans'' therefore serve also to identify the ''user requirements'' which may form the basis for the ergonomic approach to design.

XIII. ERGONOMIC EDUCATION FOR HEALTH PROFESSIONALS

The environment in which health professionals work has, historically, been the responsibility of governments, hospital administrators, and independent equipment manufacturers to whom ergonomics may be a familiar word, manifest perhaps by a few textbooks on someone else's office shelf, but little more. In recent years, this climate has shown signs of change, but with the economic recession that affected so many countries, implementation of the ergonomic approach has been limited, and so it remains. These questions therefore have to be asked: are health professionals content with this? Given the premise that ergonomic intervention should be cost-beneficial,[1-3] should they themselves not take more active steps?

In 1986, in response to a plea from the Royal College of Nursing (U.K.), a short course on ergonomics in nursing was held at the University of Nottingham with the collaboration of the Queen's Medical Center at Nottingham. (See References 16 and 51 for two books used in the course.) In addition to instruction in the basic principles, course members were given a variety of problems to solve and they were asked to inspect existing workplaces in the hospital to identify the main ergonomic problems. They took to it with enthusiasm.

Clearly, there is a need for more courses of this kind and for more extended courses to enable selected clinical teachers or nurse tutors to begin teaching ergonomics in schools of nursing and university departments of health studies. Instruction in basic ergonomic principles was included in the new system of training in patient handling for student nurses in Finland[49] and this was entirely appropriate to their education, for a qualified nurse must be prepared to adapt the working environment in order to achieve ergonomically satisfactory conditions. Increasingly, physiotherapists are receiving ergonomic instruction in the course of training, certainly those who are working in occupational health services. Ergonomics is essential to the education of an occupational therapist. It should become so for all administrators and planners in the health services, including community health.

XIV. CONCLUSION

Caring for the sick and chronically disabled presents a daunting variety of ergonomic problems. However, when health professionals begin to understand what ergonomics, as a discipline, has to offer, they then can reasonably take a more optimistic view. They can make a start in creating a safer, less uncomfortable, and more efficient working environment. However, the enlightenment that ergonomics may bring to the health professional and the improvements that will stem from ergonomic redesign of specific workplaces may be frustrated if ergonomics is not understood and encouraged throughout the hierarchy of hospital and community health services.

REFERENCES

1. **Westgaard, R. H. and Aarås, A.,** The effect of improved workplace design on the development of work-related musculoskeletal illnesses, *Appl. Ergon.,* 16, 91, 1985.
2. **Spilling, S., Aarås, A., and Eitrheim, J.,** *Kostnadanalyse av arbeidsmiljoinvesteringer ved STKs telefonfabrikk på Kons4gsvinger (STK-K),* Standard Telefon og Kabelfabrik A/S, Oslo, 1986.
3. **Silta, S., Heikkilä, S., and Kuorinka, I.,** *Ergonomia toistotyössä: rasitussairauksien ehkäisy,* Werner Söderström Osakeyhtiö, Porvoo, 1986.
4. **Jensen, R. C.,** Work-related back injuries among nursing personnel in New York, in *Proc. of the Human Factors Society,* 30th Annual Meeting, 1986.
5. **Shannon, H. S. and Manning, D. P.,** The use of a model to record and store data on industrial accidents resulting in injury, *J. Occup. Accid.,* 3, 57, 1980.
6. Régies Nationale Des Usines Renault, *Les Profiles de Postes,* Masson/Sirtes, Paris, 1976.
7. **Rohmert, W. and Landau, K.,** *Arbeitswissenschaftliche Erhebungsverfahren zur Tätigkeitsanalyse (AET),* Handbuch mit Merkalheft, Hans Huber, Bern, 1979.
8. **North, K. and Rohmert, W.,** Job analysis applied to the special needs of the disabled, *Ergonomics,* 24, 889, 1981.

9. **Karhu, O., Kansi, P., and Kuorinka, I.,** Correcting work postures in industry: a practical method for analysis, *Appl. Ergon.*, 8, 199, 1977.
10. **Holzmann, P.,** Arban — a new method for analysis of ergonomic effort, *Appl. Ergon.*, 13, 82, 1982.
11. **Foreman, T. K. and Troup, J. D. G.,** Diurnal variations in spinal loading and the effects on stature: a preliminary study of nursing activities, *Clin. Biomech.*, 2, 48, 1987.
12. **Drury, C. G.,** Task analysis methods in industry, *Appl. Ergon.*, 14, 19, 1983.
13. Työterveyslaitos, *Työpaikan Ergonominen Selvitys,* Institute of Occupational Health, Helsinki, 1984.
14. **Millard, G.,** *Commissioning Hospital Buildings,* Oxford University Press and the King Edward's Hospital Fund, London, 1982.
15. **Stanton, G.,** The development of ergonomic data for health building design guidance, *Ergonomics,* 26, 785, 1983.
16. **Pheasant, S. T.,** *Bodyspace: Anthropometry, Ergonomics and Design,* Taylor & Francis, London, 1986.
17. **Mitchell, J.,** User requirements and the development of products which are suitable for the broad spectrum of user capacities, *Ergonomics,* 24, 863, 1981.
18. Institute for Consumer Ergonomics, Seated anthropometry: the problems involved in a large scale survey of disabled and elderly people, *Ergonomics,* 24, 831, 1981.
19. **Agnew, K., Archer, B., and Patterson, G.,** *Design of Hospital Bedsteads,* King Edward's Hospital Fund, London, 1967.
20. **Mitchell, J.,** User requirements and hygienic equipment, *Ergonomics,* 24, 871, 1981.
21. **Fraser, C. and Mitchell, J.,** Evaluation of curb negotiation in a prototype and standard model of wheelchair, *J. Med. Eng. Technol.*, 4, 230, 1980.
22. **Mitchell, J.,** Wheelchairs and the discharge of personal responsibility, *Physiotherapy,* 70, 467, 1984.
23. **Dudley, H. A.,** Operative ergonomics, *Nursing Mirror,* October 28, 53, 1976.
24. **Dudley, H. A.,** Micro-ergonomics, *Nursing Mirror,* January 27, 48, 1977.
25. Styrelsen för Teknisk Utveckling, *Insatsområdet sjukvårdsmiljö,* Nr. 330, Vårdens tunga lyft, 1980, STU, Stockholm.
26. **Dehlin, O., Jäderberg, E., and Karlqvist, L.,** Mätmetod vid Patientlyft: Studier av Samband Mellan Vissa Variabler i Lyft-Situationer och Upplevd Ansträngning. En metod för att Skatta Ansträngningsupplevelse hos Vårdpersonal under Lyft och Förflyttning av Patienter, STU-Rep. 79-5689, 80-4056, Styrelsen för Teknisk Utveckling, Stockholm, 1980.
27. **Fordham, M., Appenteng, K., Goldsmith, R., and O'Brien, C.,** The cost of work in medical nursing, *Ergonomics,* 21, 331, 1978.
28. **Magora, A.,** Investigation of the relation between low back pain and occupation. I. Age, sex, community, education and other factors. II. Work history, *Ind. Med. Surg.*, 39, 465, 504, 1970.
29. **Cust, G., Pearson, J. C. G., and Mair, A.,** The prevalence of low back pain in nurses, *Int. Nurs. Rev.*, 19, 169, 1972.
30. **Stubbs, D. A., Buckle, P. W., Hudson, M. P., and Rivers, P. M.,** Back pain in the nursing profession. I. Epidemiology and pilot methodology, *Ergonomics,* 26, 755, 1983.
31. **Videman, T., Nurminen, T., Tola, S., Kuorinka, I., Vanharanta, H., and Troup, J. D. G.,** Low back pain in nurses and some loading factors of work, *Spine,* 9, 400, 1984.
32. **Jensen, R. C.,** Events that trigger disabling back pain among nurses, in *Proc. of the Human Factors Soc.*, 29th Annual Meeting, 1985.
33. **Videman, T., Nurminen, T., Cedercreutz, G., Asp, S., Rauhala, H., and Troup, J. D. G.,** Incidence of low back pain among nursing students and nurses, in The Changing Nature of Work and Workforce, Proc. of the 3rd Joint U.S.-Finnish Science Symposium (National Institute for Occupational Safety and Health, and the Institute of Occupational Health, Helsinki), 1986, 95.
34. **Takala, E.-P. and Kukkonen, R.,** The handling of patients on geriatric wards: a challenge for on-the-job training, *Appl. Ergon.*, 18, 17, 1987.
35. **Leyshon, G. E. and Francis, H. W. S.,** Lifting injuries in ambulance crews, *Public Health,* 89, 71, 1975.
36. **Kilbom, Å., Ljungberg, A.-S., and Hägg, G.,** Lifting and carrying in geriatric care: a comparison between differences in workspace layout, work organisation and use of modern equipment, in *Proc. of Int. Ergonomic Assoc. Congr.*, Taylor & Francis, London, 1985, 550.
37. **Chaffin, D. B., Herrin, G. D., and Keyserling, W. M.,** Pre-employment strength testing — an updated position, *J. Occup. Med.*, 20, 403, 1978.
38. **Keyserling, W. M., Herrin, G. D., and Chaffin, D. B.,** Isometric strength testing as a means of controlling medical incidents on strenuous jobs, *J. Occup. Med.*, 22, 332, 1980.
39. **Hansson, T., Bigos, S. J., Wortley, M. K., and Spengler, D.,** The load on the lumbar spine during isometric testing, *Spine,* 9, 720, 1984.
40. **Troup, J. D. G., Foreman, T. K., Baxter, C. E., and Brown, D.,** The perception of back pain and the role of psychophysical tests of lifting strength, *Spine,* 12, 645, 1987.
41. **Ljungberg, A.-S., Gamberale, F., and Kilbom, Å.,** Horizontal lifting — physiological and psychological responses, *Ergonomics,* 25, 741, 1982.

42. **Snook, S. H.,** The design of manual handling tasks, *Ergonomics,* 21, 963, 1978.
43. NIOSH, *Work Practices Guide for Manual Lifting,* Tech. Rep. Publ. No. 81-122, National Institute for Occupational Safety and Health, U.S. Department of Health and Human Services, Washington, D.C., 1981.
44. **Troup, J. D. G. and Edwards, F. C.,** Manual Handling and Lifting: an Information and Literature Review with Special Reference to the Back, Health and Safety Executive, Her Majesty's Stationery Office, London, 1985.
45. **Kazarian, L. E.,** Creep characteristics of the human spinal column, *Orthop. Clin. North Am.,* 6, 3, 1975.
46. **Lloyd, P., Osborne, C. M., Tarling, C., and Troup, J. D. G.,** *The Handling of Patients: a Guide for Nurse Managers,* The Back Pain Association in collaboration with the Royal College of Nursing, London, 1981.
47. **Lloyd, P., Tarling, C., Wright, B., and Troup, J. D. G.,** *The Handling of Patients: a Guide for Nurses,* The Back Pain Association in collaboration with the Royal College of Nursing, London, 1987.
48. **Troup, J. D. G., Lloyd, P., Osborne, C. M., and Tarling, C.,** *Potilaan siirron opas,* Institute of Occupational Health, Helsinki, 1985, (translated by Videman, T. and Rauhala, S.).
49. **Troup, J. D. G. and Rauhala, H. H.,** Ergonomics and training, *Int. J. Nurs. Stud.,* in press.
50. **Troup, J. D. G.,** Training in the handling of patients: the demands on education and occupational health for nurses, *Nurs. Educ. Today,* 2, 13, 1982.
51. **Grandjean, E.,** *Fitting the Task to the Man: an Ergonomic Approach,* Taylor & Francis, London, 1982.

Chapter 13

ECZEMA

Eskil Nilsson

TABLE OF CONTENTS

I. INTRODUCTION

High standards of hygiene necessitate frequent handwashing, hand disinfection, and the use of gloves for large groups of hospital workers. This inevitably involves a cumulative irritant exposure of the hands among ward orderlies, nurses, assistants, nursing assistants, physicians, and dental personnel. As the nursing staff comprises mainly women, domestic exposure to irritants may add an extra strain on the hands. Some atopic individuals run an increased risk of developing irritant hand eczema in these occupations. The complex interplay and the relative importance of "wet" hospital work, domestic work, and atopy as risk factors for hand eczema will be discussed in this chapter. Contact allergens in the hospital environment differ among countries, specialities, and occupational duties. A number of allergens listed in this chapter constitute significant risk factors for small groups, while others may involve risks for large groups of hospital workers. Basic knowledge on the etiology, diagnosis, treatment, and prevention of hand eczema, information invaluable for the management of this common disorder in hospital workers, will be presented.

II. ECZEMA

A. Definition[1]

The term eczema may be defined as an inflammatory process which takes place in the skin. External and internal factors may provoke eczema either alone or in combination. Acute eczema is characterized by erythema, edema, papules, and groups of vesicles or bullae and is often accompanied by weeping and crusting. Epidermal thickening (lichenification) and scaling are characteristic features in chronic eczema. The clinical picture of eczema may change over a few days. Thus, acute and chronic lesions can be found simultaneously in the same patient. The terms eczema and dermatitis are often used synonymously. Dermatitis is, however, used for skin diseases which do not show the clinical and histological signs of eczema. All eczemas are forms of dermatitis, but all cases of dermatitis are not eczemas. Unfortunately, there is no international agreement on the use of these terms.

B. Classification[1]

A simplified classification of the major types of eczema is as follows:

1. Exogenous eczemas
   a. Irritant contact dermatitis
   b. Allergic contact dermatitis
2. Endogenous eczemas
   a. Atopic dermatitis
   b. Seborrheic dermatitis
   c. Nummular eczema
   d. Other patterns

As the term implies, contact dermatitis is an inflammatory response to external agents. There are two major types of contact dermatitis: (1) irritant dermatitis which develops after exposure of the skin to chemically or physically damaging factors and (2) allergic contact dermatitis which is provoked by a specific immunological mechanism. In endogenous eczema, hereditary and other intrinsic factors are of great importance.

C. Irritant Dermatitis[2]

Acute irritant contact dermatitis can be provoked by a few applications of strong irritants. Eczema may also develop after repeated exposure to weak irritants. This dermatitis has been termed cumulative insult dermatitis, chronic irritant dermatitis, ''wear and tear'' dermatitis, and ''traumiterative dermatitis''. As no test can determine whether an irritant is relevant to a patient's eczema, the evaluation of the importance of irritants in dermatitis has to be a clinical decision. Many chemicals such as solvents, acids, alkalis, and detergents alone or, more commonly, in combination may provoke an irritant dermatitis. Low humidity, cold, and friction may also be important contributory factors. It is difficult to obtain scientific documentation for the widely held belief that contact with irritants is the most common cause of hand eczema. Clinical evidence is, however, overwhelming as various kinds of ''wet'' work with frequent exposure to water and detergents involve occupational hazards to the hands. Eczema of the hands among housewives is the most common irritant contact dermatitis. An irritant contact dermatitis very often develops in two phases. The first phase is characterized by dryness and slight erythema and if the exposure continues, an inflammatory reaction will develop with more pronounced erythema, edema, and vesicles.

D. Contact Allergy[2]

Allergic contact dermatitis results from sensitization to chemical antigens. The inflammatory reaction is an immunological response mediated by T-lymphocytes. Generally, allergic contact dermatitis occurs after repeated exposure to a potential sensitizer. A contact allergy is a specific reaction and involves only one or a few substances. The allergy embraces the skin over the whole body. Very small quantities of the allergen may elicit an eczematous reaction. As a rule, a contact allergy endures for a very long time, often a whole lifetime.

The responsible sensitizer in allergic contact dermatitis is generally identified by means of epicutaneous or patch testing. These are performed by applying a nonirritating concentration of the suspected allergens under occlusion to normal skin, usually on the back. In a sensitized individual, this procedure will provoke a delayed eczematous reaction with erythema, edema, and vesiculation after 48 h. Recommendations concerning the standardization of epicutaneous testing and screening test batteries for the most common contact allergens have been made by the International Contact Dermatitis Research Group (ICDRG). The ICDRG suggests the following compounds for inclusion in a standard screening series:

1. Metals
   - Potassium dichromate
   - Cobalt chloride
   - Nickel sulfate
2. Rubber chemicals
   - Mercapto-mix
   - Thiuram-mix
   - PPD-mix (black rubber mix)
   - Carba-mix
3. Medicaments
   - Neomycin sulfate
   - Benzocaine
   - Parabens mix
   - Wool alcohols
   - Ethylene diamine HCL
   - Dowicil
   - Quinoline mix
4. Balsams
   - Balsam of Peru
   - Colophony
   - Fragrance mix

5. Others
   - Formaldehyde
   - Paraphenylene diamine (PPD)
   - Epoxy resin
   - Paratertiary butylphenol formaldehyde resin
   - Primin

E. Contact Urticaria[3-5]

Contact urticaria may be defined as a wheal-and-flare response elicited from within a few minutes to up to half an hour after exposure of the skin to various substances. The term contact urticaria syndrome was proposed in 1975 to cover a broad range of clinical manifestations which may be provoked by the causative agent. von Krogh and Maibach[4] propose the following pattern of stages for this syndrome.

Cutaneous reactions only:
- Stage 1. Localized urticaria; dermatitis; nonspecific symptoms (itching, tingling, burning, etc.)
- Stage 2. Generalized urticaria

Extracutaneous reactions:
- Stage 3. Bronchial asthma; rhinoconjunctivitis; Orolaryngeal symptoms; gastrointestinal (GI) symptoms
- Stage 4. Anaphylactoid reactions

Contact urticaria may be elicited by an allergic mechanism. Immunological contact urticaria may be suspected when there has been a period of sensitization, the reaction is strong, tests on controls are negative, and the passive transfer test is positive. In nonimmunological contact urticaria, the reaction is elicited without previous sensitization in most exposed individuals. This type of contact urticaria is probably very common. When exposure is optimal, a majority of individuals will react with contact urticaria after a rather highly concentrated application to intact skin. In lower concentrations, the substances may provoke erythema, but not a true urticarial reaction. The frequency and strength of nonimmunological contact urticaria are influenced by the concentration of the chemicals and even by the chemical composition of the vehicle.

Diagnosis of contact urticaria has to be based on a careful history of immediate reactions. Diagnostic tests, guided by the case history, may be performed on intact skin as an open, occlusive, or intradermal test and may be performed on slightly affected or previously affected skin as open or occlusive tests. A stepwise test procedure has been recommended by von Krogh and Maibach.[4] Contact urticaria may elicit or aggravate a hand dermatitis.

F. Atopic Dermatitis[6,7]

The term atopy was introduced by Coca and Cooke in 1923 and means a strange disease. Today the term atopy includes allergic rhinoconjunctivitis, bronchial asthma, atopic dermatitis, and certain forms of GI allergy and urticaria. Atopic dermatitis is a genetically determined, chronically fluctuating, pruritic eczematous disorder starting in infancy or childhood. During infancy, eczematous lesions on the head and extensor surfaces of the extremities are prominent features. In older children and adults, involvement of the flexure areas of the elbows, knees, wrists, and ankles are common. The atopic skin is pale, dry, and pruritic and has a reduced resistance to irritants.

G. Occupational Aspects[8,9]

Approximately 90% of the occupational skin diseases are contact eczemas. The diagnosis of occupational irritant contact dermatitis should be based on the following criteria: the

Table 1
CRITERIA FOR OCCUPATIONAL CONTACT DERMATITIS

1. The lesion represents an eczema
2. Significant occupational exposure to irritants
3. Significant occupational exposure to an allergen positive on patch testing
4. Location corresponds to exposure
5. Improvement when exposure stops
6. Recurrence on reexposure

clinical state of the skin lesion indicates an eczema; occupational exposure to a skin irritant occurred at the time of onset; the location of the eczema corresponds to the exposure; the eczema improves considerably within days or a week once exposure stops; reexposure causes the eczema to recur. Occupational allergic contact dermatitis may be diagnosed on the basis of the same criteria, but requires a positive skin test to an occupational allergen. Generally allergic contact dermatitis is more acute and vesicular than irritant eczema. These guidelines for the diagnosis of occupational contact dermatitis summarized in Table 1 may seem simple, but in the individual case, the problem may be very complex. It can prove extremely difficult to evaluate the relative importance of an intrinsic factor such as a previous atopic dermatitis as an etiologic factor in hand dermatitis. Furthermore, it may be difficult to evaluate the importance of occupational vs. domestic exposure to irritants and allergens.

III. CONTACTANTS POTENTIALLY HAZARDOUS TO HOSPITAL WORKERS

A. Irritants[2,10-12]

Hospital work necessarily entails repeated washing and disinfection of the hands and a lot of the work has to be done in protective gloves. Exposure to water, soaps, detergents, and disinfectants such as alcohol is inevitable. This cumulative exposure may provoke a chronic irritant dermatitis of the hands. Water is hypotonic and may dissolve the hygroscopic substances in the horny layer and make the skin dry and brittle. Soaps and detergents remove the lipid film, the water-holding substances in the horny layer, and damage cell membranes.

Somewhat surprisingly, it has recently been shown that wet hospital work increased the odds (relative risk) of developing hand eczema only twice compared to dry office work.[13] The same study indicated that individual factors were of great importance for the development of hand eczema. As a part of this study, 142 wet hospital workers, of whom 91% were women, were clinically investigated because of current hand eczema. Among these patients who consulted a dermatologist for current hand eczema, risk individuals were overrepresented. Thus, a history of atopy was found in 58%, previous hand eczema was found in 67%, and a history of metal dermatitis was found in 41%. The corresponding figures for the total cohort were atopy, 22.6%; earlier hand eczema, 22.4%; and metal dermatitis, 26.3%. It was stated that the following agents had provoked the current hand dermatitis in these 142 wet hospital workers.

Agent	Number of patients
Water and cleaning agents	111
Disinfectants	26
Physical factors	24
Various foods	23
Rubber gloves	17
Oils, solvents	11
Paper towels	9
Dirt and dust	8

In 35% of the patients, it was claimed that exposure to the eliciting factors took place largely at home which highlights the importance of domestic irritant exposure in the development of hand eczema among hospital workers. In a recent textbook on dermatology, the following irritants are listed as specific for hospital workers:[8] disinfectants, quaternary ammonium compounds, hand creams, soaps, and detergents. Irritants listed as specific for dentists and dental technicians are soaps, detergents, plaster of Paris, acrylic monomer, and fluxes.

B. Sensitizers[12,14-19]

Contactants with potentially allergenic properties which can be found in the hospital environment may be listed as follows:

1. Skin protection measures
    - Chemicals in rubber gloves
    - Hand creams — vehicles; preservatives; fragrance materials; antibiotics; and topical anesthetics
2. Disinfectants — skin, instruments, etc.
    - Benzalkonium chloride and other quaternary ammonium compounds
    - Dodecyl-di-(amino-ethyl) glycine[20,21]
    - Chlorhexidine gluconate
    - Ethyl alcohol
    - Formaldehyde
    - Glutaraldehyde
    - Hexachlorphene
    - Iodine compounds
    - Isopropyl alcohol
    - Phenolic compounds
3. Antibiotics
    - Ampicillin
    - Chloramphenicol
    - Neomycin
    - Penicillin
    - Streptomycin
    - Sulfonamide
4. Other medicaments
    - Apomorphine
    - Chlorpromazine
    - Cyanamide[22]
    - Ethambutol[23]
    - Nitrogen mustard
    - Phenothiazine
    - Propranolol
5. Local anesthetics
    - Benzocaine
    - Lidocaine
    - Procaine
    - Propranidid
    - Tetracaine
6. Miscellaneous
    - Acrylic bone cement
    - Dental resin and composite

- Colophony
- Essential oils
- Mercury
- Nickel
- Piperazine

Chemicals found in rubber gloves are probably the most common source of allergic contact dermatitis among hospital personnel. In any hand dermatitis that stops at the wrist, sensitization to rubber chemicals should be suspected. Patch testing should be performed with a standard series of rubber chemicals and if the results are negative, additional rubber chemicals and the glove itself may be tested.

Hand creams for lubrication and a topical medication for dermatitis may produce contact sensitivity. Thus, various vehicle constituents, preservatives, topical anesthetics, topical antibiotics, perfumes, and other constituents in emollients and topical medicaments may cause sensitization. Approximately 80% of contact allergies due to topical medicaments can be diagnosed by means of a standard patch test series.

Isopropyl alcohol or denatured ethyl alcohol is extensively used in hospitals for cleansing the skin. Contact sensitivity to alcohol is uncommon, but may occur. Agents added to denatured alcohol are neither potent nor frequent sensitizers. Hexachlorophene, once a very popular disinfectant, rarely caused contact allergy. It is now recommended that its use be restricted because it has various systemic effects including neurotoxicity. Chlorhexidine is a common disinfectant with a low sensitizing capacity. Accidental sensitization to benzalkonium chloride used as a disinfectant for the skin and sometimes for surgical instruments has occurred, but is probably rare among hospital personnel. Iodine compounds such as tincture of iodine and iodoform are sensitizers and may give rise to systemic effects. Povidone iodine (Betadine®) is said to be a safe disinfectant, although allergic contact dermatitis has been reported in connection with its use. Hospital personnel may be exposed to formaldehyde when it is used for sterilizing instruments, as a disinfectant, or in laboratories for fixing tissues. Formaldehyde is a primary irritant and a fairly common sensitizer. Glutaraldehyde used for sterilization of instruments has caused contact sensitivity among hospital personnel. The disinfectant dodecyl-di-(amino-ethyl) glycine which is marketed under various trade names (Tego®, Tego® 103G, Desimex®, and Ampholyt G®) may sensitize hospital workers.

Penicillin and sulfonamides previously used for topical treatment may induce contact sensitivity. Accidental contact with penicillin, especially when used for injections or in syrups, may sensitize medical personnel. Allergic contact sensitivity to penicillin may be combined with an anaphylactic reaction. Previously, contact sensitivity to streptomycin was not uncommon among nurses, but improved information on the importance of skin protection and the use of amphines has reduced these problems considerably.

Chlorpromazine and related phenothiazines commonly caused contact dermatitis among personnel working in psychiatric departments in the past. Apart from allergic contact dermatitis, photosensitivity may also occur. Improved skin protection and the use of capsules and amphines instead of syrups and tablets have reduced contact allergy to these drugs in recent years. Nitrogen mustard used for the topical treatment of mycosis fungoides has sensitized several dermatologists. A glove made of polyvinyl chloride has to be used to prevent skin exposure to nitrogen mustard. Accidental sensitization to other cytostatic agents has not been reported in hospital personnel.

Dentists especially may develop contact sensitivity to local anesthetics. In the past, procaine and other compounds derived from paraaminobenzoic acid (PABA) were common sensitizers. Lidocaine (Xylocaine®) and mepivacaine (Carbocaine®) are less sensitizing.

Orthopedic surgeons may develop contact allergy to acrylate in bone cement when it is used for fixing of prostheses. As the acrylate penetrates rubber gloves, it is difficult for the surgeon to avoid skin contact, but one possible way of diminishing skin contact would be to wear two pairs of rubber gloves and to discard both immediately after acrylate exposure. Contact sensitivity to acrylic bone cement is commonly accompanied by dry and fissured fingers. A burning sensation, tingling, and paraesthesia of the finger tips is common among patients sensitized to acrylic monomer in bone cement.

Dentists and other dental personnel generally work without gloves and are routinely exposed to many sensitizers. The following are listed in a recent textbook on dermatology:[8] local anesthetics (tetracaine, procaine), mercury, rubber, UV-curing acrylates, acrylic monomer, disinfectants (formaldehyde, eugenol), nickel, epoxy resin (filling), periodontal dressing (balsam of Peru, colophony, eugenol), and catalysts methyl-*p*-toluenesulfonate and methyl-1,4-di-chlorbenzodesulfonate in impression and sealent materials. Dentists may become sensitized to local anesthetics through rubbing gums with a solution or ointment containing a local anesthetic. Exposure may also occur as a result of the solution leaking from the syringe when the compound is injected. The new, less sensitizing compounds (Xylocaine®, Carbocaine®) and improved syringes have reduced the problem considerably.

In spite of the extensive use of mercury in amalgam, contact sensitivity to metallic mercury is uncommon among dental personnel. In a survey of occupational dermatitis among health service personnel in Poland, contact sensitivity to mercury was found in four stomatologists.[15] A case report by Ancona et al. presents a dentist who developed a dermatitis on the fingers after preparing amalgam fillings by mixing the metals manually, squeezing the alloy with the fingers. Positive epicutaneous test results were obtained for metallic mercury and ammoniated mercury.[24] White and Brandt[25] patch-tested dental students to determine whether the rate of hypersensitivity to mercury increased as they were exposed to silver amalgam during their dental training. A statistically significant increase was found, but no clinical symptoms relevant for the contact sensitivity were reported.[25] Apart from metallic mercury, mercury occurs in many organic and nonorganic compounds. Most mercury compounds are toxic and may sensitize. There are various reports in the literature concerning cross-reactions between metallic mercury and organic and nonorganic mercury compounds.[26,27]

Sensitization to rubber chemicals in gloves may become an increasing problem among dental personnel as the need for personal protection against infectious agents (e.g., AIDS virus) will increase. Dental technicians and dentists may contract a contact allergy through exposure to uncured acrylic monomer which is a potent sensitizer. The self-cured resin used to repair dentures and to buildup temporary bridges, crowns, and fillings contains much more residual monomers than the heat-cured resin used in ordinary dentures.

C. Contact Urticants[3-5,28]

Contactants with potentially urticariogenic properties which can be found in the hospital environment may be listed as follows:

1. Skin protection measures
   - Rubber in gloves
   - Hand creams — vehicles; preservatives; fragrance materials (see under Miscellaneous)
2. Disinfectants
   - Chloramine
   - Ethyl alcohol
   - Formaldehyde
   - Gentian violet
   - Isopropyl alcohol
3. Antibiotics
   - Ampicillin

- Cephalosporins
- Chloramphenicol
- Neomycin
- Penicillin
- Streptomycin
- Tetracycline

4. Other medicaments
   - Chlorpromazine
   - Lindane
   - Menthol
   - Metamizole
   - Nitrogen mustard
   - Phenylbutazone
   - Promethazine hydrochloride
   - Propyphen butazone
   - Pyrazolone derivatives
   - Salicylic acid
   - Sulfur
5. Miscellaneous
   - Amyl alcohol
   - Benzocaine
   - Benzoic acid
   - Butylated alcohol
   - Butylated hydroxyanizole
   - Butylated hydroxytoluene
   - Cetyl alcohol
   - Cinnamic acid
   - Cinnamic aldehyde
   - Cinnamon oil
   - Denatonium benzoate
   - Ethylene diamine
   - Eugenol
   - *p*-Hydroxybenzoic acid
   - Lanolin alcohol
   - Parabens
   - Phenol
   - Polyethylene glycol 400
   - Polysorbate 60
   - Prophyl alcohol
   - Propylene glycol
   - Sodium benzoate
   - Sorbic acid
   - Stearyl alcohol

The contact urticaria syndrome is a relatively unexplored field. Most reports in the growing literature on contact urticaria are limited to one or a few cases. Little is known about the prevalence and the relevance in different populations with hand eczema. In a study of a selected group of 142 patients with hand eczema employed in hospital wet work, a positive prick test relevant for contact urticaria was found in 22 patients, most often obtained with rubber gloves. A few relevant prick tests were found with formaldehyde, benzalkonium chloride, and some emollients, disinfectants, and cleaning agents.[29] Contact urticaria in relation to various foods may be an occupational hazard among kitchen workers and saliva may evoke contact urticaria in dentists. It must be stressed that contact urticaria is probably much more common than has previously been suggested and the contactants listed should alert the physician to the possibility of immediate reactions as a cause of skin troubles among hospital workers.

IV. HAND ECZEMA IN HOSPITAL WORKERS

A. Definition[1,30]

The terms hand eczema and hand dermatitis are very often used synonymously. In general disorders of known etiology, such as tinea, lichen planus, and keratodermia of the palms, are not defined as hand dermatitis. As it is difficult to subclassify hand dermatitis according to the traditional morphological-etiological classifications, the nondescriptive term dermatitis serves a useful function. In typical cases of hand eczema, you may find dryness, erythema, edema, papules, vesicles, weeping, crusting, and lichenification.

B. Prevalence

Epidemiological studies of hand eczema in various populations have recorded a rather wide range of prevalence figures. In a population study in southern Sweden, the prevalence of hand dermatitis was estimated to be 1.2 to 3.4%.[31] As a part of a study on nickel allergy in Finland, the hands of 980 subjects were examined and hand eczema was found in 4%.[32] In a study in northern Norway, 14,667 adult subjects were asked about the occurrence of allergic hand eczema during the preceding 12 months. A positive reply was given by 4.9% of the men and 13.2% of the women. The age of the women influenced the prevalence. Thus, 15% of women between 25 to 34 years of age reported hand eczema, while only 9.2% of those between 45 to 49 years did so.[33] In a prevalence study in the Netherlands, an episode of eczema of the hands and forearms lasting for 3 weeks and occurring during the past 3 years was used as a criterion for classifying a case as hand eczema. In this study, the prevalence figures were 4.5% for men and 10.0% for women.[34] Of a representative sample consisting of 1961 Danish women, 22% reported a history of hand eczema.[35] In a Finnish study of 617 wet hospital workers, predominantly women, 44% had past or current hand eczema and a similar figure was found in a prospective study of wet hospital workers in Sweden.[13,36] The variety of results recorded in these investigations reflects some of the problems connected with prevalence studies of hand eczema. As hand eczema is a periodic disorder, figures based on a single examination will underrate the prevalence. Prevalence studies covering periods of time must rely on anamnestic information with its limitations regarding accuracy of diagnosis. Another aspect of the problem is how severe a dermatitis of the hands has to be before it can be diagnosed as eczema. The difference in prevalence between the two sexes and the importance of wet work is obvious from the figures.

C. Etiology and Classification[1,30]

Classification of hand eczema may be based on morphological description of the eczema using terms such as vesiculosquamous, nummular, wedding ring, finger tip, hyperkeratotic, and pompholyx. These names refer to clinical features and say very little or nothing about the etiology. A common way of classifying hand eczema is according to etiological aspects. Thus, the terms irritant or allergic contact dermatitis of the hands are used depending on the cause. As hand eczema is almost always multifactorial, simple etiological classifications are doomed to failure. In order to avoid the frustrating struggle with some of the classifications, Epstein[30] proposed that hand eczema should be analyzed for its endogenous and exogenous factors. Improved understanding of the relative importance of various endogenous and exogenous factors would be of great importance in the analysis of the various factors which contribute to the development of a hand eczema.

Table 2
HISTORY IN PATIENTS WITH HAND DERMATITIS

Family history
- Atopy
- Psoriasis
- "Skin diseases"

Personal history
- Atopic dermatitis
- Atopic skin condition (dry/itchy skin)
- Contact allergy
- Contact urticaria
- "Skin diseases"

Previous hand eczema
- Date of onset
- Location at start
- Previous trauma
- Occupation at start
- Relation to occupational or domestic factors
- Periodicity
- Progress
- Medical consultation
- Sick leave
- Change of work

Current hand eczema
- Date of onset
- Location at start
- Appearance at onset and subsequent course
- Sick leave
- Work description
- Eliciting factors (irritants, sensitizers, contact urticants)
- Occupational and/or domestic exposure
- Effects of weekends and vacations
- Topical treatment

D. Diagnosis of Hand Eczema[1,8,30,37,38]

The investigation of a patient with hand eczema is generally very time consuming. A detailed history is mandatory and should include information about the family history of atopy, psoriasis, and eczema. The patient should be questioned about his personal history of atopy, atopic skin condition with dry and pruritic skin, psoriasis, eczema, and contact allergy to common sensitizers such as metals, perfumes, and cosmetics. In order to evaluate the course and severity of earlier occurrences of hand eczema, information should be elicited about date of onset, occupation when it started, and whether the eczema was related to exposure to certain occupational or domestic factors. Information about periodicity, progress, medical consultation, sick leave, and change of employment due to hand eczema should be included.

The following information concerning the current hand eczema should be recorded: date of onset, location and appearance at onset, the subsequent course, and sick leave. The patient's own description of the eczema and his/her opinion about the eliciting factors have to be registered. A detailed work description is very important. The occurrence of irritants, contact allergens, and contact urticants in the current occupational and domestic environment should be investigated. The effects of weekends and vacations on the eczema are important items of information. Any previous treatment should be noted in the history and all kinds of topical remedies should be listed. The extensive anamnestic information necessary in patients with hand eczema is summarized in Table 2.

Table 3
ETIOLOGIC EVALUATION IN HAND ECZEMA

Facts indicating exogenous etiology
Absence of endogenous factors
Exposure to relevant irritants, sensitizers, or contact urticants
Improvement or healing by avoiding exposure
Hand only involved
Asymmetric right/left distribution
Dorsal and interdigital localization
Facts indicating endogenous etiology
Atopy
Endogenous skin disorders, i.e., psoriasis
Symmetric, palmar lesions with vesicles, pustules, or hyperkeratosis

Examination of the hand dermatitis should focus on whether the dermatitis is eczematous in nature and whether the eczema is located only on the hands or should be considered as involvement of the hands in a generalized eczema. It should be noted whether the dermatitis has the morphological presentation of an endogenous hand dermatitis, such as pustular or psoriasiform volar hand dermatitis, pompholyx, hyperkeratotic dermatitis of the palms, hereditary keratodermia of the palms, and tinea, or if the dermatitis may be a secondary spreading of some other eczematous disorder. The entire skin should be examined in order to assess whether the hand dermatitis might be a part of another skin disease.

The testing procedures for a patient with hand eczema may be very extensive. The screening patch test battery will reveal the most common contact allergens, but additional patch testing with cosmetics, topical medicaments, and other environmentals may be necessary. If there is a history of immediate reaction, a test for contact urticaria should be carried out.

Based on the history, the clinical state, and test results, one should try to evaluate the relative importance of exogenous vs. endogenous factors in the development of the hand dermatitis. The following facts suggest the importance of exogenous factors: absence of intrinsic disposition for eczema; exposure to relevant irritants, allergens or contact urticants; considerable improvement or healing of the eczema when exposure is avoided; location limited to the hands alone; asymmetric right-left distribution; eczema located on the dorsal and interdigital aspects of the hands. Facts suggesting the significance of endogenous factors are a family or personal history of atopy; atopic eczema in locations other than the hands; and symmetrical, palmar lesions with vesicles, pustules, or hyperkeratosis. Information important in the evaluation of exogenous vs. endogenous factors in the etiology of hand eczema is summarized in Table 3. Some differential diagnoses which should be considered in cases of hand dermatitis are pompholyx, psoriasis, lichen ruber planus, porphyria cutanea tarda, acropustulosis, and scabies. A dermatophyte infection of the hands may mimic hand eczema and a culture for dermatophytes should be carried out if any suspicion arises.

E. Risk Factors

A recent investigation of hospital workers has provided us with information about the relative importance of some important risk factors for hand eczema.[13] In this prospective study with a follow-up time of 20 months, hand eczema was identified by means of a questionnaire. The nursing staff investigated (''hospital wet workers'') numbered 1613, predominantly women, and comprised all kinds of staff working with patients. The main groups were: ward maids, 964; nurses/assistants/nursing assistants, 508; physicians, 55; and dental staff, 31. This study found that hospital wet work increased the odds (relative risk) of developing hand eczema only twice in comparison with ''dry'' office work. Nursing children under 4 years of age and the lack of a dishwashing machine significantly increased the risk of contracting hand eczema. When these two domestic factors occur together, the risk will rise to the same level as for wet work which underlines the importance of irritant domestic factors. Furthermore, it was found that among women, a previous manifestation of hand eczema increased the odds (relative risk) of developing current hand eczema as much as 12.9 times if they went into wet work. The hands of most adult women have been exposed to some degree of irritant domestic or occupational work which sometimes, but not always, may have caused hand eczema. Thus, earlier hand eczema may be considered a positive usage irritancy test and a major indicator of a skin vulnerability factor which predisposes the person to irritant hand dermatitis. A previous hand eczema, a positive usage test, was found in half the subjects with a history of atopic dermatitis, in one quarter of the subjects with atopic mucosal symptoms (allergic rhinoconjunctivitis and/or allergic asthma), and in one fifth of the nonatopics. Subjects with a history of atopic dermatitis were found to suffer a more severe form of hand eczema with significantly higher figures for medical consultation, sick leave, termination due to hand eczema, early debut, permanent symptoms, and vesicular lesions. One very important observation is that as many as 35 to 40% of subjects with a history of atopic dermatitis managed to work in wet hospital work without developing hand eczema. To date no tests exist which can differentiate between these high- and low-risk individuals. From the practical point of view, a domestic or occupational usage test for trivial irritants seems the most reliable.

Individuals with atopic mucosal symptoms and nonatopics who suffer from an increased risk of developing irritant hand eczema have an inherited skin vulnerability factor defined as atopic skin diathesis.[36] This condition is defined as follows:

1. Dry skin
2. A history of a low pruritus threshold for two or three of the following nonspecific irritants: sweat, dust, rough material
3. White dermographism
4. Facial pallor/infraorbital darkenings

A usage irritancy test of the hands may be the most relevant indicator of the risk of developing hand dermatitis in wet work even for this condition.

A history of metal dermatitis, defined as an itching rash related to exposure to metal buttons, cheap jewelry, or wrist watches, increases the odds (relative risk) of getting hand eczema by a factor of 1.8 in women in hospital wet work. This increase is seen as a high risk level in subjects with vulnerable skin predisposed to irritant hand dermatitis and as a low risk level in others. Metal dermatitis may develop as a cause of contact allergy and probably develops through the irritant or other effects of metals. Metal dermatitis is more common among subjects with a vulnerable skin.[39,40]

By means of simple anamnestic information about earlier hand eczema, metal dermatitis, and atopic disease, it is possible to obtain highly differentiated prognostic information about hand eczema and its consequences among women engaged in wet hospital work, as can be seen in Table 4. Figures for medical consultation, sick leave, and termination are given in the table as a percentage of the predicted probability for current hand eczema. One important finding in the study was that most hand eczemas were mild and self-limiting and could be handled by self-treatment measures.

Two epidemiological studies of hospital workers, predominantly women, found that contact sensitivity to nickel, cobalt, balsams, and rubber chemicals was the most common. Contact sensitivity to medicaments and other allergens considered specific to the hospital environment was uncommon.[13,36] The relevance of common allergens such as nickel, cobalt, and balsams for current hand eczema among wet hospital workers is hard to assess. In one of the studies, it was found that most employees with contact sensitivity were unable to

Table 4
PREDICTED RELATIVE ODDS RATIOS (OR) AND PREDICTED PROBABILITY (PP) FOR HAND ECZEMA AND ITS CONSEQUENCES AMONG WOMEN IN WET WORK

	Hand eczema		Medical consul-	Sick leave,	Changed work,
Condition	OR	PP (%)	tation, PP (%)	PP (%)	PP (%)
AD	31	91	57	14	14
AMS	31	91	42	5.6	9.1
HMD-NA	23.1	88	28	5.6	9.5
AD	17.3	84	53	6.6	10
AMS	17.3	84	37	2.6	6.5
HHE-No HMD-NA	12.9	80	25	2.6	3.9
AD	2.4	43	48	22	14
AMS	2.4	43	34	9.5	9.1
HMD-NA	1.8	36	22	9.5	5.5
AD	1.3	30	44	11	10
AMS	1.3	30	30	4.3	6.5
No HHE-No HMD-NA	1.0	24	19	4.3	3.9

Note: AD, atopic dermatitis; AMS, atopic mucosal symptoms; NA, nonatopics; HMD, history of metal dermatitis; HHE, history of earlier hand eczema; OR, odds ratio; PP, predicted probability. medical consultation: AD, $p < 0.001$; HHE, $p < 0.01$; sick leave: AD, $p < 0.01$; HMD, $p < 0.05$; changed work: AD, $p < 0.01$.

correlate the current episode of hand eczema to any obvious exposure to the allergen.[13] This observation indicates that a constitutional skin vulnerability factor occurring predominantly in atopics in combination with trivial irritant factors is of the greatest importance for the development of hand eczema in wet hospital work even among most patients with contact allergy.

F. Colonization of *Staphylococcus aureus*

It is not uncommon for hand eczema to become secondarily infected, showing such clinical signs as serous exudation and yellow crusts. Bacteriological culture in these cases very often reveals *S. aureus*. In a recent study,[41] it was found that hand eczema was commonly colonized by *S. aureus* in high counts both in atopics and in nonatopics. The density of *S. aureus* was high even if the eczema showed no signs of clinical infection. The colonization of *S. aureus* seems to parallel the severity of the hand eczema. An interesting observation in this study was that successful tropical treatment of the eczema with a potent corticosteroid without antibiotics (clobetasol propionate) significantly reduced or eliminated the colonization by *S. aureus*.

The fact that hand eczema is commonly colonized by *S. aureus* in high counts is an important observation, especially when the eczema occurs in a hospital worker. There are indeed some reports in the literature of hospital personnel who have been the source of an epidemiological spread of infections with *S. aureus* in hospital wards.[42,43]

G. Treatment[1,30,37]

Simple guidelines for the treatment of hand eczema can be listed as follows:

- Minimize irritant exposure
- Avoid allergens
- Skin lubrication
- Topical corticosteroids
- Sick leave
- Change of work

Irritant contact dermatitis of the hands in individuals without an intrinsic disposition to eczema often appears after very heavy exposure to irritant factors. The eczema is usually clearly related to exposure, and elimination of the irritant generally heals the dermatitis. Irritant hand eczema in individuals with vulnerable skin is a much greater problem. Among these patients, the inevitable daily exposure to irritants in, for example, household work, may be sufficient to maintain the eczema. For these individuals, it is sometimes impossible to eliminate the irritant factors satisfactorily.

The effect of the elimination of allergens on hand dermatitis is extremely variable and unpredictable. When the dermatitis has developed as a result of a significant exposure to a relevant allergen, elimination will cure the dermatitis. Absence of improvement in spite of the elimination of contact allergens is very common and may be due to several causes:

1. The proven contact allergy is not relevant for the current hand dermatitis.
2. Elimination of a relevant allergen is incomplete.
3. The hand dermatitis is caused mainly by endogenous factors and the contact sensitivity is only a minor contributory factor.

Even if the relevance of a positive patch test is doubtful, the patient has to be told about the positive test and informed as carefully as possible about substances containing the specific allergen.

Irrespective of the cause, a hand dermatitis will be aggravated by irritants. It is therefore important to tell a patient with hand dermatitis how to avoid exposure to irritant factors. This information should include the following recommendations:[30]

1. Avoid contact with water, soaps, cleaning agents, and other irritating chemicals.
2. Wear waterproof vinyl gloves for protection, not rubber gloves.
3. Avoid peeling oranges with the bare hands.
4. Wear protective gloves when doing dirty or dusty work and gardening.
5. Use a dishwasher and washing machine as much as possible.
6. Avoid contact with solvents.
7. Wash the hands with very little soap and rinse carefully.
8. Do not wear rings as they very often make the dermatitis worse.
9. Protect the hands from cold and wind.
10. Do not use topical medication other than that prescribed by a physician.
11. Protect the hands for several months after the dermatitis has healed.

Skin lubrication with a prescribed emollient is very important in the treatment of hand eczema. The eczematous inflammation is effectively suppressed by topical corticosteroids in most patients. It is common to use a potent steroid at the beginning of the treatment for 1 or 2 weeks and then to continue with a less potent one for long periods. Treatment with peroral antibiotics may be necessary in cases of clinically infected eczemas. Dorsal, interdigital, irritant contact dermatitis of the hands responds much better to topical steroids than does palmar, often endogenous dermatitis, which may be very resistant and need other kinds of specific therapy.

Hand eczema among women in wet hospital work is common and we have to accept that periods of sick leave may be necessary. If the eczema does not heal in spite of adequate treatment and sick leave, it may be necessary for the patient to change his/her employment. It must, however, be emphasized that most hand eczemas among wet hospital workers are mild and can be dealt with using self-treatment measures.[13,36]

H. Prevention

Occupational counseling is important in the prevention of hand eczema. The following recommendations are made based on the findings of two epidemiological studies on hand eczema in hospital workers.[13,36]

1. Individuals with a personal or family history of atopic disease but no dry and itchy skin and no earlier hand eczema may be employed in such work.
2. Individuals with a history of atopic dermatitis who have had hand eczema should be advised not to chose employment involving wet hospital work.
3. Individuals with a history of atopic mucosal symptoms or a family history of atopy who have an atopic skin condition with a dry and itchy skin should be advised not to choose such work if they have had hand eczema during the last year.

These recommendations may serve as guidelines for the physician at the preemployment examination, but they have to be used with caution. It must be emphasized that the recommendations are based on statistical data in epidemiological studies. Ambition, education, and the social situation of the applicant are also important factors to take into consideration when deciding whether to hire an applicant or not.

People employed in wet hospital work must be told about the risk of developing hand eczema and should be supplied with simple guidelines concerning self-treatment measures for mild dermatitis of the hands. If the dermatitis persists, it is important to see a physician as soon as possible. If risk individuals are employed, a clinical follow-up during the first year is valuable.

The use of rubber gloves involves the risk of contact allergy and contact urticaria and should thus be restricted as much as possible. Vinyl gloves should be used for protection and examination purposes. "Hypoallergenic" rubber gloves are an alternative for many individuals with contact allergy to rubber chemicals. As contact allergy to perfume is common among women, emollients available at the workplace should not contain fragrance materials. It is a well-known phenomenon that some individuals cannot tolerate specific hand cleaning agents or emollients for reasons which cannot be documented by any testing. These problems may be partially eliminated if alternative agents are offered at the workplace. Before new chemicals are introduced into the hospital environment, predictive tests for irritant, urticariogenic, and sensitizing properties should be carried out. If alternatives exist, the least hazardous agents should be chosen. When allergenic chemicals have to be used, adequate information about personal protection should be given.

V. SUMMARY

In two epidemiological studies of wet hospital workers, predominantly woman, approximately 40% were found to have hand eczema identified by case history. Most eczemas were mild, periodic, and self-limiting and only a few gave rise to sick leave. Many sufferers could deal with the disorder using self-treatment measures. The odds (relative risk) of developing hand eczema in wet hospital work are only approximately twice as much as in dry office work. An increase in the irritant domestic load on women's hands through nursing babies and the simultaneous lack of a dishwashing machine will increase the risk of hand eczema as much as wet hospital work.

Risk individuals are overrepresented among patients consulting a dermatologist for hand eczema. Thus, a history of atopy, previous hand eczema, and metal dermatitis are more common than in the total population. Trivial irritants in wet and domestic work played an important part in the etiology of the current hand eczema in these patients. Two recent studies of hospital workers, predominantly women, have found that the most common contact sensitivity was to nickel, cobalt, balsams, and rubber chemicals. Contact sensitivity to medicaments and other allergens considered specific to the hospital environment was uncommon. It is hard to assess the relevance of common allergens such as nickel, cobalt, and balsams for current hand eczema among wet hospital workers. Rubber chemicals are probably the most important occupational contact allergen and rubber is the most important occupational contact urticant in hospital workers.

A constitutional skin vulnerability factor occurring predominantly in atopics in combination with trivial irritant factors is of the greatest importance for the development of hand eczema in wet hospital work. From a practical point of view, a previously manifested disposition toward hand eczema, a usage irritancy test, seems to be the most reliable means of identifying risk individuals. The odds (relative risk) of developing current hand eczema are 12.9 times higher for women with previous hand eczema than for those with no earlier hand eczema if they go into wet hospital work. This increase in risk is considerable and creates a subdivision of atopics and nonatopics in a high-risk group and a normal-risk group. Half of the subjects with atopic dermatitis, one quarter of the subjects with atopic mucosal symptoms, and one fifth of the nonatopics belong to this high-risk group.

One very important observation is that as many as 35 to 40% of subjects with a history of atopic dermatitis manage to work in wet hospital work without developing hand eczema. In women, a history of metal dermatitis increases the odds (relative risk) of developing hand eczema by a factor of 1.8. This increase is seen as one high-risk level in individuals with vulnerable skin and one normal risk level in others. Taking the case history, the clinical state, and test results as a basis, one should try to evaluate the relative importance of exogenous vs. endogenous factors in the development of each case of hand dermatitis. Improved understanding of the relative importance of various exogenous and endogenous factors is very important in this evaluation. By means of simple anamnestic information about earlier hand eczema, metal dermatitis, and atopic disease, it is possible to obtain a highly differentiated prognostic information about hand eczema and its consequences in women in wet hospital work.

Treatment of hand eczema must be based on an accurate diagnosis and assessment of the etiological factors. Irrespective of the cause, a patient with hand dermatitis should minimize irritant exposure. Exposure to relevant allergens should be avoided. Topical corticosteroids suppress the eczematous inflammation effectively in most patients. Refatting the skin with safe emollients is important in the treatment of all hand eczemas. If the eczema does not heal in spite of adequate treatment and sick leave, a change of work may be necessary. Hand eczema has, however, to be accepted as a likely consequence of wet hospital work. Immediate recognition, investigation, and treatment are important in order to reduce the need of changing the work. In the prevention of hand eczema, occupational counseling according to the given recommendations is important. A preventive program for hand eczema should include simple guidelines on adequate and safe prophylaxis and self-treatment measures for mild eczema. Exposure to rubber gloves should be minimized by using vinyl gloves whenever possible. Perfumes should not be included in emollients offered at the workplace.

VI. ACKNOWLEDGMENT

Professor Sigfrid Fregert is gratefully acknowledged for his valuable support and constructive criticism.

REFERENCES

1. **Burton, J. L., Rook, A., and Wilkinson, D. S.,** Eczema, lichen simplex, erythroderma and prurigo, in *Textbook of Dermatology,* Vol. 1, 4th ed., Rook, A., Wilkinson, D. S., Ebling, F. J. G., Champion, R. H., and Burton, J. L., Eds., Blackwell Scientific, Oxford, 1986, 367.
2. **Wilkinson, J. D. and Rycroft, R. J. G.,** Contact dermatitis, in *Textbook of Dermatology,* Vol. 1, 4th ed., Rook, A., Wilkinson, D. S., Ebling, F. J. G., Champion, R. H., and Burton, J. L., Eds., Blackwell Scientific, Oxford, 1986, 435.
3. **Lahti, A.,** Non-immunologic contact urticaria, *Acta Derm. Venereol. Suppl.,* 60, 91, 1980.
4. **von Krogh, G. and Maibach, H. I.,** The contact urticaria syndrome — an updated review, *J. Am. Acad. Dermatol.,* 5, 328, 1981.
5. **Fisher, A. A.,** Contact urticaria, in *Contact Dermatitis,* 3rd ed., Fisher, A. A., Ed., Lea & Febiger, Philadelphia, 1986, 686.
6. **Rajka, G.,** *Major Problems in Dermatology,* Vol. 3, *Atopic Dermatitis,* W. B. Saunders, London, 1975.
7. **Champion, R. H. and Parish, W. E.,** Atopic dermatitis, in *Textbook of Dermatology,* Vol. 1, 4th ed., Rook, A., Wilkinson, D. S., Ebling, F. J. G., Champion, R. H., and Burton, J. L., Eds., Blackwell Scientific, Oxford, 1986, 419.
8. **Rycroft, R. J. G.,** Occupational dermatoses, in *Textbook of Dermatology,* Vol. 1, 4th ed., Rook, A., Wilkinson, D. S., Ebling, F. J. G., Champion, R. H., and Burton, J. L., Eds., Blackwell Scientific, Oxford, 1986, 569.
9. **Fisher, A. A. and Adams, R. M.,** Occupational dermatitis, in *Contact Dermatitis,* 3rd ed., Fisher, A. A., Ed., Lea & Febiger, Philadelphia, 1986, 486.
10. **Orris, L. and Tesser, M.,** Dermatoses due to water, soaps, detergents and solvents, in *Occupational and Industrial Dermatology,* Maibach, H. and Gellin, G. A., Eds., Year Book Medical Publishing, Chicago, 1982, 23.
11. **Fisher, A. A.,** Hand dermatitis due to contactants, in *Contact Dermatitis,* 3rd ed., Fisher, A. A., Ed., Lea & Febiger, Philadelphia, 1986, 258.
12. **Fisher, A. A.,** Contact dermatitis in health personnel, in *Contact Dermatitis,* 3rd ed., Fisher, A. A., Ed., Lea & Febiger, Philadelphia, 1986, 515.
13. **Nilsson, E.,** Individual and environmental risk factors for hand eczema in hospital workers, *Acta Derm. Venereol. Suppl.,* 128, 1986.
14. **Dahlqvist, I, and Fregert, S.,** Occupational dermatoses in hospital personnel, *Berufs Dermatosen,* 18, 261, 1970.
15. **Rudzki, E.,** Occupational dermatitis among health service workers, *Derm. Beruf Umwelt,* 27, 112, 1979.
16. **Foussereau, J., Benezra, C., Maibach, H. I., and Hjorth, N.,** *Occupational Contact Dermatitis, Clinical and Chemical Aspects,* Munksgaard, Copenhagen, 1982, 277.
17. **Fisher, A. A.,** Contact dermatitis in medical and surgical personnel, in *Occupational and Industrial Dermatology,* Maibach, H. and Gellin, G. A., Eds., Year Book Medical Publishing, Chicago, 1982, 219.
18. **Fisher, A. A.,** Contact dermatitis from topical medicaments, *Semin. Dermatol.,* 1, 49, 1982.
19. **Nater, J. P. and de Groot, A. C.,** *Unwanted Effects of Cosmetics and Drugs Used in Dermatology,* 2nd ed., Elsevier, Amsterdam, 1985, 44 and 109.
20. **Suhonen, R.,** Contact allergy to dodecyl-di-(aminoethyl)glycine (Desimex®), *Contact Derm.,* 6, 290, 1980.
21. **Foussereau, J., Samsoen, M., and Hecht, M. Th.,** Occupational dermatitis to Ampholyt G in hospital personnel, *Contact Derm.,* 9, 233, 1983.
22. **Conde-Salazar, L., Guimaraens, D., Romero, L., and Harto, A.,** Allergic contact dermatitis to cyanamide (carbodiimide), *Contact Derm.,* 7, 329, 1981.
23. **Holdiness, M. R.,** Contact dermatitis to ethambutol, *Contact Derm.,* 15, 96, 1986.
24. **Ancona, A., Ramos, M., Suarez, R., and Macotela, E.,** Mercury sensitivity in a dentist, *Contact Derm.,* 8, 218, 1982.
25. **White, R. R. and Brandt, R. L.,** Development of mercury hypersensitivity among dental students, *J. Am. Dent. Assoc.,* 92, 1204, 1976.
26. **Cronin, E.,** *Contact Dermatitis,* Churchill Livingstone, Edinburgh, 1980, 683.
27. **Fregert, S. and Hjorth, N.,** Increasing incidence of mercury sensitivity. The possible role of organic mercury compounds, *Contact Derm. Newsl.,* 5, 88, 1969.
28. **Nater, J. P. and de Groot, A. C.,** *Unwanted Effects of Cosmetics and Drugs Used in Dermatology,* 2nd ed., Elsevier, Amsterdam, 1985, 119.
29. **Nilsson, E.,** Contact sensitivity and urticaria in "wet" work, *Contact Derm.,* 13, 321, 1985.
30. **Epstein, E.,** Hand dermatitis: practical management and current concepts, *J. Am. Acad. Dermatol.,* 10, 395, 1984.
31. **Agrup, G.,** Hand eczema and other hand dermatoses in south Sweden, *Acta Derm. Venereol. Suppl.,* 49, 61, 1969.
32. **Peltonen, L.,** Nickel sensitivity in the general population, *Contact Derm.,* 5, 27, 1979.

33. **Kavli, G. and Førde, O. H.,** Hand dermatoses in Tromsø, *Contact Derm.*, 10, 174, 1984.
34. **Lantinga, H., Nater, J. P., and Coenraads, P. J.,** Prevalence, incidence and course of eczema of the hands and forearms in a sample of the general population, *Contact Derm.*, 10, 135, 1984.
35. **Menné, T., Borgan, O., and Green, A.,** Nickel allergy and hand dermatitis in a stratified sample of the Danish female population: an epidemiological study including a statistic appendix, *Acta Derm. Venereol.*, 62, 35, 1982.
36. **Lammintausta, K.,** *Risk Factors for Hand Dermatitis in Wet Work,* Academic dissertation, Turku, 1982.
37. **Fregert, S.,** *Manual of Contact Dermatitis,* 2nd ed., Munksgaard, Copenhagen, 1981.
38. **Adams, R. M.,** The diagnosis of occupational skin disease, in *Occupational and Industrial Dermatology,* Maibach, H. and Gellin, G. A., Eds., Year Book Medical Publishing, Chicago, 1982, 3.
39. **Nilsson, E. and Bäck, O.,** The importance of anamnestic information of atopy, metal dermatitis and earlier hand eczema for the development of hand dermatitis in women in wet hospital work, *Acta Derm. Venereol.*, 66, 45, 1986.
40. **Möller, H. and Svensson, A.,** Metal sensitivity: positive history but negative test indicates atopy, *Contact Derm.*, 14, 57, 1986.
41. **Nilsson, E., Henning, C., and Hjörleifsson, M.-L.,** Density of the microflora in hand eczema before and after topical treatment with a potent corticosteroid, *J. Am. Acad. Dermatol.*, 15, 192, 1986.
42. **Ayliffe, G. A. J. and Collins, B. J.,** Wound infections from a disperser of an unusual strain of *Staphylococcus aureus, J. Clin. Pathol.*, 20, 195, 1967.
43. **Payne, R. W.,** Severe outbreak of surgical sepsis due to *Staphylococcus aureus* of unusual type and origin, *Br. Med. J.*, iv, 17, 1967.

Chapter 14

HAZARDS IN THE DENTAL ENVIRONMENT — CURING LIGHTS AND HIGH-SPEED HANDPIECES

Arne Lervik

TABLE OF CONTENTS

I. IRRADIATION FROM DENTAL LIGHT-CURING UNITS

By nature, the human eye is protected in many ways, such as its position in the skull, the form of the face, the eyelids, etc. In spite of this, exposed to certain environments with high light intensity and short wavelengths, the human eye may suffer permanent damage.[1,2] Visible light consists of light particles (photones) emitted in waves from the sun. The white daylight is a multiple of various wavelengths ranging from approximately 320 to 700 nm. The sensitivity of the eye to the various wavelengths is highly different, showing the highest sensitivity to light in the 430- to 600-nm area (Figure 1).

The sun radiates large amounts of ultraviolet (UV) light. This is light with wavelengths below 400 nm. Fortunately, due to the presence of the ozone layer surrounding the earth, only small amounts of such light penetrate this layer. Short-term exposure to UV radiation may lead to photoconjunctivitis and photokeratitis. Photokeratitis (also called photophtalmia) may also result from visual exposure to welding. "Welders flash" is due to inflammation of the cornea and conjunctiva.

Light with a slightly longer wavelength (300 to 400 nm) travels through the ozone layer more easily. These low-energy photons are the main reason for common sunburn. Like the skin, the eye is also sensitive to this UV radiation. However, unlike the skin's ability to develop tolerance to multiple exposures, the eye does not have this ability. More than half of this short-waved light is absorbed by the lens, a quarter of it is absorbed by the cornea, and less than 1% may reach the retina and cause irreversible damage.[1] The lens may tolerate such exposures, but only to a certain point. Multiple exposures may damage the proteins in the lens itself and just as the skin develops a brown color, a yellow pigmentation is observed in the lens. This pigmentation shows filtering properties, thus allowing less radiation to reach the retina. As the accumulation of yellow pigments in the lens increases, the vision becomes more blurred. This type of light may also cause the formation of cataract.[1] Cataract disease is today the main reason for developing blindness. The pigmentation blocks out the light from entering the retina. In the U.S. more than 41 million people over the age of 40 have developed cataracts and more than 400,000 are each year afflicted with it.

Looking at light with wavelengths in the 400- to 500-nm area, we find that this blue light is about one sixth as harmful as the UV radiation, but twice as intense as incident sunlight and about 100 times more able to pass through the lens to the retina. Very little of such light is absorbed by the ozone layer. Multiple exposures to this high-energy blue light are hazardous not only because it passes through to the retina virtually without any absorption in the lens, but also because of the scattering effect. We can see this effect in the earth atmosphere. The blue light is scattered among the atmospheric molecules and radiated to space and to the ground, thus causing the blue color of the sky. The same effect occurs in the eye, although on a smaller scale. The damaging mechanism to the retina is not fully known, but hypotheses for it include the accumulation of toxic products, photooxidative membrane damage, and metabolic disorders from extended overbleach. This effect is of a photochemical rather than a thermal nature and the damage may be considered as irreversible.

Premature aging of the retina and senile macular degeneration (the decreasing ability of the macular area of the retina to provide visual acuity) sets in. It has also been discovered that exposure to the blue as well as the UV light is additive. This means that although short-term exposures may be considered safe, a multiple of such exposures during the day can add up to one potentially hazardous exposure.[3]

In the early 1970s, the first light-cured composites were introduced to dentist. These materials were cured by exposing them to UV light. This is light with very short wavelengths, typically 320 to 365 nm (Figure 1). The light-emitting apparatus filtered away light with wavelengths below 320 nm, as these were known to be harmful to the human eye. Later, to meet the demands for greater curing depths, the modern high-intensity curing units were

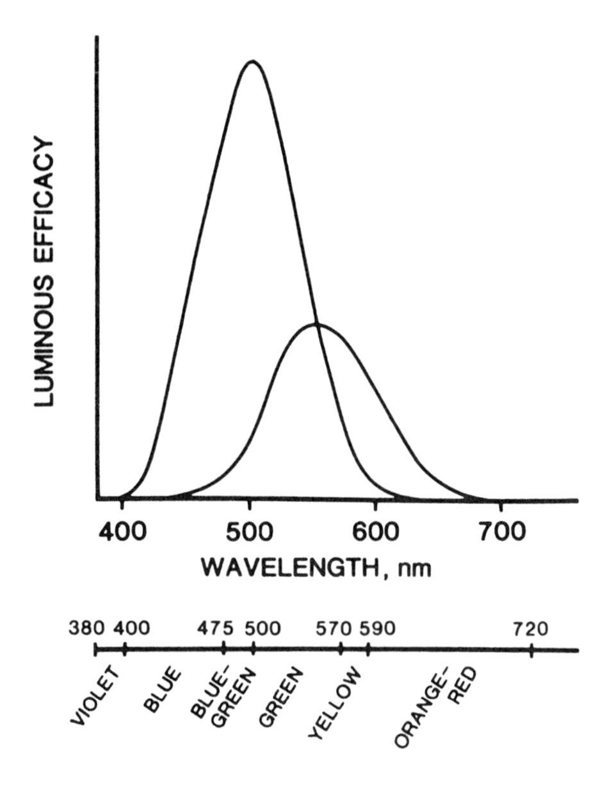

FIGURE 1. The sensitivity of the human eye to the different wavelengths of light. (Adapted from Ruyter, E. and Steien, K., *Norw. Dent. J.*, 16, 733, 1986.)

developed. They produced light with a longer wavelength than before: the high-energy blue light. The spectral energy between the different units varies as much as from 348 to 746 mW/cm^2.[4] The relative spectral energy between two different light curing units is shown in Figure 2. Composite resin systems are polymerized by light in the 470-nm area. Modern curing units generally transmit light from approximately 370 to 500 nm, some even beyond 600 nm. The specific range depends primarily on the filtering properties of the unit.[5]

More and more dentists, especially those working with bonding techniques in esthetic dentistry, are now exposed to ''blue light hazards''. Although very few cases of possible eye damage caused by the dental light-curing units have been reported, many dentists are experiencing tiredness, ''seeing spots on the walls'', and reduced night vision following extended bonding sessions.[6] To minimize the risks of retinal damage, the following steps should be acted on carefully:[7]

1. Never look directly into the exit window of the light-curing unit.
2. During long bonding sessions, dentist, auxillaries, and also the patient should wear protective glasses. Protective shields to fit onto the tip of the curing unit to minimize the blue light-scattering effect are also recommended.
3. Be sure to wear the right kind of protection. Ordinary sunglasses are unsuitable because of their poor filtering properties. Even good sunglasses filter up to only 30% of the UV and blue light emission. Protective filtering glasses should reduce this transmission by 99 to 100% for all wavelengths below 510 nm. The filtering effect should not be reduced during the years due to bleaching of the glasses. The tinted layer which produces the filtering properties should be molded into the material itself and not left as a coating on the surface only to be abraded in a very short time.

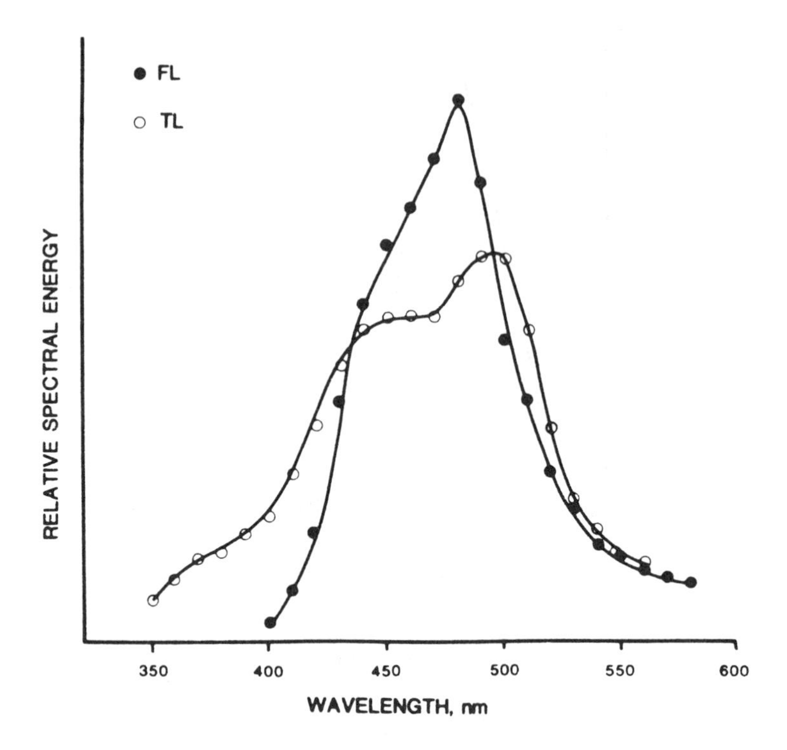

FIGURE 2. The relative spectral energy from two different light curing units: TL, Translux; FL, Fotofil. (Adapted from Ruyter, E., *Norw. Dent. J.*, 6, 271, 1985.)

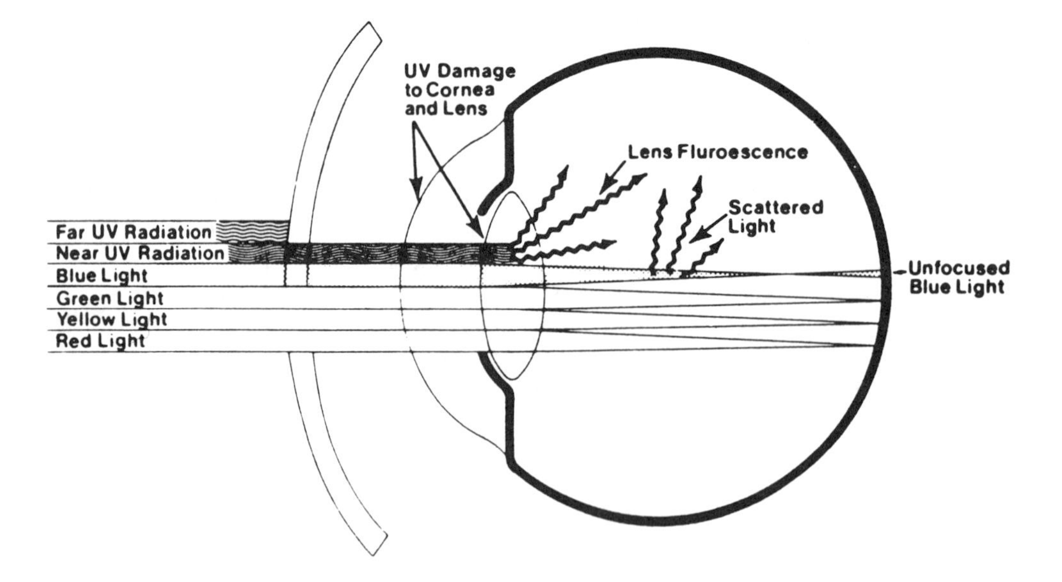

FIGURE 3. Transmission through ordinary sunglasses.

Figures 3 and 4 show light transmission through sunglasses and through what we may consider ''ideal'' protective glasses.

II. NOISE-INDUCED HEARING LOSS FROM THE USE OF HIGH-SPEED HANDPIECES

The human ear is exposed to various kinds of noise. Young adults may normally be able

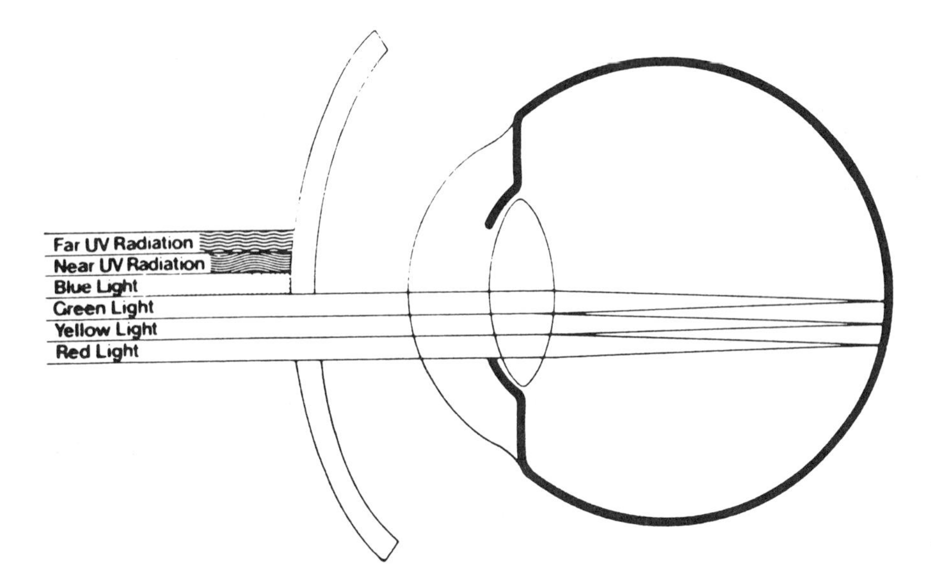

FIGURE 4. Transmission through glasses with ideal filtering properties.

to hear frequencies from 20 to 20,000 Hz. Most of the sounds relevant to the proper hearing of speech are in the range of 500 to 6000 Hz. The cochlear hair cells may be easily damaged when exposed to high-intensity noise such as gunfire, high-pressure drills, music from discothèques (which may exceed 100 dB) or airplane engines. The components involved in noise-induced hearing loss are the reduction of frequency resolution which means the ability to hear one sound in the presence of others; the difficulty to understand what is being said against a background of other sounds; and last a feature called tinnitus which is most commonly described as a singing or whistling sound.

The noise from high-speed handpieces has, since their introduction in the late 1950s, been suspected to be a hazard to the hearing of dentists using such handpieces. Early studies of what we call noise-induced hearing loss were inconclusive; some studies indicated that actual damage to the hearing may have resulted,[10-12] while others gave no hint of any such effect.[13,14] In 1965 Taylor et al.[15] made a carefully controlled study indicating conclusive evidence that damage to the hearing may result from such noise. Other studies made by Ward and Holmberg[16] and by Flottorp[17] confirmed this possibility of damage.

The early high-speed handpieces produced high-intensity noise of 95 dB or more at frequencies up to 8000 Hz, and as they became worn, they produced even higher frequencies.[21,22] These were handpieces of the ballbearing type. As the air bearing types were introduced, both the noise and frequency level dropped to 85 dB and 2000 to 6000 Hz, although some of these still maintained noise levels of up to 95 dB after being used for some time.[17]

What is the background for the potential hazard of noise-induced hearing loss? To understand this, we must look at the various elements of which the phenomenon called noise is composed:

1. Noise intensity is measured in decibels on a scale from 0 dB which means no noise at all up to 120 dB which is accepted to be hazardous even for a short period of time.
2. The frequency of the noise is measured in cycles per second (hertz). The ear is most vulnerable to frequencies in the 2000- to 5000-Hz area.
3. The duration of the noise is important. It has been shown[17] that intermittent noise is more hazardous that continuous noise.

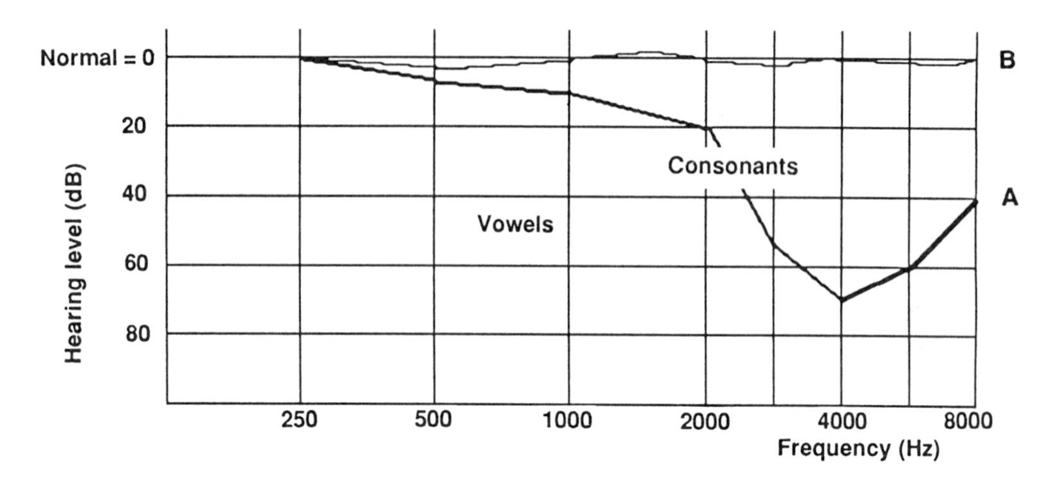

FIGURE 5. Audiogram of a typical case of noise-induced hearing loss (curve A) compared with a case with normal hearing (curve B).[17,19]

Noise dosage is composed of the above three elements. We may assume that the possible damaging effects result from the total amount of noise absorbed by the ear in a specified amount of time. Various countries may have different values for what is considered the safe value of daily noise dosage.

In the U.K.,[18] the maximum daily dosage considered acceptable is at present 90 dB for an 8-h working day. This is the same noise dosage as 93 dB for 4 h or 96 dB for 2 h or 102 dB for 30 min. In Norway the values are set to a maximum of 85 dB for 8 h, with 75 dB recommended for the same duration as absolutely safe. These different acceptable levels for different durations of exposure are relevant to the dentist in view of the fact that noise sources in the dental office are intermittent, with perhaps a total daily time from 30 min to 2 h of exposure to handpiece noise.[19]

Although 90 dB for an 8-h day may be considered safe, we should rather consider 75 to 80 dB as a more relevant limit as the individual tolerance for noise varies so much. Once damage by noise is done, there is no way of undoing it. The cochlear hair cells do not regenerate, and no known treatment has been demonstrated to alleviate this condition. Hearing aids can help to some extent by amplifying some of the speech sounds, but they also have the drawbacks of making speech inaudible in areas with high background sounds. Figure 5 shows a typical case of noise-induced hearing loss.

Modern high-speed handpieces are shown to have better properties in reducing the noise energy level than older ones (Figure 6). Recent studies[20] show this is mainly due to the reduction in speed from 450,000 to 300,000 rpm, to higher torque from larger turbine rotor fans, and in using rubber or silicone-insulated bearings. However, these properties are only fulfilled in new and not in used and mistreated handpieces. To prolong these properties, a proper and careful maintenance procedure is called for. This also includes a more frequent change of the endpiece itself when it becomes noisy.

More recently, some doctors have introduced new preparation techniques to save time and thus increase productivity. This calls for higher air pressure to increase rotating speed, which again produces more noise due to increased speed and faster wear of the endpiece. For these doctors, proper ear protection should be compulsory. Excellent ear protectors which filter out the higher frequencies and still maintain the ability to hear normal speech are available. In addition to this, frequent hearing tests are recommended. Regarding the noise emitted from high-speed handpieces, appropriate data are missing from the International Organization for Standardization (ISO) standard. The manufacturers should give more information about the noise levels and frequencies at which their handpieces operate.

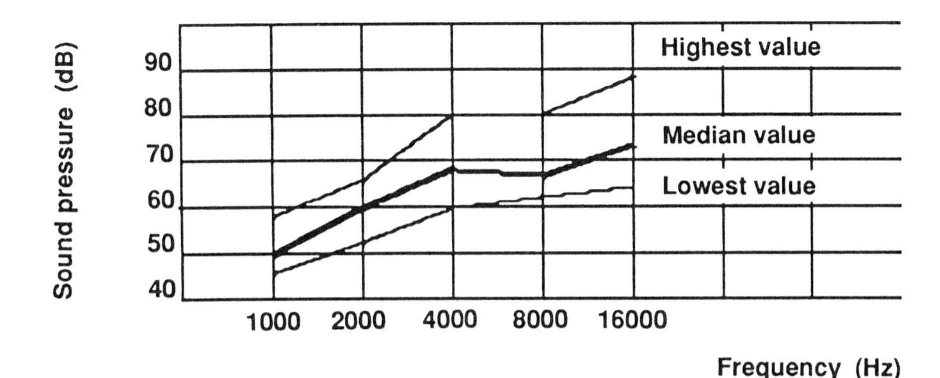

FIGURE 6. Spectrum of ten high-speed handpieces. (Adapted from Coles, R. R. A. and Hoare, N. W., *Br. Dent. J.*, 159, 209, 1985.)

REFERENCES

1. **Ham, W. T.,** Ocular hazards of light sources, *J. Occup. Med.*, 25, 1983.
2. **Lerman, S.,** *Ann. Opthalmol.*, 1982.
3. **Sliney, D. H.,** U.S. Army Environmental Hygiene Agency: standards for use of visible- and non-visible radiation on the eye, *Am. J. Optom. Physiol. Opt.*, 60, 1983.
4. **Cook, W. D.,** Spectral distributions of dental photopolymerization sources, *J. Dent. Res.*, 61, 1436.
5. Counsel on Dental Materials, Instruments and Equipment, *J. Am. Dent. Assoc.*, 112, 533, 1986.
6. **Lervik, A.,** Herdelamper gir Stråleskader, SPA-Info 59, Oslo, Norway, 1985.
7. **Ham, Mueller, Ruffolo, Clark, Moon,** *Invest. Opthalmol.*, 1974.
8. **Ruyter, E. and Steien, K.,** Lysherding av komposítter: karakterisering av beskyttelsesglass (briller), *Norw. Dent. J.*, 16, 733, 1986.
9. **Ruyter, E.,** Aktiveringslampers betydning for kvaliteten på komposítters herding, *Norw. Dent. J.*, 6, 271, 1985.
10. **Robin, I. G.,** Effect of noise made by the dental turbine drill, *Dent. Practitioner,* 10, 148, 1960.
11. **Bernier, J. L. and Knapp, M. J.,** Methods used in evaluation of high speed dental instruments and some results, *Oral Surg.*, 12, 234, 1959.
12. **Brenman, A. K., Brenman, H. S., Erulkar, S., and Ackerman, J. L.,** The effect of noise producing dental instruments on the auditory mechanism, *J. Dent. Res.*, 39, 738, 1960.
13. **Hopp, E. S.,** Acoustic trauma in high-speed dental drills, *Laryngoscope,* 72, 821, 1962.
14. **Weston, H. R.,** Severity of noise from high speed dental drills and hearing conservation, *Aust. Dent. J.*, 7, 210, 1962.
15. **Taylor, W., Pearson, J., and Mair, A.,** The hearing thresholds of dental practitioners exposed to air turbine drill noise, *Br. Dent. J.*, 118, 206, 1965.
16. **Ward, W. D. and Holmberg, C. J.,** Effects of high-speed drill noise and gunfire on dentists' hearing, *J. Am. Dent. Assoc.*, 79, 1383, 1969.
17. **Flottorp, G.,** Kan tannlegenes boremaskin fremkalle hørselsskade?, *Norw. Dent. J.*, 69, 362, 1959.
18. Health and Safety Executive, Code of Practice for Reducing the Exposure of Employed Persons to Noise, Her Majesty's Stationery Office, London, 1972.
19. **Coles, R. R. A. and Hoare, N. W.,** Noise-induced hearing loss and the dentist, *Br. Dent. J.*, 159, 209, 1985.
20. **Blomquist, G.,** Hørselsskador hos tandlakare, *Swed. Dent. J.*, 5, 227, 1986.
21. **Smith, A. F. J. and Coles, R. R. A.,** Auditory discomfort associated with the use of the air turbine dental drill, *J. R. Nav. Med. Ser.*, 15, 259, 1970.
22. **Wark, C.,** Deafness in dentists, *J. Oto. Laryngol. Soc. Aust.*, 79, 1383, 1969.

Chapter 15

EFFECTS OF VIBRATION IN THE HANDS

Ronnie Lundström

TABLE OF CONTENTS

I. INTRODUCTION

A. General Aspects of the Vibration Syndrome

The vibrations in many hand-held tools or work pieces may cause a complex of neurological, vascular, and musculoskeletal disturbances. These phenomena, occurring together or independently, are becoming widely recognized as an important occupational disease known as the vibration syndrome.[1] The best documented and most easily observed condition connected to hand-arm vibration exposure is a vasoconstrictive phenomenon, appearing as episodes of finger blanching together with tingling and numbness in the exposed hands. However, similar symptoms of finger blanching were first described by Raynaud, as early as the middle of the 19th century, on healthy nonvibration-exposed individuals (primary Raynaud's disease).[2] Ever since, many causes of finger blanching have been recognized, and these are usually referred to as "secondary Raynaud's phenomenon". It was not until the beginning of this century that vibration-induced finger blanching was first described in stone cutters and rock miners in Italy.[3] Scientific papers relevant to this specific injury are today quite numerous and this disease has been denoted by several different names, among which "Raynaud's phenomenon of occupational origin", "traumatic vasospastic disease", and "vibration-induced white fingers" (VWF) are the most commonly used.[1]

In the early stages of VWF, a person usually reports symptoms of neurological disturbances, such as slight tingling and numbness of his fingers. Later, after regular prolonged exposure, the tips of one or more fingers suffer attacks of finger blanching. The episodic vasoconstrictions are usually precipitated by an exposure to cold. With further exposure, in the magnitude of years, the area of finger blanching extends proximally to the base of the exposed fingers. During the periods of blanching, the fingers feel swollen, painful, and inflexible. Moreover, there is often in parallel a reduction in tactile sensibility, manual dexterity, and grip strength.[1,4,5] The vasoconstriction terminates after some minutes up to about 1 h, being due to the degree of severity of the VWF attack and the possibility of local warming of the hands. During this period, a reactive hyperemia occurs, usually painful and unpleasant.

The mechanisms behind these vasomotor reactions are still not known, but several theories have been suggested. All of these are based primarily on changes in peripheral vessels and/or in the peripheral nervous system.[6-10] Hyvärinen et al.[10] has suggested that a sympathetic reflex, triggered by an exposure to cold, loud noise, and/or vibration, entails vasoconstriction. As regards vibration, the activity originating in Pacinian corpuscles was suggested to elicit the sympathetic reflex. This type of mechanoreceptor has been shown to be very sensitive to vibrations.[11] At about 250 Hz, the stimulus amplitude required for evoking a response could be as low as 10^{-6} m peak to peak.

Apart from vasomotor reactions, other types of injuries seem to be associated with the use of vibrating hand-held tools. Peripheral neuropathy, changes in nerve conduction velocity, and changes in vibrotactile perception in the hands have been reported.[12-14] Musculoskeletal changes such as abnormally strong muscle fatigue and vacuoles, cysts, and decalcification of bone tissues are less commonly reported.[15,16]

B. Risk Assessment

The international standard, ISO 5349,[17] specifies methods for measuring and guidelines for the assessment of hand-transmitted vibration with frequencies up to about 1500 Hz. The guidelines for the assessment, specified in terms of frequency weighted acceleration for the dominant axes and daily exposure time, are based on our present knowledge on the dose-effect relationship.[18] The nominal gain of the frequency weighting network is to be 0 from approximately 6 to 16 Hz, and further on, up to about 1250 Hz, the acceleration signals will be attenuated by 6 dB per octave. The attenuation at still higher frequencies should be at least 12 dB per octave. The frequency weighted value can also be determined from one-third octave band data. In these calculations it seems very common that all contributions from bands with higher center frequencies than 1250 Hz are totally neglected. For any of these methods, the risk of getting vibration injuries is thus considered to decrease with increasing frequency, assuming that man is less sensitive to high-frequency components.

II. HEALTH EFFECTS OF VIBRATIONS IN CLINICS

At the end of the 1970s, complaints about numbness, stiffness, and reduced tactile sensibility in the fingers were noticed among dentists after practicing their profession for many years.[33] The inconvenience was above all of a subjective nature and was very similar to that expected among industrial workers using vibratory hand-held tools. Some of them had in fact been forced to resign due to a reduction in motor skills. The degree of ability for precision work could not for this reason be maintained. These experiences gave rise to the idea that vibrations in the handpieces used by dentists in their daily work might be the cause of the trouble.

Against this background, vibration measurements on a few high-speed tools (100,000 to 150,000 rpm) commonly used in modern dentistry were taken.[19] The outcome of this investigation showed levels as high as about 100 m/s^2 within the frequency range of 1 to 50 kHz.[19] Similar results have also been obtained in a later investigation.[20] To my knowledge, no data on vibration levels for the older and lower-speed types of drills can be found in the literature.

As earlier mentioned, all risk assessment shall be limited to vibrational frequencies below approximately 1.5 kHz, i.e., those covered by the ISO standard.[17] According to previous measurements, it seems like most of the vibrational energy in dental handpieces contains frequencies outside this range. Interestingly, the number of scientific papers attributed to this specific topic is also quite small. One reason might be that high vibration frequencies (above 1 to 2 kHz) are above the upper limiting frequency for man's tactile sensibility, i.e., we can't feel the vibrations.[11] These vibrations have therefore not been associated with the risk of getting vibration injuries. Another reason might be that accurate measurements of high-frequency vibration are considered to be quite difficult.[19-21] Against this background, most investigators have restricted their measurements and assessments to the lower frequency range, today covered by the ISO standard. Many of these investigators have, however, been studying health effects caused by tools containing very high-frequency components as well, such as pedestal grinders and percussive tools.[22,23]

It is known that the greater part of mechanical energy, at high frequencies, is absorbed by superficial tissues in direct contact with or close to the vibration source.[24,25] A risk of adverse effects on these tissues can therefore not be excluded. Cutaneous mechanoreceptors located in the glabrous skin of the hand are among the tissues that can be affected in this case, with a loss in sensibility and motor precision performance as a consequence.[11,26] Furthermore, these disturbances indicate an influence on the nervous system and may be an early sign in the development of more serious effects, such as VWF (see Section I, General Aspects of the Vibration Syndrome).[1]

In light of the suspicions that hand-transmitted vibrations at frequencies above 1.5 kHz may have a detrimental effect on man's peripheral nervous system, a group consisting of ten dentists has been tested with regard to the vibration perception in both of their hands.[27] They had all been using high-speed handpieces professionally during the last 5 years. None of them had any complaints indicating symptoms of vibration injuries. For the sake of comparison, a reference group was also examined. The perception threshold was defined as the amplitude at which the vibration was perceived by the subject. Ten test frequencies within the range of 40 to 400 Hz were used. For the dentists, all right-handed, a 3-dB-higher (approximately 1.4 times higher) vibration amplitude in average was needed for perception for the exposed hands compared to the others. For the corresponding reference group, no difference between hands could, as expected, be found. Thus, these findings support the idea that a long-term exposure to very high-frequency vibrations may have an untoward effect on man's tactile sensibility. However, the sensory losses seem to be quite small and the affected persons probably fully adapt to the loss at this stage. With prolonged vibration exposure, which may entail further reduction in tactile sensibility, the motor skills of the hand may be impaired, and disturbances may occur in the regulation of grip forces.[26] This is not only a serious handicap in most activities, but it may also imply an increased risk for accidents, for instance, those caused by undesired slipping. The mechanism behind the observed sensory losses is not at all clear. There are, of course, other factors than vibration in the dentist's working environment, such as an exposure to different kinds of chemicals like mercury and the use of heavy handgrips, that might be the cause of the trouble.

Interestingly, similar symptoms of numbness and loss in tactile sensibility have also been observed lately among dental technicians.[34] Recent measurements have confirmed that even this group is professionally exposed to high-frequency vibrations.[28] Vibration intensities between 10 to 30 m/s^2_{rms} were found within the frequency range of 5 to 20 kHz. Dental technicians are in general using tools with much lower rotational speed than the dentists, usually close to about 20,000 rpm. On the other hand, it is clear that most of the dental technicians use their tools for a much longer period of time during the working day. These observations give additional support to health effects caused by high-frequency vibrations.

In a study conducted later by Andersson et al.,[20] no increased prevalence of Raynaud's phenomenon could be found among 193 dentists compared to a group of 144 physicians. A prevalence of about 10 and 12.5% was found for the dentists and the physicians, respectively. However, among those suffering from Raynaud's phenomenon, the prevalence in their "working hands" was slightly greater in dentists than physicians.

The most common type of vibratory hand-held tools used by professionals in the medical service are probably ultrasonic devices for diagnostics or therapy. Devices for therapy usually have an operating frequency of around 1 MHz.[29] By using a special measuring technique, based on ultrasonic transducers, the radiation against the therapists hand has been estimated to be about 10 to 50 mW/cm^2.[29] This is below the recommendation of 100 mW/cm^2 considered acceptable by many.[30] For therapists, these recommendations may, however, be exceeded if the therapist's fingers are in contact with the metal part of the ultrasonic head or if the treatment is carried out in water.[29] In these particular cases, the radiation intensity can be as much as 100 times higher than the recommended limit. Ultrasonic devices designed for diagnostic purposes use radiation intensities which are much lower compared to the therapy devices.

In a follow-up study to the one on dentists mentioned above,[27] a group of ten therapists, using ultrasonic therapy devices, has also been tested with regard to vibro-tactile perception in their hands.[31] For comparison, a sex- and age-matched reference group with similar working tasks but not giving ultrasonic therapy was also examined. The outcome suggested that a certain reduction of tactile sensibility occurs for the therapists' hands compared with those of the references. From this study, together with the earlier one on dentists,[27] it can be concluded that both groups showed an impairment on the tactile sense for vibration-exposed hands. Even if these reductions are considered as quite small, they support the idea that vibration with high frequencies ($>$1000 Hz) has a detrimental effect on man.

III. VIBRATION ISOLATION

For hand-held tools, vibrations are probably generated through the following mechanisms: oscillations generated in mechanical parts of the motor and transmission, turbulence in the air flow through the tool, and frictional drags between tools and material worked on. In principle, there are three ways to solve a vibration isolation problem:

1. Reduce the source of vibration which usually implies mounting of a more or less well-balanced and vibration-free motor.
2. Reduce the transmission of vibration from the source through the tool, usually done by mounting different kind of dampers or clothing the pathway with viscoelastic and energy-absorbing materials.
3. Reduce vibration at or within the vibration receiver, normally difficult to carry out when the receiver is a human being.

Therefore, we have to concentrate our efforts to the first two alternatives.

A final solution to this problem is, of course, the development of vibration-free tools. Many of the manufacturers of dental and ultrasonic equipment are today trying hard to develop better tools in this matter. The possibility of offering vibration-free tools has become important in marketing. However, we must agree on the fact that vibration damping in this way may be a very difficult task and will certainly cost a great deal of money, both for the manufacturer and for the purchaser of such equipment.

Clothing the surface of vibration transmission with an isolating and damping material may sometimes be very efficient. In traditional vibration isolation activities, for instance in industry, this is done by mounting vibration dampers, usually consisting of metal springs or blocks of rubber. At this stage, it is important to note that dampers must be chosen correctly, otherwise there is a certain risk for making the situation worse.

The result of vibration isolation depends on several factors, such as the stiffness and viscoelastic properties of the damping material, vibration frequency, and the contact pressure on the material itself. In general, damping at high frequencies is easier to carry out compared to lower ones. For dental and ultrasonic equipment, vibrating at high frequencies, a soft and rubber-like material would work satisfactorily in most cases.

A material suitable for this application seems to be a silicon-rubber material named COLGRIP®. The material was first developed and designed for mounting on nonvibrating instruments, held by dentists and technicians in their daily service, for ergonomical purposes. As a consequence of reported risks for getting symptoms of vibration injuries as mentioned above, the material has also been tested as regards its ability to isolate vibrations with high frequencies. The results from these measurements are shown in Figure 1.[32] As can be seen, the transfer of vibration depends first of all on the frequency, but it also depends on the static pressure applied. The static pressures were chosen to correspond to those normally applied when using dental and ultrasonic equipment. Furthermore, it can be seen that the material behaves as a traditional mass-spring system. This means that damping occurs only at frequencies higher than the natural frequency of the systems, in this case above approximately 1000 Hz for the highest pressure. Moreover, at low frequencies, no damping at all occurs, i.e., the material will follow the vibrational source perfectly. In a frequency region between low and high frequencies, the material will behave in a resonant way, leading to an undesired amplification of the vibration amplitude. Consequently, for vibration isolation purposes, the material can only be used for equipment vibrating with higher frequencies than at least 2000 to 3000 Hz. Once again, the vibration characteristics should be carefully checked before mounting the COLGRIP® material on any tool; otherwise there is a risk of making the situation even worse.

Finally, while awaiting more vibration-free tools, it is worth using the material preventionally, at least for ultra-high-speed dental handpieces and for ultrasonic devices. It should also be noted that the equipment may be cleaned and sterilized keeping this COLGRIP® clothing on. Another advantage with COLGRIP® clothing is the increased friction against

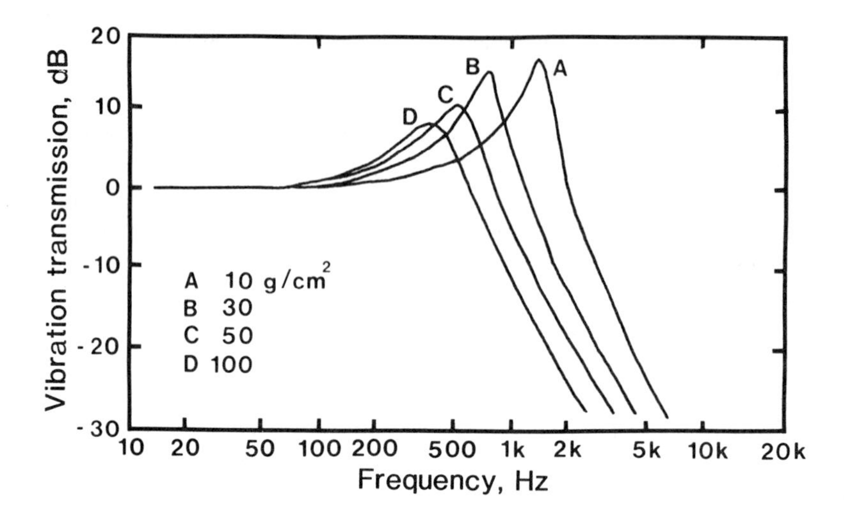

FIGURE 1. Transmission curves for COLGRIP® at different contact pressures applied on the material. The transmission (T) is expressed in decibels according to the formula: $T = 20 \cdot \log_{10}(a_r/a_t)$. Indexes r and t indicate that the vibration acceleration amplitude is measured at the receiving and transmitting side of the material, respectively. Therefore, transmission figures below 0 dB indicate damping of the vibration and figures above indicate vice versa. (For more details, see text and Reference 32.)

the skin which mostly entails a looser handgrip applied when using the tool. A looser handgrip reduces the mechanical coupling between the tool and the user which in turn leads to a lower amount of absorbed vibration energy in the hand and arm.

REFERENCES

1. **Taylor, W. and Brammer, A. J.,** Vibration effects on the hand and arm in industry: an introduction and review, in *Vibration Effects on the Hand and Arm in Industry,* Brammer, A. J. and Taylor, W., Eds., John Wiley & Sons, New York, 1982, 6.
2. **Raynaud, M.,** On local asphyxia and symmetrical gangrene of the extremities, Selected monograph, New Sydenham Society, London, 1888.
3. **Loriga, G.,** Il laboro coir mertelli pneumatica, *Boll. Ispett.,* lavoro 2, 35, 1911.
4. **Bannister, P. A. and Smith, F. V.,** Vibration-induced white fingers and manipulative dexterity, *Br. J. Ind. Med.,* 29, 264, 1972.
5. **Färkkilä, M. A., Pyykkö, I., Starck, J. P., and Korhonen, O. S.,** Hand grip force and muscle fatigue in the etiology of the vibration syndrome, in *Vibration Effects on the Hand and Arm in Industry,* Brammer, A. J. and Taylor, W., Eds., John Wiley & Sons, New York, 1982.
6. **Ashe, W., Cook, W. T., and Old, J. W.,** Raynaud's phenomenon of occupational origin, *Arch. Environ. Health,* 5, 333, 1962.
7. **Magos, L. and Okos, G.,** Raynaud's phenomenon, *Arch. Environ. Health,* 7, 341, 1963.
8. **Stewart, A. M. and Goda, D. F.,** Vibration syndrome, *Br. J. Ind. Med.,* 27, 19, 1970.
9. **Gurdjian, E. S. and Walker, L. E.,** Traumatic vasospastic disease of the hand, *JAMA,* 129, 668, 1945.
10. **Hyvärinen, J., Pyykkö, I., and Sundberg, S.,** Vibration frequencies and amplitudes in the aetiology of traumatic vasospastic disease, *Lancet,* 1, 791, 1973.
11. **Lundström, R.,** Responses of mechanoreceptive afferent units in the glabrous skin of the human hand to vibration, *Scand. J. Work Environ. Health,* 12, 413, 1986.
12. **Seppäläinen, A. M.,** Peripheral neuropathy in forest workers. A field study, *Work Environ. Health,* 9, 106, 1972.
13. **Seppäläinen, A. M.,** Nerve conduction in the vibration syndrome, *Work Environment Health,* 7, 82, 1970.
14. **Lidström, I.-M., Hagelthorn, G., and Bjerker, N.,** Vibration perception in persons not previously exposed to local vibration and in vibration exposed workers, in *Vibration Effects on the Hand and Arm in Industry,* Brammer, A. J. and Taylor, W., Eds., John Wiley & Sons, New York, 1982, 59.

15. **Kumlin, T., Wiikeri, M., and Sumari, P.,** Radiological changes in carpal and metacarpal bones and phalanges caused by chain saw vibration, *Br. J. Ind. Med.*, 30, 71, 1973.
16. **Färkkilä, M., Pyykkö, I., Korhonen, O., and Starck, J.,** Vibration-induced decrease in the muscle force in lumberjacks, *Eur. J. Appl. Physiol.*, 40, 1, 1980.
17. ISO 5349, Guidelines for the Measurements and the Assessment of Human Exposure to Hand-Transmitted Vibration, International Organization for Standardization, Geneva, 1985.
18. **Brammer, A. J.,** Relation between vibration exposure and the development of the vibration syndrome, in *Vibration Effects on the Hand and Arm in Industry,* Brammer, A. J. and Taylor, W., Eds., John Wiley & Sons, New York, 1982, 283.
19. **Lundström, R. and Liszka, L.,** High Frequency Vibrations — Ultrasound in Hand-Held Tools, Invest. Rep. No. 8, The Swedish National Board of Occupational Safety and Health, 1979.
20. **Andersson, R., Bergström, B., Dandanell, R., Engström, K., Grave, B., Magnusson, L., Nilsson, L., and Sjöstrand, O.,** Undersökning av Vibrationsmiljön och Skaderisker vid Arbete med Tandläkarborr med Höga Varvtal och Ultraljudsverktyg, Invest. Rep. TH 83:64, Saab-Scania, Sweden, 1983.
21. **O'Connor, D. E. and Lindquist, B.,** Method for measuring the vibration of impact pneumatic tool, in *Vibration Effects on the Hand and Arm in Industry,* Brammer, A. J. and Taylor, W., Eds., John Wiley & Sons, New York, 1982, 97.
22. **Dandanell, R. and Engström, K.,** Vibrations from riveting tools in the frequency range 6 Hz to 10 MHz and Raynauds phenomenon, *Scand. J. Work Environ. Health,* 12, 338, 1986.
23. **Seppäläinen, A. M., Starck, J., and Härkönen, H.,** High frequency vibration and sensory nerves, *Scand. J. Work Environ. Health,* 12, 420, 1986.
24. **Suggs, C. W.,** Vibration of machine handles and controls and propagation through the hands and arm, in *4th Int. Ergonomics Congr.*, France, July 6 to 10, 1970.
25. **Reynolds, D. D. and Angevine, E. N.,** Hand-arm vibration. II. Vibration transmission characteristics of the hand and arm, *J. Sound Vib.*, 51, 255, 1977.
26. **Westling, G. and Johansson, R. S.,** Factors influencing the force control during precision grip, *Exp. Brain Res.*, 53, 277, 1984.
27. **Lundström, R. and Lindmark, A.,** Effects of local vibration on tactile perception in the hands of dentists, *J. Low Frequency Noise Vib.*, 1, 1, 1982.
28. **Brändström, E. and Liszka, L.,** Measurements and Analysis of High Frequency Noise and Vibration, Invest. Rep. No. 43, The Swedish National Board of Occupational Safety and Health, 1985.
29. **Sävenstrand, U.,** Ultrasound radiation towards the therapist's hand, *Sjukgymnasten,* 10, 22, 1981.
30. WHO, Environmental Health Criteria. Ultrasound, World Health Organization, Geneva, 1981.
31. **Lundström, R.,** Effects of local vibration transmitted from ultrasonic devices on vibrotactile perception in the hands of therapists, *Ergonomics,* 28, 793, 1985.
32. **Lundström, R.,** Undersökning av Silikonmaterialet COLGRIP® ur Vibrationsisolerande Synpunkt (Investigation of the Vibration Isolating Properties of COLGRIP®, *Protetica Dental,* Invest. Rep., Stockholm, 1982.
33. Personal communication, 1978.
34. Unpublished data, 1984.

Chapter 16

MERCURY IN DENTISTRY

L. Gerhardsson and D. K. Brune

TABLE OF CONTENTS

I. INTRODUCTION

Ever since the first reported use of amalgam in dental restorations in 1818, the concern over toxic effects of mercury has continued to be an active issue in dentistry. The concern regards both the patient who is exposed to mercury from dental fillings and dental personnel who may be exposed, both from their own dental fillings and from relatively high levels of mercury vapor and mercury-containing particulates in the working environment. During the last decades much attention has been paid to contamination control in dental surgeries, resulting in a substantial improvement. As mercury-based amalgams continue to be the most common filling in current dental practice, there is a need to increase the knowledge and observance of the potential hazards.

II. PROPERTIES OF DENTAL MERCURY

Mercury is a heavy, silvery-white metal with a vapor pressure of 0.002 mmHg at 25°C. The equilibrium concentration of mercury vapor in air at 25°C is calculated to be 20 mg/m^3.[1] Its vapor pressure increases rapidly as the temperature rises. The solubility of mercury in water at 25°C is rather low, 56.2 μg/l.[2] Inorganic mercury has three oxidation stages: metallic (Hg°), mercurous (Hg_2^{2+}), and mercuric (Hg^{2+}) mercury.

Metallic mercury appears in three forms: solid, liquid, and vapor. Mercury is the only common metal liquid at room temperature. It easily forms alloys with many metals, such as gold, silver, and tin, which are called amalgams. Dental amalgam is made by mixing alloy powder and mercury. Conventional amalgam is formed by mixing metallic mercury (ratio of approximately 1:1) with an alloy powder containing about 70% silver, 25% tin, and a small amount of copper and zinc. Later a nongamma-2 amalgam with a higher copper concentration was developed. Copper amalgam alloy containing approximately 65% mercury, 30% copper, and a few percent cadmium and zinc is also available. This alloy is in limited use due to its susceptibility to corrosion.

The most important organometallic compounds connected with human exposure are monomethyl (CH_3Hg^+) and dimethyl [$(CH_3)_2Hg$] mercury. Organic mercury compounds have been used for seed treatment and are still in use as fungicides and algaecides in paints and plastics and as disinfectants.

III. EXPOSURE IN DENTAL OFFICES

A. Mercury Contamination

Mercury is emitted from several sources in dental surgeries, such as from accidental spills, mechanical amalgamators, ultrasonic amalgam condensers, amalgam mulling in hand, squeezing out excess mercury from freshly mixed amalgam, hot air sterilizers (amalgam-contaminated instruments), storage of mercury and amalgam scrap, and removal of old fillings.[3,4] Diffusion is generally reduced by general ventilation. Local exhaust ventilation is, however, far more efficient.

Table 1
MERCURY THRESHOLD LIMIT VALUES (TLVs)

Mercury	TWA ($\mu g/m^3$)	STEL ($\mu g/m^3$)
Alkyl compounds	10	30
All forms except alkyl	—	—
Vapor	50	—
Aryl and inorganic compounds	100	—

Note: TLV-TWA, threshold limit value-time-weighted average; TLV-STEL, threshold limit value-short-term exposure limit; TLV time-weighted average adopted by the ACGIH for a normal 8-h workday and 40-h work week; STEL, a 15-min time-weighted average.

Modified from American Conference of Governmental Industrial Hygienists, Threshold Limit Values and Biological Exposure Indices for 1986-1987, ACGIH, Cincinnati, OH, 1986.

Mercury vapor arising from spills on the floor is a major cause of mercury contamination.[2] Mercury droplets collected in carpets and floor cracks are not easily removed by routine cleaning. Elevated room temperature increases vaporization. General ventilation is not flexible enough to ventilate point sources. Dust in dental laboratories is emitted from various operations, such as cutting, grinding, polishing, centrifuge casting (gold), soldering, and preparation of gypsum models. The dust contains several substances other than mercury, e.g., gold, chromium, cobalt, nickel, porcelain, and denture base materials.

B. Reported Mercury Levels

The National Institute for Occupational Safety and Health (U.S.) (NIOSH) standard for mercury is currently 50 $\mu g/m^3$ (Table 1).[5] In 1980, a World Health Organization (WHO) Study Group recommended a time-weighted average of 25 $\mu g/m^3$ for long-term exposure.[6] It has been estimated that at least 10% of all dental offices in the U.S. exceed a level of 50 $\mu g/m^3$.[7-9] In a Canadian study[10] of 375 operatories, 56% of the measurements in the general location of the dental chair showed mercury levels between 10 and 49 $\mu g/m^3$. Only about 3% of the measurements showed concentrations higher than 50 $\mu g/m^3$. In the U.K., Lenihan et al.[11] reported that in about 10% of 23 dental surgeries investigated, mercury vapor levels exceeded 50 $\mu g/m^3$. About 10% of 200 dental surgeries studied by Nixon et al.[12] displayed ambient air levels greater than 20 $\mu g/m^3$. In three operatories (1.5%), the mercury vapor levels exceeded 50 $\mu g/m^3$. In a recent study from 82 dental clinics in northern Sweden,[13] the average amount of mercury vapor in the breathing zone was below 25 $\mu g/m^3$. The mercury concentrations, however, varied considerably over the room area. The highest values were registered at the floor level of the amalgam preparation area, above the amalgam preparation area, above and at the floor level of the amalgam scrap container, and above the waste container. The exposure was higher in private dental care than in public dental care (see also Chapter 18).

C. Mercury Levels During Removal of Old Fillings

Substantial amounts of mercury vapor and mercury-containing particulates are released during the removal of old amalgam fillings. The exposure close to the dental chair varies considerably with the technique used (dry or wet cutting, etc.). Appropriate ventilation combined with the use of water spray kept the exposure well below 50 $\mu g/m^3$.[14] On the other hand, dry high-speed drilling with improper ventilation may temporarily raise the exposure levels to several hundred micrograms per cubic meter. In some instances the peak values have exceeded 1 mg/m^3.[15,16]

IV. METABOLISM

A. Inorganic Mercury

1. Elemental Mercury

The daily intake of inorganic mercury from food and water is low and will normally not exceed 10 μg/d. Recent data indicate that release of mercury vapor from amalgam fillings may be a dominant source of nonoccupational exposure. Occupational exposure varies within the wide limits and must be related to the working conditions. Inhaled mercury vapors are readily absorbed to about 80%.[17] Ingested liquid metallic mercury is poorly absorbed from the gastrointestinal (GI) tract. Skin absorption is low, but no quantitative data are available.

Dissolved mercury vapor in the body is transported by the blood to the brain[18] and other organs. Mercury vapor also penetrates the placental barrier.[19] In blood, the major part is dissolved in red blood cells which have a much larger capacity to dissolve elemental mercury than plasma. Elemental mercury in blood cells and tissues is rapidly oxidized to mercuric mercury under the influence of the hydrogen peroxidase-catalase complex.[20,21] Ethanol and aminotriazole inhibit catalase mercury oxidation,[22,23] resulting in decreased uptake of mercury vapor by inhalation. In the brain, mercury is predominantly distributed to grey matter and accumulated in some nuclei in the brain stem and in some parts of the cerebellum.[24] The retention time varies widely among different organs, with a biological half-life from a few days to months. Brain, kidneys, and the testicles have the longest retention times. The biological half-life for the brain may exceed several years. After oxidation, elemental mercury is mainly excreted as mercuric mercury in urine and feces. Exhalation of small quantities of mercury vapor have been demonstrated in animals[25] and humans.[26]

2. Mercuric Mercury

Quantitative data from the uptake after inhalation of mercuric mercury are still lacking. Divalent mercuric mercury is absorbed to 5 to 10% in the adult GI tract. Animal experimentation data indicate that a higher percentage may be absorbed in neonates and infants. Some skin absorption may occur, but the mechanism is unclear.

In the blood, mercuric mercury is distributed between red cells and plasma, a somewhat higher percentage has been reported in plasma.[27] In erythrocytes mercury is bound to the sulfhydryl groups of hemoglobin and glutathione[28] and in plasma it is bound to albumin and macromolecules. Contrary to elemental mercury, mercuric mercury does not readily pass the blood-brain barrier or the placental barrier. The dominating pool is the kidney cortex holding about 90% of the body burden at steady state. Second is the liver with the highest concentration in the periportal areas.

Urine and feces are the dominating excretion routes. Mercury is excreted by the liver through the bile.[29] Partition between urine and feces is dose dependent. At higher dose levels, a larger fraction is excreted by urine. A minor fraction is excreted by sweat and saliva. A small amount of mercuric mercury may be reduced to mercury vapor and exhaled. Few reports are available on the biological half-life of the total body burden in humans. Rahola et al.[30] reported a half-life of 40 to 45 d for the whole body, after an oral tracer dose. Ingested mercurous or monovalent mercury is rapidly transformed into mercuric mercury and elemental mercury in the body tissues.

B. Organic Mercury

In food a large part of the mercury is present in the form of methyl mercury. The main source is fish. The daily fish consumption ranges from none to 500 g or more per day. In Sweden the daily intake of methyl mercury from normal fish varies between 1 and 20 μg/d.[31] The methyl mercury in food and seafood is explained by the biotransformation of elemental and mercuric mercury in the terrestrial and aquatic environments. High fish consumption from mercury-contaminated waters has caused intoxications (e.g., Minamata and

Niigata, Japan, 1953-1966). A high consumption of fish containing up to 20 mg of methyl mercury per kilogram may result in a daily intake of up to 5 mg.[31]

Inhaled methyl mercury is absorbed to about 80% in the lung. Ingested methyl mercury is absorbed to more than 95% from the GI tract. Skin absorption can occur. More than 90% of the methyl mercury in the blood is bound to hemoglobin of the red blood cells. The rest is bound to plasma proteins. Methyl mercury penetrates the blood-brain barrier and accumulates in the brain. The brain contains about 10% of the body burden.[32] Methyl mercury also concentrates in liver and kidneys. In other body tissues, methyl mercury is rather evenly distributed. Methyl mercury penetrates the placental barrier[33] and particularly accumulates in the brain of the fetus. Methyl mercury is eliminated via the liver into the bile (enterohepatic circulation) and via the kidney into the urine. The intestinal microflora demethylates a fraction of the methyl mercury secreted into the intestinal lumen; most of inorganic mercury is excreted in the feces. As inorganic mercury is absorbed only to 5 to 10%, the demethylation contributes to an increased excretion. About 90% of the total human excretion of methyl mercury is by feces.

Methyl mercury is also excreted in breast milk. The milk concentration is about 3 to 5% of the maternal blood levels.[34] The amount of methyl mercury in hair is proportional to the blood concentration at the time of incorporation. The net excretion in humans, about 1% of the body burden, corresponds to a biological half-life of 70 d under nontoxic conditions.[35]

C. Biotransformation of Inorganic Mercury

Inorganic mercury is methylated *in vivo* in the gut of rats. *In vitro* experiments have shown methylation in the human gut.[36] Oral bacteria (streptococci) methylate inorganic mercury as shown in *in vitro* studies.[37] The degree of mercury methylation in the oral cavity, however, is so far not evaluated.

V. HEALTH EFFECTS

A. Inorganic Mercury

1. Elemental Mercury

The lung is the critical organ after acute accidental exposure to high concentrations of mercury vapor. The vapor causes erosive bronchitis and bronchiolitis with interstitial pneumonitis. The central nervous system (CNS) is the critical organ after extended exposure to toxic levels of mercury vapor. At low levels, increasing doses may develop an unspecific, asthenic vegetative syndrome (so-called micromercurialism) with weakness, fatigue, anorexia, loss of weight, and disturbance of GI functions.[38] At higher exposure levels, tremor appears, starting as a fine trembling of peripheral parts like the fingers, eyelids, and lips. It may progress into generalized tremor with violent spasms of the extremities. These symptoms are accompanied by severe behavioral and personality changes, increased excitability, loss of memory, and insomnia, which may develop into depression. Delirium and hallucination may occur. After severe intoxication, persistent disturbances of the nervous system are common.[39] Repeated occupational exposure to mercury concentrations in the air exceeding 100 μg/m³ may produce mercurialism.[38] Micromercurialism has been reported to appear at concentrations between 10 and 100 μg/m³.[31]

2. Mercuric Mercury

The acute and long-term effects of mercuric salts are primarily GI disturbances and renal damage manifested as a tubular dysfunction and in severe cases tubular necrosis. Another type of renal damage may also occur. In animal experiments, inorganic mercury has induced antibodies against the basal membrane of the glomeruli.[40,41] This autoimmune reaction, possibly with a genetic component, develops a nephrotic syndrome with proteinuria[42-44] and signs of glomerular nephritis. Sensitization may occur with allergic symptoms such as dermatitis and bronchial asthma. The lethal dose in man is about 1 g of mercuric salt. Mercury concentrations in the kidney between 10 and 70 mg/kg have been reported from intoxicated humans with renal injury. Normal kidney levels are usually below 3 mg/kg.[31]

B. Organic Mercury

The brain is the critical organ after methyl mercury exposure. Prenatal intoxication leads to infantile cerebral palsy[35] characterized by ataxic motor disturbances and mental symptoms. In less severe cases, dose-related psychomotor retardation has been noted in infants.[45] Autopsy findings show a symmetrical atrophy of the cerebrum and cerebellum. An incomplete or abnormal migration of neurons to the cerebellar and cerebral cortices and deranged cortical organization of the cerebrum have been reported.[46] Postnatal poisoning results in symptoms characterized by sensitivity disturbances with paresthesia in the distal extremities, the tongue, and around the lips. After severe intoxications, ataxia, concentric constriction of the visual field, impairment of hearing, and extrapyramidal symptoms may occur. Clonic seizures have been reported from very severe cases. Histological findings in the CNS are mainly general neuronal degeneration in the cerebral cortex with gliosis accompanied by atrophy of the cerebral cortex.

Body burdens below 0.5 mg/kg of body weight, which corresponds to blood levels of less than 200 μg/l and mercury values in hair below 50 mg/kg, are not likely to give detectable neurological disturbances in adults.[31] However, such levels of methyl mercury exposure in pregnant females may result in inhibited brain development of the fetus with psychomotor retardation of the child.

VI. ENVIRONMENTAL MONITORING

Environmental monitoring quantifies exposure by measurements outside the human body. Humans are exposed from air, water, food, and soil; the exposure varies over time. By environmental monitoring, specific sources of exposure are identified and exposure levels are related to preset permissible levels. The effectiveness of environmental or industrial control measures may also be evaluated. Quantified exposure doses are used in epidemiological studies and are related to health effects. They are also used for risk assessment.[47]

The respiratory tract, the GI tract, and the skin are the routes for uptake of metals through air, food, water, and product handling. The body burden depends on the difference between total uptake and excretion by exhaled air, urine, feces, and sweat. As air is the dominant exposure source, air contaminants — gaseous and particulates — must be evaluated. Both fixed sampling stations and mobile personal samplers or monitors are used. Stationary monitors — both at the work place and for ambient monitoring — sample air representative of a certain area. Personal samplers carried by the exposed individual sample breathing air. Size-selective samplers simulate deposition of particles as inspirable particulate mass, thoracic particulate mass, and respirable particulate mass. Less soluble particles cleared through mucociliary action contribute to intake by ingestion. Food can be contaminated by improper personal hygiene. Smoking contributes to inhalation exposure.

Laboratory mercury analyses are undertaken by various techniques. Colorimetric, cold vapor atomic absorption (CVAA), neutron activation analysis (NAA), and various types of plasma atomic emission spectrometry (AES) are in current use. On-site analyses are undertaken by mobile instruments. Small hand-carried direct reading instruments are used for measurements of mercury vapors. In the mercury vapor detector, mercury atoms absorb the characteristic UV wavelengths from a mercury lamp. The reduction of the intensity is registered by a photocell. The response is proportional to the mercury concentration. The instrument is calibrated for direct reading. Other types of direct-reading instruments are available, e.g., colorimetric indicators. Gas detectors measure trace concentrations of gases and vapors such as mercury.[48] The cost for the measurements should be observed. A practical rule is to not use more expensive analytical tools than are needed for receiving the wanted information.

VII. BIOLOGICAL MONITORING

A. Inorganic Mercury

The mercury concentrations in blood and urine correlate to prolonged exposure when compared on a group basis. To eliminate the impact of methyl mercury from fish consumption, it is necessary to distinguish between total mercury and inorganic mercury in blood. An alternative is to analyze the mercury concentration in plasma and blood cells separately as methyl mercury primarily accumulates in the red cells.

Total mercury in urine mainly reflects the excretion of inorganic mercury; methyl mercury is only excreted in urine to a very limited extent. The individual urinary excretion varies considerably from day to day and during the day. During ongoing exposure, blood and urine mercury levels are related to recent exposure. After cessation of exposure, urine mercury concentrations are related to the kidney concentrations of mercury. To adjust for variation in urine flow rates, urine mercury levels are related to the creatinine concentration. No methods are at present available for routine biological monitoring in a critical organ such as the brain.[47]

B. Organic Mercury

Under steady-state conditions, mercury levels in the blood are correlated both to the daily intake of methyl mercury and to the accumulated concentrations of methyl mercury in the brain. The mercury concentration in red cells is the most reliable indicator of methyl mercury body burden and brain concentration because more than 90% of the methyl mercury in blood is bound to the erythrocytes. Hair is a good indicator which reflects past interval blood mercury levels when analyzed longitudinally segment by segment. Hair samples may be analyzed by the use of nondestructive X-ray fluorescence (XRF) or particle induced X-ray emission analysis (PIXE) and may be used to estimate changes in brain levels.[47]

In humans the blood to brain ratio of mercury is usually between 5 or 10:1, the ratio of blood to hair is about 1:250, and the ratio between red cell and plasma is about 20:1.[47] There are considerable individual differences as well as large species differences. Blood concentrations reflect both recent exposure and body burden of methyl mercury. Hair monitoring can be used for retrospective exposure evaluations.

C. Levels in Human Tissues and Body Fluids

1. Blood

Blood mercury levels vary considerably according to several investigators.[49-51] Normal blood total mercury levels usually range from 1 to 7 μg/l. Blood mercury concentrations are dependent of the daily fish consumption. In subjects with an extremely high consumption of contaminated fish, the blood mercury values may be 10 to 100 times higher. Workers occupationally exposed to metallic mercury at the threshold limit value have values about 30 μg/l.[52] Dentists had elevated whole blood concentrations of both total mercury and methyl mercury compared to controls in a study by Cross et al.[53]

2. Urine

Occupational mercury exposure after World War II was mainly checked by urine sampling. Gradually, urine analyses were replaced by air monitoring. Normal urine total mercury levels are about 5 μg/l (5 μg/g of creatinine). The values in urine are relatively independent of

Table 2
MERCURY LEVELS (μg/g) IN HAIR AND NAILS

	Dentists (n = 87)		Dental assistants (n = 80)		Controls (n = 16)	
Specimen	**Median**	**Range**	**Median**	**Range**	**Median**	**Range**
Head hair	9.6	1.2—160	9.4	1.8—288	4.0	1.2—9.9
Pubic hair	3.5	0.7—404	2.9	0.6—27	1.4	0.7—6.5
Fingernails	68.0	2.7—3070	16.9	1.3—2580	3.4	1.4—15.0
Toenails	3.0	0.8—181	6.5	0.8—102	1.4	0.6—4.5

Modified from Lenihan, J. M. A., Smith, H., and Harvey, W., *Br. Dent. J.*, 135, 365, 1973.

the fish consumption. Workers occupationally exposed to metallic mercury at the threshold limit value reach urine mercury levels around 80 μg/l.[52] In a large U.S. study, the mean urine mercury level of the dental personnel was 15.3 μg/l compared to 3.4 μg/l for the controls.[2] Recently, a Swedish study[54] showed higher urine mercury levels for dental personnel (n = 505) than for the control group (n = 41). However, the values were all low and well below the proposed occupational exposure limit of 28 nmol of mercury per millimole of creatinine[6] for subjects exposed to mercury vapor.

3. Hair and Nails

Analyses of mercury levels in hair and nails are increasingly used for exposure control. Examples of measured levels are given in Table 2. In the study by Lenihan et al.,[11] the dental personnel displayed higher mercury concentrations in hair and nails compared to controls. In head hair, pubic hair, and fingernails, the surgery assistants showed lower mercury levels than the dentists. However, the toenail levels were higher in the former group, probably due to a more common use of sandals and open-toed shoes.

4. Saliva

Saliva is not routinely used for biological monitoring of mercury. However, studies report higher mercury levels in the saliva of dental staff members than in controls.[2] It has been found that persons with amalgam fillings show somewhat higher mercury saliva levels than persons without fillings.[55]

5. Brain

The total mercury concentration in the brain in Swedes with no known occupational exposure is about 10 μg/kg of wet weight. In subjects with occupational exposure to metallic mercury or to methyl mercury, the brain concentrations may be ten times higher or more.[52]

VIII. HAZARDS TO DENTAL STAFF

The risks to be considered are external and internal exposure to mercury from the working environment and from their own dental fillings. Raised mercury levels have been reported in urine,[54,56,57] blood,[49] saliva,[58] and hair and nails[11] from asymptomatic dental personnel. In a survey of 501 male general practitioners,[59] no difference in hand steadiness (visual and motor coordination) was found between dentists with blood mercury levels below 10 μg/l and in those with increased levels (15 to 57 μg/l). In a study of 111 dentists and assisting personnel,[60] no correlation was noted between blood mercury levels and handwriting tremor.

Mercury levels in different body tissues have also been determined. In a study[61] of 298 dentists, 23 had head and wrist mercury concentrations exceeding 20 μg/g of tissue. Of the latter group, 30% showed electrophysiological evidence of a subclinical polyneuropathy not observed in the control group. Conduction velocity along the sensory nerve fibers was delayed in an individual intoxicated by elemental mercury.[62] Behavioral changes (erethism) have also been reported.[63,64]

Nylander et al.[65] observed considerably higher total mercury concentrations in the pituitary glands of three deceased dentists when compared with four controls. However, the clinical significance of these findings was not clear nor was it known if these high values were related to recent or earlier exposure. Compared to controls, a twofold risk of developing brain tumors (glioblastoma) has been observed for dentists and dental nurses.[66] Mercury, chloroform, and radiography were the suspected environmental factors.

The relationship between mercury exposure and pregnancy outcome among dental professionals has been studied by Brodsky et al.[67] There were no increased rates of spontaneous abortions or congenital abnormalities in the children of men and women who were occupationally exposed to low vs. high levels of mercury. Occupational mercury exposure may in certain cases cause allergic reactions among practicing dentists and assistants (Chapter 13). However, a complete reversal of the allergic symptoms usually occurs upon removal of mercury contamination. Fatal intoxications of dental personnel are rare. However, a rapidly fatal nephrotic syndrome of a 42-year-old dental surgery assistant is reported.[68]

Simple clinical diagnostic tests are not available for the evaluation of mercury hazards. Brain levels of inorganic mercury which are of prime interest for risk assessment cannot be calculated with ease. A rather time-consuming combination of biological and environmental monitoring is needed to quantify the accumulated body burden.

IX. HAZARDS TO PATIENTS

Dental amalgam is a major contributor to the daily nonoccupational intake of mercury. Mercury is ingested and inhaled in connection with placement, removal, and wear of the restoration.[4] Mercury is released during chewing and toothbrushing.[69,70] Mercury concentrations up to 100 $\mu g/m^3$ have been measured in intraoral and exhaled air after chewing and toothbrushing.[69-71] Chewing stimulation by humans with amalgam fillings increased the mercury concentration sixfold from unstimulated conditions (from a mean value of 4.9 to 29.1 $\mu g/m^3$). This meant a 54-fold increse over levels observed in control subjects (0.7 $\mu g/m^3$).[71] Lack of quality control data and uncertainties as to the conditions studied, such as mouth to nose breathing ratio, time interval to preceding meal, sampling techniques, etc., hamper the evaluation. Most probably, the inhaled mercury intake in these studies is overestimated.[52]

Abraham et al.[51] noted statistically significant differences in the mouth air mercury levels before and after chewing in a group with amalgams, but not in a group without amalgams. Blood mercury concentrations were positively correlated to the number and surface area of amalgam restorations and were significantly lower in the group without dental amalgams. The daily intake dose has been calculated for a different number of fillings. Vimy and Lorscheider[72] estimated the daily mercury dose to be 29 μg for subjects with 12 or more occlusal amalgam surfaces and to be 8 μg in subjects with four or fewer occlusal amalgam surfaces. In comparison, industrial workers inhaling 8 to 10 m^3/d at a threshold limit value of 50 μg of mercury per cubic meter get a repeated daily dose of about 300 μg (calculated absorption rate of 80%).

Mercury concentrations in the CNS have been determined in autopsy studies of 17 persons, with no known occupational exposure to mercury.[73] A statistically significant positive correlation was found between the concentration of total mercury in the occipital cortex (range of 41.8 to 28.7 ng of mercury per gram of wet weight) and the number of amalgam restorations. Other studies show somewhat higher brain mercury concentrations of 50 to 100 ng/g of wet weight.[74,75] Mercury miners in Yugoslavia had brain levels around 700 ng/g of wet weight.[76] Fatal intoxications of animals raised the brain tissue concentrations to over 1000 ng/g of wet weight.[77] Data are lacking for brain concentrations in cases of so-called micromercurialism.

The release of mercury in the oral cavity is complex and depends on the composition and treatment of the alloys, pH, and oxygen variations in the oral cavity, presence of proteins, masticatory functions, etc. The mercury release from freshly prepared amalgams (vapor and particulates) during solely static conditions may be of the same magnitude as the intake from food and drink the first day after insertion.[78] During long-term static release conditions, the total mercury release seems to be far below the food and drink intake.[79] Chewing considerably increases the release.

Systemic uptake in animals and humans and urinary clearance of mercury from dental fillings have been traced with radioactive ^{203}Hg.[80] This investigation concludes that the body uptake from dental amalgam is small. There is little or no difference in the urinary mercury excretion from patients before and after placement of amalgam fillings.[81] Silver amalgam containing radioactive mercury (^{203}Hg) was inserted in the teeth of five male volunteer patients.[82] Each patient got four to five occlusal fillings with a total surface area ranging from 0.6 to 0.9 cm^2. Later (7 d), the fillings were substituted by nonradioactive amalgam. Scintillation counting was used to determine the amounts of mercury in urine and feces. In urine the concentrations slowly increased to a mean value of 2.5 μg/l on the fifth day, after which the level decreased to nondetectable levels. A second maximum of 4 μg/l was reached the day after the fillings had been removed. The highest value in urine was 5 μg/l. The excretion with feces reached its first maximum level (mean value about 7000 μg/kg of wet weight) the first day after the insertion of the amalgam fillings. A second maximum was obtained 1 to 2 d after the removal of the fillings. Nilsson and Nilsson[54] found a statistically significant correlation between the number of amalgam surfaces and the urine mercury values in a group of nonoccupationally exposed subjects with amalgam fillings.

In conclusion, available evidence suggests that the health hazards from amalgam as compared to occupational mercury exposure are small. Early signs and symptoms from the CNS (so-called micromercurialism) caused by low exposure to mercury vapor are unspecific and other possible causes should also be considered. As the potentially exposed population is large, individuals with a high susceptibility to mercury can at present not be excluded.[52] Further studies are ongoing. It should be noted that allergic reactions (e.g., rash, dermatitis, and edema) may occur.[4,82-84] Several of such cases report a previous mercury sensitization from medical or occupational exposure.[4] Allergies are further dealt with in Chapter 13. Inorganic mercury penetrates the placenta and accumulates in the fetus, who has a higher susceptibility to mercury. In order to avoid this risk, comprehensive amalgam restorations should be avoided during pregnancy.

X. PREVENTIVE MEASURES

Preventive measures follow established strategies and should be systematized. Prevention includes three steps: appropriate choice of materials and working methods, adequate measures to prevent diffusion of vapors and particulates, and proper action to avoid exposure.

A. Handling

As mercury is a toxic metal, mercury-containing materials should when possible be replaced with less toxic alternatives. Diffusion is reduced by good working practices and proper ventilation. Exposure is prevented by appropriate routines and personal protection devices when necessary. Mercury should be handled with care and stored in unbreakable, tightly sealed containers. An amalgamator with completely enclosed arms and capsule during amalgamation should be used. Ultrasonic amalgam condensers should be avoided.[2] Scrap amalgam material should be stored under water or in a bacteriostatic solution in tightly sealed containers. To avoid spreading of mercury, heating of mercury-containing material should be avoided. Mercury-contaminated instruments should not be placed in heat sterilizers. Spilled mercury should be properly cleaned up. Inaccessible spillage can be covered up by powdered sulfur, calcium polysulfide, or activated alumina. When necessary, rubber gloves should be worn. If mercury contacts the skin, thorough washing with soap and water should be undertaken. Cracked and seamed counter tops, floors with cracks, and carpets should be avoided in dental operatories and supportive areas as they collect spills.[4] Water spray combined with appropriate evacuation decreases exposure during removal of old amalgam restorations.[3] Drinking, eating, or smoking should not be permitted in the amalgam processing area.

B. Filter Masks

The use of face masks requires proper choice of type, information, and training of the user. Specifications should be assured as there are many inferior trademarks. The retention of respirable dust (particle fractions below 5 μm) should be specified.

C. Local Exhausts

Airborne contaminants should be collected as close to the point of origin as possible. Local exhaust ventilation should be complemented by general ventilation. There are several types of local exhausts available: hoods or tubes placed near the point of operation, exhaust nozzles integrated in the tools, and others.

D. General Ventilation

General ventilation dilutes the air contaminants and is expressed in number of air changes per time unit, normally number of air changes per hour. General ventilation is more expensive than local exhaust ventilation due to the large air volumes and heating of the air needed. The proper number of air changes must be related to the emission of air contaminants. Placement of inlet and outlet is important.

E. Air Filters

The contaminated air from the ventilation systems must be filtered. There are different types of filters: cloth, electrostatic, and others.[85] The efficiency of the filters should be specified and particular attention should be given to particulates smaller than 5 μm. In some cases, air from general ventilation systems is recirculated after cleaning. In such cases it is important to avoid accumulation in the room air of toxic gases and particulates. Maintenance and control of the ventilation system are important; filters should be checked at predetermined intervals to avoid overloading.

XI. CONCLUSION

Several investigations have shown that mercury vapor is released from amalgam fillings, particularly after chewing. Insertion and removal of amalgam fillings by drilling increase the excretion of mercury in urine for several days. The concentration of mercury in urine is correlated to the number of amalgam fillings in the mouth. A correlation has been observed between the number of amalgam surfaces in the mouth and the concentration of mercury in brain and kidneys. Different calculations indicate that the daily uptake of mercury from a moderate amalgam loading increases urinary mercury excretion by 2 to 4 μg/l. There are considerable individual differences.

Discrete effects on the CNS and the kidneys are reported at urine mercury levels of 50 μg/l or higher. A urine mercury concentration of 50 μg/l corresponds to an air concentration of about 30 $\mu g/m^3$ and a blood mercury level of about 20 μg/l.[52] Total mercury concentrations in whole blood below 10 μg/l and urine concentrations lower than 10 μg/l are considered to be permissible. A total mercury concentration in whole blood exceeding 20 μg/l or urine levels higher than 50 μg/l indicate an exposure which may lead to mercury intoxication.[52]

Mercuric mercury interacts with other metals. Animal experiments have shown that selenium counteracts the toxic effects of inorganic mercury. Data are lacking for the detailed interaction between mercury vapor and selenium in humans. During the last decade, the working environment of dental surgeries has considerably improved. The exposure to mercury has decreased due to various preventive measures, such as increased use of less toxic materials in dental fillings, changed handling routines, improved general ventilation, and the introduction of local exhausts. However, measurements of air contaminants show that permissible levels for mercury and mercury compounds may be exceeded during shorter periods. A systematic surveillance of the working conditions is important. Such programs include several steps: working routines should be checked, air measurements should be undertaken at intervals, and a program for ventilation maintenance and control should be enforced.

Exposure control combines biological and environmental monitoring. In biological monitoring, both blood and urine measurements are used. Air monitoring may be performed by direct reading instruments such as mercury vapor detectors or detector tubes. In periodic health examinations, special attention should be paid to symptoms ascribed to so-called micromercurialism which is supposed to appear at exposure levels between 10 and 100 μg of mercury per cubic meter of air. This syndrome, however, is unspecific, and the symptoms may have other causes than mercury. At exposure levels higher than 100 $\mu g/m^3$, there is a risk for mercurialism with tremor.

REFERENCES

1. **Steere, N. V.,** Mercury vapor hazards and control measures, *J. Chem. Educ.*, 42, A529, 1965.
2. **Rao, G. S. and Hefferren, J. J.,** Toxicity of mercury, in *Biocompatibility of Dental Materials,* Vol. 3, *Biocompatibility of Dental Restorative Materials,* Smith, D. C. and Williams, D. F., Eds., CRC Press, Boca Raton, FL, 1982, 19.
3. **MacDonald, G.,** Occupational hazards in dentistry, *J. Calif. Dent. Assoc.*, 12, 17, 1984.
4. **Bauer, J. G.,** Action of mercury in dental exposures to mercury, *Oper. Dent.*, 10, 104, 1985.
5. American Conference of Governmental Industrial Hygienists, Threshold Limit Values and Biological Exposure Indices for 1986-1987, ACGIH, Cincinnati, OH, 1986.
6. WHO, Report of a WHO Study Group, 1980, Recommended Health-Based Limits in Occupational Exposure to Heavy Metals, Tech. Rep. Ser. No. 647, World Health Organization, Geneva, 1980.
7. **Gronka, P. A., Bobkoskie, R. L., Tomchick, G. J., Bach, F., and Rakow, A. B.,** Mercury vapor exposures in dental offices, *J. Am. Dent. Assoc.*, 81, 923, 1970.
8. **Mantyla, D. G. and Wright, O. D.,** Mercury toxicity in the dental office: a neglected problem, *J. Am. Dent. Assoc.*, 92, 1189, 1976.
9. **Gordon, H., Tsujii, D., and Breysse, P.,** Mercury body burdens and office vapor levels in Seattle practices, *J. Dent. Res.*, 56 (Abstr. 118), B81, 1977.
10. **Jones, D. W., Sutow, E. J., and Milne, E. L.,** Survey of mercury vapour in dental offices in Atlantic Canada, *J. Can. Dent. Assoc.*, 6, 378, 1983.
11. **Lenihan, J. M. A., Smith, H., and Harvey, W.,** Mercury hazards in dental practice, assessment and control by activation analysis, *Br. Dent. J.*, 135, 365, 1973.
12. **Nixon, G. S., Whittle, C. A., and Woodfin, A.,** Mercury levels in dental surgeries and dental personnel, *Br. Dent. J.*, 151, 149, 1981.
13. **Nilsson, B. and Nilsson, B.,** Mercury in dental practice. I. The working environment of dental personnel and their exposure to mercury vapor, *Swed. Dent. J.*, 10, 1, 1986.

14. **Brune, D., Hensten-Pettersen, A., and Beltesbrekke, H.,** Exposure to mercury and silver during removal of amalgam restorations, *Scand. J. Dent. Res.*, 88, 460, 1980.
15. **Cooley, R. L. and Barkmeier, W. W.,** Mercury vapor emitted during ultraspeed cutting of amalgam, *J. Ind. Dent. Assoc.*, 57, 28, 1978.
16. **Richards, J. M. and Warren, P. J.,** Mercury vapour released during the removal of old amalgam restorations, *Br. Dent. J.*, 159, 231, 1985.
17. **Nielsen Kudsk, F.,** Absorption of mercury vapour from the respiratory tract in man, *Acta Pharmacol. Toxicol.*, 23, 250, 1965.
18. **Nordberg, G. F. and Serenius, F.,** Distribution of inorganic mercury in the guinea pig brain, *Acta Pharmacol. Toxicol.*, 27, 269, 1969.
19. **Clarkson, T. W., Magos, L., and Greenwood, M. R.,** The transport of elemental mercury into fetal tissues, *Biol. Neonate,* 21, 239, 1972.
20. **Halbach, S. and Clarkson, T. W.,** Enzymatic oxidation of mercury vapor by erythrocytes, *Biochim. Biophys. Acta,* 523, 522, 1978.
21. **Magos, L., Halbach, S., and Clarkson, T. W.,** Role of catalase in the oxidation of mercury vapor, *Biochem. Pharmacol.*, 27, 1373, 1978.
22. **Nielsen Kudsk, F.,** The influence of ethyl alcohol on the absorption of mercury vapour from the lungs in man, *Acta Pharmacol. Toxicol.*, 23, 263, 1965.
23. **Magos, L., Sugata, Y., and Clarkson, T. W.,** Effects of 3-amino-1,2,4-triazole on mercury uptake by in vitro human blood samples and by whole rats, *Toxicol. Appl. Pharmacol.*, 28, 367, 1974.
24. **Berlin, M., Fazackerley, J., and Nordberg, G.,** The uptake of mercury in the brains of mammals exposed to mercury vapor and to mercuric salts, *Arch. Environ. Health,* 18, 719, 1969.
25. **Clarkson, T. and Rothstein, A.,** The excretion of volatile mercury by rats injected with mercuric salts, *Health Phys.*, 10, 1115, 1964.
26. **Hursh, J. B., Cherian, M. G., Clarkson, T. W., Vostal, J. J., and vander Mallie, R.,** Clearance of mercury (Hg-197, Hg-203) vapor inhaled by human subjects, *Arch. Environ. Health,* 31, 302, 1976.
27. **Suzuki, T., Miyama, T., and Katsunuma, H.,** Mercury contents in the red cells, plasma, urine and hair from workers exposed to mercury vapour, *Ind. Health,* 8, 39, 1970.
28. **Nielsen Kudsk, F.,** Factors influencing the in vitro uptake of mercury vapour in blood, *Acta Pharmacol. Toxicol,* 27, 161, 1969.
29. **Klaassen, C. D.,** Biliary excretion of mercury compounds, *Toxicol. Appl. Pharmacol.*, 33, 356, 1975.
30. **Rahola, T., Hattula, T., Korolainen, A., and Miettinen, J. K.,** Elimination of protein bound and ionic mercury — Hg-203— in man, *Scand. J. Clin. Lab. Invest.*, 27(Suppl. 116), 77, 1971.
31. **Berlin, M.,** Mercury, in *Handbook on the Toxicology of Metals,* Vol. 2, 2nd ed., Friberg, L., Nordberg, G. F., and Vouk, V. B., Eds., Elsevier, Amsterdam, 1986, 387.
32. **Aberg, B., Ekman, L., Falk, R., Greitz, U., Persson, G., and Snihs, J.-O.,** Metabolism of methyl mercury (^{203}Hg) compounds in man, excretion and distribution, *Arch. Environ. Health,* 19, 478, 1969.
33. **Reynolds, W. A. and Pitkin, R. M.,** Transplacental passage of methylmercury and its uptake by primate fetal tissues, *Proc. Soc. Exp. Biol. Med.*, 148, 523, 1975.
34. **Bakir, R., Damluji, S. F., Amin-Zaki, L., Murtadha, M., Khalidi, A., Al-Rawi, N. Y., Tikriti, S., Dhahir, H. I., Clarkson, T. W., Smith, J. C., and Doherty, R. A.,** Methylmercury poisoning in Iraq, an interuniversity report, *Science,* 181, 230, 1973.
35. Swedish Expert Group, Methyl mercury in fish, a toxicologic-epidemiologic evaluation of risks. Report from an expert group, *Nord. Hyg. Tidskr.*, Suppl. 4, 1971.
36. **Rowland, I. R., Grasso, P., and Davies, M. J.,** The methylation of mercuric chloride by human intestinal bacteria, *Experientia,* 31, 1064, 1975.
37. **Heintze, U., Edwardsson, S., Dérand, T., and Birkhed, D.,** Methylation of mercury from dental amalgam and mercuric chloride by oral streptococci in vitro, *Scand. J. Dent. Res.*, 91, 150, 1983.
38. **Friberg, L. and Nordberg, G. F.,** Inorganic mercury — relation between exposure and effects, in *Mercury in the Environment, an Epidemiological and Toxicological Appraisal,* Friberg, L. and Vostal, J., Eds., CRC Press, Boca Raton, FL, 1972, 113.
39. **Baldi, G., Vigliani, E. C., and Zurlo, N.,** Il mercurialismo cronico nei cappellifici, *Med. Lav.*, 44, 161, 1953.
40. **Sapin, C., Druet, E., and Druet, P.,** Induction of antiglomerular basement membrane antibodies in the brown-Norway rat by mercuric chloride, *Clin. Exp. Immunol.*, 28, 173, 1977.
41. **Shull, R. M., Stowe, C. M., Osborne, C. A., O'Leary, T. P., Vernier, R. L., and Hammer, R. F.,** Membranous glomerulo-nephropathy and nephrotic syndrome associated with iatrogenic metallic mercury poisoning in a cat, *Vet. Hum. Toxicol.*, 23, 1, 1981.
42. **Kazantzis, G., Schiller, K. F. R., Asscher, A. W., and Drew, R. G.,** Albuminuria and the nephrotic syndrome following exposure to mercury and its compounds, *Q. J. Med.*, 31, 403, 1962.
43. **Strunge, P.,** Nephrotic syndrome caused by a seed disinfectant, *J. Occup. Med.*, 12, 178, 1970.

44. **Lindqvist, K. J., Makene, W. J., Shaba, J. K., and Nantulya, V.,** Immunofluorescence and electron microscopic studies of kidney biopsies from patients with nephrotic syndrome, possibly induced by skin lightening creams containing mercury, *East Afr. Med. J.*, 51, 168, 1974.
45. **Marsh, D. O., Myers, G. J., Clarkson, T. W., Amin-Zaki, L., Tikriti, S., and Majeed, M. A.,** Fetal methylmercury poisoning: clinical and toxicological data on 29 cases, *Ann. Neurol.*, 7, 348, 1980.
46. **Choi, B. H., Lapham, L. W., Amin-Zaki, L., and Saleem, T.,** Abnormal neuronal migration, deranged cerebral cortical organization, and diffuse white matter astrocytosis of human fetal brain: a major effect of methylmercury poisoning in utero, *J. Neuropathol. Exp. Neurol.*, 37, 719, 1978.
47. **Elinder, C. G., Gerhardsson, L., and Oberdoerster, G.,** Biological monitoring of toxic metals, Overview, in *Biological Monitoring of Toxic Metals,* Clarkson, T. W., Friberg, L., Norberg, G. F., and Sager, P. R., Eds., Rochester Series on Environmental Toxicity, Plenum, NY, 1988, 1.
48. ACGIH, Air Sampling Instruments for Evaluation of Atmospheric Contaminants, American Conference of Governmental Industrial Hygienists, 5th ed., Cincinnati, 1978, Sections I, S, and U.
49. **Battistone, G. C., Hefferren, J. J., Miller, R. A., and Cutright, D. E.,** Mercury: its relation to the dentist's health and dental practice characteristics, *J. Am. Dent. Assoc.*, 92, 1182, 1976.
50. **Iyengar, G. V., Kollmer, W. E., and Bowen, H. J. M.,** *The Elemental Composition of Human Tissues and Body Fluids,* Verlag Chemie, Weinheim, New York, 1978.
51. **Abraham, J. E., Svare, C. W., and Frank, C. W.,** The effect of dental amalgam restorations on blood mercury levels, *J. Dent. Res.*, 63, 71, 1984.
52. Swedish Expert Group, Health Risks from Amalgams, Swedish National Board of Health and Welfare, Rep. 1987:10, Allmanna Forlaget AB, S-10647, Stockholm Sweden (in Swedish).
53. **Cross, J. D., Dale, I. M., Goolvard, L., and Lenihan, J. M. A.,** Methyl mercury in blood of dentists, *Lancet,* 2, 312, 1978.
54. **Nilsson, B. and Nilsson, B.,** Mercury in dental practice. II. Urinary mercury excretion in dental personnel, *Swed. Dent. J.*, 10, 221, 1986.
55. **Ott, K. H. R., Loh, F., Krönche, A., Schaller, K. H., Valentin, H., and Weltle, D.,** Zur Quecksilberbelastung durch Amalgamfüllungen, *Dtsch. Zahnaerztl. Z.*, 39, 199, 1984.
56. **Battistone, G. C., Sammons, D. W., and Miller, R. A.,** Mercury excretion in military dental personnel, *Oral Surg.*, 35, 47, 1973.
57. **Marks, V. and Taylor, A.,** Urinary mercury excretion in dental workers, *Br. Dent. J.*, 146, 269, 1979.
58. **Beste, L., Hefferren, J. J., Muller, T., Bejnarowicz, E., Rubin, A., and Zimmerman, M.,** Atmospheric and biological mercury levels of dental offices and personnel, *J. Dent. Res.*, 55 (Abstr. 719), B241, 1976.
59. **Ayer, W. A., Getter, L., Machen, J. B., and Haller, G. R.,** Hand steadiness and mercury blood levels among practicing dentists: preliminary findings, *J. Am. Dent. Assoc.*, 92, 1208, 1976.
60. **Vandenberge, J., Moodie, A. S., and Keller, R. E., Jr.,** Blood serum mercury test report, *J. Am. Dent. Assoc.*, 94, 1155, 1977.
61. **Shapiro, I. M., Sumner, A. J., Spitz, L. K., Cornblath, D. R., Uzzell, B., Ship, I. I., and Bloch, P.,** Neurophysiological and neuropsychological function in mercury-exposed dentists, *Lancet,* 1, 1147, 1982.
62. **Iyer, K., Goodgold, J., Eberstein, A., and Berg, P.,** Mercury poisoning in a dentist, *Arch. Neurol.*, 33, 788, 1976.
63. **Merfield, D. P., Taylor, A., Gemmell, D. M., and Parrish, J. A.,** Mercury intoxication in a dental surgery following unreported spillage, *Br. Dent. J.*, 141, 179, 1976.
64. **Smith, D. L., Jr.,** Mental effects of mercury poisoning, *South. Med. J.*, 71, 904, 1978.
65. **Nylander, M.,** Mercury in pituitary glands of dentists, *Lancet,* 1, 442, 1986.
66. **Ahlbom, A., Norell, S., Rodvall, Y., and Nylander, M.,** Dentists, dental nurses, and brain tumours, *Br. Med. J.*, 292, 662, 1986.
67. **Brodsky, J. B., Cohen, E. N., Whitcher, C., Brown, B. W., Jr., and Wu, M. L.,** Occupational exposure to mercury in dentistry and pregnancy outcome, *J. Am. Dent. Assoc.*, 111, 779, 1985.
68. **Cook, T. A. and Yates, P. O.,** Fatal mercury intoxication in a dental surgery assistant, *Br. Dent. J.*, 127, 553, 1969.
69. **Svare, C. W., Peterson, L. C., Reinhardt, J. W., Boyer, D. B., Frank, C. W., Gay, D. D., and Cox, R. D.,** The effect of dental amalgams on mercury levels in expired air, *J. Dent. Res.*, 60, 1668, 1981.
70. **Patterson, J. E., Weissberg, B. G., and Dennison, P. J.,** Mercury in human breath from dental amalgams, *Bull. Environ. Contam. Toxicol.*, 34, 459, 1985.
71. **Vimy, M. J. and Lorscheider, F. L.,** Intra-oral air mercury released from dental amalgam, *J. Dent. Res.*, 64, 1069, 1985.
72. **Vimy, M. J. and Lorscheider, F. L.,** Serial measurements of intra-oral air mercury: estimation of daily dose from dental amalgam, *J. Dent. Res.*, 64, 1072, 1985.
73. **Friberg, L., Kullman, L., Lind, B., and Nylander, M.,** Mercury in the central nervous system in relationship to dental amalgam fillings, *Lakartidningen,* 83, 519, 1986 (in Swedish).

74. **Mottet, N. K. and Body, R. L.,** Mercury burden of human autopsy organs and tissues, *Arch. Environ. Health,* 29, 18, 1974.
75. **Kitamura, S., Sumino, K., Hayakawa, K., and Shibata, T.,** Mercury content in human tissues from Japan, in *Effects and Dose-Response Relationships of Toxic Metals,* Nordberg, G. F., Ed., Elsevier, Amsterdam, 1976, 290.
76. **Kosta, L., Byrne, A. R., and Zelenko, V.,** Correlation between selenium and mercury in man following exposure to inorganic mercury, *Nature (London),* 254, 238, 1975.
77. **Fukuda, K.,** Metallic mercury induced tremor in rabbits and mercury content of the central nervous system, *Br. J. Ind. Med.,* 28, 308, 1971.
78. **Brune, D. and Evje, D. M.,** Man's mercury loading from a dental amalgam, *Sci. Total Environ.,* 44, 51, 1985.
79. **Brune, D.,** Metal release from dental biomaterials, *Biomaterials,* 7, 163, 1986.
80. **Frykholm, K. O. and Odeblad, E.,** Studies on the penetration of mercury through the dental hard tissues, using Hg^{203} in silver amalgam fillings, *Acta Odontol. Scand.,* 13, 157, 1955.
81. **Hoover, A. W. and Goldwater, L. J.,** Absorption and excretion of mercury in man. X. Dental amalgams as a source of urinary mercury, *Arch Environ. Health,* 12, 506, 1966.
82. **Frykholm, K. O.,** On mercury from dental amalgam. Its toxic and allergic effects, and some comments on occupational hygiene, *Acta Odontol. Scand. Suppl.,* 15(22), 7, 1957.
83. **Fernström, A. I. B., Frykholm, K. O., and Huldt, S.,** Mercury allergy with eczematous dermatitis due to silver-amalgam fillings, *Br. Dent. J.,* 13, 204, 1962.
84. **Feuerman, E. J.,** Recurrent contact dermatitis caused by mercury in amalgam dental fillings, *Int. J. Dermatol.,* 14, 657, 1975.
85. **Brune, D. and Beltesbrekke, H.,** Dust in dental laboratories. III. Efficiency of ventilation systems and face masks, *J. Prosthet. Dent.,* 44, 211, 1980.

Chapter 17

RISKS AND PREVENTION OF MUSCULOSKELETAL DISORDERS AMONG DENTISTS

Mats Hagberg and Catharina Hagberg

TABLE OF CONTENTS

I. INTRODUCTION

Musculoskeletal disorders are common in the general population and are the most common grounds for paying disablement pensions.[1,2] The prevalence of neck pain in the U.S. population is 5% for males and 7% for females, increasing with age.[2] For the low back, the corresponding figures are 9 and 11% for males and females, respectively.[2] In a sample of a population aged 25 to 74 years, 5% of males and 8% of females are receiving treatment for musculoskeletal disorders.[2] Epidemiological studies have revealed that occupational factors may account for more than 90% of the musculoskeletal disorders in certain occupational groups.[3] Dentists are an occupational group among whom up to every second practicing dentist has been reported to have had back problems related to occupation.[4] Initially, the workstation for the dentist was simply a chair in which the patient was seated. Chairs that could be elevated and tilted came into use at the turn of the century. In the 1920s and 1930s, the modern dental workstation (dental unit) was designed consisting of a patient chair, spittoon, a table on a hinged arm, and a drill powered by an electric motor. The dental mirror for use in the mouth of the patient gave the dentist better access to posterior teeth and to the palatal surfaces of the teeth in general. The modern dental workstation is for four-handed, low-seated dentistry, meaning that the dentist is seated on a low stool and the assistant, also seated, provides continuous chair-side aid.[5] In this workstation design, the dentist is seated to the right of the patient; the dental assistant is seated to the left. A job analysis of a dentist's activities from 22,000 random readings using a work sampling sheet showed that the 3 most common activities were general practice operative dentistry (38%), prophylaxis (8%), and fixed bridgework (5%).[6] The dentist is confined to a small working area, i.e., the mouth of the patient, and in this working area uses hand tools with great precision. This precision work will determine posture, movement, and the musculoskeletal load.

II. POSTURE, MOVEMENT, AND MUSCULOSKELETAL LOAD IN DENTISTRY

A. Preferred Chair-Side Positions in Sit-Down Dentistry

Sit-down dentistry has been advocated sine the mid-1960s in preference to stand-up dentistry. Today, as questionnaire studies indicate,[7,8] sit-down dentistry is probably the most common work technique for most dentists. In sit-down dentistry, there are different dental operating concepts. There was an international standard published by the International Organization for Standardization (ISO)[9] concerning the definitions and principles of the working space of a dentist, which has now, however, been withdrawn. Nevertheless, a postural classification can be made according to the ISO concepts.[9,10] If the patient's teeth are divided into four quadrants, the sit-down working posture may be changed within local positions according to the quadrant the dentist is working in (Figure 1A and B). There are two general positions from which the dentist can work in all four quadrants of the patient's mouth. In one the dentist sits at 9:00 to 11:00 o'clock; the other alternative is that the dentist sits at 10:30 to 12:30 and works in all four quadrants of the patient mouth (Figure 1C and D).

Sit-down dentistry work puts major sites of posture strain on the neck, shoulder, and low back. In a survey of Irish dentists, 73% operated in a seated position the majority of the time.[7] Only 10% of the dentists operated standing up the majority of the time.[7] The preferred chairside positions varied evenly between 3:00, 9:00, and 12:00 o'clock.[7] However, when these figures are evaluated, the low response rate of 14% in the Irish study has to be considered.[7] The claimed optimum position for the patient's head was 80° from the vertical and the principal chair-side position was at 11:00 to 7:00 o'clock in a study which was said to be based on electro-myography (EMG) recordings and ratings.[11] However, no information

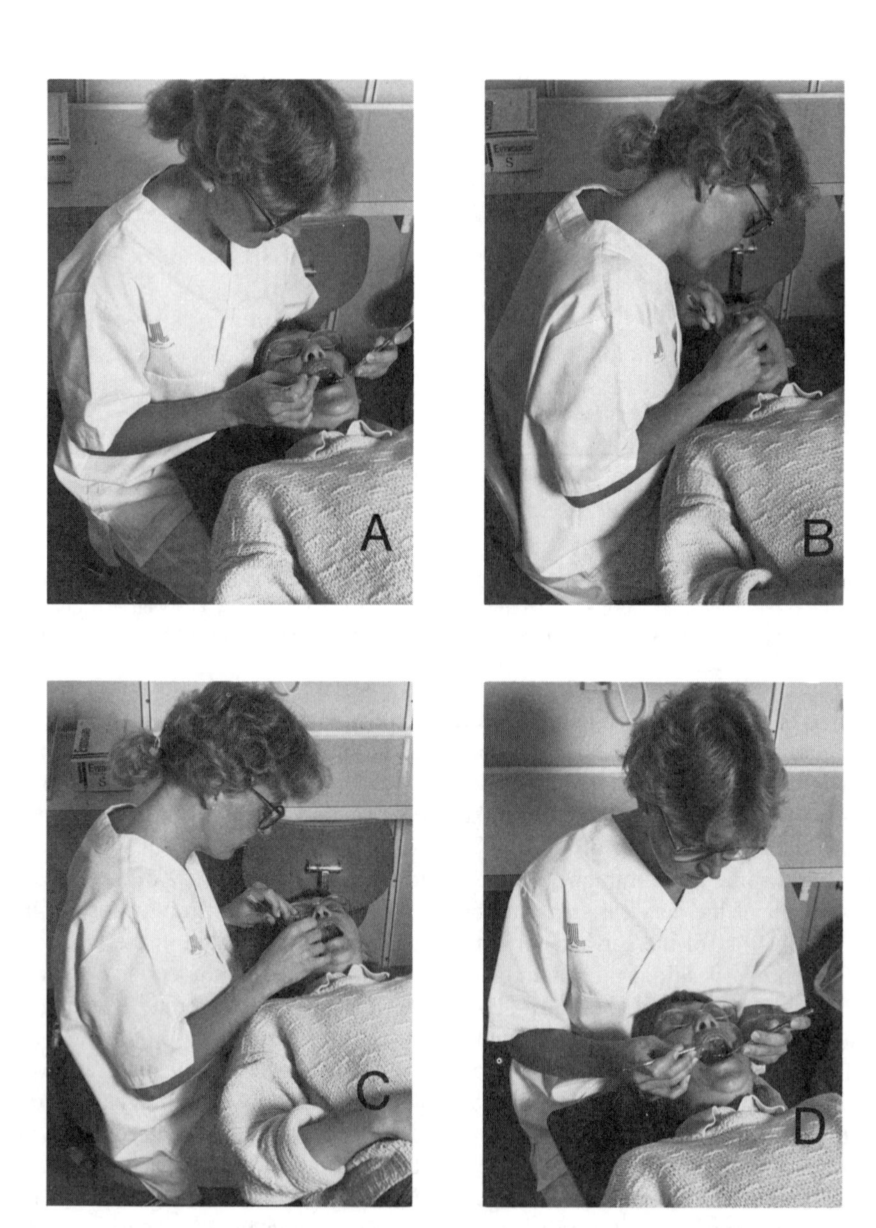

FIGURE 1. (A) The 11:00 o'clock position where the dentist works in the first quadrant; (B) the 8:30 o'clock position where the dentist works in the fourth quadrant. Two general positions from which the dentist can work all four quadrants of the patient's mouth are (C) 9:00 to 11:00 o'clock and (D) 10:30 to 12:30 o'clock. Note the head-down posture of the dentist in all four positions.

about the number of subjects or EMG techniques, etc. appeared in the study.[11] In a study of the patients' attitudes to different chair positions (0°, horizontal; 15°, 30°, and 60°, nearly upright), the patients preferred the 30° position.[12] The operator in the same study preferred the nearly horizontal position (15°).

Eccles and Davies[13] compared chair, head, and operator positions for access time and posture using a patient simulator consisting of a phantom head. They found that it was better to have the chair in the horizontal position than at 30° since the dentist could achieve better posture.[13] Furthermore, they concluded that the operator should work in the 9:00 or 12:00 o'clock position and not in the 3:00 o'clock position when considering both access time and posture.[13]

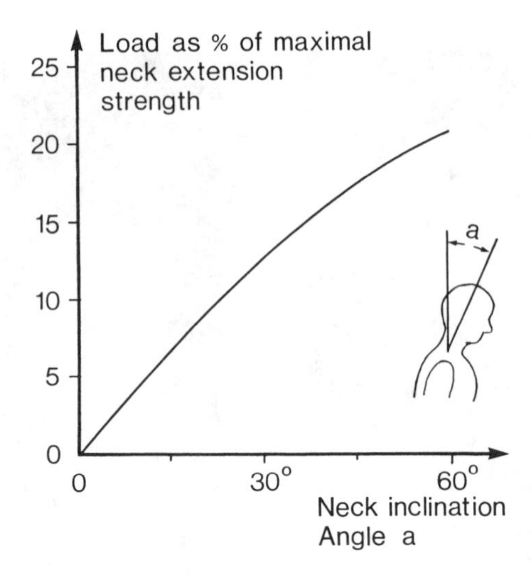

FIGURE 2. The load on the neck muscles at different angles of cervical spine flection. On the vertical axis, the load is given as a percentage of maximal voluntary contraction (%MVC) for women. (Modified from Chaffin, D. B. and Andersson, G. B. J., *Occupational Biomechanics*, John Wiley & Sons, New York, 1984.)

B. Posture of Cervical Spine and Musculoskeletal Load

Green and Brown[14] visited the offices of "four successful dentists" and made notes and sketches of common positions and methods. They observed that the dentist's head was kept in a head-down position in 58 to 83% in the frames studied.[14] The head-down position was defined as a cervical bending of 45 to 90°, the extreme position with 90° forward flection of the cervical spine was common.[14] The weight of the head is about 7% of the body weight and the weight of the cervical spine in about 1%.[15] The center of gravity for the head is located just in front of the ear and for the cervical spine it is located in the middle of the spine. In a normal position, the center of gravity for the head and the cervical spine lies in front of the vertical line of the body. Thus, a flexing moment or torque is acting on the head and the cervical spine. The magnitude of the torque is dependent on the force (weight of the head and cervical spine) and the lever arm. Thus the torque increase and with the angle of flection. Since the head and the cervical spine in a normal position have a forward-acting torque, the shoulder/neck muscles must be active to keep the head in an upright position. In a position of 30° of neck flection, the load on the neck muscles is about 15% of the maximal voluntary contraction (Figure 2).[16] Furthermore, if the angle of neck flection is 30°, the time taken to reach significant fatigue is approximately 2 h (Figure 3).[17] Among accounting machine operators, it was found that the angle of neck flection during work correlated to neck pains.[18]

C. Shoulder Posture and Musculoskeletal Load

In studying postures of dentists, Green and Brown[14] have described postures with 90° of abduction in the glenohumeral joint, although there was no information given about the frequency or duration of the postures during work. In seated light manual work, the arms are slightly abducted and forward flexed. If the working area is higher than elbow height, such as in sit-down dentistry, the shoulder girdle may be elevated. This position may cause strain on both the shoulder/neck angle and the glenohumeral joint. At an angle of 60° of abduction, the blood flow to the supraspinatus tendon may decrease since the intramuscular pressure exceeds the circulatory pressure and one of the main arteries runs through the supraspinatus muscle.[19] Also static contraction of the rotator cuff muscle may cause impairment of the blood flow in the tendon by static tension in the tendon, since increased

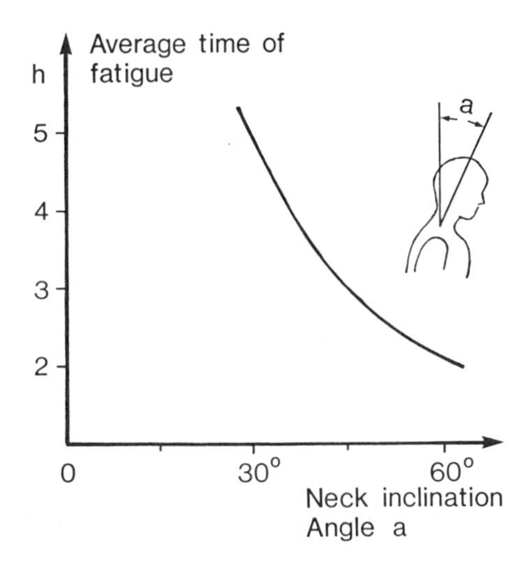

FIGURE 3. The relationship between forward flection of the neck and the duration until severe fatigue in the neck is perceived. (Modified from Chaffin, D. B. and Andersson, G. B. J., *Occupational Biomechanics,* John Wiley & Sons, New York, 1984.)

tension impairs blood flow in a tendon.[20] Decreased blood flow in the tendons to the rotator cuff muscle causes an accelerated degeneration that is an important station on the way to shoulder tendinitis. Static contractions of the shoulder elevators such as the descending part of the trapezius muscle may cause chronic myalgia often referred to as tension neck syndrome. Tension neck syndrome may be caused by local ischemia or metabolic disturbances.[21]

D. Low Back Posture, Movements, and Musculoskeletal Load

Trunk movement during work was studied by Nordin and co-workers[22] with a back flection analyzer. The occupations studied were, besides dentistry, nurses' aide work and warehouse work. Ten dentists with "regular, normal outpatient work" were studied; the dentists had modern equipment that allowed them to treat the patients in a supine position.[22] The dentists were free to sit or to stand during work. It was found that the dentists spent most of their time (mean of 52.2 min/h) in a semiflexed position, not exceeding 36° of back flection.[22] Furthermore, the dentistry was regarded as mainly static with the dentists performing a mean of eight deep forward flections of more than 73°/h.[22]

In an electromyographical study of 20 dental students and faculty members, the effect of stool height and lumbar support was investigated in sit-down dentistry.[23] It was found that lumbar support was a significant factor in reducing muscle activity in the upper and lower back for the dentists.[23] They found that stool height had no significant influence, but suggested that there was a trend for low stool heights (knee angles of 90 and 75°) to produce less activity than high stool heights (knee angle of 105°) if the back were supported.[23] Postures associated with low back strain are those involving forward flection, lateral bending, and rotation.[24] In a forward flexed position, the torque will cause both a high muscular load on the errector spinae muscle and increase the intradiscal pressure.[16]

III. PREVALENCE AND RISKS OF MUSCULOSKELETAL DISORDERS

In a study of 147 public health service dentists, disorders of the musculoskeletal system were reported by 48% of younger dentists and 64% of older dentists in comparison to 26 and 37% of younger and older clerks, respectively, serving as referents.[25] In a meta-analysis, these figures translate into a Mantel-Haenzel odds ratio (MOR) (relative risk) of 2.8 (a 95%

confidence interval of 1.6 to 4.9). A higher prevalence of absence from work due to musculoskeletal disorders was reported by the younger and older dentists — 17 and 26% — compared to the referents (6% and 6%, respectively).[25] A meta-analysis gives a MOR (relative risk) of 2.5 and a 95% confidence interval of 1.2 to 5.0.

A. Low Back Pain in Dentists

In an Irish study of dentists, 48% had experienced pain in the lower back (only 87 respondents out of 650).[7] The overall prevalence of back pain in a sample of 172 Canadian dentists with a mean age of 41.4 years was 36% (the response rate in this questionnaire study was 50%).[26] Furthermore, they found no relation between back pain incidence and working posture (standing practitioners vs. sitting practitioners).[26] In a study of Swedish public health service dentists, the prevalence of back pain was 20 and 24%, respectively, in younger and older dentists.[25] The relative risk (MOR) of back pain was 2.3 (95% confidence interval of 1.1 to 5.1) when compared to clerks. In a study of 68 dentists in Finland (62 females and 6 males) it was reported that 35% of the dentists had aches during the 6 months preceding the survey.[8] In South Wales, 54% of the male dentists reported that they had experienced back pain since starting clinical work.[27] The number of dentists in the questionnaire study was 261 with a response rate of 89%.

B. Cervical Spine Disorders in Dentists

The prevalence of musculoskeletal disorders in the back of the head and neck was 8 and 18%, respectively, in younger and older dentists; there was no difference from clerks.[25] Aches in the neck during the 6 months preceding a survey were found in 27% of dentists (62 females and 6 males).[8] In the study of male dentists from South Wales, 24% reported disorders in the neck occurring since they had started clinical dentistry.[27] The findings of a radiographic examination of the cervical spine in 119 male dentists from Finland showed that 50% had cervical spondylosis compared to 31% in the reference group consisting of 192 male farmers.[28] The relative risk (odds ratio) of cervical spondylosis was 4.0 with a 95% confidence interval of 2.0 to 7.9.[3] However, the correlation between symptoms from the cervical spine and radiographic cervical spondylosis is poor or nonexistent.[29,30] Neither cervical disk degeneration nor encroachment of the intervertebral foramina of the cervical spine are correlated to symptoms.[29]

C. Shoulder Disorders in Dentists

A relative risk of 2.9 (95% confidence interval of 1.1 to 7.8) was found for right shoulder musculoskeletal disorders when Swedish dentists (16% males and 84% females) were compared with clerks. The given prevalence was 15 and 16% in younger and older dentists, respectively.[25] In a study of 68 dentists (6 males and 62 females) in Finland, the part of the body in which aching most frequently occurred was the shoulder, with 42% of dentists experiencing aching during the 6 months preceding the survey.[8] Glenohumeral joint osteoarthrosis was found in 13% of 119 male Finnish dentists.[28] When the dentists were compared to farmers, a standardized odds ratio of 4.2 was found, with a 95% confidence interval of 2.1 to 8.5.[3] However, glenohumeral osteoarthrosis is poorly correlated to shoulder pain.[21]

D. Hand and Wrist Problems in Dentists

Among the dentists from South Wales, 23% reported musculoskeletal pain in the hands occurring since they started clinical dentistry.[27] In the study where 147 dentists working in the public health service were compared to 95 clerks, 10% of younger dentists and 22% of the older dentists had musculoskeletal disorders of the right hand.[25] A meta-analysis of these figures gives a MOR (relative risk) of 2.4 (95% confidence interval of 0.96 to 6.1). In the left hand, the prevalence of musculoskeletal disorders was similar for both dentists and clerks.[25] The prevalence was 7% for younger dentists and 8% for older dentists.[25] The corresponding MOR is 1.0 (95% confidence interval of 0.37 to 2.6).

IV. PREVENTION OF MUSCULOSKELETAL DISORDERS IN DENTISTRY

To prevent musculoskeletal disorders in dentistry, the occupational factors promoting or modifying such disorders must be identified. Possible important occupational factors are not only traumatogens such as extreme cervical spine flection during work, but also constitutional, psychosocial, and behavioral factors. Constitutional factors that are important include height, muscle strength, and endurance. Psychosocial factors are mental stress, attitudes toward the work, and perceived fulfillment in the work. Behavioral factors include physical fitness, training, and diet.

A. Optimizing Working Postures in Dentistry

The increased risk of low back pain among dentists may be correlated to their semiflexed working position (see above). Thus it is important to provide a more upright working posture for the dentist and low back support. An easily adjustable stool is essential for the dentist. Gas spring-operated controls should facilitate adjustment of the chair height and lumbar support. Also the patient chair should have a sufficient range of height variation that allows the working position of the dentist to alternate between sit-down and stand-up dentistry. To minimize the stress on the shoulder, the arms should be kept close to the body. Work technique training is probably one way of teaching dentists to avoid elevation of the arms. The value of support as a means to reducing the work load on the shoulder is doubtful. The use of support may even increase the muscular load.[31] Even when a correct working height is achieved for the low back, the operator may bend the neck down to achieve optimal vision in the precision work. Magnifying glasses corrected for an optimal working height distance may reduce this stress.[10]

B. Optimizing the Musculoskeletal Load by Work Organization Changes

Since the work of Rohmert,[32] the importance of interrupting static contractions in order to enhance local work capacity has been recognized. The muscular load level that can be sustained for 1 h in an isometric (static) contraction by the elbow flexors is as low as 8.8% of the maximal voluntary contraction (MVC);[33] whereas the mean load level that can be sustained in an intermittent isometric contraction for 1 h is 24% of the MVC even if the rests are only of 2-s duration.[33] For a working posture with continuous repetitive forward flections in the shoulder, EMG fatigue development in the descending part of the trapezius muscle was correlated to the load as glenohumeral torque and similar to isometric exercise of the elbow flexors.[34] This implies that repetitive arm elevations may induce a static load on the shoulder/neck muscles. As the work pace increases in dentistry, the more "static" is the musculoskeletal workload exposure for the dentist. One commonly held early theory of the development of muscle fatigue concerned the accumulation of metabolites and a drop in the pH level inhibiting glycolytic enzymes. The theory has been questioned by Mills and Edwards,[35] who reported muscular fatigue development despite normal pH levels in patients with McArdles syndrome. Another possible available mechanism is depletion of intracellular potassium.[36] It is, however, beyond dispute that the recovery of muscle function through relaxation is dependent on the perfusion of the muscle.

Job rotation is often proposed as a compensatory measure for poor working postures, but it is still not known whether job rotation reduces occupational musculoskeletal stress. One possibility is that a work task producing a low static load on muscles may lead to selective fatigue of low threshold motor units. If a change in posture or task is made with a change in muscle load, this may cause a favorable recruitment of other motor units and fatigue rate. If this hypothesis is correct, even a change of work involving an increase of the load would reduce the muscular strain. In dentistry a variation of musculoskeletal load may be achieved by having a work organization which allows for both sit-down and stand-up dentistry. Furthermore, the work organization should aim at giving the dentist as many tasks as possible. Work pace must allow for short rests (micropauses) in the work.

C. Optimizing Individual Resistance to Musculoskeletal Strain

Individual resistance to musculoskeletal strain may be decreased by disease. In silent ankylosing spondylitis, there is an increased susceptibility to muscular fatigue, probably due to a generalized myopathy.[37] Acute viral infections, e.g., the common cold, may affect muscle function.[38] It is possible that acute infections may predispose a subject to an occupational musculoskeletal disorder in two ways. Firstly, during acute infections, when the muscle is infiltrated with viruses, it may have a reduced tolerance to stress. Secondly, after an acute infection, reactive tendinitis or myalgia may occur due to prior occupational degenerative changes.[31]

It is not known whether muscle strength training reduces musculoskeletal susceptibility to strain. Selection of workers through isometric strength tests was found in one report to be a means of controlling medical incidents in strenuous jobs in tire manufacturing.[39] Good isometric endurance in the back was found to be a factor that may prevent first-time occurrence of low back pain in men in a prospective population study.[40]

Epidemiological studies point to an increased risk of musculoskeletal disorders among dentists. High prevalence rates and high relative risks of musculoskeletal disorders indicate that prevention may have a great impact on the health of dentists. However, we lack good strategies for either musculoskeletal health or hazard surveillance. Furthermore, there are no controlled intervention studies to evaluate different work-station designs or different work organizations in dentistry.

V. SUMMARY

A majority of dentists practice sit-down dentistry. Considering both access time and posture, the optimum chair-side position was at 9 or 12 o'clock. The dentists spent most of their time in a semiflexed back position. Among dentists working in the public health service, the prevalence of back pain was 20 to 25%. The relative risk was 2.3 when dentists and clerks were compared. The head-down position (cervical bending 45 to 90°) was reported to have a duration of 58 to 83% of the working time. An increased risk of cervical spondylosis of 4.0 in dentists contrasted to farmers may also be due to the load on the cervical spine. The prevention program for occupational musculoskeletal disorders in dentists includes optimizing working postures, changing work organization, and improving individual resistance to musculoskeletal strain. There is a need for a musculoskeletal health and hazard surveillance program and controlled intervention studies.

REFERENCES

1. **Cunningham, L. S. and Kelsey, J. L.,** Epidemiology of musculoskeletal impairments and associated disability, *Am. J. Public Health,* 74, 574, 1984.
2. **Maurer, K.,** Basic Data on Arthritis, DHEW Publ. No. (PHS) 79-16611979, U.S. Department of Health, Education and Welfare, Washington, D.C., 1979.
3. **Hagberg, M. and Wegman, D. H.,** Prevalence rates and odds ratios of shoulder neck diseases in different occupational groups, *Br. J. Ind. Med.,* 44, 602, 1987.
4. **Norris, C.,** Is your back biting back?, *Dent. Manage.,* 17, 69, 1977.

5. **Eccles, J. D.,** Dental practice — a field for ergonomic research, *Appl. Ergonomics,* 7.3, 151, 1976.
6. **Green, E. S., Lynam, W. A., and Cleveland, B. S.,** Work simplification: an application to dentistry, *J. Am. Dent. Assoc.,* 57, 242, 1958.
7. **Hope-Ross, A. and Corcoran, D.,** A survey of dentists' working posture, *J. Ir. Dent. Assoc.,* 32, 13, 1985.
8. **Murtomaa, H.,** Work-related complaints of dentists and dental assistants, *Int. Arch. Occup. Environ. Health,* 50, 231, 1982.
9. International Organization for Standardization, Dentistry — Working Space of the Dentist — Definitions and Principles, International Standard ISO 3246-1977 (E),
10. **Coburn, D. G.,** Vision, posture and productivity, *Oral Health,* 74, 13, 1984.
11. **Fox, J. G. and Jones, J. M.,** Occupational stress in dental practice, *Br. Dent. J.,* 123, 465, 1967.
12. **Davies, M. H. and Eccles, J. D.,** Attitudes of dental patients to conservation treatment in different chair positions, *J. Dent.,* 6, 294, 1978.
13. **Eccles, J. D. and Davies, M. H.,** A study of operating positions in conservative dentistry, *Dent. Practitioner,* 21, 221, 1971.
14. **Green, E. J. and Brown, M. E.,** An aid to the elimination of tension and fatigue: body mechanics applied to the practice of dentistry, *J. Am. Dent. Assoc.,* 67, 679, 1963.
15. **Gowitzke, B. A. and Milner, M.,** *Understanding the Scientific Bases of Human Movement,* Williams & Wilkins, Baltimore, 1980.
16. **Chaffin, D. B. and Andersson, G. B. J.,** *Occupational Biomechanics,* John Wiley & Sons, New York, 1984.
17. **Chaffin, D. B.,** Localized muscle fatigue — definition and measurement, *J. Occup. Med.,* 15, 346, 1973.
18. **Maeda, K., Hunting, W., and Grandjean, E.,** Localized fatigue in accounting machine operators, *J. Occup. Med.,* 22, 810, 1980.
19. **Järvholm, U., Palmerud, G., Styf, J., Herberts, P., and Kadefors, R.,** Intramuscular pressure in the supraspinatus muscle, *J. Orthop. Res.,* 6, 230, 1988.
20. **Schatzker, J. and Branemark, P. I.,** Intravital observation on the microvascular anatomy and microcirculation of the tendon, *Acta Orthop. Scand. Suppl.,* 126, 3, 1969.
21. **Hagberg, M.,** Shoulder pain — pathogenesis, in *Clinical Concepts in Regional Musculoskeletal Illness,* Hadler, N. M., Ed., Grune & Stratton, Orlando, FL, 1987, 191.
22. **Nordin, M., Örtengren, R., and Andersson, G. B. J.,** Measurements of trunk movements during work, *Spine,* 9, 465, 1984.
23. **Hardage, J. L., Gildersleeve, J. R., and Rugh, J. D.,** Clinical work posture for the dentist: an electromyographic study, *J. Am. Dent. Assoc.,* 107, 937, 1983.
24. **Andersson, G.,** Low back pain in industry: epidemiological aspects, *Scand. J. Rehabil. Med.,* 11, 163, 1979.
25. **Kajland, A., Lindvall, T., and Nilsson, T.,** Occupational medical aspects of the dental profession, *Work Environ. Health,* 11, 100, 1974.
26. **Diakow, P. R. P. and Cassidy, J. D.,** Back pain in dentists, *J. Manipulative Physiol. Ther.,* 7, 85, 1984.
27. **Eccles, J. D. and Powell, M.,** The health of dentists. A survey in south Wales 1965/1966, *Br. Dent. J.,* 123, 379, 1967.
28. **Katevuo, K., Aitasalo, K., Lehtinen, R., and Pietila, J.,** Skeletal changes in dentists and farmers in Finland, *Community Dent. Oral Epidemiol.,* 13, 23, 1985.
29. **Friendenberg, Z. B. and Miller, W. T.,** Degenerative disc disease of the cervical spine, *J. Bone Jt. Surg.,* 45-A, (6), 1171, 1963.
30. **Lawrence, J. S.,** Disc degeneration, its frequency and relationships to symptoms, *Ann. Reum. Dis.,* 28, 121, 1969.
31. **Hagberg, M.,** Optimizing occupational muscular stress of the neck and shoulder, in *The Ergonomics of Working Postures,* Corlett, N., Wilson, J., and Manenica, I., Eds., Taylor & Francis, London, 1986, 109.
32. **Rohmert, W.,** Ermittlung von Erholungspausen fur statiche Arbeit des Menchen, *Int. Z. Angew. Physiol. Einschl. Arbeitsphysiol.,* 18, 123, 1960.
33. **Hagberg, M.,** Muscular endurance and surface electromyogram in isometric and dynamic exercise, *J. Appl. Physiol.,* 51, 1, 1981.
34. **Hagberg, M.,** Work load and fatigue in repetitive arm elevations, *Ergonomics,* 24, 543, 1981.
35. **Mills, K. R. and Edwards, R. H. T.,** Muscle fatigue in myophosphorylase deficiency: power spectral analysis of the electromyogram, *Electroencephalogr. Clin. Neurophysiol.,* 57, 300, 1984.
36. **Sjögaard, G.,** Electrolytes in slow and fast muscle fibers of humans at rest and with dynamic exercise, *Am. J. Physiol.,* 245, R25, 1983.
37. **Hagberg, M., Hagner, I.-M., and Bjelle, A.,** Shoulder muscle strength, endurance and electromyographic fatigue in ankylosing spondylitis, *Scand. J. Rheumatol.,* 16, 161, 1987.
38. **Friman, G.,** Effect of acute infections disease on human isometric muscle endurance, *Upsala J. Med. Sci.,* 83, 105, 1978.

39. **Keyserling, W. M., Herrin, G. D., and Chaffin, D. B.,** Isometric, strength testing as a means of controlling medical incidents on strenuous jobs, *J. Occup. Med.,* 5, 332, 1980.
40. **Biering-Sörensen, F.,** Physical measurements as risk indicators for low-back trouble over a one-year period, *Spine,* 9, 106, 1984.

Chapter 18

DENTAL LABORATORIES

Bodil Persson and Dag Brune

TABLE OF CONTENTS

I. INTRODUCTION

Dental technicians are exposed to several agents in the working environment that may be potentially hazardous. In the dental laboratory, a lot of different chemicals are used. Exposure levels are, however, often low compared to those in the industrial environment. Also other potential risk factors such as noise, vibrations, and stress occur.

Only a few studies are published concerning mortality and cancer morbidity among dental technicians. In the Cancer-Environmental Registry of Sweden,[1] all forms of cancer among dental technicians occurred slightly more frequently than expected. Excess risk for breast cancer among females and for cancer in the urinary tractus (except the kidneys) among the male technicians was observed. A slight excess risk for lung cancer was observed. In another study, lung cancer appeared more often than expected among dental technicians.[2] This may be due to differences in smoking habits or to exposure in the working environment.

This survey will describe various agents and concomitant exposure levels in the working environment associated with reported or suspected hazards and make suggestions for improvements. The potential hazards in the environment of the dental laboratory are related to ergonomics, chemicals, dust, noise, light, etc. Also, the stress factor seems to prevail in this work area. Generally, headache is a common symptom among dental technicians[3] and could be related to stress factors.

II. GASES

A. General Aspects

Dental technicians are exposed to various gases, mostly from solvents, e.g., methylmethacrylate (MMA). Solvents may cause acute or chronic effects on the central nervous system (CNS). The acute symptoms are headache, fatigue, and drowsiness. These acute symptoms are reversible if exposure is discontinued. Long-term solvent exposure is connected with chronic effects on the CNS. The core symptoms are fatigability, bad memory, difficulties in concentration, and loss of initiative. Other symptoms are depression, dysphoria, emotional lability, headache, irritability, paresthesity, sleep disturbances, and vertigo.[4] Similar symptoms could also appear due to other reasons, e.g., psychiatric disorders. Also irritative effects on the mucous membranes in the nose, eyes, and respiratory tracts could be associated with exposure to solvents. In dental laboratories, MMA is widely used and this compound will be paid special attention.

B. Methylmethacrylate (MMA)

1. Production

Commercial production of MMA in the U.S. was first reported in 1937. The production in western Europe in 1976 was about 0.2 millions tons.[5]

2. Use

In dentistry, MMA is specially employed in the removable dentures, orthodontic appliances, occlusal bite planes and splints, veneer crowns, tooth-colored fillings (composites), and pit and fissure sealants.

3. Chemical Properties

MMA is a clear, flammable liquid with a strong acrid odor. In the atmosphere MMA odor may be recognized at a concentration of 0.210 ppm.[6] MMA polymerizes easily and it forms clear, ceramic-like resins and plastics. The present Swedish TLV (threshold limit value) constitutes 50 ppm or 200 mg/m^3. The Occupational Safety and Health Administration (OSHA) and the American Conference Governmental Industrial Hygienists (ACGIH) have recommended 100 ppm or 410 mg/m^3 as the TLV.

4. Toxicity

MMA is irritative to the eyes and respiratory tract and is a potential sensitizer, especially when used in connection with hydroquinone or tertiary amines.[7] The polymer is inert and nontoxic.[8] MMA has been reported to be mutagenic for *Salmonella typhimurium* bacteria.[9] Also an increased incidence of chromosome aberrations were found in rats exposed to a mixture of MMA and chloroprene.[10,11] The carcinogenic potential of MMA is evaluated by the International Agency for Research on Cancer (IARC). Due to the experimental data from rats, the studies were considered inadequate by their working group for an evaluation of the carcinogenicity.[5]

5. Health Effects

Several reports have been published on the occurrence of dermatitis associated with the medical and industrial use of acrylic monomers. Both allergic skin reactions and nonallergic dermatitis are reported among dental technicians.[12] It is estimated that 10% of technicians have acquired hypersensitivity to MMA after between 2 and 14 years of contact with the material.[13] The frequency of current hand eruptions among technicians was 19% according to a Finnish investigation.[12] In another investigation,[3] 8% had skin reactions, mostly non-allergic dermatitis, that could be due to work. It is important to note that the skin after repeated stress from washing is not a very good barrier against foreign chemicals.

As a solvent, MMA monomer may effect the nervous system. MMA has been suggested to cause mild axonal degeneration of the digital nerves when MMA is handled bare handed before polymerization.[14] MMA exposure in dental laboratories has been suggested to cause chronic effects on the nervous system.[15] Other investigators do not agree with that conclusion.[3] The IARC considered that the human data available did not permit an evaluation of the carcinogenicity to humans.[5] Further aspects on health effects of MMA are given in Chapter 8.

6. Exposure Levels

Exposure levels in dental laboratories are often low compared to TLV. In orthopedic surgeries, exposure levels appear to be higher. Due to the relatively low exposure levels in dental laboratories, it should not cause chronic cerebral effects. Exposure levels during various operations are presented in Table 1. The level of MMA in the breathing air can be considerably reduced by using local ventilation (see Table 1).

C. Formaldehyde

1. Properties and Toxicity

These aspects are described in Chapter 9.

2. Release of Formaldehyde

During finishing and polishing denture base resin, small amounts of MMA and formaldehyde are released into the environment of the dental technician.[17] Levels of formaldehyde in the breathing air compared to the present Swedish TLV are presented in Figure 1. During finishing and polishing denture base resin, a local temperature increase will arise in the surface of the prepared specimen, while unpolymerized MMA is released together with MMA which is produced during depolymerization due to heat transfer. Release of formaldehyde from denture base polymers has also been observed.[18]

D. Mercury Vapor

Mercury exposure has occurred to some extent in dental laboratories.[19,21] Dental technicians trimming dies or mulling amalgam in their palms may have been exposed to mercury vapor levels essentially exceeding the present TLV of 50 μg/m^3. The mercury levels arising

Table 1
EXPOSURE LEVELS OF METHYLMETHACRYLATE (MMA) IN DENTAL LABORATORIES AND IN ONE ORTHOPEDIC SURGERY

Exposure level (mg/m³)			
Mean	**Range**	**Working procedure**	**Local exhaust**
127	17—347	Polymerization in dental laboratories[a]	Not used
7	3—31	Grinding in dental laboratories[a]	Not used
13	6—20	Finishing and polishing denture base resin[b]	Not used
2	—	Finishing and polishing denture base resin[b]	Used
1064	824—1356	Polymerization in orthopedic surgery[c]	Not used
121	53—254	Polymerization in orthopedic surgery[c]	Used

[a] Data from five dental laboratories in Stockholm.[16]
[b] Data from measurements of the levels of MMA in dental workroom air.[17]
[c] Data from an orthopedic surgery in Oslo. The exposure levels were measured during the mixing of MMA.

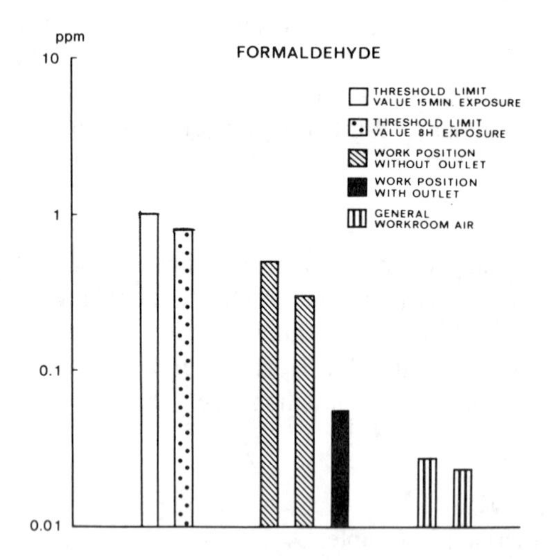

FIGURE 1. Levels of formaldehyde in breathing air close to the work piece as well as in the general workroom air of a dental laboratory during finishing and polishing denture base resin. (From Brune, D. and Beltesbrekke, H., *Scand. J. Dent. Res.*, 89, 113, 1981. With permission.)

during grinding amalgam dies can be considerably reduced using appropriate outlet systems. However, such equipment was previously not in frequent use. Nowadays, the procedure of trimming amalgam dies is rather uncommon. Exposure to mercury vapor at levels of 10 to 100 μg/m³, relevant nowadays only for a minor group of dental technicians, may be associated with micromercurialism (c.f. Chapter 16). For comparison it should be mentioned that a substantial part of dental staff members, i.e., dentists and their assistants, may have been exposed to mercury at the 10 to 100 μg/m³ level during their working day even during the 1980s (c.f. Chapter 16).

E. Precautions

Using a special local exhaust during work with solvents is recommended. Chemicals not

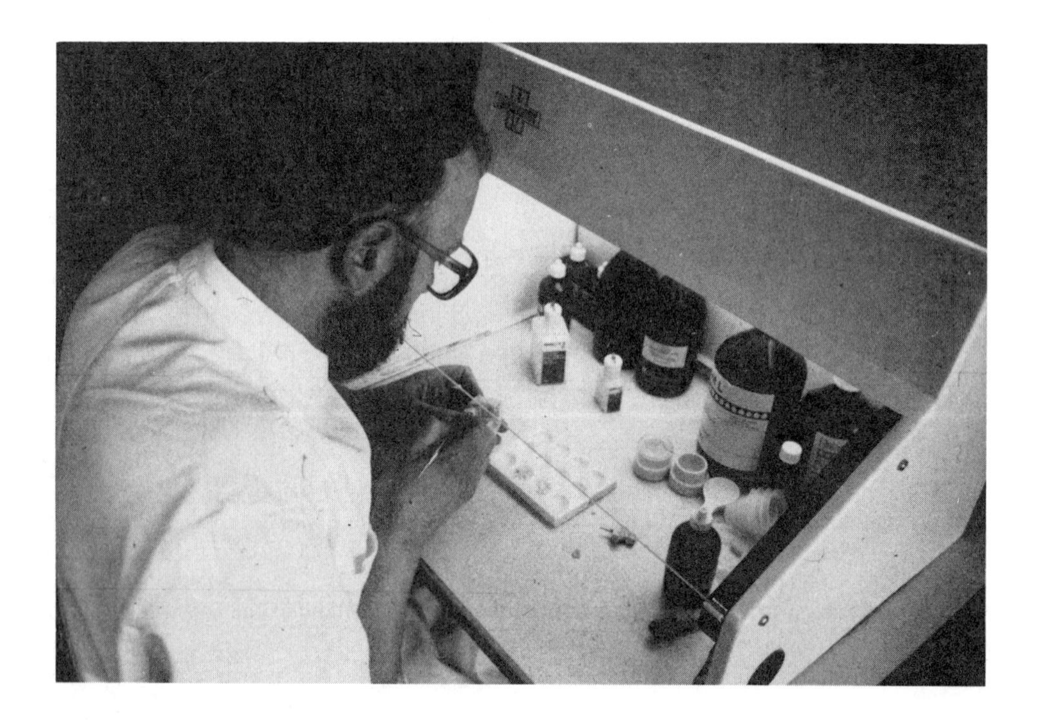

FIGURE 2. Work inside a closed hood. (From Andersson, I., Bornberger, S., Persson, B., and Åkerstedt, K., *Tandteknikern,* 3, 92, 1984. With permission.)

used should be sealed and all waste containing MMA should be kept inside a closed hood (Figure 2) or under a special exhaust. Chemicals not used should be removed and all chemicals should be clearly labeled. Since MMA is a potential sensitizer, it is important to reduce the skin contact and to develop nontouch techniques. The use of protective gloves is dubious. All commercial surgeon's gloves with sufficient flexibility and durability are permeable to MMA monomer.[20] It is suggested that the use of disposable polyethylene gloves with a separate rubber finger hood would be effective enough to prevent MMA penetration during quick procedures (up to 10 min).[12]

III. DUST

Dust in dental laboratories arises from different operations, e.g., cutting, coarse and fine grinding, polishing, centrifuge casting (gold), soldering, and gypsum and investment works. Specimens involved in such procedures constitute gypsum models; porcelain; denture base materials; and alloys for restorative and prosthetic dental procedures, such as gold, chromium, cobalt-, and nickel-based alloys, as well as amalgam dies. During cutting, grinding, or polishing various dental materials, dust arises from the prepared specimen as well as from the tools used, i.e., cut wheels or abrasives. The main components of cut wheels and stones are silicon carbide (SiC) and corundum (Al_2O_3). In addition to these macrocomponents, wheels and stones contain several minor constituents. According to semiquantitative analyses of wheels and stones by emission spectrography, the content of B, Co, Cr, Cu, Mn, Na, Ni, Ti, V, W, Zn, or Zr in the range of 0.01 to 1% by weight was revealed.[21]

Gypsum has a wide application in dental laboratories. Not only pure gypsum ($CaSO_4$), but also modifications with various additives occur. The embedding masses consist of gypsum and different forms of silica (SiO_2), e.g., quartz, tridymite, and cristobalite. These different forms of silica also arise when heated.

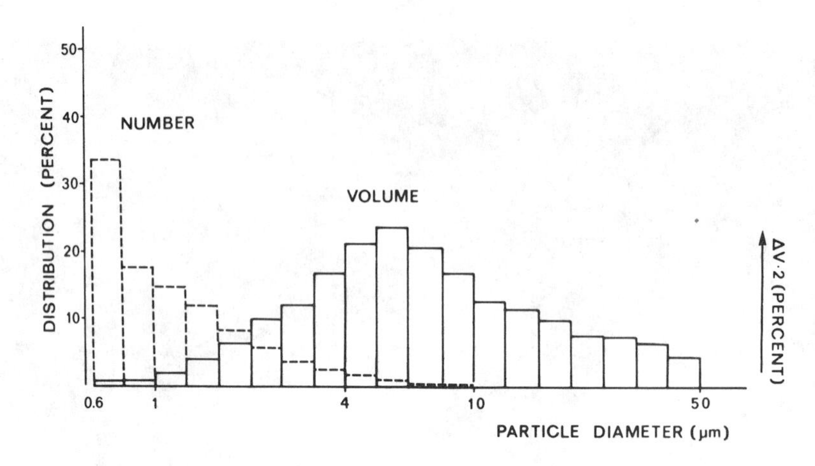

FIGURE 3. Particle size distribution in grinding of porcelain, Coulter-counter measurements. (From Brune, D., Beltesbrekke, H., and Strand, G., *J. Prosthet. Dent.*, 44, 82, 1980. With permission.)

In various casting procedures, an asbestos liner is used to allow expansion when heated. During heating, the asbestos liner becomes brittle and fibers will be released to breathing air in minor quantities during dismantling the mold.[17]

A. Dust Levels

Measurement of particulate matter collected in breathing air, i.e., at a distance of about 30 cm from the work piece during the above mentioned operations, has been conducted.[21] In laboratories with no or improper local ventilation systems, dust levels considerably exceeding the TLV were recorded. Levels in excess of ten times the TLV were observed for solid mercury and silver present in amalgam dust, for cobalt present in nonprecious alloys, and for gypsum. A slight excess of the TLV was measured for copper, nickel, tin, and porcelain in cases of insufficient ventilation.

B. Particle Size Distribution

Deposition of particles in alveolar and tracheobronchial compartments of the lung is influenced by the physical characteristics of the particles, such as their size, shape, and density. Particle deposition in the alveolar compartments increases with decreasing particle size. Knowledge of physical parameters, such as levels of various airborne elements or compounds, as well as their size distributions in the polluted breathing air, is essential in the evaluation of health risks. It is evident that a great part of the dental material dust is composed of particles with dimensions of less than 5 μm.[22] These small particles are defined as respirable according to the Johannesburg convention (Figures 3 and 4).[23]

C. Health Effects

Pneumoconiosis is defined as the nonneoplastic reaction in the lungs to inhaled mineral or organic dust and the resultant alteration in their structure. The fibrotic reactions in the lungs cause the pneumoconiosis. It is the small respirable particles that may cause pneumoconiosis. Silicosis and asbestosis are caused by the inhaled dust of free silica or asbestos. Silica as tridymite or cristobalite appears to have a greater fibrogenic potency than quartz. In cases with pneumoconiosis, dust particles are found in the lungs. In silicosis the fibrosis may progress long after the exposure has ceased. The symptoms could be breathlessness during pronounced effort and later with lesser degrees of effort. In silicosis the presence and severity of dyspnea and impairment of lung function correlates poorly with radiographic appearances.

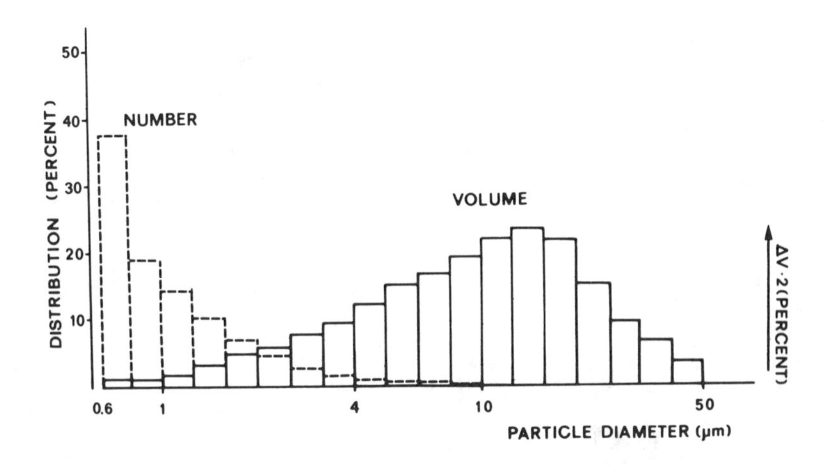

FIGURE 4. Particle size distribution in grinding dust of a chromium-cobalt alloy. (From Brune, D., Beltesbrekke, H., and Strand, G., *J. Prosthet. Dent.*, 44, 82, 1980. With permission.)

During the last 10 years, pneumoconiosis among dental technicians has been described. Lob and Hugonnaud[24] diagnosed 1 case of pneumoconiosis among 25 dental laboratory technicians. This man had been working for 22 years, manufacturing prostheses containing chromium, cobalt, and nickel. In 1980 a clinical study of dental laboratory technicians was reported[25] 70 dental technicians participated; 53 of these had abnormal radiological results and 45 had abnormal function results. Two cases of serious fibrosis were noted. In another study,[26] 178 dental technicians were examined and 8 were diagnosed as having a simple pneumoconiosis by chest radiograph. These technicians had been grinding nonprecious metal alloys. The morphology of pneumoconiosis among dental technicians has also been studied by systematic light and electron microscopical investigation of 30 lung preparations.[27] Diffuse streaky fibrosis and nodular fibrosis were observed together with occasional formation of silicotic nodules. Determination of the elemental components of dust particles deposited in pulmonary tissue was performed by energy-dispersive X-ray microanalysis on histologic sections under a scanning electron microscope (SEM). The combination of aluminium, silicon, chromium, and cobalt was the most frequently observed. From Belgium[28] two cases of pneumoconiosis among dental technicians are reported. The patients had worked 31 and 8 years, respectively, with molding, cutting, and polishing or sandblasting Vitallium® prosthetic appliances. Vitallium® is a chromium-cobalt-molybdenum alloy. Mineralogic analyses were performed on bronchoalveolar fluid from both patients and from lung tissue from one of them. The energy-dispersive analyses of the BAL fluid from the patients showed chromium, cobalt, and molybdenum. In addition, bare asbestos fibers as well as typical asbestos bodies were found in the BAL fluid from one of the patients. From Denmark[29] a report of a case of silicosis in a dental laboratory was published. The dental technician was exposed to silica dust for about 10 years. The dental technician's pneumoconiosis is suggested to be a real entity differing from other pneumoconiosis such as silicosis, hard metal disease, or asbestosis. The Cr-Co-Mo alloys are suspected to play a central role in the genesis. Also components from the tools, e.g., cut wheels and abrasives, may take part in the genesis.

D. Precautions

To prevent chronic effects to the lungs, local exhaust should be used at all dust-generating procedures. Ventilation with a local exhaust system possesses the ability to reduce dust levels in dental laboratories. A nozzle on the tube improves the efficiency of the ventilation system. The effect is most pronounced very close to the tube inlet. Face masks, especially those protecting against dust, can be used if local exhaust is not available (Figures 5 to 8).

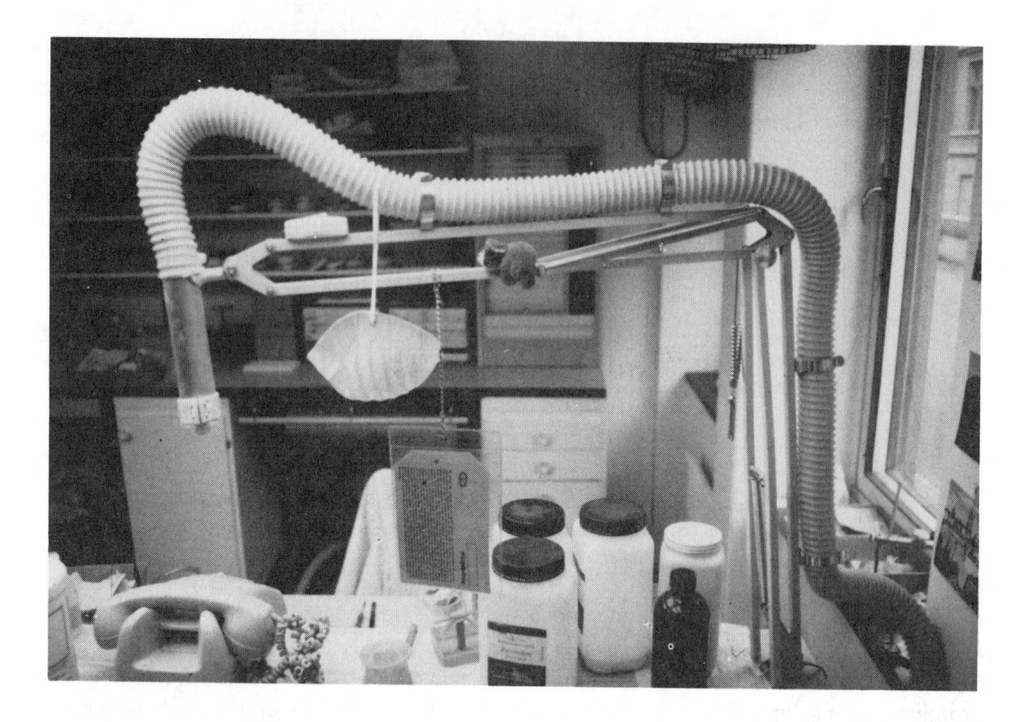

FIGURE 5. Mobile local exhaust. (From Andersson, I., Bornberger, S., Persson, B., and Åkerstedt, K., *Tandteknikern,* 3, 92, 1984. With permission.)

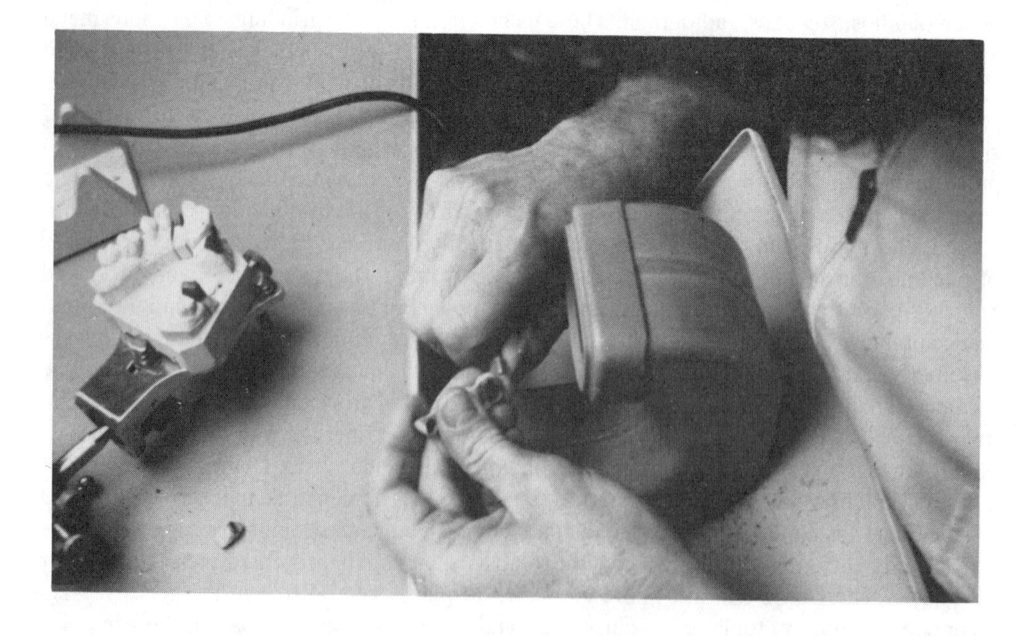

FIGURE 6. Immobile local exhaust. (From Andersson, I., Bornberger, S., Persson, B., and Åkerstedt, K., *Tandteknikern,* 3, 92, 1984. With permission.)

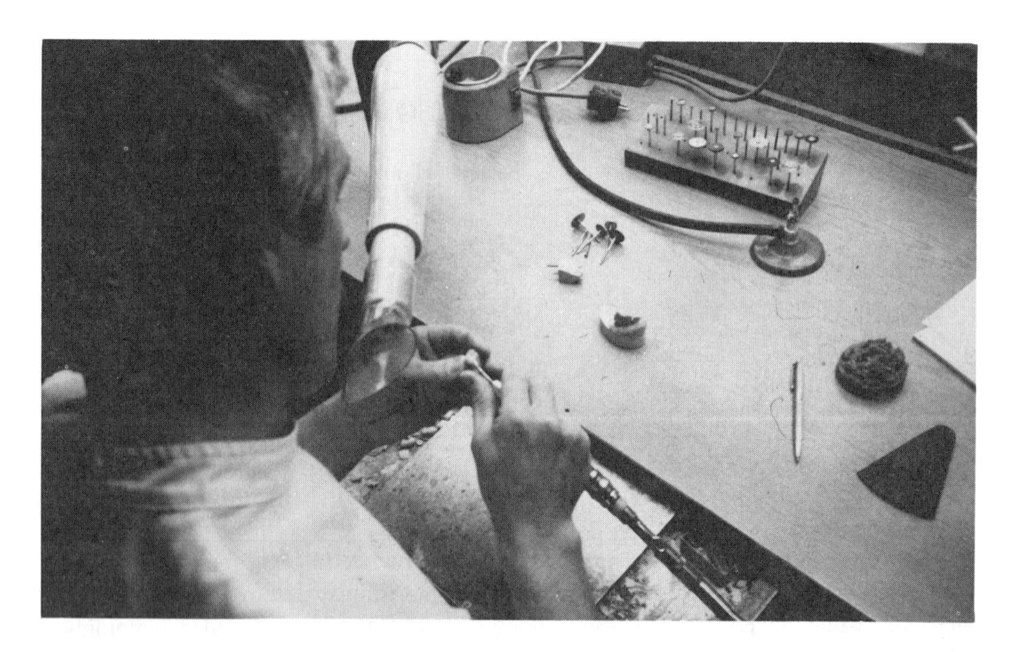

FIGURE 7. Local exhaust with a nozzle. (From Andersson, I., Bornberger, S., Persson, B., and Åkerstedt, K., *Tandteknikern,* 3, 92, 1984. With permission.)

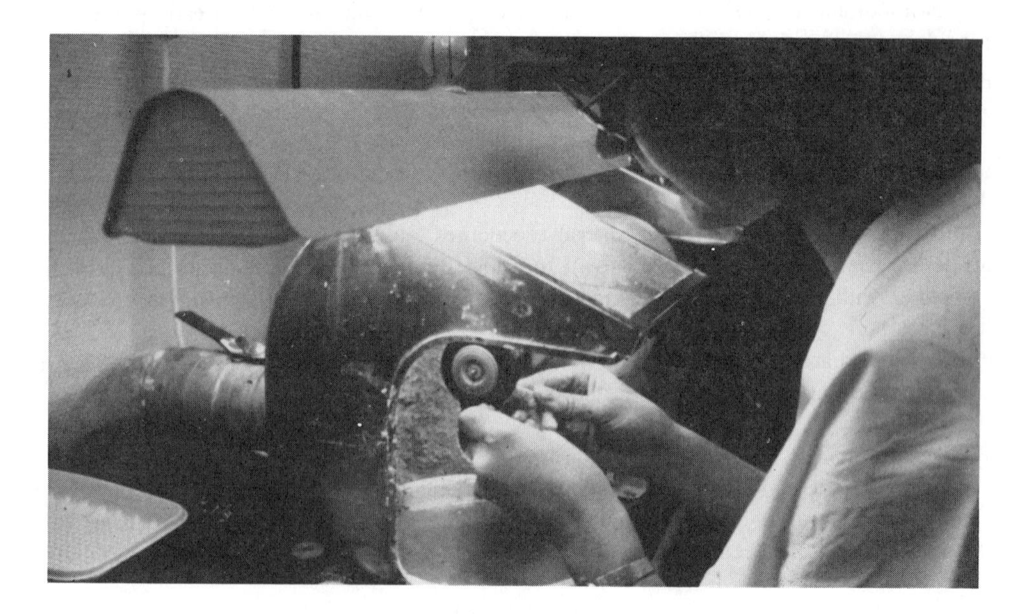

FIGURE 8. Polishing with local exhaust. (From Andersson, I., Bornberger, S., Persson, B., and Åkerstedt, K., *Tandteknikern,* 3, 92, 1984. With permission.)

IV. VIBRATIONS

Grinding and polishing with vibrating tools are time-consuming moments in a dental technician's work. The effects of local vibrations among dentists have been described in Chapter 15. The local vibrations generated by the grinding tools in dental work are of higher frequency than those generated by, for example, industrial pneumatic grinding tools or drilling machines.

Table 2
SOUND LEVELS AT DIFFERENT PROCEDURES (MEASURED DURING 60 S)[a]

Procedure	Sound level (dB A)	Range (dBA)
Cleaning with pneumatic air	96	85—105
Gypsum trimming	90	83—100
Cleaning with ultrasound	86	70—93
Grinding		
Denture base	86	81—90
Metal	87	82—94
Gypsum	88	85—90
Local dust exhaust	74	65—83

[a] Results from a survey among 21 dental laboratories.[3]

A. Health Effects

Local vibrations may cause vasospastic symptoms of white fingers (traumatic vasospastic disease — TVD). Effects due to local vibrations comprising symptoms of numbness, fumbling, and white fingers are described among dental technicians.[30] Similar symptoms are observed among dentists (see Chapter 15).

B. Precautions

It is important to develop tools that generate less vibrations and it is already possible to provide the tools with insulating coverings in order to moderate the vibrations. Such coverings are available on the commercial market.

V. NOISE

A. Sound Levels and Health Effects

In dental laboratories there are several procedures that generate noise. Measured sound levels at different procedures are shown in Table 2 below. These sound levels should be compared with the level (85 dB A) during continuous work that is considered to cause hearing loss.

Effects of hearing loss are shown among a group of Swedish dental technicians.[3] Noise from the working environment could be the reason.

B. Precautions

It is important to reduce the noisy procedures in dental laboratories, e.g., place the equipment on vibration-damping ground. When other noise-reducing measures are imperfect, the use of hearing protection should be considered.

VI. LIGHT

Working procedures in the dental laboratories could in this aspect be compared with jewelers or watchmakers. The illumination in dental laboratories is often insufficient and inadequate. Injuries associated with certain light wavelengths pertinent to the dental profession have been described in Chapter 14.

VII. ERGONOMICS

This topic is in general described for health professionals in Chapters 11 and 12. However, various symptoms specific to dental technicians are described below.

Table 3
FREQUENCY OF MUSCULOSKELETAL SYMPTOMS DURING THE LAST 12 MONTHS AMONG DENTAL TECHNICIANS AND REFERENT GROUPS FROM DIFFERENT TRADES

	Dental technicians[a]		Referents[b]	
Localization	**Male (n = 39)**	**Female (n = 11)**	**Male (n = 7569)**	**Female (n = 5158)**
Back of the neck	26	64	24	40
Shoulder(s)	26	46	24	41
Elbow(s)	18	18	10	11
Hand(s) or wrist(s)	23	45	13	22
Lower back	26	36	41	41
Hip(s)	8	9	11	14
Knee(s)	23	9	25	22
Foot/feet or ankle(s)	5	9	13	15

[a] Results from a Swedish survey.[3]
[b] Referent data are from the Department of Occupational Medicine, Örebro.

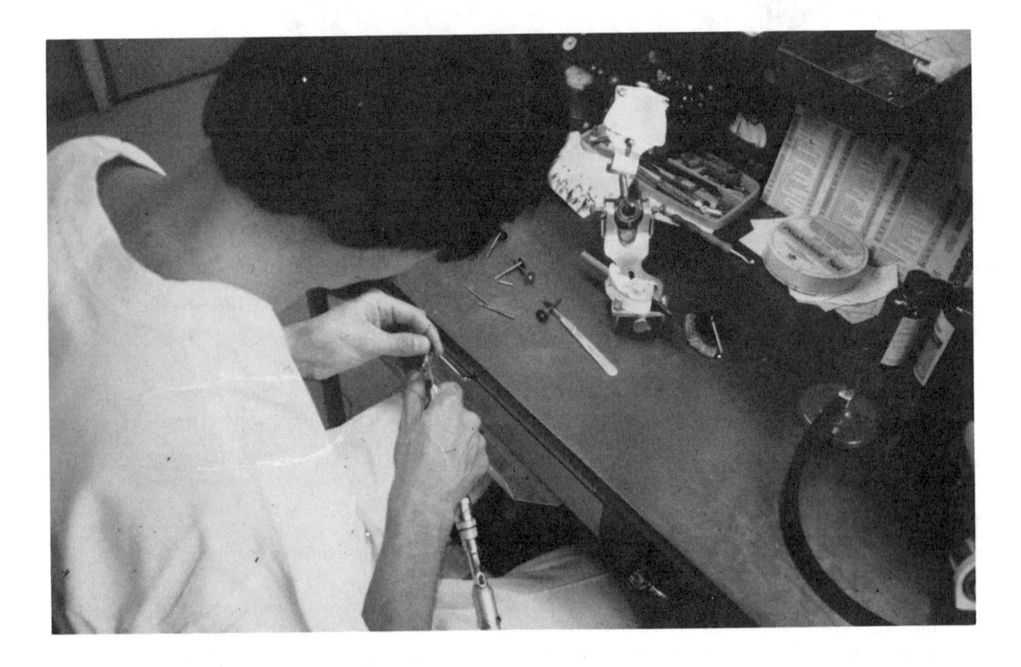

FIGURE 9. Working position during grinding and polishing. (From Andersson, I., Bornberger, S., Persson, B., and Åkerstedt, K., *Tandteknikern,* 3, 92, 1984. With permission.)

A. Health Effects

In a Swedish investigation, the ergonomic problems were found to be by far the most prominent in the trade. Musculoskeletal symptoms with different localizations are presented in Table 3. For comparison, the frequencies among workers in some other trades are given.Symptoms from the upper extremities dominate. During grinding and polishing, dental technicians work with their heads bent forward and in static working positions. The grinding tools used for work by hand are also often heavy. The design of the work place in dental laboratories is more often arranged traditionally than according to ergonomic properties (Figures 9 and 10).

FIGURE 10. Working position during grinding with the head bent forward. (From Andersson, I., Bornberger, S., Persson, B., and Åkerstedt, K., *Tandteknikern,* 3, 92, 1984. With permission.)

B. Precautions

It is important to reduce the static load due to work and to be aware of the working positions. These aspects are described in Chapters 11, 12, and 17.

REFERENCES

1. **Malker, H. and Weiner, J.,** Cancermiljöregistret. Exempel på utnyttjande av registerepidemiologi inom arbetsmiljöområdet, *Arbete och Hälsa,* 9, 50, 1984.
2. **Menck, H. R. and Henderson, B. E.,** Occupational differences in rates of lung cancer, *J. Occup. Med.,* 18, 797, 1976.
3. **Andersson, I., Bornberger, S., Persson, B., and Åkerstedt, K.,** Tandteknikerprojektet 1982-1983, *Tandteknikern,* 3, 92, 1984.
4. Organic Solvents and the Central Nervous System and Diagnostic Criteria: Report on a Joint WHO/Nordic Council of Ministers Working Group, Environmental Health Series, World Health Organization, Geneva, 1985, 5.
5. IARC Monographs on the Evaluation of the Carcinogenic Risk of Chemicals to Humans, 19, 187, 1979.
6. **Stahl, W. H.,** Compilation of Odor and Taste Threshold Values Data, ASTM Data Ser. DS48, American Society for Testing Materials, Baltimore, MD, 1973.
7. **Fisher, A. A.,** *Contact Dermatitis,* 2nd ed., Lea & Febiger, Philadelphia, 1973, 358.
8. **Clayton, G. D. and Clayton, F. E.,** *Patty's Industrial Hygiene and Toxicology,* Vol. 2A, John Wiley & Sons, New York, 1981, 2298.
9. **Poss, R., Thilly, W. G., and Kaden, D. A.,** Methylmethacrylate is a mutagen for *Salmonella typhimunium, J. Bone Jt. Surg.,* 61 (Abstr.), 1203, 1979.
10. **Bagramjan, S. B. and Babajan, E. A.,** Cytogenic study of the mutagenic activity of chemical substances isolated from Nairit latexes MKH and LNT-1, *Biol. Zh. Arm.,* 27, 102, 1974.

11. **Bagramjan, S. B., Pogosjan, A. S., Babajan, E. A., Ovanesjan, R. D., and Charjan, S. M.,** Mutagenic effect of small concentrations of volatile substances, emitted from polychloroprene latexes LNT-1 and MKH, during their combined uptake, *Biol. Zh. Arm.*, 29, 98, 1976.
12. **Estlander, T., Rajaniemi, R., and Jolanki, R.,** Hand dermatitis in dental technicians, *Contact Dermatitis,* 10, 201, 1984.
13. **Spiechowicz, E.,** Experimental studies on the effect of acrylic resin on rabbit skin, *Berufs Dermatosen,* 19, 132, 1971.
14. **Seppäläinen, A. M. and Rajaniemi, R.,** Local neurotoxicity of methylmethacrylate among dental technicians, *Am. J. Ind. Med.*, 5, 471, 1984.
15. **Christiansen, M. L., Adelhardt, M., Kjaergaard Jörgensen, N., and Gyntelberg, F.,** Metylmetakrylat en årsag til toksisk hjerneskade?, *Ugeskr. Laeg.*, 24, 1491, 1986.
16. **Anker, K., Ekenvall, L., Gustavsson, P., and Göthe, C.-G.,** Expositionsförhållanden och symtom hos tandtekniker vid arbete med kallhärdande metylmetakrylat, MMA, *Tandteknikern,* 50, 444, 1981.
17. **Brune, D. and Beltesbrekke, H.,** Levels of methylmethacrylate, formaldehyde, and asbestos in dental work air, *Scand. J. Dent. Res.*, 89, 113, 1981.
18. **Ruyter, I. E.,** Release of formaldehyde from denture base polymers, *Acta Odontol. Scand.*, 38, 17, 1980.
19. **Brune, D. and Beltesbrekke, H.,** Levels of mercury and silver in dust from trimming of amalgam dies, *Scand. J. Dent. Res.*, 87, 462, 1979.
20. **Waegemaekers, T. H. J. M., Seutter, E., den Arend, J. A. C. J., and Malten, K. E.,** Permeability of surgeon's gloves to methylmetacrylate, *Acta Orthop. Scand.*, 54, 790, 1983.
21. **Brune, D. and Beltesbrekke, H.,** Dust in dental laboratories. I. Types and levels in specific operations, *J. Prosthet. Dent.*, 43, 687, 1980.
22. **Brune, D., Beltesbrekke, H., and Strand, G.,** Dust in dental laboratories. II. Measurement of particle size distributions, *J. Prosthet. Dent.*, 44, 82, 1980.
23. **Orenstein, A. J.,** *Proc. of the Pneumoconiosis Conf.*, J. and A. Churchill, London, 1960, 120.
24. **Lob, M. and Hugonnaud, C.,** Pulmonary pathology. Risk of pneumoconiosis due to hard metals and berylliosis in dental technicians during modelling of metal prosthesis, *Arch. Mal. Prof. Hyg. Toxicol. Ind.*, 38, 543, 1977.
25. **von Kronenberger, H., Morgenroth, K., Tuengerthal, S., Schneider, M., Meier-Sydow, J., Riemann, H., Kroidl, R. F., and Amthor, M.,** Pneumokoniosen bei einem Zahntechnikerkollektiv, *Atemwegs Lungenkrankheiten,* 4, 279, 1980.
26. **Rom, W. N., Lochey, J. E., Lee, J. S., Kimball, A. C., Bang, K. M., Leaman, H., Johns, R. E., Penota, D., and Gibbons, H. L.,** Pneumoconiosis and exposures of dental laboratory technicians, *Am. J. Public Health,* 74, 1252, 1984.
27. **Morgenroth, K., Kronenberger, H., Michalke, G., and Schnabel, R.,** Morphology and pathogenesis of pneumoconiosis in dental technicians, *Pathol. Res. Pract.*, 179, 528, 1985.
28. **de Vuyst, P., van de Weyer, R., de Coster, A., Marchandise, F. X., Demortier, P., Ketelbant, P., Jedwab, J., and Yernault, J. C.,** Dental technicians pneumoconiosis. A report of two cases, *Am. Rev. Respir. Dis.*, 133, 316, 1986.
29. **Hansen, H. M.,** Silikose hos en tandtekniker, *Ugeskr. Laeg.*, 145, 2378, 1983.
30. **Hjortsberg, U., Jonsson, K., Larsson, B., Necking, L.-E., Nise, G., and Örbaek, P.,** Neuropati och vasospasm i fingrar hos tandtekniker exponerade för högfrekventa vibrationer, *Sven. Laekaresaellsk.*, Handlingar, Hygiea, Stockholm, 1986, 120.

Chapter 19

THE PSYCHOSOCIAL WORKING ENVIRONMENT

Töres Theorell

TABLE OF CONTENTS

I. A CASE HISTORY

One of the patients who consulted me recently is a dentistry nurse who had recently developed hypertension and had even had a small stroke when she was 30 years old. This had started during a period of marital conflict which later resulted in divorce. The stroke had caused a partial paralysis of her right hand and also a partial aphasia. These symptoms soon disappeared when she was in a calm condition, but they reappear when she is tired or rushed, even after many years have passed. This may result in difficulties in her work as a dentistry nurse because when she feels rushed she may drop expensive or sterilized instruments on the floor or be unable to find the right words when talking to patients.

After the stroke episode, the patient consulted hospitals because she needed to control her hypertension. When she started monitoring her own blood pressure by means of a self-triggered blood pressure measurement instrument, she observed that her blood pressure was normal without medication during her summer vacation, but that it rose slowly morning after morning when she started working. She also noticed that the levels became dangerously high in the middle of the day. At this stage she was able to regulate her blood pressure by means of small doses of medication and rest in the middle of the day.

The dentist she worked for decided to buy a new clinic. Interest rates were very high and this meant that he and his nurses had to work much harder. The lunch intermission was taken away. During this period my patient's condition deteriorated rapidly. Her blood pressure was high despite medication and during the most rushed periods she started dropping things on the floor and made erroneous comments and notes. The dentist wanted to get rid of her (due to Swedish laws protecting employment, this is very difficult) and constantly made critical remarks. This had a bad effect on the patient's symptoms and a vicious circle was started. This case illustrates several points:

1. In subjects with a pronounced "blood pressure vulnerability", the working environment may have serious effects — in this case it endangered her cerebral function. The effect of the interaction between a "strained" psychosocial work and blood pressure vulnerability on the blood pressure levels at work has recently been shown in an epidemiological study of 28-year-old men.[1]
2. In the caring occupations, contacts with patients/clients may create a kind of emotional strain which does not exist in most noncaring occupations.

One way of describing the interplay between the environment and the individual is the illustration in Figure 1.

In the case described above, the condition started deteriorating when the environmental (financial) pressure increased. It is obvious that the vicious circle that started when the dentist started criticizing the nurse created bad experiences for her in rushed situations. This in itself created deteriorating symptoms in rushed situations. The nurse was "programmed" in a harmful way. Social support was nonexistent. There were no other workmates whom she could talk to. If she had had the opportunity of discussing daily problems with someone at work, her experiences might not have been as bad, and fewer tensions might have arisen. I shall now use the labels in the diagram in an effort to describe the main currents in international research on the psychosocial environment of the carring occupations.

II. REACTIONS

Most of the research in this area has been devoted to individual reactions to an adverse working environment. Psychological as well as physiological reactions have been described. Of psychological reactions, "burnout" has been the most common concept during recent

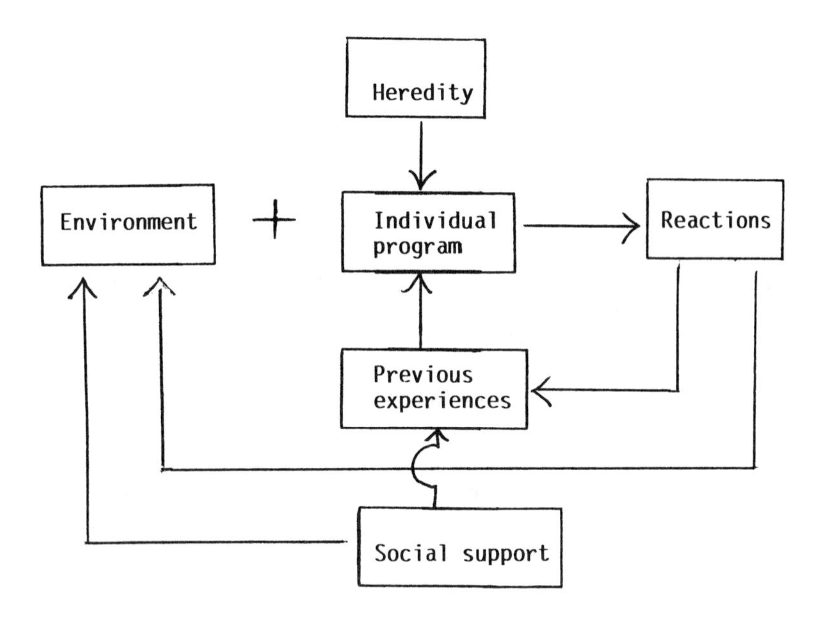

FIGURE 1. Interplay between individual and environment in generating reactions. (Modified from Kagan, A. R. and Levi, L., *Society, Stress and Disease,* Part 1, Levi, L., Ed., Oxford University Press, London, 1971.)

years. Burnout has three main components, namely, emotional exhaustion, depersonalization, and lack of personal accomplishment.[3]

Of these, emotional exhaustion has been the most extensively studied one. Emotional exhaustion relates to a state of "emotional emptiness", like a "battery that has been worn out and cannot be charged". Most studies have indicated that long exposure to a caring job and a high degree of emotional devotion to clients/patients are predisposing factors in relation to emotional exhaustion. Emotional exhaustion has been related to a number of work variables. Eisenstat and Felner have studied 168 workers in a range of human service occupations.[4] They conclude that job stressors increase the risk of emotional exhaustion. As pointed out by Maslach, it is common among workers in caring occupations to put the blame on "personality" and other personal characteristics for psychosocial problems, whereas in fact work organization and work load are frequently the most important factors.[3] It has been suggested that those nurses who want to work in an intensive care unit (ICU) should be tested with regard to "hardy personality" — a "hardy" person (according to an inventory measurement) runs less risk of burnout.[5] In my opinion, no personality inventories have such precision that this testing procedure before employment would be warranted! Burnout in general and emotional exhaustion in particular have been very helpful concepts in the discussions about the psychosocial working environment in the caring occupations. In Sweden numerous discussions have been started in many cancer words, psychiatric hospitals, etc. The concepts are immediately recognizable and therefore constructive discussions may take place.

III. BURNOUT AND GENDER

In a recent study, female and male physicians were compared in interviews. In general, female physicians reported more burnout than male physicians.[6] We do not know the reason for this. It could be, for instance, that female caregivers demand more emotional involvement from themselves than a male caregiver does and also that patients expect more "emotional" caregiving from women than they expect from men. A study was made of a cohort of

Swedish physicians which was compared to the total Swedish working population and to academically educated working Swedes, in particular. This showed that female physicians had a markedly elevated age-adjusted relative risk of committing suicide, both compared to the female Swedish working population and to the academically educated one.[6] The mortality of the male physicians did not differ significantly from that of the general working population. Compared to other male academic occupations, however, male physicians committed more suicides, but men with academic occupations committed fewer suicides than the total male working population. This observation is consistent with observations of male physician mortality in the U.S. and Canada.[7,8]

In a recent study of randomly selected male and female physicians in the greater Stockholm area (all of them active in routine in- and outpatient care; participation rate of 61%, an ''emotion protocol'' was utilized. Every subject was asked to mark appropriate adjective alternatives (four grades from ''not at all'' to ''very much'') approximately once an hour during an ordinary working day. Figure 2 shows the average ''percentage'' of certain states during the day. In most subjects, four different days during a year were studied. The average of each subject is thus based on a large number of occasions. The percentage is based on the existence of at least ''rather much''.

1. ''Rushed'' or ''nervous'' or ''stressed''
2. ''Sad'' or ''depressed''
3. ''Angry'' or ''irritated''

The figure clearly shows that the physicians when compared with other service groups reported more ''rush'' and more ''anger'' than other men. They also tended to report more ''sadness'', although that difference was not statistically significant. The female physicians even reported more ''sadness'' and ''anger'' than the male physicians.[9]

IV. BEHAVIORAL REACTIONS

Another interesting kind of response to an adverse environment is increased cigarette smoking. Physical and emotional stress reactions at work as well as dissatisfaction with job rewards were associated with increased smoking in hospital nurses.[10]

V. PHYSIOLOGICAL REACTIONS

Physiological reactions in the work environment could aim at mobilizing energy (catabolic) or protecting the organism (anabolic). Catecholamines, heart rate, and blood pressure are parameters that have been studied in caregivers during working hours. They could all be seen as indicating degree of sympathoadrenal arousal. Cortisol in plasma and urine could be regarded as a parameter mainly measuring adaptation to novel ''distress'' induced in situations with some element of stress.

Catecholamines have been studied along with feelings of ''stress'' in emergency situations during ambulance transports involving physicians and drivers. In these situations, marked elevations of adrenaline and heart rate have been observed.[11] Anxiety states, systolic blood pressure, and heart rate have been studied during ''strain periods'' in nursing students. The findings clearly show that rushed unpredictable situations are also associated with sympathoadrenal arousal in this group.[12] On the other hand, studies have shown that surgeons mostly do not have a high heart rate or high catecholamine output when they operate — if no unusual emergency situation arises.[13-14] This could probably be interpreted in the following way: in most of the work taking place during a surgeon's usual working day, there is no unusual sympathoadrenal arousal taking place. However, the more often unpredictable emergency situations occur the more sympathoadrenal arousal takes place.

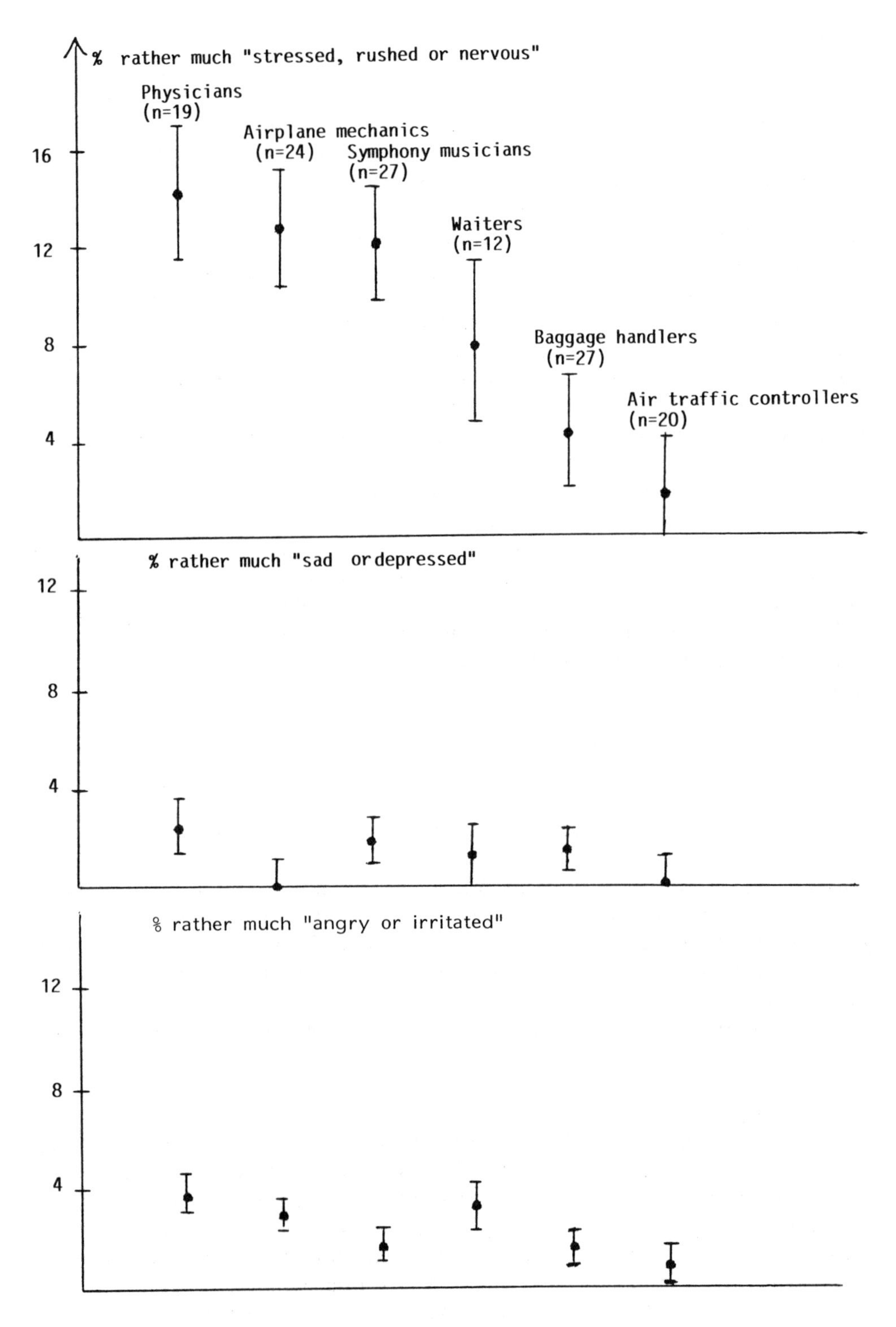

FIGURE 2. Reported emotional states during ordinary workdays in men (means and standard errors).

In the Swedish cohort study of physician mortality, general surgeons had higher cardiac mortality than other male physicians. This may be consistent with the sympathoadrenal arousal theory.[15,16] In the study of emotions among physicians working in routine care referred to above, blood pressure was measured during activity at work and leisure. Physicians, male as well as female, had low average blood pressures at work and small differences between blood pressure at work and during leisure, possibly indicating a low average degree of sympathoadrenal arousal. At the same time, however, plasma cortisol levels in the morning were higher than in the other service occupations in which cortisol levels were studied (air traffic controllers, symphony musicians, baggage handlers, and airplane mechanics).[9] Only one other group that we have studied — prison personnel — also a ''caregiving'' group, had as high plasma cortisol levels in the morning as the physicians.[17] One possible interpretation of this is that the confrontation with serious life problems in other human beings is associated with pronounced distress reactions resulting in adrenocortical arousal. As has been pointed out by several authors, the distressing effect of the confrontation with other people's serious life problems is mostly denied or regarded as an inevitable part of work.[18]

The way in which burnout reactions are regarded among caregiving colleagues varies markedly between institutions. A recent survey of internal medicine programs in the U.S. was made during 1979 to 1984.[19] A total of 63% of the directors responded to the survey; 56% of the programs granted leaves of absence to medical residents because of emotional impairment; 0.9% of the male residents demanded such leaves. The incidence in female residents was about twice the male incidence. Of those who left the programs for this reason, 79% continued in medical work, whereas 10% abandoned medicine completely; 2% had committed suicide and a further 3% had tried (but failed) to commit suicide. It is likely that burnout is frequently unrecognized.

VI. INDIVIDUAL FACTORS

A. Age

One of the groups of researchers who showed that young age increases the risk of burnout and who also studied personnel taking care of the mentally retarded was Livingston and Livingston.[20] Härenstam et al. found that morning plasma cortisol levels in prison personnel were very high, particularly in young personnel.[17] Thus, at least in some kinds of care, young age (and lack of experience?) seems to be associated with increased risk. Maslach[3] has pointed out that massive exposure and a high degree of involvement in clients/patients increase the risk. Thus, the highest risk is found in unexperienced caregivers who start working at a young age and spend a lot of time with patients who demand emotional engagement.

B. Personality

One personality trait that has been studied in relation to burnout and other long-lasting reactions to care is the ''hardy personality''. This is an individual program characterized by active coping in crisis situations. Such a personality trait is probably not stable, but rather is influenced to a great extent by the experiences that the individual has had in stressful situations (mastery or failure). The hardy personality has been associated with lowered risk of burnout among nurses and with lowered risk of stress reactions in intensive care nurses.[5,21]

Another personality factor that has been discussed in this research area is ''locus of control'' — the individual's tendency to regard himself as able/unable to change his environment. Is the environment controlling me (external locus) or am I controlling the environment (internal locus)? Obviously this is theoretically related to hardy personality — ''active copers'' are likely to regard themselves as able to change the environment. Vredensburgh and Trinkaus, in a study of 566 nurses in four city hospitals, showed that internal locus of control protected the nurses against ''role stress''.[22]

C. Education

Education has also been mentioned as an individual factor of importance. Vredensburgh and Trinkaus found that a higher level of education was associated with lower level of role stress.[22]

D. Expectations

Role expectations in caregivers entering their occupation have been discussed in the literature. Maslach[3] has pointed out that high expectations may increase the risk of burnout, whereas Seabold, who made a cross-sectional and a 1-year predictive study of burnout in 139 first and second year nurses, states that a high level of occupational expectation protects against burnout.[19,23] Thus, expectation in general does not have a clear role in relation to burnout. Maybe unrealistic expectations increase the risk of burnout. No one has, however, studied expectations in this way.

E. Interactions

Several authors have pointed out that the environment interacts with individual characteristics in generating occupational stress reactions. Arsenault and Dolan have pointed out that the ''job context'' (organizational aspect) interacts with ''personality'' in generating job stress, and Paredes, who studied 98 nurses in medical wards, found that social support (in particular, the one from supervisors) interacted with psychological resources in protection against burnout and occupational stress — the effect of social support was better in nurses with good psychological resources than in nurses with poor ''resources''.[24,25]

F. Job Environment

A lot of attention has been paid to the acute arousal reactions taking place in emergency situations. From studies of physicians, ambulance drivers, and nurses in such situations, we may conclude that crisis situations have powerful arousal effects. On the other hand, such situations are relatively uncommon for most caregivers. In the daily routines, other factors are much more important in generating adverse health effects for most caregivers. Lack of intellectual discretion and lack of authority over decisions are common among all subordinates in the caregiving occupations. Gardell and Gustafsson,[26] in a large-scale study of caregivers in Swedish hospitals in 1976, found that the work environment in hospitals was authoritarian. Nursing aides reported a small authority over decisions as well as low levels of feedback and intellectual discretion. Physicians reported considerably higher levels, whereas nurses, physiotherapists, and other paramedical personnel were in an intermediate position.[26] These observations were particularly interesting because Sweden was considered to be a democratically advanced country at the time. The hierarchical structure in caregiving occupations may be even more pronounced in other countries. Ehrenreich and Ehrenreich have pointed out that there are pronounced class conflicts in hospital care.[27]

Also when compared to other occupations on the female labor market, several of the caregiving occupations stand out as occupations in which many workers report a rushed tempo and lack of intellectual discretion. In many of these occupations, a high proportion of the workers also report fatigue. Table 1 shows how men and women in nine different caregiving occupations have described their working conditions and their degree of fatigue. The percentages have been compared with all other occupations on the Swedish labor market. The interviews were done with randomly selected Swedes from 1977 to 1981 (9051 men and 4191 women) — ''4'' means that the occupation belongs to the 25% worst occupations and ''1'' that it belongs to the 25% most favored occupations. The table shows that physicians, male as well as female, report a lot of rush and frequently that they have felt ''tired since two weeks''. Mental nurses frequently report monotony and being tired in the morning. Female nursing aides report more than average rush, lack of influence over working hours,

Table 1
DISTRIBUTION IN QUARTILES OF THE TOTAL SWEDISH MALE AND FEMALE WORKING POPULATION OF CERTAIN CHARACTERISTICS ACCORDING TO THE NATIONAL SWEDISH INTERVIEW SURVEYS (1977 TO 1979)

	Rush		Monotony		Lack of influence over working hours		Lack of possibility to learn new things		Tired since 2 weeks		Tired in the morning		Tired in the evening	
Subject	**Men**	**Women**	**Men**	**Women**	**Men**	**Women**	**Men**	**Women**	**Men**	**Women**	**Men**	**Women**	**Men**	**Women**
Physicians	4	4	1	1	3	2	1	1	3	3	1	2	2	2
Dentists	3	—	1	—	1	—	2	—	1	—	1	—	1	—
Registered nurses	—	2	—	1	—	2	—	1	—	1	—	1	—	1
Obstetrical nurses	2	1	—	1	—	3	—	1	—	1	—	1	—	1
Psychiatric nurses	2	1	3	3	3	1	2	2	2	3	3	3	1	1
Nursing aides	2	3	3	2	4	3	3	2	4	2	4	3	3	2
Dentistry nurses	—	3	—	2	—	4	—	2	—	1	—	2	—	1
Medical tech. assistants	—	4	—	2	—	2	—	2	—	4	—	4	—	4
Physiotherapists	—	1	—	1	—	1	—	2	—	4	—	3	—	3

Note: 4, The occupation belongs to the "worst" 25% of all occupations (Swedish Central Bureau of Statistics); 1, the occupation belongs to the "best" 25% of all occupation; —, too few interviewed subjects.

and being tired in the morning. Physiotherapists, despite a rather favorable description of their jobs, frequently report tiredness.

Gardell and Gustafsson[26] summarized their findings in the concept "hospital care on the assembly line". A paradoxical picture of the hospital world emerges: physicians traditionally have more power than other caregivers. They also report a high level of intellectual discretion and in general relatively good authority over decisions, although some aspects of the decision authority are, of course, poor (possibility to influence work tempo, for instance). In a previous section we observed that physicians, in particular female physicians, commit many suicides and also that their everyday life is characterized by many emotional reactions and in men we observed high cortisol levels. It could be that physicians as a group might benefit psychologically from sharing power with other caregivers. The other caregivers, on the other hand, may also benefit from this because of less organizational frustration. Jackson studied a practical experiment in increased participation in decision-making.[28] The experiment was performed with an experiment and a control group, with pretests in the experiment group; 95 persons participated. A significant positive effect was observed, with reduced job-related strain arising after the increased participation in decision-making had started. Amaral et al. have described a pilot project in a cancer ward.[29] A practical experiment was performed aimed at increasing information and support to the personnel. The personnel stated high levels of satisfaction with the program, but there was no statistical evidence that it had affected the outcome variables.

Another experiment by Arnetz et al. originated in the following goal formulations.[30] Elderly tenants in a service house were observed to be socially extremely passive and a program based on the tenants' own interests was started aimed at increasing social interaction. The program was evaluated by means of concomitant observations in the experimental ward and in a control ward. During 6 months, the tenants in the experimental ward became socially more active as well as psychologically more active and less tense than those in the control ward. Physiologically they improved their anabolism and their carbohydrate metabolism compared to the control tenants.[31] The personnel in the experimental ward changed their attitude — becoming more interested in social activation than before the experiment. Absenteeism from work and the personnel turnover decreased. The change in attitudes has lasted even after 5 years.

G. Social Support

Social support has been stressed as a protective factor by many authors. Paredes, who studied 98 nurses in medical wards, found that support at work, in particular supervisory support, was more important than social support outside the job concerning protection against burnout.[25] Norbeck studied 164 emergency nurses comparing married and unmarried nurses. Lack of support at work "explained" 24% of reported job stress in married nurses.[32] In unmarried nurses, lack of support from relatives explained 10% of psychological symptoms. It is obvious that different sources of social support may have different meanings for married and unmarried caregivers. Vredensburgh and Trinkaus have pointed out that the workmates' acceptance of the worker's behavior protects against role stress. This profound aspect of social support — reassurance of personal worth — has been stressed by many researchers.[22]

Cassileth has studied doctors and nurses in a ward treating acute leukemia.[33] The research technique was social-anthropological. This study indicated that the profound insecurity regarding treatment and prognosis ties the team together. The team "becomes a family". The author regards this process as evidence that a humanistic attitude to patients is developing in highly specialized care.

H. Relatives of Patients

The patients' relatives are often regarded as complicating factors in the giving of care. Recently a study was performed aimed at increasing the participation of relatives in the care of cancer patients.[34] The study was controlled — relatives involved in the experimental program of caring for cancer patients were compared with comparable relatives who were not actively involved in caring for patients, but were involved only in routine care. The results indicated that the activation program for relatives of cancer patients triggered an accelerated psychological and physiological grief process in these relatives before the death of the patient, but that they benefited psychologically from this after the death of the patient. Among the personnel this initially created increased role conflicts. However, the attitude toward relatives became more favorable. This study indicates that changes in the role of relatives may create initial tensions in the personnel.

I. Violence

A somewhat neglected research area has been violence in the caring environment. In their review of psychological stress experienced by health care personnel, Leppänen and Olkinuora point out that workers in some kinds of care are frequently exposed to violence from patients.[35] This is particularly true in mental hospitals and emergency units. More research is needed in order to clarify to what extent such violence could be due to lack of communication, too rushed a tempo, rigid rules, etc., factors that are potentially amenable to improvement. So far, however, the research in this area has mainly described the extent of the problem.[36,37] In future care, the problems arising due to the personnel's fear of human immunodeficiency virus- (HIV) contaminated blood will grow into a major psychosocial problem. No systematic research in this field has been published.

J. Turnover

Increasing turnover of personnel is an increasing problem in the caring occupations in Scandinavian countries. It is possible that psychosocial problems in the work place may account for part of this problem, although this has not been sufficiently analyzed. Turnover is mostly regarded as a technical research problem that creates sources of error in the analyses. Our own study of the elderly showed that active discussions with personnel leading to changed goal formulations and improved social integration of the elderly was also associated with a decreased turnover of personnel.[30] This may be an important area for future research.

VII. CONCLUSION

A review of research on the psychosocial work environment in caregiving shows that there is rapid development in this field. Large representative samples are studied and standardized questionnaires as well as observation techniques and physiological stress analyses are being utilized. There is evidence that emotional tensions are common in most kinds of care, that many of these tensions result in burnout, and also that some of the burnout remains unrecognized. Systematic studies have shown that lack of support and information are common organizational factors, resulting in adverse consequences. Evaluations of interventions aiming at improved support and information as well as improved goal formulations have shown promising results.

REFERENCES

1. **Theorell, T., Knox, S., Svensson, J., and Waller, D.,** Blood pressure variations during a working day at age 28: effects of different types of work and blood pressure level at age 18, *J. Hum. Stress,* 11, 36, 1985.
2. **Kagan, A. R. and Levi, L.,** Adaptation of the psychosocial environment to man's abilities and needs, in *Society, Stress and Disease,* Part 1, Levi, L., Ed., Oxford University Press, London, 1971.

3. **Maslach, C.,** *Burnout — The Cost of Caring,* Prentice-Hall, Englewood Cliffs, N.J., 1982.
4. **Eisenstat, R. A. and Felner, R. D.,** Toward a differentiated view of burnout: personal and organizational mediators of job satisfaction and stress, *Am. J. Comm. Psychol.,* 12, 411, 1984.
5. **Keane, A., Ducette, J., and Adler, D. C.,** Stress in ICU and non-ICU nurses, *Nurs. Res.,* 34, 231, 1985.
6. **Arnetz, B., Andreasson, S., Strandberg, M., Eneroth, P., Kallner, A., and Theorell, T.,** The Working Environment of Physicians, Stress Res. Rep. No. 187, National Institute for Psychosocial Factors and Health, Stockholm, 1986 (in Swedish).
7. **Goodman, L. J.,** The longevity and mortality of American physicians, 1969-1973, Millbank Memorial Fund Quarterly, *Health Soc.,* 53, 353, 1975.
8. **Elliott, R.,** Mortality among Canadian physicians, *Can. Med. Assoc. J.,* 96, 1178, 1967.
9. **Theorell, T., Preski, A., Sigala, F., and Ahlberg-Hultén, G.,** Six service occupations, psychosocial, psychological and endocrinological observations, in preparation (in Swedish).
10. **Tagliacozzo, R. and Vaughn, S.,** Stress and smoking in hospital nurses, *Am. J. Public Health,* 72, 441, 1982.
11. **Lehman, M., Dörges, V., Huber, G., Zöllner, G., Spöri, U., and Keul, J.,** Zum Verhalten der freien Katecholamine in Blut and Harn bei Sanitätern und Ärzten während des Einsatzes, *Int. Architectural Occup. Environ. Health,* 51, 209, 1983.
12. **Eden, D.,** Critical job events, acute stress and strain, a multiple interrupted times series, *Organ. Behav. Performance,* 30, 312, 1982.
13. **Becker, W. G. E., Ellis, H., Goldsmith, R., and Kaye, A. M.,** Heart rates of surgeons in theatre, *Ergonomics,* 26, 803, 1983.
14. **Payne, R. L., Rick, J. T., Smith, G. H., and Cooper, R. G.,** Multiple indications of stress in an "active" job — cardiothoracic surgery, *J. Occup. Med.,* 26, 805, 1984.
15. **Arnetz, B., Hörte, L.-G., Hedberg, G., Theorell, T., Allander, E., and Malker, H.,** The Mortality Pattern (Coronary Heart Disease and Suicide) in Swedish Physicians, Stress Res. Rep. No. 185, National Institute for Psychosocial Factors and Health, Stockholm, 1986 (in Swedish).
16. **Arnetz, B., Hörte, L.-G., Hedberg, A., Theorell, T., Allander, E., and Malker, H.,** Suicide patterns among physicians related to other academics as well as to the general population, *Acta Psychiatr. Scand.,* 75, 139, 1987.
17. **Härenstam, A., Theorell, T., and Palm, U.-B.,** The working environment in prisons. IV. Summary, Statshälsan and National Institute for Psychosocial Factors and Health, Stockholm, 1987 (in Swedish).
18. **Hellström, L. and Uddenberg, N.,** Both master and servant — the physician and the work mates, *Lakartidningen,* 82, 3672, 1984 (in Swedish).
19. **Smith, J. W., Denny, W. F., and Witzke, D. B.,** Emotional impairment in internal medicine house staff. Results of a national study, *JAMA,* 255, 1155, 1986.
20. **Livingston, M. and Livingston, H.,** Emotional distress in nurses at work, *Br. J. Med. Psychol.,* 57, 291, 1984.
21. **Maloney, J. P. and Bartz, C.,** Stress-tolerant people: intensive care nurses compared with non-intensive care nurses, *Heart Lung,* 12, 389, 1983.
22. **Vredensburgh, D. J. and Trinkaus, R. J.,** An analysis of role stress among hospital nurses, *J. Vocation. Behav.,* 22, 82, 1983.
23. **Seabold, F. C.,** The Relationship of Psychological Resources and Social Support to Occupational Stress and Burnout in Hospital Nurses, Doctoral thesis, University of Houston, Texas, 1982.
24. **Arsenault, A. and Dolan, S.,** The role of personality, occupation and organization in understanding the relationship between job stress, performance and absenteeism, *J. Occup. Psychol.,* 56, 227, 1983.
25. **Paredes, F. C.,** The Relationship of Psychological Resources and Social Support to Occupational Stress and Burnout in Hospital Nurses, Doctoral thesis, University of Houston, Texas, 1982.
26. **Gardell, B. and Gustaffson, R.,** *Hospital Care on the Assembly Line,* Prisma, Stockholm, 1979 (in Swedish).
27. **Ehrenreich, B. and Ehrenreich, J. H.,** Hospital workers: class conflicts in the making, *Int. J. Health Serv.,* 5, 43, 1975.
28. **Jackson, S. E.,** Participation in decision-making as a strategy for reducing job-related strain, *J. Appl. Psychol.,* 68, 3, 1983.
29. **Amaral, P., Nehemkis, A. M., and Fox, L.,** Staff support group on a cancer ward: a pilot project, *Death Educ.,* 5, 267, 1981.
30. **Arnetz, B., Eyre, M., and Theorell, T.,** Social activation of the elderly. A social experiment, *Soc. Sci. Med.,* 16, 1685, 1982.
31. **Arnetz, B., Theorell, T., Levi, L., Kallner, A., and Eneroth, P.,** An experimental study of social isolation of elderly people. Psychoendocrine and metabolic effects, *Psychosom. Med.,* 45,395, 1983.
32. **Norbeck, J. S.,** Types and sources of social support for managing job stress in critical care nursing, *Nurs. Res.,* 34, 225, 1985.
33. **Cassileth, B. R.,** Surviving: Staff Adaptations to Stress on a Cancer Ward, Doctoral thesis, University of Pennsylvania, Philadelphia, 1978.

34. **Theorell, T., Häggmark, C., and Eneroth, P.,** Endocrinological reactions in women with a relative suffering from cancer: effects of an activation programme on the cancer ward and correlation to psychosocial reaction, *Acta Oncol. Scand.*, 26, 419, 1987.
35. **Leppänen, R. A. and Olkinuora, M. A.,** Psychological stress experienced by health care personnel, *Scand. J. Work Environ. Health,* 13, 1, 1987.
36. **Mindus, P. and Struwe, G.,** Violent acts in a psychiatric emergency unit: a retrospective study, *Lakartidningen,* 74, 4606, 1977 (in Swedish).
37. **Stymne, I.,** Risks of violence in the work environment, Arbetarskyddsstyrelsen, Solna, 1981 (in Swedish).

INDEX

A

B

C

D

E

F

G

H

I

J

K

L

M

N

O

P

S

T

U

V